Food Flavor and Chemistry
Explorations into the 21st Century

Food Flavor and Chemistry
Explorations into the 21st Century

Edited by

A.M. Spanier
US Dept of Agriculture, Rockville, Maryland, USA
F. Shahidi
Memorial University of Newfoundland, St John's, NF, Canada
T.H. Parliment
Parliment Consulting, New City, NY, USA
C. Mussinan
IFF R&D, Union Beach, NJ, USA
C.-T. Ho
Rutgers University, New Brunswick, NJ, USA
E. Tratras Contis
Eastern Michigan University, Ypsilanti, MI, USA

RSC | Advancing the Chemical Sciences

The proceedings of the 11th International Flavor Conference: Recent Advances in Food Flavor Chemistry, 3rd George Charalambous Memorial Symposium held on 29 June to 2 July 2004 in Samos, Greece.

Special Publication No. 300

ISBN 0-85404-653-4

A catalogue record for this book is available from the British Library

© The Royal Society of Chemistry 2005

All rights reserved

Apart from any fair dealing for the purpose of research or private study for non-commercial purposes, or criticism or review as permitted under the terms of the UK Copyright, Designs and Patents Act, 1988 and the Copyright and Related Rights Regulations 2003, this publication may not be reproduced, stored or transmitted, in any form or by any means, without the prior permission in writing of The Royal Society of Chemistry, or in the case of reprographic reproduction only in accordance with the terms of the licences issued by the Copyright Licensing Agency in the UK, or in accordance with the terms of the licences issued by the appropriate Reproduction Rights Organization outside the UK. Enquiries concerning reproduction outside the terms stated here should be sent to The Royal Society of Chemistry at the address printed on this page.

Published by The Royal Society of Chemistry,
Thomas Graham House, Science Park, Milton Road,
Cambridge CB4 0WF, UK

Registered Charity Number 207890

For further information see our web site at www.rsc.org

Printed by Athenaeum Press Ltd, Gateshead, Tyne and Wear, UK

Preface

The Third George Charalambous Memorial Symposium was the 11th International Flavor Conference, held July 1-4, 2004 on Samos Island, Greece. The meeting was sponsored by the Agricultural and Food Chemistry Division of American Chemical Society and was attended from around the globe by professionals and scientific leaders belonging to academia, government laboratories and industry. Presentations were mainly in the flavor science and technology area, but advances in food science and nutrition were also highlighted.

This book represents primarily the proceedings of the presentations made at the conference. It covers articles of overview nature (Chapter 1), flavor of wine and dairy products (Chapters 2-9), composition of volatile aroma compounds of fruits, vegetables and meat (Chapters 10-16) as well as aroma generation (Chapters 17-21) and analytical issues in food flavors (Chapters 22-25). In addition, analysis of phenolic glycosides and migration of chemicals in food packaging is included (Chapters 26 and 27). A section is also devoted to discussions about antioxidants and specialty lipids with respect to their role in health promotion (Chapters 28-35). Finally, quality of food stuff as affected by processing is presented (Chapters 36-46).

The organizers acknowledge the generous contributions towards conference's expenses from International Flavors and Fragrances as well as Eastern Michigan University for setting up the website for the conference and production and printing of the abstracts of the conference. We, the editors, also thank all contributors for their excellent presentations and proceeding chapters.

The Editors
April 2005

Contents

Overview

Halal Issues in Flavor Industry   3
M. N. Riaz, M. M. Chaudry and M. Sadek

Dairy and Wine Flavors

Real Time Release of Flavor Compounds and Flavor Perception: An Application to Cheese   13
E. Pionnier, J.L. Le Quéré and C. Salles

Effects of Aroma Profiles of Piacentinu and Ricotta Cheese Using Different Tool Materials during Cheesemaking   23
S. Mallia, S.Carpino, L.Corallo, L.Tuminello, R. Gelsomino and G. Licitra

Volatile Profiles of Piacentinu Ennese Cheese Produced with Raw and Pasteurized Milk   35
S. Carpino, T. Rapisarda, J. Horne, A.i Falco and G. Licitra

Heat Stability of Ca-caseinate/Whey Emulsions. Effect of pH, Salts and Protein Ratios   42
E. A. Theologou, P. G. Demertzis, M. Minor

Flavor of Wines Produced by Cells Immobilized on Various Supports   51
Y. Kourkoutas, M. Kanellaki, A. Bekatorou, A.A. Koutinas and M. Iconomopoulou

Changes in the Profile of Major Volatile Compounds in Greek Wines Stored under Cellar Temperature Conditions   61
Vasileios C. Siaravas, Panagiotis G. Demertzis and Konstantoula Akrida-Demertzi

Role of Anthocyanins in the Differentiation of Tempranillo Wines   72
E. Revilla, M.J. González-Reig, P. Garcinuño and E. García-Beneytez

Organic Acid Composition of Ume Liqueur   82
Rie Kuramitsu and Shoji Furukawa

Composition

The Volatile Components of Indian Long Pepper, *Piper longum* Linn.   93
L. Trinnaman, N.C. Da Costa, M.L. Dewis, T.V. John

Investigation on Aroma Volatiles from Fresh Flowers of Saffron (*Crocus sativusv*   104
L.)
M. Bergoin, C. Raynaud, G. Vilarem and T. Talou

GC/MS Analysis of the Volatile Compounds of *Tuber Melanosporum* from   115
Tricastin and Alpes de Haute Province F (FRANCE)
Gaston Vernin, Cyril Párkányi and Hervé Casabianca

Characterization of Off-odor of Local Duck Meat   136
Apriyantono, R. Hustiany, J. Hermanianto, P.S. Hardjosworo

Glycosidically-bound Aroma Compounds Present in Green and Cured Vanilla   145
Beans
D. Setyaningsih, A. Apriyantono, M. T. Suhartono

Flavor Studies on Some Amazonian Fruits 1. Free and Bound Volatles of Cocona   156
(*Solanum sessiliflorum* Dunal) Pulp Fruit
Alberto Fajardo, Alicia L. Morales and Carmenza Duque

Flavor Studies on Some Amazonian Fruits. 2. Free and Bound Volatile of Cocona   164
(*Solanum sessiliflorum* Dunal) Pulp Fruit
Alberto Fajardo, Alicia L. Morales and Carmenza Duque

Formation of Flavors

Generation of Potentially New Flavoring Structures from Thiamine by a New   175
Combinatorial Chemistry Program
René M. Barone, Michel C. Chanon, Gaston A. Vernin and Cyril Párkányi

Generation of Aldehydes from Maillard Reaction of Glucose and Amino Acids   213
Jiangang Li and Chi-Tang Ho

The Biosynthesis of Furanones in Strawberry: Are the Plant Cells All Alone?   219
Kyriacou and I. Zabetakis

Evolution of Volatile Compounds and Sensory Rancidity in Purified Olive Oil   224
during Storage under Normal and Accelerated Conditions (25-75 °C)
M.D. Salvador, S. Gómez-Alonso, V. Mancebo-Campos and G. Fregapane

Dewatering-impregnation-Soaking in Nonconventional Solutions as Source of   231
Natural Flavorants of Colombian Azúcar Varieties of Mango (*Mangifera indica*)
and Perolera Pineapple (*Ananas comosus*)
L. Morales, M. P. Castaño, D. C. Sinuco, G. Camacho and C. Duque

Analysis

Application of GCxGC (Comprehensive 2DGC) in Flavor Analyses 243
Hajime Komura and Mineko Kawamura

Aroma Active Norisoprenoids in Orange and Grapefruit Juices 252
Kanjana Mahattanatawee, Russell Rouseff, Kevin Goodner and Michael Naim

Black Truffle Flavor: Investigation into the Impact of High-Boiling-Point Volatiles by GC-Olfactometry 260
Jensen, T. Talou, C. Raynaud and A. Graset

Evidence of the Presence of (S)-Linalool and of (S)-Linalool Synthase Activity in *Vitis vinifera* L., cv. Muscat de Frontignan 271
G. M. de Billerbeck, F. Cozzolino and C. Ambid

Isolation and Identification of Phenolic Glycosides from Quinoa Seeds (*Chenopodium quinoa* Willd) 278
Nanqun Zhu, Shengmin Sang, Robert T. Rosen and Chi-Tang Ho

Potential Migration of Organic Pollutants from Recycled Paperboard Packaging Materials into Dry Food 283
V.I. Triantafyllou, K. Akrida-Demertzi and P.G. Demertzis

Antioxidants and Health

Antioxidant Capacity of Phenolic Extract from Wild Blueberry Leaves and Fruits 293
Marian Naczk, Ryszard Amarowicz, Ryszard Zadernowski, Ron Pegg, and Fereidoon Shahidi

Antioxidative Activity of Cruciferous Vegetables and the Effect of Broccoli on Edible Oil Oxidation 304
Ryoyasu Saijo, Rong Wang, Keiko Saito, Reiko Nakata, Satoko Ofuji, Tomoko Inoue, Yuko Mori, Miwa Motoki and Yoko Tabata

Biological Studies and Antioxidative Activity for White Truffle Fungus (*Tuber borchii*) 312
Emad S. Shaker

Structured Lipids Containing Long-Chain Omega-3 Polyunsaturated Fatty Acids 323
S.P.J.N. Senanayake and F. Shahidi

Hepatic Acute-Phase Response to 3-Alkyl-2-phenyl-2-hydroxymorpholinium Cations 335
F.M. Fouad, O. A. Mamer, F. Sauriol, A. Lesmple, F. Shahidi and G. Ruthenstroth-Bauer

Stimulation of Heptic Acute-Phase Response by Stress, Sucrose Polyester and   341
Zocor in Animal Model
F.M. Fouad, O. Mamer, F. Sauriol, A. Lesimple, F. Shahidi, M. Khayyal, M. Hasseeb and G. Ruthenstroth-Bauer

The Functionality of Buckwheat Sour Juice   350
Kozo Nakamura, Chiho Nakamura, Shigeyoshi Maejima, Masataka Maejima, Michiaki Watanabe, Mayumi Shiro, and Hiroshi Kayahara

Functionality Enhancement in Germinated Brown Rice   356
Kozo Nakamura, Su Tian, and Hiroshi Kayahar

Quality

Comparison of the Influence of Hydrodynamic Pressure (HDP)-treatment and   375
Aging on Beef Striploin Protein
A.M. Spanier, T.M. Fahrenholz, E.W. Paoczay and R. Schmukler

Hydrodynamic Pressure (HDP)-treatment: Influence on Beef Striploin Proteins   391
A.M. Spanier and T.M. Fahrenholz

Use of Capillary Electrophoresis (CE) to Assess the Influence of Hydrodynamic   405
Pressure (HDP)-treatment and Aging of Beef Striploin Proteins. A Method for
Assessment of Meat Tenderness
A.M. Spanier and T.M. Fahrenholz

Changes in Protein Distribution in Beef *Semitendinosus* Muscle (ST) in Samples   418
Showing Varying Response to Hdrodynamic Pressure (HDP)-treatment
A.M. Spanier and T.M. Fahrenholz

Multiquality Enhancement of Muscle Food: A Hypothesis Explaining How   431
Hydrodynamic Pressure (HDP)-Treatment Leads to Meat Tenderness
A.M. Spanier and R.D. Romanowski

Flavor and Quality Characteristics of Bakery Products from Frozen Dough with   447
Various Added Ingredients
V. Giannou and C. Tzia

Shelf-life Prediction and Management of Frozen Strawberries with Time   459
Temperature Integrators (TTI)
E.D. Dermesonlouoglou, M.C. Giannakourou and P.S. Taoukis

Stability of methanolic extract activity for leaves, peels and Citrus seeds under UV   472
Irradiation
Emad S. Shaker

Effect of Different Initial and Supplementary Brining Treatments on the   484
Fermentation of cv. Conservolea Green Olives
E.Z. Panagou, C.C. Tassou, and K.Z. Katsaboxakis

Shelf-life Determination of Untreated Green Olive cv. Conservolea Packed in   492
Polyethylene Pouches under Different Modified Atmospheres
E.Z. Panagou, and C.C. Tassou

Microbiological, Physicochemical and Sensorial Changes of Marinated Fish   500
Products
J. S. Arkoudelos, C.C. Tassou, P. Galiatsatou, and F.J. Samaras

Subject Index   507

Overview

Overview

HALAL ISSUES IN THE FLAVOR INDUSTRY

M. N. Riaz[1]. M. M. Chaudry[2] and M. Sadek[3]

[1]Food Protein Research & Development Center, 2476 Texas A&M University, College Station, TX 77843-2476, USA
[2]Islamic Food and Nutrition Council of America, 5901 N. Cicero Avenue, Suite 309; Chicago, IL 60646, USA
[3]Islamic Food Council of Europe; 4 Rue De La Presse; 1000 Brussels, Belgium

1. ABSTRACT

Halal is a term that describes the foods permitted for consumption by Muslim, who makes up more than 20 percent of the world population. Food products imported by Muslim countries must meet health and Halal requirements. With the advancement of ingredient technology, flavor industry has become quite complex. Use of alcohol and animal derived ingredients in the development of flavors presents challenges for the producers of Halal Foods. Guidelines and requirements for Halal food productions are discusses in this article, to assist those interested in dealing with Muslim and Muslim markets.

2. INTRODUCTION

Food consumed by Muslims which meets the Islamic dietary code is called Halal food. Muslims use two major terms to describe food: Halal and Haram. Halal means permitted or lawful and Haram means forbidden or unlawful. Other terminologies used are makrooh, mashbooh and dhabiha. Makrooh is an Arabic word meaning religiously 'discouraged' or 'disliked'. It covers any foods and liquids which are disguised or harmful to the body. Mash-Booh is also an Arabic word and means 'suspected'; it covers the gray area between the Halal and Haram. Dhabiha means 'slaughtered' it implies that the animal has been slaughtered by a Muslim, according to the Islamic method of slaughter.

The food industry, like any other industry, responds to the needs and desires of the consumer. People all over the world are now more conscious about foods, health and nutrition. They are interested in eating healthy foods that are low in calories, cholesterol, fat, and sodium among others. Many people are interested in foods that are organically produced without the use of synthetic pesticides and other non-natural chemicals. The ethnic and religious diversity in America and Europe has encouraged the food industry to prepare products which are suitable to different groups, Chinese, Japanese, Italian, Indian, Mexican, Seventh Day Adventist, Vegetarian, Jewish, Muslim, etc.

2.1 What are Halal Foods?

By definition, Halal foods are those that are free from any component that Muslims are prohibited from consuming. According to the Quran (the Muslim scripture), all good and clean foods are Halal. Consequently, almost all food of plant and animal origin are considered Halal except those that have been specifically prohibited according to the Quran and the Sunnah (Tradition of Muhammad).

Accordingly, all foods pure and clean, are permitted for consumption by the Muslims except the following categories, including any products derived from them or contaminated with them:

- Carrion or dead animals.
- Flowing or congealed blood.
- Swine, including all by-products.
- Animals slaughtered without pronouncing the name of God on them.
- Animals killed in a manner that prevents their blood from being fully drained from their bodies.
- Animals slaughtered while pronouncing a name other than God.
- Intoxicants of all types, including alcohol and drugs.
- Carnivorous animals with fangs, such as lions, dogs, wolves, tigers, etc.
- Birds with sharp claws (birds of prey), such as falcons, eagles, owls, vultures, etc.
- Land animals such as frogs, snakes, etc.

2.2 How does one translate major prohibitions into practice in today's industrial environment? Let us look at how the laws are translated into practice:

2.2.1 Carrion and Dead Animals. It is generally recognized that eating carrion is offensive to human dignity and probably nobody voluntarily consumes carrion in the modern civilized society. However, there is a chance of an animal dying from the shock of stunning before it is properly slaughtered. This is more common in Europe than in North America. The meat of such dead animals would not be proper for Muslim consumption.[1]

2.2.2 Proper Slaughtering. There are strict requirements for the slaughtering of animals: the animal must be of a Halal species, i.e., cattle, lamb, etc.; the animal must be slaughtered by a Muslim of proper age; the name of God must be pronounced at the time of slaughter; and the slaughter must be done by cutting the throat of the animal in a manner that induces rapid, complete bleeding and results in the quickest death.

Certain other conditions should also be observed. These include considerate treatment of the animal, giving it water to prevent thirst, using a sharp knife, etc. These conditions ensure the humane treatment of animals before and during slaughter. Any by-products or derived ingredients must also be from duly slaughtered animals to be good for Muslim consumption.

2.2.3 Swine. Pork, lard, and their by-products or derived ingredients are categorically prohibited for Muslim consumption. All chances of cross-contamination from pork into Halal products must be thoroughly prevented. In fact, in Islam, the prohibition extends beyond eating. For example, a Muslim must not buy, sell, raise, transport, slaughter, or in any way directly derive benefit from swine or other Haram media.

Overview 5

2.2.4 Blood. Blood that pours forth (liquid blood) is generally not offered in the marketplace or consumed, but products made from blood and ingredients derived from it are available. There is general agreement among religious scholars that anything made from blood is unlawful for Muslims.

2.2.5 Alcohol and Other Intoxicants. Alcoholic beverages such as wine, beer, and hard liquors are strictly prohibited. Foods containing added amounts of alcoholic beverages are also prohibited because such foods, by definition, become impure. Non- medical drugs and other intoxicants that affect a person's mind, health, and overall performance are prohibited, too. Consuming these directly or incorporated into foods is not permitted.

2.3 Foods are broadly categorized into four groups for the ease of establishing their Halal status and formulating guidelines for the industry.

2.3.1 The meat and poultry group contains four out of five Haram (prohibited) categories. Hence higher restrictions are observed here. Animals must be Halal. One cannot slaughter a pig the Islamic way and call it Halal. Animals must be slaughtered by a sane Muslim, while pronouncing the name of God. A sharp knife must be used to severe the jugular veins, carotid arteries, trachea, and esophagus, and blood must be drained out completely. Islam places great emphasis on humane treatment of animals, so dismemberment must not take place before the animal is completely dead, as described earlier.

2.3.2 Fish and Seafood. To determine the acceptability of fish and seafood, one has to understand the rules of different schools of Islamic jurisprudence, as well as the cultural practices of Muslims living in different regions. All Muslims accept fish with scales; however, some groups do not accept fish without scales such as catfish. There are even greater differences among Muslims about seafood, such as molluscs, and crustaceans. One must understand the requirements in various regions of the world. For example for exporting products containing seafood flavors.

2.3.3 Milk and Eggs from Halal animals are also Halal. The predominant source of milk in the West is from cow, and the predominant source of eggs is the chicken. All other sources are required to be labeled accordingly. There are a variety of products made from milk and eggs. Milk is used for making cheese, butter, and cream. Most of the cheeses are made with various enzymes, which could be Halal if made with microorganisms or Halal slaughtered animals. The enzymes could be Haram if extracted from porcine sources or questionable when obtained from non-Halal slaughtered animals. Similarly, emulsifiers, mold inhibitors and other functional ingredients from non-specified sources may make milk and egg products doubtful to consume.

2.3.4 Plant and Vegetable materials are generally Halal except alcoholic drinks or other intoxicants. However, in the modern day processing plants, vegetables and meats may be processed in the same plant and on the same equipment, increasing the chance of cross-contamination. Certain functional ingredients from animal sources may also be used in the processing of vegetables, which make the products doubtful. Hence processing aids and production methods have to be carefully monitored to maintain the Halal status of foods of plant origin.

From the above discussion on the laws and regulations it is clear that there are several factors determining the Halal/Haram status of a particular foodstuff. It depends on its nature, how it is processed and how it is obtained. As an example, any product from pig would be considered as Haram because the material itself is Haram. Similarly, beef from an animal that has not been slaughtered according to Islamic rites would still be considered unacceptable. Food and drink that are poisonous or intoxicating are obviously Haram even in small quantities because they are harmful to health.

Producing Halal food is similar to producing regular foods, except for certain basic requirements. Halal foods can be processed by using the same equipment and utensils as regular food, with few changes.

2.4 Halal and Kosher:

Many of the Muslim rules are similar to Kosher laws in terms of slaughtering rituals of animals and the complete avoidance of pork and all pork derivatives. So, often times consumers tend to assume Kosher is similar to Halal. Whereas Kosher and Halal are two different entities carrying a different meaning and spirit. Kosher certification is not compatible with complex Muslims' dietary laws. There is a philosophical difference between Halal and Kosher certification and Halal may never overlap with Kosher.[3] Alcohol is prohibited for the Muslims, whereas all the wines are Kosher and are part of the meal, Several gelatins are being marketed as Kosher regardless of the origin, and there are some Kosher gelatins whose origin is pork.[2] Similarly, enzymes' source is not a big issue in Kosher but the source of enzymes in cheese is a big issue for Muslims.

2.5 Flavor Issues in Halal Food Production:

The major uses of alcohol today are for alcoholic beverages and as a solvent in the food, cosmetics and pharmaceutical industries. Alcoholic beverages legally can contain between 0.5% and 80% ethyl alcohol by volume. Pure industrial alcohol may be 95% alcohol. Alcoholic beverages can be consumed directly or added to foods, either as an ingredient during formulation or during cooking. When alcohol is an added ingredient, the ingredient label of the food product must list the specific alcoholic beverage that has been added, if the final amount of alcohol is greater than 0.5%. Examples of this would be liqueur-flavored chocolates, cakes and meals containing wine, such as beef stroganoff in wine sauce.

Foods are cooked in alcohol to enhance the flavor or to impart a distinctive flavor. Wine is the most common form of alcohol used in cooking. While it may seem that all of the added alcohol evaporates or burns off during cooking, it does not. Rena Cultrufelli of the USDA has prepared a table listing the amount of retained alcohol in foods cooked in alcohol. The retained alcohol varies depending upon the cooking method. The following gives some of the retained alcohol content of foods prepared by different cooking methods.[4]

- Added to boiling liquid and removed from the heat   85%
- Cooked over a flame   75%
- Added without heat and stored overnight   70%
- Baked for 25 minutes without stirring   45%
- Stirred into a mixture and baked or simmered for 15 minutes   40%
- Stirred into a mixture and baked or simmered for 30 minutes   35%
- Stirred into a mixture and baked or simmered for 1 hour   25%
- Stirred into a mixture and baked or simmered for 2 hours   10%

- Stirred into a mixture and baked or simmered for 2½ hours   5%

Two of the major uses of pure alcohol are as a solvent and raw material. As a solvent, it is used to extract flavoring chemicals from plant materials such as vanilla beans. Dilute ethyl alcohol is almost universally used for the extraction of vanilla beans. After the extraction, vanilla flavor, called natural vanilla flavoring, is standardized with alcohol. By the Food and Drug Administration's) FDA's standard of identity, natural vanilla flavoring must contain at least 35% alcohol by volume, otherwise it may not be called natural vanilla flavoring.[5]

One important function of alcohol is to facilitate the mixing of oil-based ingredients into water-based products or water-based ingredients into oil-based products. This is an important use in the production of flavors. Most flavors are oils. For example, orange flavor is oil derived from orange skins. Orange flavor would not dissolve in water but will dissolve in alcohol. The mixture of alcohol and orange flavor will then dissolve in water. So to produce an orange flavored carbonated drink, alcohol is used to make sure the orange flavor is fully mixed and dissolved in the carbonated water and remains dissolved over the expected shelf life of the product.[6]

Alcoholic beverages of any type are prohibited for Muslims. The use of alcoholic beverages in preparing or producing food items or drinks is also prohibited. Hence eating or drinking products made with alcoholic beverages, such as spiked punch, or cakes containing brandy are not permitted. Grain alcohol or synthetic alcohol may be used in the production of food ingredients, as long as it is evaporated to a final level of less than 0.5% in food ingredients and 0.1% in consumer products. These guidelines are practiced by some of the Halal certification organizations, while others follow a somewhat stricter guidelines.

The following points may be helpful for the use of alcohol in Halal food production.

1. Natural products containing a small amount of intrinsic alcohol do not present a Halal issue.

2. Alcohol contained in a natural product may be concentrated into its essence, thereby concentrating the amount of alcohol. Most Halal certifying bodies would accept a small amount of such inherent alcohol, generally less than 0.1%, and sometimes up to 0.5%.

3. Use of alcohol in any concentration in an industrial process is acceptable due to technical reasons, where other viable alternatives are not available. The final alcohol content in the product of such industrial application must be reduced to less than 0.5% by evaporation or conversion to acetic acid. This means flavors that will be used in food production must not contain more than 0.5% alcohol to qualify as Halal. Some countries, however, do accept amounts higher than 0.5%, while others have an even lower cut-off.

4. Addition of any amount of fermented alcoholic drinks such as beer, wine, liquor, etc., to any food product or drink renders the product Haram. However, if the essence is extracted from these products and alcohol is reduced to negligible amount, most Halal certifying agencies and importing countries accept the use of such essences in food products. Consultation with proper authorities or end users can clarify this issue.

5. Consumer products with added ingredients that contain alcohol must have less than 0.1% alcohol including both added and any natural alcohol to qualify as Halal. At this level one cannot taste the alcohol, smell the alcohol, or see the alcohol, a criterion generally applied for the impurities. This reasoning has been established by the Islamic Food and Nutrition Council of America. Other groups may accept more lenient or stricter guidelines than these. The food industry should consult their customer companies or Halal approval agencies for

their exact stander.

3. HALAL MARKETS AND DEMOGRAPHICS:

The estimates of the number of Muslims in 2002 vary from 1.2 to 1.5 billion. According to a conservative estimate, one-sixth to one-seventh of the world's population is Muslim.[7] According to the Center for American Muslim Research and Information (New York), one-fourth to one-fifth of the world is Muslim. A figure of 1.3 billion quoted by Chaudry (2002)[8] seems to be the best estimate. The Muslim population is estimated to reach 12.2 million in the year 2018 in the U.S.[9] In 1976, fifty-seven (57) countries were reported to have a Muslim majority accounting for about 800 million people in 1980,[10] which would be over one billion in the year 2002.

More than 400 million Muslims are estimated to live as minorities in different nations of the world, forming a part of many different cultures and societies. In spite of their geographic and ethnic diversity, Muslims all follow their belief and the religion of Islam. Halal is a very important and integral part of religious observance for all Muslims. Hence, Halal constitutes a universal standard for a Muslim to live by.

With expanding global markets, innovative food companies are leading the charge by carving a new niche to gain a competitive edge in the marketplace. For the domestic marketer, the question is "How can my company get the additional edge in the marketplace" whereas for the global marketer, the question is "How can I comply with the import requirements of the Muslim countries?" The Halal food market potential in the world is not limited to Muslim countries. Countries like Singapore, Australia, New Zealand, and South Africa (with very small Muslim populations) have become significant contributors to the world Halal trade.

The world is now much more accessible because of improved transportation and communications systems and has become truly a global supermarket. The demand for Halal foods and products in countries around the world is on the increase, as Muslim consumers are creating an educated demand for Halal foods and products. In the past many Muslim countries produced most of their food requirements domestically or imported them from other Muslim countries. However, population increases are outpacing the food supply, and Muslim countries now import food from agriculturally advanced countries. Changes in the food habits of people, where western style franchised food is becoming popular, is another factor in the changes taking place in international marketing.

4. GLOBAL HALAL MARKET:

There are around 1.3 billion Muslims in the world and 1.5 billion Halal consumers, which means that one out of every four human beings consumes Halal products. The difference of 0.2 billion between the Halal consumers and Muslims is accounted for by non-Muslims living in Muslim majority countries where most foods are Halal such as Indonesia, Bangladesh and others. Presently there are two strong markets for Halal products - Southeast Asia and the Middle East.[11] All major U.S. poultry processors export to these markets, while secondary suppliers provide beef. The primary sources of beef in those markets are imports from Australia and New Zealand, whose governments are very supportive of Halal programs. Worldwide the Muslim consumer base is estimated to be

$150 billion per year and is spread over more than 112 countries.[12] Marketing efforts to supply certified Halal products throughout the world are gaining momentum.

Southeast Asia is home to more than 250 million Halal consumers. Indonesia, Malaysia, and Singapore have had regulations to control the import of Halal certified products for a number of years. Recently, Thailand, the Philippines and other countries have realized the value of Halal certified products and their governments are formulating regulations to promote both export and import of Halal certified products. For export to many of the Association of South East Asian Nations (ASEAN), even the simplest of vegetable products must be certified. In this region, even non-Muslim consumers perceive Halal as a symbol of quality and wholesomeness.

Middle Eastern countries are net importers of processed foods both for the food service and retail markets. Saudi Arabia, United Arab Emirates and other Middle Eastern countries have been importing food for decades. North Africa and other African countries also offer opportunities for export of processed food, as their economies and political conditions improve. South Asia comprising India, Pakistan, Bangladesh, and Sri Lanka, is home to almost 1.3 billion people, out of which over 400 million are Muslim. Although this region is an agricultural economy, these countries do import certain processed items, especially for food service.

In the late 80's and early 90's, the potential of the Southeast Asian and Middle Eastern markets for Halal foods started to be realized, leading to an increase in the production and certification of Halal foods. This has expanded into South Asia, the Mediterranean, Europe, and Central Asia. The benefits of trade for the Western corporations with the Muslim majority countries are clear. Even Muslim minority countries like Singapore, South Africa and others have shown that the Halal food business is a good one. Although the Muslim community forms only 16% of the 3.8 million population of Singapore, the Halal food industry is big business in this cosmopolitan city. McDonald's, A & W, Kentucky Fried Chicken, and Taco Bell are some international brands that have gone 100% Halal in Singapore.

The opportunities available to a corporation able to supply Halal products are continuously growing. Muslims are starting to blend the best of Western attitudes with their generally Eastern cultures. Additionally, the large addition of Westerners to the faith of Islam is resulting in some changes in the behavior of the Muslim community. Whereas in the past Muslims simply avoided foods that did not meet the dietary standard of Halal, today Muslims are making their presence felt, socially and politically. Muslims are now requesting food products that meet their dietary needs. They are offering services and cooperating with those producers with the foresight and wisdom to cater to the Muslim consumer.

5. HALAL CERTIFICATION

Most of the food companies look at the Halal symbol as a seal of quality over and above the religious meaning. The Halal seal is also considered a proxy for quality. Since Halal food is perceived by many as a food specially selected and supervised at all stages of preparation and processing to achieve the highest standards of wholesomeness and hygiene. A food company that wants to introduce Halal in their product line should obtain Halal certification of their food product. The situation with regard to Halal food certification has come of age, and several major U.S. food corporations are seriously considering getting into Halal marketing for the American Muslim consumers. The Halal certification is an authoritative, reliable and independent testimony to support the food manufacturer and that their products or food meets the Halal requirements. Muslim customers will have greater confidence in consuming such products or food. Halal certification involves an intention of the company to go Halal, inspection of the production facility, review of sanitation, review of ingredients and labels, and training the company personal is understanding and meeting the Halal requirements.[12]

References

1. M. M. Chaudry. Islamic food laws: Philosophical basis and practical implications. *Food Technology*, 1992, **46(10)**, 92-93, 104.
2. M. Regenstein. The Hook -R- How Kosher supervision can serve the needs of other religious group better. Special Forum "The Flavor Industry and Religious Foods" IFT annual meeting at Atlanta, GA .1998, June 20-24.
3. J. Byron. Kosher rabbis could assist Muslims, vegetarian and others with special diets. Food Chemical News. 1998. July 6. page 4.
4. J. Larsen. Ask the Dietitian. Hopkins Technology, LLC.421 Hazel Lane, Hopkins, MN. 1995. (http://www.dietitian.com/alcohol.html).
5. Food and Drug Adminstration. 2000. 21 CFR 169.3 (c)
6. R. Othman and M. N. Riaz. Alcohol – A drink/ A chemical: Halal Consumer Fall Issue. 2000. pp. 17-21.
7. C. I. Waslien. Muslim dietary laws, nutrition, and food processing. In *Encyclopedia of Food Science and Technology*, Ed. Hui Y. H. Vol. 2. John Wiley and Sons Inc. 1992. pp1848-1850.
8. M. M. Chaudry. Halal Certification Process. Presented at "Market Outlook: 2002 Conference; Toward Efficient Egyptian Processed Food Export Industry in a Global Environment. Cairo, Egypt, 2002.
9. D. Merill, Facts, figures about Islam. Muslim future: Projected Muslim Population in the USA. By USA Today, 1999. June, 25.
10. J. B. Noss, *Man's Religion*, 6th Ed. Macmillan, New York, N.Y., 1980, p 15.
11. M. N. Riaz, Halal Food - An insight into a growing food industry segment. *International Food Marketing and Technology*.1998, **12(6)**, 6-9.
12. M. M. Chaudry, Islamic foods move slowly into marketplace. (Press article). *Meat Processing* 1997, **36(2)**, 34-38.

Dairy and Wine Flavors

REAL TIME RELEASE OF FLAVOR COMPOUNDS AND FLAVOR PERCEPTION. AN APPLICATION TO CHEESE

E. Pionnier, J.L. Le Quéré and C. Salles

Institut National d la Recherche Agronomique, Unité Mixte de Recherche INRA-ENESAD sur les Arômes, 17 rue Sully, BP86510, F21065 Dijon Cedex, France

1. ABSTRACT

Relating flavor (aroma and taste) compounds to flavor perception experienced by a consumer is still a challenge. Real time release of aroma compounds may be monitored in vivo through breath by breath analysis using Atmospheric Pressure Chemical Ionization - Mass Spectrometry. Off-line analysis of the time course release of non-volatiles in the saliva is achievable using a sampling method such as the cotton bud technique followed by subsequent MS or HPLC analyses. Simultaneous time-intensity (TI) measurements allow time course evaluation of sensory attributes (aroma and taste) to be undertaken. An application on a model flavored cheese tries to relate aroma and taste release in the mouth during consumption to time-intensity curves of target sensory attributes and to oral parameters. Larger individual differences were observed for most measured parameters. The aroma release parameters Cmax and AUC (In-nose maximum concentration and area under the curve) could be related to respiratory and masticatory parameters. For non volatile compounds, high Tmax (time to reach Cmax) and AUC values could be related to high chewing time, low saliva flow rates, low chewing rates, low masticatory performance and low swallowing rates. Concerning temporal perception, Tmax of salty, sour and moldy attributes was positively related to Tmax of sodium, citric acid and 2-heptanone and to oral parameters.

2 INTRODUCTION

Flavor release has been extensively studied in recent years.[1,2] Most of the studies were carried out on very simple models consisting of one or a few aroma compounds within a simple aqueous medium. The first experiments were conducted using static and dynamic headspace[3], then, later, breath by breath analyses were developed to follow flavor release in air expired by people eating food.[4] However, few studies were conducted on flavor release from real solid or semi-solid food. In vivo, the aroma stimulus depends on the concentration of aroma compounds in the nosespace. It is affected by release rate of compounds in the mouth which is mainly a function of food matrix composition and the mastication process. Food matrix composition is particularly important in this process because interactions between the matrix and volatile compounds are significant factors for aroma release.[5] Considering mastication, the phenomena occurring during the stay of food in the mouth affect the release of aroma. The movement of food in the mouth then during swallowing, provides information about the transfer of volatile compounds.[6] To study the impact of the mastication process on flavor release *in vitro*, devices simulating more or less precisely eating events have been proposed.[7] Though important information has been learned from these apparatus such as the effect of saliva, shearing, temperature and fat, they are far from reproducing the complexity of the overall phenomenon occurring in a real mouth. Concerning non volatile compounds, few studies were carried out during food mastication. In-mouth sensors were used to measure conductivity and pH during chewing.[8] Davidson et al. recorded sucrose release during chewing in analyzing saliva samples at different times during the chewing sequence by API mass spectrometry.[9]

Individual differences were already observed in numerous Time-Intensity studies. Investigations on the possible influence of some oral parameters including salivary flow and chewing rate were made on aroma release using some model mouth systems.[7] However, very few studies investigated the global effect of several oral parameters on in-vivo flavor release and flavor perception.

The aim of this paper is to present a synthesis of recent results obtained in our laboratory. We have tried to explain the individual differences observed in flavor perception by the individual differences observed during *in vivo* flavor release and in oral parameters.

3 MATERIALS AND METHODS

A general scheme of the procedures used for this study is presented on Figure 1.

3.1 Model Cheese

The food product was a model cheese made with casein rennet, anhydrous milk fat, water, nonvolatile compounds and aroma compounds using analogue cheese technology. The nonvolatile compounds consisted of organic acids (citric and lactic acids), minerals (sodium, potassium, calcium chloride, sodium dihydrogenophosphate, trisodium citrate) and amino acids (L-leucine, L-phenylalanine, L-glutamic acid). The cheese was flavored with 6 aroma compounds: propionic acid, butyric acid, diacetyl, 2-heptanone, ethyl hexanoate and 2-heptanol. The procedure of flavored cheese making is detailed in Pionnier et al.[10]

3.2 Flavor Release

In vivo flavor release was studied both for aroma and non-volatile compounds. Eight panelists took part in oral, physico-chemical and sensory experiments. They were asked to eat 5 g of model cheese without any constraint.

The individual aroma release profiles were obtained by performing nosespace analyses for 3 min while eating cheese. Expired air from subjects' noses was sampled and analyzed simultaneously by Atmospheric Pressure Ionization Mass spectrometry (in the gaseous phase) and by Solid Phase Microextraction Gas chromatography Mass spectrometry according to the nature of the compounds as described by Pionnier et al.[10]

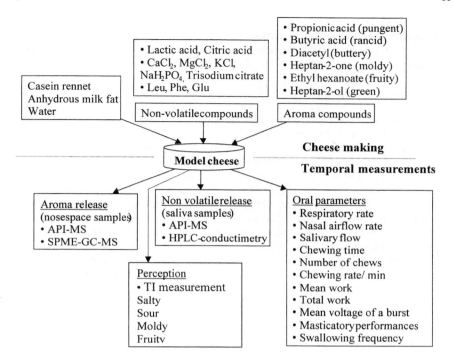

Figure 1 *Procedure for model cheese making and physicochemical, sensory and oral parameters measurements.*

The individual non-volatile release profiles were obtained in a discontinuous way by sampling saliva at several times while eating cheese. Saliva was sampled on the tongue and analyzed by API-MS after extraction (in the liquid phase) for free amino acids and HPLC using ionic chromatography for minerals and organic acids as described by Pionnier et al.[11]

To summarize the information contained in the flavor release curves, some parameters were defined and studied by analysis of variance. These parameters were Cmax (maximum concentration), Tmax (time when Cmax is reached), slope (initial slope of the curve measured between 0 and 10 sec) and finally, AUC (Area Under the Curve) corresponding to the total flavor release over a 3-min period.

3.3 Oral Parameters

Several oral parameters were measured during the mastication of the cheese. First, breath parameters including respiratory rate and nasal air flow were monitored with a flow meter. Simultaneously, chewing parameters including chewing time, number of chews, chewing rate/min, mean and total work and mean voltage of a burst were determined via Electromyography recording masseter and temporal activities.[12]

Swallowing events were controlled using a necklace strain gauge (which provides a deviation in the baseline when a swallow was triggered). Masticatory performance and salivary flow rate during mastication of parafilm were also determined as described in Pionnier et al.[13]

3.4 Temporal Perception

Concerning sensory analysis, a modified Time Intensity procedure was used to investigate perception changes during the consumption of the cheese.[13] The eight subjects did not know when and what attributes they would be asked to note. Two mastication sequences were used to obtain the 9 notation times from 5 to 180 sec. They assessed 5 attributes over a 3-min period of chewing: fruity, moldy and rancid for aromas and salty and sour for tastes. The 5 attributes were mixed within each sequence and each mastication. For each attribute, all points were obtained within the same session. Once the attribute appeared on the screen, the subjects had 3 sec to read it and note their perception on a 10 cm unstructured horizontal scale. The experiments were done in triplicate. In order to reduce individual variation due to the way of using the notation scale, raw T-I data were first calibrated with a visual calibration using a series of gray surfaces. After obtaining all the intensities, all the points were joined. We obtained the perception of each attribute over a 3-min period. In the same way as in the flavor release study, four parameters were extracted from individual T-I curves: Imax (maximum intensity), Tmax, slope, and finally, AUC.

3.5 Statistical Treatment

Analysis of variance, correlation and regression analysis were performed with SAS software version 8.01 (SAS Institute Inc. Cary NC). Statistical models used are detailed in Pionnier et al.[10,11,13]

4 RESULTS AND DISCUSSION

To investigate relationships between temporal perception (time intensity study), flavor release and oral parameters, we proceeded in two phases. First, we related flavor release to oral parameters and, secondly, time intensity to flavor release taking into account the results found in the first phase. Indeed, it was easier to relate flavor release to oral parameters than it was to relate sensory analysis to oral parameters as there is no integration step in flavor release measurement.

Our main goal was to know if the individual differences observed in flavor release or temporal perception were explained by the individual differences observed in oral parameters.

4.1 Flavor Release

Aroma release profiles illustrating great individual variation obtained for two aroma compounds (2-heptanone and ethyl hexanoate) and four panelists are presented in Figure 2.

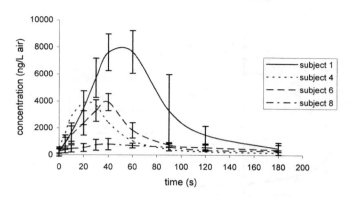

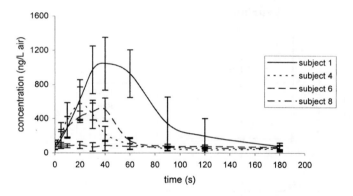

Figure 2 *Volatile release patterns observed for heptan-2-one and ethyl hexanoate (both analysed continuously by API-MS) for subjects 1, 4, 6 and 8 over a 3-min period. Data were smoothed and the standard deviations were reported for some sampling times (5, 10, 20, 30, 40, 60, 90, 120, 180 min).*

Subject 1 released significantly more 2-heptanone and ethyl hexanoate than the other subjects. On the contrary, subject 8 released the least 2-heptanone and ethyl hexanoate. It can be noticed that even if there were differences in release profiles between the subjects, significant and positive correlations are observed between the release parameters for 2-heptanone, ethyl hexanoate and 2-heptanol. Subjects with high values of Tmax, Cmax, slope and AUC for 2-heptanone also showed high values of release parameters for ethyl hexanoate and 2-heptanol.

The same observations could be made from the nonvolatile release profiles. Figure 3 presents two examples, for four panelists, illustrating the great individual variation obtained for calcium and L-leucine release.

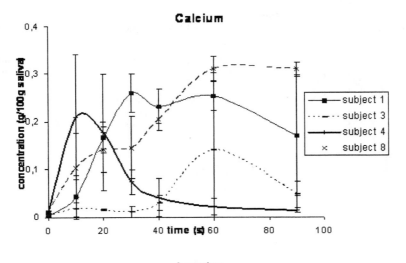

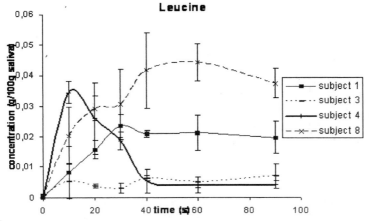

Figure 3 *Variation in the concentrations of calcium and L- leucine in saliva during the eating of the processed cheese. Data obtained were not smoothed. Standard deviations were reported for the times corresponding to the saliva sampling times.*

Different patterns of release could be observed according to the panelist. In particular, a great difference between subjects 4 and 8 for the Tmax parameter was observed. Nonetheless, for a given panelist, the shapes of the curves were similar for most of the 12 non volatile compounds (NVC). Numerous significant and positive correlations were found between the release parameters of the NVC (not shown, see[11]).

In order to explain individual differences observed in flavor release by the individual differences observed in oral parameters, regression analyses between release parameters and oral parameters were performed. Table 1 shows the significant relationships between the oral parameters and some aroma compounds and NVC. Subjects with high Cmax and AUC values for 2-heptanol, 2-heptanone and ethyl hexanoate were also the subjects with high values of chew number, muscle activity and respiratory rate. Concerning NVC, as no significant result was observed at the 5 % level, all the variables with a P value lower than 15 % were reported in Table 1. For several taste compounds, subjects with high Tmax values were also subjects with high values of chewing times and with low values of salivary flow rate, chewing rate, swallowing rate and masticatory performance.

Table 1 *Relationships between oral parameters and flavor compounds release parameters.*

	Respiratory rate	Chew number	Muscle activity	Chew time	Salivary flow	Chew rate	Swallow rate	Masticatory performances
Cmax Hol	NS (P>0.05)	+	+			NS (P>0.05)		
Cmax Hone	+	+	+					
Cmax EH	+	+	+					
AUC Hone	+	+	+					
AUC EH	+	+	+					
Tmax Ca		NS (P>0.15)		+	-	-	-	-
Tmax K				+	-	-	-	-
Tmax Leu				+	-	-	-	-
Tmax Na				+	-	-	-	-

Cmax : maximum concentration ; AUC : area under the curve ; Tmax : time at the maximum concentration ; Hol : 2-heptanol ; Hone : 2-heptanone ; EH : ethyl hexanoate
NS : non significant ; + : positive correlation ; - : negative correlation

Thus, different oral parameters are involved in aroma and NVC release. Aroma release is essentially explained by respiratory rate, the number of chews and the activity of masticatory muscles. NVC release is essentially explained by chewing time and by salivary flow rate, swallowing rate and masticatory performance.

4.2 Temporal Perception

Concerning sensory analysis, two examples of TI curves illustrating the great individual variation obtained for moldy and salty attributes are presented in Figure 4.

In particular, subjects 7 and 3 perceived significantly more moldy and salty attributes than the other subjects.

Regression analyses between temporal perception (TI) and flavor release were performed to try to explain individual differences observed in perception by the individual differences observed in flavor release taking into account relationships between flavor release and oral parameters. Two conditions were fixed to relate sensory data to physicochemical data. First, as we used a model cheese with a known composition, each perceived attribute could be directly related to one or several chemicals incorporated in the cheese. For instance, the moldy descriptor was related to 2-heptanone, fruity to ethyl hexanoate, rancid to butyric acid, salty to the sodium ion and sour to citric acid. Then, for each attribute, we chose to perform regression analyses between each T-I parameter and the corresponding physico-chemical parameter obtained during the study of kinetics of aroma and non-volatile release. Indeed, we tried to relate the Tmax obtained during the sensory study with the Tmax obtained during the flavor release study. We also tried this for the other parameters. The results of the regression analyses (P>0.1) are presented in Table 2. The significant relationships concerned the Tmax parameter. Tmax of moldy, Tmax of sour and Tmax of salty attributes are positively related to the Tmax of 2-heptanone, Tmax of citric acid and Tmax of Na, respectively. So, subjects with high values of Tmax for the moldy and sour attributes are also the subjects with high values of Tmax of 2-heptanone and Tmax of citric acid. Subjects with high values of Tmax of salty attributes also have high values of Tmax of Na.

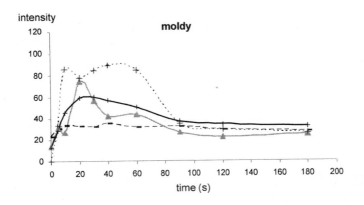

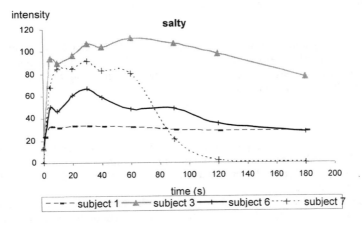

Figure 4 *T-I curves obtained for moldy and salty descriptors considering subjects 1, 3, 6 and 7 over a 3-min period. Each point of the curve represents a measurement time. Measurements are the mean of three replicates.*

Table 2 *Relationships between Tmax of flavor attributes and Tmax of flavor compounds.*

T max	Heptan-2-one		Citric acid		Sodium	
	R^2	P value	R^2	P value	R^2	P value
Moldy	0.38	0.10				
Sour			0.43	0.08		
Salty					0.47	0.06

R^2: regression coefficient ; Tmax: time at maximum concentration

We had previously shown that individual variation in Tmax of Na could be explained by oral parameters. Thus, the Tmax of the salty attribute is directly related to the Tmax of Na and indirectly related to oral parameters. However, no relationship with Imax, AUC and slope could be observed.

The number of relationships with perception is rather low. That can be explained by the high complexity of the mechanisms of perception. These include adaptation, integration steps and the relationships including perception, flavor release, texture and oral parameters that should be taken into account. Moreover, the phenomenon in the mouth occurring during chewing, already investigated by several authors,[14-16] are not well known yet. Their study should contribute to better understanding of how flavor release can be related to perception.

5 CONCLUSION

The methodologies used in this study allowed us to highlight relationships between flavor release and oral parameters. Aroma release is related to respiratory rate, the number of chews and the activity of masticatory muscles; NVC release to chewing time and to salivary flow rate, swallowing rate and masticatory performance. Concerning flavor perception, we could only show 3 significant relationships between temporal perception, flavor release and oral parameters out of the twenty possible relationships. Perception is directly related to flavor release and indirectly to oral parameters for Tmax of the salty attribute. It is also directly related to flavor release for Tmax of moldy and Tmax of sour attributes.

The relationships observed should be confirmed. The use of a higher number of panelists could increase the strength of the statistical treatment and give more significant relationships. Also, as generally several parameters change at the same time while chewing food, the use of a chewing simulator able to reproduce *in vitro* most of the physiological function of the mouth could allow us to see independently the effect of each oral parameter on flavor release.

Acknowledgements

We would like to acknowledge: Dr Sophie Nicklaus and Claire Chabanet (UMR Arômes, INRA-ENESAD, Dijon, France) for help in perception measurement and statistical treatment, respectively, Dr Laurence Mioche (SRV, INRA, Theix, France) for electromyography measurements and Pr Andrew J. Taylor (Samworth flavor laboratory, Nottingham University, UK) for help in API-MS of non volatile compounds.

References

1. A.J. Taylor, *Crit.Rev. Food Sci. Nut.*, 1996, **63**, 765.
2. S.M. Van Ruth, C.H. O'Connor and C.M. Delahunty, *Food Chem.*, 2000, **71**, 393.
3. J. Bakker, N. Boudaud and M. Harrison, *J. Agric. Food Chem.*, 1998, **46**, 2714.
4. A.J. Taylor, R.S.T. Linforth, B.A. Harvey and B. Blake, *Food Chem.*, 2000, **71**, 327.
5. E. Guichard, *Food Rev. Int.*, 2002, **18**, 49.
6. A. Buettner and P. Schieberle, *Food Chem.*, 2000, **71**, 347.
7. K.D. Deibler, E.H. Lavin, R.S.T. Linforth, A.J. Taylor and T.E. Acree, *J. Agric. Food Chem.*, 2001, **49**, 1388.
8. J.M. Davidson, R.S.T. Linforth and A.J. Taylor, *J. Agric. Food Chem.*, 1998, **46**, 5210.
9. J.M. Davidson, R.S.T. Linforth, T.A. Hollowood and A.J. Taylor, *J. Agric. Food Chem.*, 1999, **47**, 4336.
10. E. Pionnier, C. Chabanet, L. Mioche, J.L. Le Quéré and C. Salles, *J. Agric. Food Chem.*, 2004, **52**, 557.
11. E. Pionnier, C. chabanet, L. Mioche, A.J. Taylor, J.L. Le Quéré and C. Salles, *J. Agric. Food Chem.*, 2004, **52**, 565.
12. L. Mioche, P. Bourdiol, J.F. Martin and Y. Noël, *Arch., Oral Biol.*, 1999, **44**, 1005.
13. E. Pionnier, S. Nicklaus, C. Chabanet, L. Mioche, A.J. Taylor, J.L. Le Quéré and C. Salles, *Food Qual. Pref.*, 2004, in press.
14. A. Buettner, A. Beer, C. Hannig, M. Settles and P. Schieberle, *Food Qual. Pref.*, 2002, **13**, 497.
15. R.S.T. Linforth, F. Martin, M. Carey, J. Davidson and A.J. Taylor, *J. Agric. Food Chem.*, 2002, **50**, 1111.
16. M. Hodgson, R.S.T. Linforth and A.J. Taylor, *J. Agric. Food Chem.*, 2003, **51**, 5052.

EFFECTS ON AROMA PROFILE OF PIACENTINU AND RICOTTA CHEESE USING DIFFERENT TOOL MATERIALS DURING CHEESEMAKING

S. Mallia[1], S. Carpino[1], L. Corallo[1], L. Tuminello[1], R. Gelsomino[1] and G. Licitra[1,2]

[1]*CoRFiLaC,* Regione Siciliana, S.P. 25 km 5, 97100 Ragusa, Italy.
[2]Dipartimento di Scienze Agronomiche, Agrochimiche e delle Produzioni Animali, Catania University, 95100 Catania Italy.

1 ABSTRACT

Solid phase microextraction (SPME) coupled to gas chromatography/mass spectrometry/olfactometry (GC/MS/O) was used to determine volatile profile of Piacentinu and Ricotta cheese, produced by traditional (wood/copper) and stainless steel tools. Cheeses made by wood's tools showed a richer aroma profile, contening 2,3-butanedione, 3-hydroxy-2-butanone, 1,2-dimethyl hydrazine, limonene, copaene, α-caryophyllene, 3-methyl butanal, 2-furanmethanol, 2-hexene, dimethyl sulfone, and 2,6 dimethyl-3-ethyl- pyrazine, that were not found in copper-cheese and steel-cheese. Natural microflora of the wooden vat, may influence and enrich the volatile profile of these cheeses.

2 INTRODUCTION

Piacentinu Ennese is a traditional pressed ewe's cheese, produced in the central area of Sicily, in the province of Enna. Saffron and black pepper, added during cheese-making, give to this cheese spicy and floral aroma, as well as a fruity and piquant taste. The rind is yellow and wrinkled, due to the rush basket shape that contains the curd before salting; the body is yellow and the texture is compact, with small round eyes. The cheeses are aged from one month (fresh cheese), two-four months (semi-aged) and over four months (aged cheese). During traditional Piacentinu cheese-making, the raw milk is warm up at 35°C and is curdled, without adding any starter culture, in a special wooden or copper vat, called *tina*; the breaking of the curd is done using the "*ruotula*", wooden stick with one rounded club end. Ricotta cheese is also produced during Piacentinu cheese-making, from the whey; it is a fresh cheese, characterized by a fruity and spicy aftertaste, due to the water soluble compounds from saffron.

The aim of this study was to compare the effects of traditional (wood/copper) and modern (stainless steel) tools used during cheese-making process on the volatile fraction of Piacentinu and Ricotta cheese. So far, little work has been done to study the effects of traditional tools on aroma and sensory properties of cheese.

Recently, Licitra[1] and Corallo[2] demonstrated that the surface of the wood used for traditional tools during Ragusano cheese-making develops a natural microflora, that serves as a starter culture and reservoir for the growth of desirable bacterial. The bacterial flora are necessary to develop acidity in the curd and for its ripening. The microflora composes a natural biofilm of the wooden *tina*, that may influence and enrich the volatile profile and the sensory characteristics of Ragusano cheese. The present work is an initial attempt to define the best conditions and the best available materials to be used during cheese-making process that will yield the highest aromatic, microbiological and sensory qualities of Piacentinu and Ricotta cheese.

3 MATERIALS AND METHODS

3.1 Cheese Samples

The experiments were carried out in a traditional farm located in the province of Enna, in the period February-March 2004. Twenty-seven Piacentinu cheese samples were produced from raw ewe's milk, without starter addition, using artisan lamb rennet made on the same farm. The cheese processing was carried out using 3 different materials for the *tina* (container where the milk is curdled): wood, copper and stainless steel (Figure 1).

Each cheese processing used 50 L of milk to produce 3 cheeses: 3 Wood-cheeses (W), 3 Copper-cheeses (C) and 3 stainless Steel-cheeses (S). The experiment was replicated 3 times. Nine cheese samples (3 W, 3 C, 3 S), one from each cheese-making, were ripened during two months and then analyzed. The other cheeses will be sampled at 4 and 6 months for future analyses. Ricotta cheese was produced from the whey and all these samples were immediately analyzed.

3.2 Volatile Organic Compounds Analysis

3.2.1 Sample Preparation
The Piacentinu cheese samples were finely grated, mixed thoroughly and frozen at −20 °C until analysed. Fresh Ricotta cheese was homogenized by blender and used immediately for analysis.

3.2.2 Cheese Volatile Sampling Conditions
Five grams of Piacentinu or Ricotta cheese were placed in a 20 mL headspace vial stopped with a PTFE (Polytetrafluoroethylene)/silicone septum (Supelco, Bellefonte, PA).

After equilibration for 30 min at 35°C, in a thermostatic bath, a SPME (solid phase microextraction) manual holder equipped with a 1 cm x 50/30 μm Stableflex divinylbenzene/carboxen/polydimethylsiloxane (DVB/CAR/PDMS, Supelco, Bellefonte, PA) was inserted through the PTFE cap. The fiber coating was exposed for 1 h to the sample headspace and then desorbed in the injector port of a gas chromatograph. The same fiber was used for all the extractions.

3.2.3 Gas Chromatography/Mass Selective Detection/Olfactometry (GC/MSD/O)
Volatile compounds extracted by SPME fiber were injected, in triplicate for each sample, into a HP 6890 gas chromatograph (Hewlett-Packard, Palo Alto, CA) equipped with a HP-5 MS capillary column (30 m x 0.25 mm x 0.25 μm film thickness, HP, Palo Alto, CA).

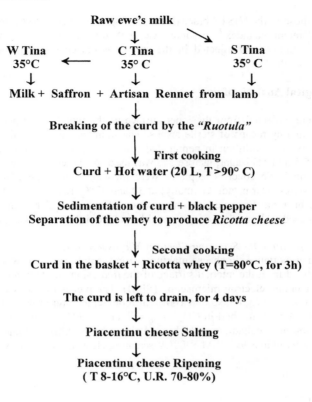

Figure 1. *Flowchart of the experimental protocol used for processing of Piacentinu and Ricotta cheese*

Mass selective detection (HP 5973) and olfactometric analysis (Datu Inc., Geneva, NY) were carried out simultaneously, splitting the flow at the end the capillary column with split ratio (MSD/O) = 3:1. Chromatographic conditions were as follows: temperature 1, 30°C (hold 3 min); rate 6°C/min; temperature 2, 200°C (hold 0); rate 2, 30 °C/min; final temperature, 240°C. Injection temperature, 250°C; mode: splitless. The mass spectrometer was operated in the electron impact ionization mode (70 eV). Mass range was from 33 to 300 amu; scan rate was 5.4 scan/sec. GC/O analysis was performed using the single sniff method, where the sniffer was trained using a procedure and a group of standard compounds designed for GC/O subject selection.[3] The sniffer was exposed during 30 min to a stream of humid air mixed with the effluent from the column and described the odor perceptions.[4] The odor information was converted to retention index, calculated relative to a series of normal alkanes (C7-C20) previously injected into the same gas chromatograph.[5,6] Volatile organic compounds were identified comparing the mass spectra of sample with those in the NIST library (NIST, 1998), using FlavorNet Internet database [7] and published retention index,[8] in some case, authentic standard compounds (Sigma-Aldrich, Milano, Italy) were injected in the same gas chromatograph under the same conditions.

3.3 Microbiological Analysis of *tina*

The samples were collected from the inner surface of the wooden, copper and stainless steel *tina* (500 cm^2) by means of sterile swabs. The samples were immediately analyzed after collection by serial dilution in peptonated water and plating on agar media. MRS acidified to pH 5.5 and M17 agar, láctose 5% was used for the enumeration, respectively, of lactobacilli and lactic cocci. The plates were incubated under anaerobic conditions (Anaerocult A, Merck, Darmstadt, Germany) at 30 and 42°C, respectively, for mesophilic end termophilic bacteria. For selective enumeration of enterococci Kanamycin Aesculin Azide Agar was used. All media were procured from Oxoid (Basingstoke, UK).

3.3.1 Sample Preparation for Scanning Electron Microscopy (SEM)
Because of their importance, in cheese making process, small pieces of wood material (4 mm^2) were removed from the internal surface of the *tina* using sterile blades and used for observation at scanning electron microscopy (SEM). The pieces of *tina* were dehydrated in graded ethanol series (75, 85, 95 and 100%) at room temperature for 12 h in each bath, dried by the critical point method in CO_2 using a Polaron CPD7501 instrument (Polaron, Watford, UK), and gold palladium coated in a Polaron SC7620 mini sputter coater. The samples were observed in a Jeol JSM 5900LV scanning electron microscopy (Jeol, Tokyo, Japan).

4 RESULTS

4.1 Cheese Volatile Organic Compounds

4.1.1 Piacentinu Ennese cheese
The analyses showed that Piacentinu W cheeses were richer in impact aroma volatile compounds than C and S cheeses (Table 1). Forty two volatile compounds were detected by GC/MSD/O in W cheeses, whereas 36 were found in C cheeses and 37 in S cheeses.
The W cheeses were characterized by the presence of aldehydes, aromatic compounds, ketones and terpenes. Twenty-one odor-active compounds were smelled by GC/O in W samples. Among those, 9 odorant compounds were "unique", that were not found in C and S cheeses: butanol ("solvent"), 1-octen-3-ol ("mushroom"), (Z)-3-nonenal ("green"), (Z)-2-nonenal ("green leaves"), p-anisaldehyde ("minty"), 2,3-butanedione ("butter"), 2,6-dimethyl-3-ethyl pyrazine ("roasty"), α-ionone ("earthy"), and an unknown compound (RI=813, "earthy" odor).

The analyses showed C and S Piacentinu cheeses having similar volatile and odor profiles; in fact in both of cheese types were identified 2-methyl-1-propanol ("fruity"), ethyl-2-methyl butyrate ("fruity"), propyl butyrate ("apple"), methional ("potato"), 2-phenyl ethanol ("sweet/honey") and ethyl decanoate ("grape/fruity"), which were not found in W cheeses. The C and S cheeses contained more esters and alcohols than W cheeses. Compounds found in C and S cheeses such as 3-methyl-1-butanol and 2-methyl-1-propanol indicate the reduction of the aldehyde produced from leucina.[9] The presence of

Table 1 *Volatile compounds identified in W, C and S Picentinu Cheese by GC/MS/O*

Compounds GC/MS	Chemical class	Odor	[a] LRI	[b] Identification	W	C	S
hexane	hydrocarbon	nd	600	MS, PI		x	
2,3-butanedione	ketone	buttery	617	MS, PI	x		
unknown		buttery	661	PI		x	
ethanol	alcohol	nd	668	MS, PI		x	x
acetic acid	acid	nd	702	MS, PI	x	x	x
3-hydroxy-2-butanone	ketone	nd	718	MS, PI	x		
3-methyl-1-butanol	alcohol	nd	736	MS, PI		x	x
2-methyl-1-propanol	alcohol	fruity	752	MS, PI		x	x
2-pentanone	ketone	nd	754	MS, PI	x	x	x
1,2-dimethyl hydrazine	amine	nd	755	MS	x		
toluene	aromatic	nd	773	MS, PI	x		
ethyl butanoate	ester	apple	796	MS, PI		x	x
butanol	alcohol	solvent	798	MS, PI	x		
unknown		earthy	813	nd	x		
ethyl-2-methyl butyrate	ester	fruity	850	PI		x	x
butanoic acid	acid	butyric	861	MS, PI	x	x	x
2-heptanone	ketone	nd	895	MS, PI	x	x	x
propyl butyrate	ester	apple	900	PI		x	x
methional	aldehyde	potato	905	PI, ST		x	x
methoxy-phenyl oxime	aromatic	nd	913	MS	x		
bicyclo(3,1,0)hex-2-ene,2-methyl-5-(1-methylethyl)*	hydrocarbon	nd	926	MS	x	x	x
α-pinene+	terpene	fresh	932	MS, PI	x	x	x
unknown		floral	966	nd		x	x
camphene	terpene	nd	974	MS, PI	x	x	x
bicyclo(3,1,0)hexane,4-methylene-1(1-methylethyl)*	hydrocarbon	nd	974	MS	x	x	x
1-octen-3-one	ketone	mushroom	979	PI, ST	x	x	x
1-octen-3-ol	alcohol	earthy/mushroom	983	PI, ST	x		
sabinene+	terpene	earthy/wood	991	MS, PI	x		x
β-mircene+	terpene	nd	992	MS, PI	x	x	x
β-pinene+	terpene	nd	1002	MS, PI	x	x	x
ethyl hexanoate	ester	fruity	1003	MS, PI	x	x	x
α-phellandrene+	terpene	green apple	1005	MS, PI	x	x	x
hexanoic acid	acid	nd	1019	MS, PI	x	x	x
1-methyl-4-(1-methylethyl)-benzene*	aromatic	nd	1027	MS	x		x
limonene+	terpene	nd	1033	MS, PI	x		
3-carene+	terpene	nd	1034	MS, PI	x	x	x
2,6-dimethyl-3-ethyl pyrazine	pyrazine	roasty	1084	MS, PI	x		

Table 1. Continued

Compounds GC/MS	Chemical class	Odor	[a] LRI	[b] Identification	W	C	S
cyclohexene,1-methyl-4-(1-methylethylene)*	hydrocarbon	nd	1089	MS	x	x	x
2-nonanone	ketone	nd	1093	MS, PI	x	x	x
(Z)-3-nonenal	aldehyde	green	1096	PI	x		
3,7-dimethyl-1,6-octadien*	hyrocarbon	nd	1102	MS	x	x	x
nonanal	aldehyde	fresh	1103	MS, PI	x	x	x
2-phenyl ethanol	alcohol	sweet/honey	1119	MS, PI		x	x
unknown		fruity	1136	PI		x	x
(Z)-2-nonenal	aldehyde	green/hay	1149	PI	x		
2-nonenal	aldehyde	green	1156	MS, PI	x	x	x
(E)-2-nonenal	aldehyde	hay	1162	PI	x	x	x
unknown		plastic/solvent	1172	nd	x		x
ethyl octanoate	ester	nutty	1197	MS, PI	x	x	x
(2,6,6-trimethyl)-1,3-cyclohexadiene-1-carboxaldehyde* (safranal)	aldehyde	almond	1207	MS	x	x	x
p-anisaldehyde	aldehyde	minty	1246	PI	x		
copaene	terpene	nd	1387	MS, PI	x		
ethyl decanoate	ester	grape/fruity	1391	MS, PI		x	x
α-ionone	terpene	wood	1411	MS, PI	x		
β-caryophyllene+	terpene	nd	1469	MS, PI	x		
					42	36	37

[a] LRI, linear retention index using a DB-5 column. [b] Identification: MS=spectra comparison using NIST library; ST=authentic standard injection; PI=comparison with published LRI; nd=not detected.* volatile compounds from saffron (D'Auria, M. and al., 2004) + volatile compounds from black pepper (Tairu, A. O. and al., 1999)

higher esters in C and S Piacentinu, due to the reaction between short- to medium-chain fatty acid and alcohols derived from lactose fermentation or from amino acid catabolism,[10] give to these cheeses fruity and floral notes.

Interesting is the presence of 2,3-butanedione and 3-hydroxy-2-butanone, that occurred only in the W Piacentinu cheeses; these compounds are mainly due to the activity of lactic acid bacteria,[11] from raw milk and from wooden *tina*. The *tina* may play an important role in determining the odor quality of cheeses, in fact the microorganisms naturally present in the wood contribute to the formation of volatile compounds.[1] The terpenes found in the cheeses come from the black pepper,[12] added to Piacentinu, and from the plants of pasture.[13] Hydrocarbons and safranal are mainly derived from saffron,[14] added to the milk during coagulation.

4.1.2 Ricotta Cheese
The W Ricotta cheese showed different quality and a higher number of volatile compounds (43) than in C (36) and S (33) samples (Table 2). W Ricotta cheese was characterized by the presence of aldehydes and hydrocarbons. Then "unique" odorants were identified in W Ricotta cheese: methional ("potato"), 2-nonenal ("green"), (*E,E*)-2,4-decadienal ("fried oil"), 3-methyl butanoate ("fruity"), 3-methyl butyl acetate ("fruity"), 2,3-butanedione ("butter"), γ-nonalactone ("coconut"), 2,6-dimethyl-3-ethylpyrazine ("roasted"), and two unknown compounds having respectively RI= 683 ("garlic") and RI=863 ("caramel"). The C and S samples showed similar aroma profiles and were characterized by the presence of alcohols, as found in Piacentinu cheese. In C cheeses were found three "unique" odorants: furaneol ("creamy"), 5-ethyl-(5H)furan-2-one ("spicy") and 2-ethyl-5-methylpyrazine ("fruity"). Ethyl propionate ("fruity") and 1-octen-3-ol ("mushroom") were smelled and identified only in S samples. C and S Ricotta cheeses were characterized by fruity, sweet and vanilla notes.

4.2 Microbiological results and SEM observation

The microbiological results on wood, copper and stainless steel *tina* are shown in Figure 2. This figure shows that the LAB presence is very different comparing the type of *tina* analyzed (wood, copper, stainless steel). Enterococci were absent in all type of *tina*.

The LAB detected in the inner surface of *tina* decreases going from the wooden to the copper to the stainless steel. In detail the number of lactic cocci, lactobacilli mesophilic and termophilic in wood is 3-log units greater then lactic acid bacteria present on copper *tina*. Lactic cocci, lactobacilli and enterococci were instead absent in the stainless steel vat. The observation of pieces of wooden *tina* by scanning electron microscope (SEM) showed a relevant presence of lactic acid bacteria colonized inner surface of wooden *tina*, composing a natural biofilm. Biofilms are microbial communities that exist on abiotic and sometimes biotic surface.[15] A big and close group of cocci were observed confirming the presence of natural micro flora in the porous wood structure, as showed in Figure 3 (7,000X). This observation on Piacentinu cheese-making confirms previous studies on Ragusano cheese,[1,2] demonstrating that natural micro flora on the wooden surface of *tina* is released to the milk, having the role of starter culture in the traditional cheese-making process.

Table 2 Volatile compounds identified in W, C and S Ricotta by GC/MS/O

Compounds GC/MS	Chemical class	Odor	aLRI	bIdentification	W	C	S
2-propanone	ketone	nd	503	MS, PI		x	x
hexane	hydrocarbon	nd	600	MS, PI	x	x	x
2,3 butanedione	ketone	buttery	617	MS, PI	x		
3-methyl butanoate	ester	fruity	659	MS	x		
3-methyl butanal	aldehyde	nd	663	MS, PI	x		
ethanol	alcohol	nd	668	MS, PI	x	x	x
unknown	-	garlic	683	-	x		
pentanal	aldehyde	nd	697	MS, PI	x		x
heptane	hydrocarbon	nd	700	MS, PI	x	x	x
acetic acid	acid	nd	702	MS, PI	x		
ethyl propyonate	ester	fruity	704	MS			x
3-methyl-1-butanol	alcohol	nd	729	MS, PI			x
toluene	aromatic	solvent	754	MS, PI	x	x	x
2-pentanone	ketone	nd	754	MS, PI	x	x	x
1-pentanol	alcohol	nd	763	MS, PI		x	x
2-octene(Z)	hydrocarbon	nd	792	MS	x	x	x
hexanal*	aldehyde	nd	795	MS, PI	x		
ethyl butanoate	ester	fruity	798	MS, PI	x	x	x
ethyl 2-methyl butyrate	ester	fruity	848	MS, PI		x	x
unknown	-	caramel	863	-	x		
1,2-dimethyl benzene	aromatic	nd	867	MS		x	
2-furanmethanol*	alcohol	nd	871	MS	x		
3-methyl butyl acetate	ester	fruity	878	MS	x		
2-heptanone	ketone	nd	890	MS, PI	x	x	x
nonane	hydrocarbon	nd	896	MS, PI	x		
heptanal	aldehyde	soapy	900	MS, PI	x	x	x
methional	aldehyde	potato	905	PI, ST	x		
methoxy-phenyl-oxime	aromatic	nd	918	MS	x	x	x
2-hexene (Z)*	hydrocarbon	nd	925	MS	x		
2(5H)-furanone*	ketone	spicy	926	MS	x		
a-pinene+	terpene	nd	932	MS, PI	x	x	x
dimethyl sulfone*	sulfur	nd	940	MS	x		
benzaldehyde*	aldehyde	nd	960	MS, PI	x	x	
b-pinene+	terpene	nd	978	MS, PI		x	
1-octen-3-ol	alcohol	mushroom	981	PI, ST			x
2-ethyl-5-methyl pyrazine	pyrazine	fruity	991	PI		x	
b-mircene+	terpene	nd	993	MS, PI		x	x
octanal	aldehyde	nd	1002	MS, PI		x	
ethyl hexanoate	ester	fruity	1003	MS, PI	x	x	x

Table 2. Continued

Compounds GC/MS	Chemical class	Odor	[a]LRI	[b]Identification	W	C	S
a-phellandrene+	terpene	nd	1004	MS, PI	x	x	x
bicyclo(2.1.1)-heptane 7,7-dimethyl-2-methylene*	hydrocarbon	nd	1005	MS	x		
3-carene+	terpene	nd	1011	MS, PI		x	
4-methyl-1-(1-methylethyl) cyclohexene	hydrocarbon	nd	1018	MS		x	x
1-methyl-4-(1-methylethyl)-benzene*	aromatic	nd	1025	MS	x	x	x
d-limonene+	terpene	nd	1029	MS, PI	x	x	x
bicyclo(3.1.1)-heptane 6,6-dimethyl-2-methylene*	hydrocarbon	nd	1031	MS	x		
3-methyl decane	hydrocarbon	nd	1035	MS			x
2,5,5 trimethyl hexane	hydrocarbon	nd	1055	MS	x		
furaneol	alcohol	creamy	1060	PI, ST	x		
2,6 dime-3-ethyl-pyrazine	pyrazine	roasted	1082	PI, ST	x		
2-nonanone	ketone	nd	1092	MS, PI	x	x	x
nonanal	aldehyde	soapy	1104	MS, PI	x	x	x
3,5,5 trimethyl 2-cyclohexen-1-one*	ketone	nd	1126	MS	x	x	x
(Z)-2-nonenal	aldehyde	green	1147	PI	x	x	x
2-nonenal	aldehyde	green	1154	MS, PI	x		
(E)-2-nonenal	aldehyde	hay	1160	PI	x	x	x
safranal*	aldehyde	nd	1204	MS	x	x	x
decanal	aldehyde	nd	1207	MS, PI			x
(E,E) 2,4 decadienal	aldehyde	fried oil	1318	MS, PI	x		
g-nonalactone	lactone	coconut	1366	MS	x		
unknown	-	vanilla	1382	-		x	x
5-ethyl-(5H)furan-2one	ketone	nd	1395	MS		x	
					43	36	33

[a] LRI, linear retention index using a DB-5 column. [b] Identification: MS=spectra comparison using NIST library; ST=authentic standard injection; PI=comparison with published LRI; nd=not detected.* volatile compounds from saffron (D'Auria, M. and al., 2004) + volatile compounds from black pepper (Tairu, A. O. and al., 1999).

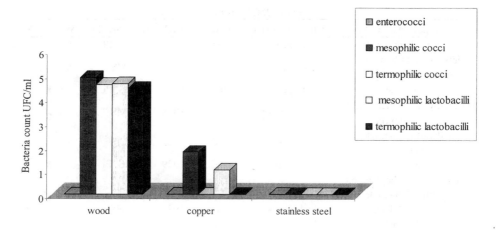

Figure 2 *Lactic acid bacteria colonized inner surface of wooden, copper and stainless steel tina*

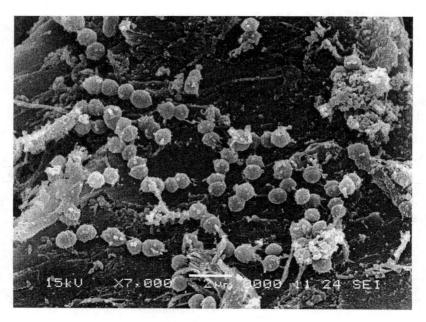

Figure 3 *Cocci community colonizing inner surface of wooden tina*

5 CONCLUSIONS

Piacentinu Ennese and Ricotta cheese made using wooden *tina* clearly showed different quality and higher numbers of volatile and odor compounds from those made using copper and stainless steel vat. The W cheeses were characterized by "buttery", "earthy", "woody", "roasty", "green" and "minty" notes, whereas C and S cheeses by "fruity" and "sweet" odors.

The microbiological analysis and SEM observation of the natural biofilm in the inner surface of wooden *tina* confirmed that lactic acid bacteria are released to milk during cheese-making, enriching the natural micro flora of raw milk. The technology and the tools used during cheese-making, therefore, clearly influenced the aroma profile of the final product. It is evident the role and the importance of wood tools in maintaining the traditional cheese-making and determining the good quality of Piacentinu Ennese and Ricotta cheeses. For the small number of samples analyzed, this is considered a preliminary study; future research will investigate other peculiar characteristics, considering: the effects of the use of wooden *tina* influencing aromatic and sensory properties of Piacentinu, at different ripening times; the release of bacterial cells from *tina* to milk; the identification of strains of lactic cocci in the biofilm, adhered to the inner wood surface of *tina*. These studies may be useful to identify the best Piacentinu cheese-making conditions, for future recognition of this cheese as a Protected Denomination of Origin.

ACKNOWLEDGEMENTS

Financial support was provided by the Ministero dell'Istruzione, dell'Università e della Ricerca (MIUR). A special thanks to *CoRFiLaC* laboratory technicians whose assistance was invaluable in the data collection steps.

References

1. G. Licitra and V. Bottazzi, *Il Ragusano, DDL srl (Ed.)*, 2001.
2. L.Corallo, P.S. Cocconcelli, R. Gelsomino, P. Campo, S. Carpino, and G. Licitra, *J. Dairy Sci.*, 2002, Vol. **85**, *Suppl. 1*, 258.
3. A.B. Marin, T.E. Acree, J. Barnard, *Chem. Senses*, 1988, **13**, 435.
4. T.E. Acree and J. Barnard, *Proceedings of the 7th Weurman Flavour Research Symposium, Noordwijkerhout, Netherlands*, 1994, Vol. **35**, 211.
5. E. Kovats, *In Adv.in Chromat.. Giddings J. C. and Keller R. A. Eds., Marcel Dekker, New York*, 1966, **41**, 1915.
6. H. Van Den Dool and P. Kratz Dec., *J. Chromatography*, 1963, **11**, 463.
7. H. Arn, and T. E. Acree,. *In Food Flavors: Formation, Analysis and Packaging Influences, Elsevier, Lemnos*, 1998.
8. N. Kondjoyan and J. Berdagué, *A compilation of relative retention indices for the analysis of aromatic compounds. Clermont-Ferrand: Edition du Laboratoire Flaveur*, 1996.
9. P.M.G. Curioni and J.O. Bosset, *Int. Dairy J.*, 2002, **12**, 959.
10. J.O. Bosset and R. Liardon, *Lebensm.-Wiss.U.-Technol*, 1984, **17**, 359.

11 F. W. Welsh., W. D. Murray, and R. E. William, *CRC Critical Reviews in Biotechnology*, 1989, **9**, 105.
12 A.O. Tairu, T. Hofmann, and P. Schieberle, *Perspectives on new crops and new uses. ASHS Press, Alexandria, J. Janick (ed.)*, 1999, 474.
13 R. G. Mariaca, T. F. H. Berger, R. Gauch, M. I. Imhof, B. Jeangros and J. O. Bosset,. *J. of Agric. and Food Chem.*, 1997, **45,** 4423.
14 M. D'Auria, G. Mauriello, G.L. Rana, *Flav. and Fragr. J.*, 2004, **19**, 17.
15 V.L. Poulsen, *Lebensm.-Wiss.U.-Technol.*, 1999, **32,** 321.

VOLATILE PROFILE OF PIACENTINU ENNESE CHEESE PRODUCED WITH RAW AND PASTEURIZED MILK

Stefania Carpino[1], Teresa Rapisarda[1], John Horne[1], Antonio Di Falco[1], and Giuseppe Licitra[1,2]

[1] *CoRFiLaC*, Regione Siciliana, S.P. 25 km 5, 97100 Ragusa Italy
[2] Dipartimento di Scienze Agronomiche, Agrochimiche e delle Produzione Animali, Catania University, 95100 Catania Italy

1 ABSTRACT

Piacentinu is one of the many traditional cheeses of Sicily. Produced in the province of Enna, it is a pressed ewes' milk cheese with peppercorns and saffron that gives it a typical spicy/floral flavors and a characteristic yellow color. The objective of this study was to determine whether changing the traditional conditions under which Piacentinu cheeses are made affects their volatile organic compounds (VOC). Cheeses were obtained from seven different farms, three of which used raw milk, no starters, and artisanal lamb and/or kid rennets (R). Four other farms used pasteurized milk, commercial LAB starter cultures and commercial lamb rennets (P+S). Cheeses were ripened from 2 to 6 months. VOC fractions were extracted onto SPME fibers using a retronasal aroma simulator (RAS). Gas chromatography/olfactometry (GC/O) and mass spectrometry (GC/MS) were used to identify VOC. R cheeses contained a more diverse group of VOC, especially among terpenes which points to the importance of pasture in promoting flavor development of traditional cheeses. Differences among other groups of volatiles were likely due to the unique microbial communities and enzymes found in R cheeses that differed from those found in P+S cheeses. R cheeses also had significantly stronger aroma intensities in most categories except fruity. Principal components analysis on VOC and sensory data separated R and P+S cheeses by their VOC "fingerprints."

2 INTRODUCTION

Raw milk is frequently pasteurized to eliminate pathogens and/or spoilage bacteria.[1] However, pasteurization also negatively affects many of the endogenous milk flora. Licitra *et al* [2] recently reported that traditional cheeses derive many of their desired and unique properties from their production areas and from methods specific to the cheese-maker handed down from generations. Previous studies have shown that raw-milk cheeses are more intensely flavored and ripen faster than cheeses made from pasteurized milk.[3,4,5] The endogenous microflora are responsible for the differences in flavor between raw and pasteurized milk cheese.[6] The present study represents a first attempt to determine whether non-traditional cheese-making methods, using pasteurized milk, alter the volatile profile of Piacentinu Ennese cheese, made traditionally from raw milk.

3 MATERIALS AND METHODS

3.1 Samples

Twenty-one cheeses were produced by seven different farms in the province of Enna, in Sicily, between March and April 2003. Three farms produced cheeses from raw milk using no starter cultures and artisanal lamb and/or kids' rennet ("R" samples). Four other farms produced cheeses from pasteurized milk (70-72°C) with commercial lamb rennets ("P+S" samples) and with mixed strain commercial lactic acid bacteria (LAB) starter cultures. All cheeses were ripened for 50 to 55 days at the farm. Subsequent ripening took place on-site at *CoRFiLaC*.[1]

3.2 Extraction of volatile organic compounds

Volatile organic compounds (VOC) were extracted by dynamic headspace with a retronasal aroma simulator (RAS) in duplicate and adsorbed onto a solid phase micro extraction (SPME) fiber with a 65 μm carbowax/divinylbenzene coating (Supelco, Bellefonte, PA). Fibers were pre-conditioned before initial use by inserting them into the injector port of a gas chromatography olfactometer (GC/O) for 1 h at 225°C, and reconditioned between extractions at the same temperature for 5 min, followed by 10 min at ambient temperature. The RAS apparatus consisted of a water-jacketed stainless steel blender with a screw-top lid having a regulated gas source attachment to provide airflow (laboratory purified air) and another stoppable attachment with a self-sealing silicone septum to hold the SPME fiber in place. For each extraction, 130 g of cheese was cut in cubes and combined with 26 ml of an odorless artificial saliva, previously described by Roberts and Acree.[7] This mixture was placed into a blender and agitated at 60% power with the lid and stopcock closed for 10 min (water temperature 37°C; airflow 20 ml/min). These conditions simulated the mastication and breathing that take place while eating at the natural temperature of the mouth. A syringe holding the SPME fiber was then fit into place, the stopcock opened and 1 cm of the fiber was exposed to the dynamic headspace of the sample for an additional 10 min. The syringe was removed from the septum and the volatiles analyzed by GC/O. The dynamic headspace of the sample was maintained during this time by closing the stopcock and leaving the blender and airflow running.

3.3 Gas Chromatography Olfactometry

GC/O analysis was performed by a single sniffer, previously trained using the procedure and the standard compounds described by Marin et al.[8] These standards consisted of a group of eight compounds used to evaluate olfactory acuity and to determine if a sniffer has specific anosmia for certain odors. After extraction of volatiles, the fiber was desorbed into a modified Hewlett Packard 6890 gas chromatograph (Datu Inc., Geneva, NY) (fused-silica capillary column HP-1, 12 m x 0.32 mm x 0.52 μm film thickness; splitless injection at 250°C; oven temperature program: 35°C for 3 min, 6°C/min to 190°C, then 30°C/min to 225°C and 225°C for 3 min; He carrier gas, column flow rate 1.9 ml/min). The eluted compounds were mixed with humidified air in a method described by Acree and Barnard[9] and the sniffer was continuously exposed to this source for 30 min. The time of response to individual odors perceived by the sniffer was recorded by Charmware software (v.1.12, Datu Inc., Geneva, NY). These times were converted into retention indices (RI) for each VOC and displayed by the software as a series of peaks in an "aromagram." RI values were calculated relative to a series of normal alkanes (C_7–C_{18}) previously injected into the FID port of the same gas chromatograph.

3.4 Gas Chromatography Mass Spectrometry

The SPME fiber was re-exposed to the dynamic headspace of the same sample into the RAS and then desorbed into a gas chromatograph mass spectrometer (Hewlett Packard 6890, crosslinked methyl siloxane capillary column HP-1, 25 m x 0.20 mm x 0.11 μm film thickness) using the same GC/O temperature program. Retention times (RT) of volatile compounds were calculated relative to the same series of normal alkanes (C_7–C_{18}) used in GC/O, that had been previously injected into the GC/MS. This procedure permitted a direct comparison between RI values obtained from GC/O and RT values obtained from GC/MS.

3.5 Statistical analysis

Counts of identified and unidentified VOC were tabulated for each cheese, with a 1 representing the presence of a particular VOC in a sample and a 0 representing its non-presence. Sums were calculated for each sample across all VOC and among the following categories: alcohols, aldehydes, free fatty acids, fatty acid esters, ketones, lactones, pyrazines, sulfurous components, terpenes and unidentified components. Counts of total VOC, VOC within chemical categories were all treated as dependent variables in a two-way split-plot ANOVA design. Method of production (R or P+S) and ripening time (2, 4 or 6 months) were treated as fixed main effects, while farm was treated as random and nested within method of production. Hypothesis testing for the method of production main effect used the farm (method) term as error, while the ripening time effect and method X ripening time interaction were tested using the farm X ripening time (method) interaction as error. Student's t-tests ($\alpha = 0.05$) were used to determine differences between individual ripening time means when significant differences were found among that effect. Univariate analyses were performed using proc GLM in SAS v. 8.2 (The SAS Institute).[10]

4. RESULTS

4.1 Cheese Volatile Organic Compounds

A total of 35 VOC were identified in the different cheeses evaluated and another six were present but were unable to be identified. All compounds are shown by chemical class in Table 1.

Table 1 *Identified and unidentified volatile organic components by chemical class* ● = *component was found in all cheeses in a given category;* ○ = *component was found in some cheeses in a given category;* -- = *component was not found in a given category; Tot. (total) = percent of cheeses in each milk treatment group where an individual component was found.*

Kovats RI [a]	Component	Odor	P+S (N = 12)				R (N = 9)			
		ripening time (months)	2	4	6	Tot.	2	4	6	Tot.
Free fatty acids										
710	acetic acid	sour	●	●	●	100	●	●	●	100
861	butanoic acid	rancid	●	●	●	100	●	●	●	100
890	hexanoic acid	sweaty	○	●	○	67	○	○	○	25
Alcohols/Aldehydes										
1074	guaiacol	burnt	--	○	--	11	--	--	--	--
1169	lavandulol	herbaceous	--	○	--	11	--	--	○	8
1043	(Z)-2-octenal	green leaf	--	--	--	--	○	--	--	8
1079	nonanal	soapy	●	●	●	100	●	●	○	92
1128	(E)-nonenal	green	●	●	●	100	●	●	●	100
1133	(Z)-nonenal	hay	●	●	●	100	○	●	●	92
1186	decanal	soapy	●	●	●	100	○	●	●	92
1234	(E)-2-decenal	pungent	--	--	--	--	○	--	--	8
1283	(E,E)-2,4-decadienal	nutty	--	--	○	11	--	○	○	25
1346	vanillin	vanilla	○	●	○	78	○	●	○	67
Fatty acid esters										
785	ethyl butanoate	apple	●	●	●	100	○	○	●	75
983	ethyl hexanoate	orange	●	●	●	100	○	●	●	92
1307	methyl decanoate	wine	--	--	--	--	○	--	--	8
1318	benzyl butanoate	floral	--	●	--	33	○	--	--	8
1371	butyl octanoate	fruity	--	--	--	--	○	--	--	8
Ketones/Lactones										
655	diacetyl	buttery	○	○	○	67	○	●	●	75
956	1-octen-3-one	mushroom	●	○	○	67	●	○	○	58
1407	δ-octalactone	floral	--	--	--	--	○	○	--	17
1424	γ-decalactone	sweet	--	--	--	--	○	--	--	8

Inspection of these data suggests that R cheeses tended to contain certain chemical compounds more frequently than P+S cheeses. Hexanoic acid, for example, was present in more R than P+S cheeses. Fatty acids in cheeses originate from lipolysis of milk fat, and the lipases responsible for this process originate from the milk itself, moulds, LAB and/or propionic bacteria (PAB).[11] The greater frequency of hexanoic acid likely indicates enhanced hexanoic-specific lipase activities in R cheeses. These activities might have arisen from a wild LAB strain from the raw milk or lipases in artisanal rennets that were not present in the starters or the commercial rennets. The apparent reduction in frequency of this acid with ripening time is somewhat curious, especially given that short-chain fatty acids tend to increase with ripening time.[12] One explanation for our results could be that the hexanoic acid in Piacentinu cheeses was metabolized to aldehydes or esters (e.g., ethyl hexanoate) as cheeses were ripened. A group of five aldehydes: nonanal, (E)-nonenal, (Z)-nonenal, decanal and vanillin, and two ethyl esters: ethyl butanoate and ethyl hexanoate were found in almost all R and slightly fewer P+S cheeses. Larger differences were seen between R and P+S cheeses for dimethyl disulfide (DMDS) and dimethyl trisulfide (DMTS). These volatiles were found in all of the R cheeses but only in about half of the P+S cheeses. DMDS and DMTS are formed principally by PAB-mediated conversion of methionine.[11] Their greater frequencies in R cheeses therefore suggests that this group of bacteria or other wild strain bacteria were present in greater numbers in R than in P+S cheeses.

Each of the six identified terpenes, while not found universally in R cheeses, were found more frequently in R cheeses than in P+S cheeses. Curioni and Bosset[13] in a recent review noted that terpenes in cheeses are of plant and not microbial origin. The greater incidence of terpenes in R cheeses studied here then suggests that the influence of pasture on cheese flavor was reduced by the use of non-traditional methods in cheese production. Reducing the influence of local pasture may in turn reduce or eliminate the ability to trace a cheese to a specific location or farm[14] causing the cheese to lose some of its identity.

Lastly, with the exception of a single compound (RI = 1332, nutty aroma), all of the unknowns were found more frequently, and two of the six were found exclusively in R cheeses. Three terpenes: (E)-rose oxide, citronellol and α-terpineol were likewise exclusive to R cheeses, while (Z)-2-octenal, (E)-2-decenal, methyl decanoate, butyl octanoate, δ-octalactone, γ-decalactone, and one of the unknowns were exclusive to P+S cheeses. This last group of molecules, having predominantly sweet, floral, fruity and herbaceous aromas, may have resulted from the high temperature milk treatment and were found almost exclusively in cheeses ripened two months. Lactones are also formed from esterification and lactonization of lipids often carried out by LAB.[15] Medium-chain aldehydes and esters are formed by similar processes.[13] These compounds might have been formed by starter LAB strains not found among the wild LAB strains in the raw milk or from enzymes specifically found in commercial rennets.

The volatile "fingerprint" of Piacentinu Ennese cheese tended to change somewhat with ripening time. (E,E)-2,4-decadienal, 2-isopropyl-3-methoxypyrazine, 2-acetyl-thiazoline, (E)-rose oxide, L-carvone, citronellol and four of the six unknowns were found exclusively in cheeses ripened four or more months. There were also a few compounds that tended to increase in frequency with increasing ripening time. These included terpinolene, one of the unknowns (RI = 1327, floral aroma), and in P+S cheeses, DMDS and DMTS. A different group of molecules, including hexanoic acid, 1-octen-3-one, 2,6-dimethyl-3-ethylpyrazine, 5-ethyl-2,4-dimethylthiazole and one unknown (RI = 1029, sweaty aroma), were found most frequently in cheeses ripened two or four months. The last component was found in all of the R cheeses aged two and four months and then disappeared. The off-odor associated with it might have been destroyed by the high temperature milk treatment in the P+S cheeses, while it dispersed only as a function of time in R cheeses.

Formal statistical analyses confirmed many of the above observations with respect to the mean numbers of VOC per sample in each of the chemical categories (Table 2). R cheeses contained over 30% more VOC per sample than P+S cheeses ($F_{1,5}$ = 11.3). This same trend was present for most of the individual chemical categories, but was significant for only sulfurous compounds, terpenes and the unknowns ($F_{1,8}$ > 9.8). In two cases, this trend was not observed. Mean numbers of ketones were identical between R and P+S cheeses, and P+S cheeses contained significantly more lactones per sample ($F_{1,5}$ = 6.5). However, in the latter case, no lactones were found in any of the R cheeses, thus violating one of the assumptions of the ANOVA model. No significant differences were found among mean numbers of VOC between ripening times. This last finding is in general agreement with other research on ewes' and cow's milk cheeses [12,16,17] where increases in concentrations but not numbers of VOC have been reported with increased ripening times.

Table 2 *Mean numbers of volatile odor components in cheese samples by chemical class*

Class	n [c]	Prod. Method		Ripening time (months)		
		R (N=9)	P+S (N=12)	2 (N=7)	4 (N=7)	6 (N=7)
Free fatty acids	3	2.67	2.25	2.46	2.63	2.29
Alcohols	2	0.22	0.08	--	0.33	0.13
Aldehydes	8	4.89	4.83	4.58	5.25	4.75
Fatty acid esters	5	2.33	1.92	2.00	2.38	2.00
Ketones	2	1.33	1.33	1.45	1.29	1.25
Lactones	2	-- [b]	0.25 [a]	0.25	0.13	--
Pyrazines	2	1.11	0.92	0.75	1.25	1.04
Sufurous	5	3.89 [a]	2.92 [b]	3.25	3.21	3.75
Terpenes	6	1.78 [a]	0.33 [b]	0.33	1.42	1.42
Unknowns	6	1.89 [a]	0.42 [b]	0.67	1.29	1.50
TOTAL	41	20.11 [a]	15.25 [b]	15.75	19.17	18.13

[a,b] *Means with different superscript letters in the same row were significantly different ($P < 0.05$) within that main effect*
[c] *n = number of compounds in a given chemical class*

5. CONCLUSIONS

Piacentinu Ennese cheeses made using non-traditional methods clearly had different volatile "fingerprints" both in numbers and types of VOC present from those made using traditional means. Differences between the cheeses were further evidenced among the aromatic sensory profiles, with R cheeses having stronger aromas in most categories. These findings were in agreement with other previous studies and were likely due to differences in the types and diversities of microbial communities in the two cheese types. Lastly, cheeses ripened longer periods of time tended to have stronger aromas from those ripened for shorter periods. Aroma differences were probably due to increased volatile concentrations, which were not measured here.

Acknowledgements

Financial support was provided by the Ministero dell'Istruzione, dell'Università e della Ricerca (MIUR). A special thanks to *CoRFiLaC* laboratory technicians whose assistance was invaluable in the data collection steps.

References

1. H. Burton, *International Dairy Federation,* 1986, Bulletin 200, Brussels.
2. G. Licitra, G. Leone, F. Amata and D. Mormorio, *F. Motta (Ed)*, 2000.
3. K.Y. Lau, D.M. Barbano and R.R. Ramussen, *J. Dairy Sci.* 1991, **74**, 727.
   P.L.H. McSweeney, P.F. Fox, J.A. Lucey, K.N. Jordan and T.M. Cogan, *Int. Dairy J.* 1993, **3**, 613.
4. Shakeel-Ur-Rehmann, P.L.H. McSweeny, J.M. Banks, E.Y. Brechany, D.D. Muir and P.F. Fox, *Int. Dairy J.* 2000, **10**, 33.
5. J.M. Banks and A.G. Williams, *Int. J. of Dairy Tech.*, 2004, Vol. 57, **2-3**, 145.
6. D.D. Roberts and T.E. Acree, 4th Wartburg Aroma Symposium, Eisenach 1994, M. Rothe & H.-P. Kruse (Eds.), 1995, 619.
7. A.B. Marin, T.E. Acree and J. Barnard, *Chem. Senses*, 1988, **13**, 435.
8. T.E. Acree and J. Barnard, 7th Weurman Flavor Research Symposium, Noordwijkerhout, H. Maarse (Ed.) Amsterdam, Elsevier 1994, Vol. 35, 211.
9. The SAS Institute, Statistical Analysis Software, version 8.2., 2001, Cary, NC.
10. P.L.H. McSweeney and M.J. Sousa, *Le Lait* 2000, **80**, 293.
11. A. Mulet, I. Escriche, C. Rossello and J. Tarrazo, *Food Chem.* 1999, **65**, 219.
12. P.M.G. Curioni and J.O. Bosset, *Int. Dairy J.* 2002, **12**, 959.
13. C. Bugaud, S. Buchin, A. Hauwuy and J.B. Coulon, *Le Lait* 2001, **81**, 757.
14. R.C. Jolly and F.V. Kosikowski, *J. of Agric. and Food Chem.* 1975, **23**, 1175.
15. J.A. Gomez-Ruiz, C. Ballesteros, M.A. Gonzalez Vinas, L. Cabezas and I. Martinez-Castro, *Le Lait* 2002, **82**, 613.
16. A.M. Partidario, M. Barbosa and L. Vilas-Boas, *Int. Dairy J.* 1998, **8**, 873.

HEAT STABILITY OF CA-CASEINATE/WHEY EMULSIONS. EFFECT OF PH, SALTS AND PROTEIN RATIOS.

E. A. Theologou[1], P. G. Demertzis[1], M. Minor[2]

[1]Department of Chemistry, University of Ioannina-GR-45110 Ioannina, Greece
[2]Technology Development and Quality, Numico Research B.V., Bosrandweg 20, 6700 CA Wageningen, P.O. Box 7005, the Netherlands

1 ABSTRACT

The effect of compositional factors, such as pH and mineral composition, on heat stability of both whey and Ca-caseinate emulsions are investigated. The divalent ions seem to play a crucial role on the heat coagulation time (HCT) of both emulsions. In addition, the effect of citrate concentration on the integrity of the micelle and on the thermal coagulation of these emulsions is also examined. In Ca-caseinate/whey emulsions the HCT-pH curve is more complicated. It is concluded that the presence of whey proteins is the reason of both the maximum and the minimum of the curve, strongly dependent on the contribution of the minerals. Based on the above findings, a new hypothesis is established to explain the heat stability profile of the studied emulsions.

2 INTRODUCTION

Milk is something with which we all are familiar. From a physico-chemical point of view, milk can be described as a colloid. It can be also considered as an oil-in-water emulsion of fat droplets stabilized by protein. But milk is also a colloidal dispersion. The dispersed particles, known as casein micelles, are aggregates of the casein family of phosphoproteins with mineral calcium phosphate.[1]

Milk (liquid or dried) and milk constituents (e.g. caseinate and whey proteins) find use as ingredients in several food emulsions. The manner in which these molecules interact with each other can ultimately determine the structure and stability of the product formed. The physico-chemical properties of milk and its components have been investigated by many researchers.[2-6]

The heat stability of the emulsions is markedly affected by pH. The mechanism of the coagulation of caseinate-whey emulsion is similar to that of milk. Tessie and Rose (1964) showed that β-lactoglobulin (β-lg) has a very specific effect on the heat stability of milk and furthermore in milk protein stabilized emulsions.[6]

Generally, the mechanism of the thermal coagulation of milk has been determined and largely accepted. A very extensive literature has accumulated which has been reviewed periodically.[7-16] A new hypothesis on the pH-dependence and coagulation mechanism of milk is also reported by O' Connell.[9]

In this work, the heat stability of Ca-caseinate/whey emulsions was studied in relation to pH, concentration of salts and protein ratio. It is suggested that whey proteins and salt concentration are responsible for the maximum of the heat coagulation time (HCT)-pH curve. Divalent ions seem to play a crucial role on HCT of Ca-caseinate/ whey emulsions. A combination of mechanisms was used to explain in depth the heat stability of the above emulsions.

3 METHOD AND RESULTS

3.1 Constructive Models

The general composition of the emulsions was the following: 50 g/kg protein powder, 50 g/kg canola oil, 100 g/kg malto-dextrins (DE19) and 800 g/kg simulated milk ultra filtrate (SMUF) (or SMUF/distilled water mixture). Using different ratio of isolac powder [commercial whey protein isolate (WPI), supplied by Milei] and calcium-caseinate powder (DMV International) six emulsion models were prepared; model A: 100/0 Ca-caseinate/isolac, model B: 80/20 Ca-caseinate/isolac, model C: 60/40 Ca-caseinate/isolac, model D: 40/60 Ca-caseinate/isolac, model E: 20/80 Ca-caseinate/isolac and, model F: 0/100 Ca-caseinate/isolac.

3.1.1 Preparation of SMUF. SMUF solution was prepared by a modification of the method of Jenness et al. as described in Lavrijsen's report.[17,18] Three solutions with different salt composition were prepared in 1 L volumetric flasks. The composition of these solutions is shown in Table 1. The SMUF solution was prepared in a 1 L volumetric flask by mixing equal amounts (20 ml) of each solution and diluting to 1 L in distilled water. The composition of SMUF solution is shown in Table 2.

Table 1   *Composition of salts in three solutions (I, II, III)*

Solution I		Solution II		Solution III	
Component	Weight (g)	Component	Weight (g)	Component	Weight (g)
KH_2PO_4	79.0	$CaCl_2*2H_2O$	66.03	K_2CO_3	15.0
$K_3Citr.*H_2O$	60.0	$MgCl_2*6H_2O$	32.5	KCl	30.0
$Na_3Citr.*2H_2O$	89.5				
K_2SO_4	9.0				

Table 2 *Composition of SMUF solution*

Compound	Concentration (mmol/l)
K	29.8
Na	6.1
H	11.6
Mg	3.2
Ca	9
Cl	32.5
CO_3	2.2
PO_4	11.6
SO_4	1
Citrate	9.8

3.1.2 pH adjustment. The pH was adjusted just before the heat treatment by slowly adding hydrochloric acid or potassium hydroxide solutions. The pH values ranged between 6.4 and 7.4.

3.1.3 Salts addition. A number of salts (tri-potassium citrate, potassium chloride, sodium chloride, magnesium chloride and calcium chloride) were separately added to investigate their effect on the heat stability of Ca-Caseinate/whey emulsions.

3.1.4 Heat treatment. Sterilization process has been simulated using a falling-ball viscometer (Klaro-graph). HCT was measured at 127 ^{0}C.

3.2 Stability of Whey and Ca-Caseinate Emulsions during Heating

Experiments in model F (100% whey protein) showed that the increase of salt concentration caused destabilization of the emulsion on heating. The changes were more pronounced for divalent than monovalent ions due to the high charge of the counter-ions. Instability of whey emulsion is probably due to the increase of free thiol groups and the decrease of electrostatic repulsion between protein molecules. On the other hand, as pH was raised (from 6.4 to 7.4), whey emulsions were more stable, at constant salt concentration.

On the contrary, in model A, Ca-Caseinate emulsion was more stable as SMUF concentration was increased. Furthermore, HCT was increased by increasing pH.

Our experiments showed that the heat stability of Ca-Caseinate emulsions is strongly dependent on the ion strength of the emulsion and on the charge and the structure of the micelle. The integrity of the micelle is very important for heat stability. Changes of salt equilibrium, especially of divalent ions, result in alterations in the micelle structure.

In Ca-Caseinate emulsion prepared in distilled water, the addition of monovalent ions (K^+) caused instability (low HCT values) at low pH values. Figure 1 shows more clearly the effect of the concentration of potassium chloride on HCT values as a function of pH.

After the addition of 20 mmol/l of potassium chloride, HCT was rapidly increased from 4 to 23 minutes. More salt addition resulted in a HCT decrease.

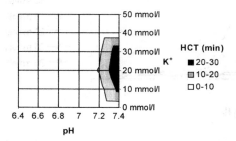

Figure 1 *Influence of monovalent ions (K^+) on HCT-pH profile of Ca-Caseinate emulsion at 127 ^0C.*

The effect of the addition of divalent ions (Ca^{2+}, Mg^{2+}) on the heat stability of Ca-caseinate emulsion prepared in distilled water was also investigated. By adding citrate, the amount of divalent ions in the serum was rapidly reduced. Moreover, HCT was increased by increasing citrate concentration and pH.

When the Ca-Caseinate emulsion was prepared in SMUF, the concentration of salts was higher. The effect of citrate addition was more pronounced as shown in Figure 2, namely the addition of citrate in low concentrations (5 and 10 mmol/l) results in a rapid increase in HCT by raising pH. On the contrary, addition of citrate in higher concentrations (15 and 20 mmol/l) results in a gradual decrease in HCT by raising pH. Addition of even higher citrate amounts (25 and 30 mmol/l) destabilizes the emulsion in the whole pH range. It is possible that the increase of citrate concentration, especially at higher pH values, caused dissociation of the micelle by breaking the calcium phosphate bridges, resulting in a swelling or even destruction of the micelle.

Whether aggregation of micelles occurs or not mainly depends on the interaction forces between the micelles and on the balance between attractive and repulsive forces.[19] In conclusion, it can be said that the salt composition in and out of the micelle plays an essential role in HCT of Ca-Caseinate emulsions.

3.3 Effect of SMUF-Ratio on Heat Stability of Ca-Caseinate/Whey Emulsions

Emulsions (models B to E) were prepared in SMUF or SMUF/distilled water mixtures. After homogenization, pH was adjusted to between 6.4 and 7.4.

Model D (40/60 Ca-caseinate/whey protein) in 10% SMUF was used to test the hypothesis of the pH-dependence and coagulation mechanism.[9] The HCT-pH profile of the emulsion can be divided into 4 regions (Figure 3).

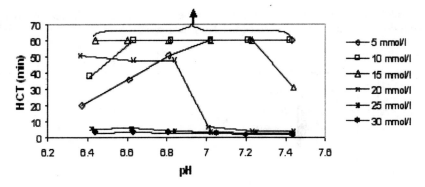

Figure 2   *Influence of the concentration of citrate on HCT-pH profile of Ca-Caseinate emulsion at 127 °C*

In region 1, the emulsion, probably due to the low pH and high Ca^{2+}-activity, is unstable and coagulates rapidly by heating at 127 °C.[9] On the other hand, casein micelles are more compact (low micelle charge) and therefore more susceptible to aggregation.

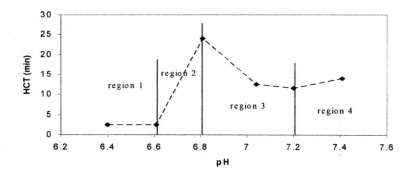

Figure 3   *HCT-pH profile of model D (40/60 Ca-Caseinate/whey) in 10 % of SMUF at 127 °C.*

In region 2, stability increases as a function of pH. Casein micelles have higher charge, and a more stable micelle is formed by heating. Moreover, whey proteins, which exist as dimers with the thiol groups inaccessible and unreactive to serum phase constituents, are denatured through a calcium-mediated mechanism and enhance the thermal stability of the casein micelles by chelating calcium.[9,20]

In region 3, Ca-Caseinate/whey emulsion has relatively low heat stability. It was found that the amount of the hairs (κ-casein) on the micelle is reduced due to very high charge and β-casein protrudes from the surface of the micelle (increased swelling). Moreover, in this pH range, β-lactoglobulin, which exists as a monomer, interacts with itself, κ-casein and $α_{s2}$-casein through disulfide interchange reactions, resulting in a decrease of HCT.[6,21]

In region 4, heat stability is increased by increasing pH. This could be attributed to a decrease in Ca^{2+}-activity and an increase in both charge and hydration at higher pH values.[22] Van Boekel et al. proposed that increased stability in this region is partly due to re-association of κ-casein molecules with the micelles.[23]

Salt concentration strongly affects HCT-pH profile of the emulsions (Figures 4, 5, 6 and 7). It can be seen that by increasing the SMUF ratio, the maximum HCT of the emulsions is shifted toward higher pH values. Moreover, HCT of emulsions where whey proteins are the predominant component (20/80 and 40/60 Ca-Caseinate/whey proteins), is initially increased by increasing SMUF ratio and, after reaching a maximum value, is decreased (Figures 4 and 5). On the other hand, HCT of emulsion with high Ca-Caseinate concentration (80/20 Ca-Caseinate/whey proteins) is steadily increased by increasing SMUF ratio (Figure 7).

Variations of Ca-Caseinate/Whey proteins concentration ratio also influence the heat stability of emulsions. By changing the protein ratio an alteration of the optimum pH (pH where HCT receives its maximum value) occurs. The increase of Ca-Caseinate concentration causes an increase of both optimum pH and HCT of the emulsions. This can be possibly explained by the fact that Ca-Caseinate at more stable in higher pH where Ca^{2+}-content in the serum is decreased.

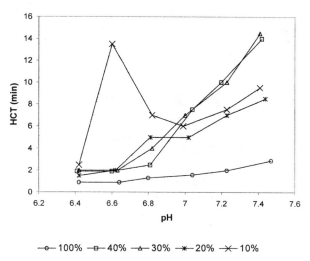

Figure 4 *HCT-pH profile of model E (20/80 Ca-Caseinate/whey) in SMUF ratios between 10 and 100% at 127 ^{0}C.*

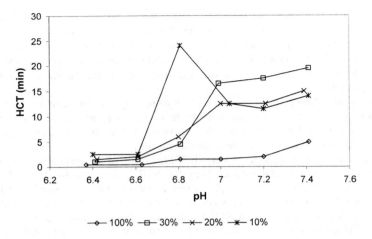

Figure 5 *HCT-pH profile of model D (40/60 Ca-Caseinate/whey) in SMUF ratios between 10 and 100% at 127 ^0C.*

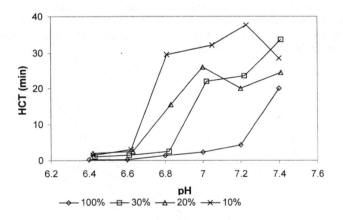

Figure 6 *HCT-pH profile of model C (60/40 Ca-Caseinate/whey) in SMUF ratios between 10 and 100% at 127 ^0C.*

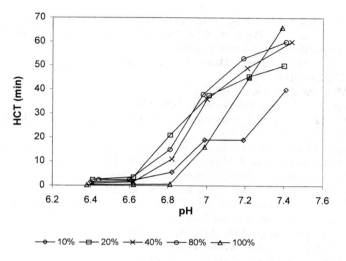

Figure 7 *HCT-pH profile of model B (80/20 Ca-Caseinate/whey) in SMUF ratios between 10 and 100% at 127 ^0C.*

4 CONCLUSION

HCT of pure whey and Ca-Caseinate emulsions is strongly affected by salt concentration. In whey emulsion, HCT is decreased by increasing salt concentration. The opposite effect was observed for Ca-Caseinate emulsions. HCT of both emulsions is strongly affected by divalent ions. However, for Ca-caseinate/whey protein emulsions, the HCT-pH profile is more complicated. It was found that HCT-pH profile is strongly influenced by changes in salt equilibrium and Ca-Caseinate/whey protein ratios. At higher Ca-caseinate and salt concentration the optimum pH and HCT values are increased.

References

1. C.V. Morr, 'Heat induced whey proteins-casein micelle interactions in milk and gelation of whey protein concentrate' in: *Protein Interactions*, ed. H. Visser, 1992, pp. 100-116.
2. H.J.M. van Dijk, *Neth. Milk Dairy J.*, 1990, **44,** 125.
3. J.E. O'Connell and P.F. Fox, *J. Dairy Sci.*, 1999.
4. T. Aoki, T. Umeda and T. Nakato, *Milchwissenschaft*, 1999, **54** (2), 91.
5. T.H.M. Snoeren, J. Koops and D. Westerbeek, *Neth. Milk Dairy J.*, 1978, **32**, 255.
6. J.E. O'Connell, and P.F. Fox, PhD Thesis, University College, Cork, Ireland, 1999.
7. D. Rose, *Dairy Sci. Abstr.*, 1963, **25**, 45.
8. P.F. Fox and P.A. Morrissey, *J. Dairy Res.*, 1977, **44**, 627.

9. J.E. O'Connell and P.F. Fox, 'Heat-induced coagulation of milk' in *Advanced Dairy Chemistry volume 1. Proteins, 3rd edition,* eds., P.F. Fox and P.L.H. McSweeney, Aspen Publications, Gaithersburg, 2000.
10. P. Walstra and R. Jenness, Dairy chemistry and physics, ed. Wiley & Sons, New York, 1984, pp. 162-185.
11. D.D. Muir, *Int. J. Biochem.*, 1985, **17**, 291.
12. H. Singh, *N.Z. J. Dairy Sci. Technol.*, 1988, **23**, 257.
13. M.A.J.S. van Boekel, J.A. Nieuwenhuijse, and P. Walstra, *Neth. Milk Dairy J.*, 1989, **43**, 147.
14. H. Singh and L.K. Creamer, In *Advanced Dairy Chemistry, 1 Proteins*, ed. P.F. Fox, Elsevier Applied Science Publishers, London, 1992, pp. 621-656.
15. C.H. McCray and D.D. Muir, In *Heat-induced Changes in Milk*, Special Issue 9501, International Dairy Federation, Brussels, 1995, pp. 206-230.
16. H. Singh, L.K. Creamer and D.F. Newstead, In *Heat-induced Changes in Milk*, Special Issue 9501, International Dairy Federation, Brussels, 1995, pp. 256-274.
17. R. Jenness and J. Koops, *Neth. Milk Dairy J.*, 1962, **16**, 153.
18. B.W.M Lavrijsen, *Numico Rescearch Report*, 2000.
19. D.S. Horne and J. Leaver, *Food Hydrocolloids*, 1995, **9** (2), 91.
20. C.A. Zittle, E.S. Dellamonicia, R.K. Rudd and J.H. Custer, *J. Amer. Chem. Soc.*, 1957, **79**, 4661.
21. N.K.D. Kella and J.E. Kinsella, *Int. J. Peptide Proteins Res.*, 1988, **32**, 396.
22. H. Singh, In *Heat-induced Changes in Milk*, Special Issue 9501, International Dairy Federation, Brussels, 1995, pp.86-99.
23. M.A.J.S. van Boekel, J.A. Nieuwenhuijse and P. Walstra, *Neth. Milk Dairy J.*, 1989, **43**, 97.

FLAVOR OF WINES PRODUCED BY CELLS IMMOBILIZED ON VARIOUS SUPPORTS

Y. Kourkoutas, M. Kanellaki, A. Bekatorou, M. Iconomopoulou, A. Malouhos and A.A. Koutinas,

Food Biotechnology Group, Department of Chemistry, University of Patras, Patras, GR 26500, Greece.

1 ABSTRACT

Five supports (kissiris, γ-alumina, gluten pellets, delignified cellulosic material and fruit pieces) were used for immobilization of baker's yeast, an industrial yeast strain and a psychrotolerant and alcohol-resistant *Saccharomyces cerevisiae* strain isolated from the Greek agricultural area for wine making. The supports are abundant, cheap and the immobilization process is simple, easy, no time consuming and of negligible cost. The immobilized biocatalysts were used for repeated batch fermentations in a wide range of temperatures (30-0°C) and lasted for a long time period without any significant loss of the biocatalyst activity. Batch fermentations of must were faster than those of free cells at various temperatures. Volatile by-products were examined by GC and GC/MS. A reduction of the percentage of higher alcohols and an increase of ethyl acetate in the total volatiles and a general decrease in the ratio of alcohols to esters were observed in the products particularly at low temperatures.

2 INTRODUCTION

Cell immobilization in alcoholic fermentation is a rapidly expanding research area because of its attractive and economic advantages compared to the conventional free cell system.[1,2]
 Although many immobilization supports have been proposed for use in wine making,[3-6] studies concerning full industrial application of the technology are scarce in literature. For industrial wine production, it is important to identify a suitable support for cell immobilization that is of food-grade purity, readily available, abundant, of non-degradable nature, cost-effective and which may contribute to an overall improvement in the sensory characteristics of the final product. The immobilized system should also be suitable for low temperature wine fermentations, since low temperature fermentations lead to improved quality.
 The aim of this investigation, therefore, was to identify and propose a number of biocatalysts suitable for low-temperature wine making, of low cost, abundant in nature and which could improve wine quality.

3 MATERIALS & METHODS

3.1 Yeast Strains

Commercial baker's yeast, *S. cerevisiae* AXAZ-1 strain isolated from the Greek agricultural area[7] and dried *S. cerevisiae* of *Uvaferm* 299 were used in the present study.

3.2 Supports and Cell Immobilization

Cell immobilization on kissiris,[8] γ-alumina,[9] gluten pellets,[10] delignified cellulosic (DC) material[11] and fruit pieces[12-15] were carried out separately as described in previous studies.

3.3 Fermentations

Fermentations using immobilized cells on the above supports were carried out as described earlier.[8-15]

3.4 Analyses

Volatile by-products were determined by gas chromatography[16,17,18] and GC/MS headspace technique.[12,13]

3.5 Preliminary Sensory Evaluation

Samples of the wines were tested for their aroma and taste characteristics. Ten testers, familiar with wine tastes, were asked to give scores on a 0-10 scale using locally approved protocols in our laboratories for taste and aroma. The sensory evaluation was a blind test in colored glass without light.

4 RESULTS & DISCUSSION

Yeast immobilization was carried out on various supports and the immobilized biocatalysts were used for repeated batch wine fermentations at room and low temperatures. Fermentations were continued up to ~7 months without any significant loss of the biocatalytic activity. Cell immobilization was proven by repeated batch fermentations and by electron microscopy (Figure 1). The high number of repeated batch fermentations showed a tendency for high operational stability. Fermentation times were significantly reduced compared to traditional fermentations, which is useful for scale-up of the processes.

4.1 Volatile By-products

In order to evaluate product quality, the produced wines were analyzed for the formation of the major volatile by-products (Table 1). Concentrations of 1-propanol, isobutanol and amyl alcohols (2-methyl-1-butanol and 3-methyl-1-butanol) were low. Low temperature generally led to a reduction of higher and amyl alcohol concentrations, which is considered a positive factor in product quality.

Dairy and Wine Flavors

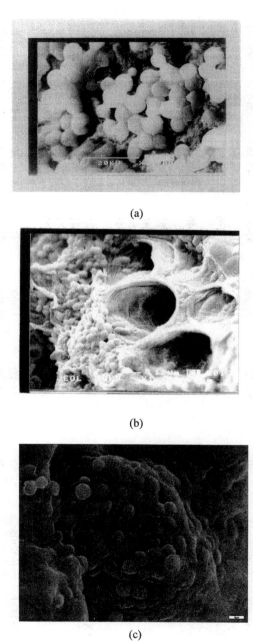

Figure 1 *Electron micrographs of immobilized cells S. cerevisiae AXAZ-1 on (a) apple, (b) quince and (c) pear pieces.*

Table 1 *Major volatile by-products in wines produced by immobilized yeast cells on various supports*

Immobili zation support	Yeast strain	Grape variety	Fermentation temperature (°C)	Acetaldehyde (ppm)	Ethyl acetate (ppm)	1-propanol (ppm)	Isobutanol (ppm)	Amyl alcohols (ppm)	Methanol (ppm)
Kissiris	Baker's yeast	Chondrada	30	61-222	20-38	32-66	42-56	106-139	285-471
γ-alumina	Baker's yeast	Trechumena	30	31-88	27-53	Tr-76	35-58	42-130	154-261
Gluten pellets	AXAZ-1	Sideritis	30	10-68	53-84	23-46	38-63	118-147	34-83
			15	8-75	41-55	32-55	36-55	104-169	30-53
			10	33-60	42-123	43-48	34-43	149-177	35-66
			5	8-57	25-89	35-68	23-30	136-195	17-78
			0	14-31	19-25	32-45	20-25	122-135	9-21
Delignified cellulosic material	AXAZ-1	Sideritis	30	18-25	117-132	37-48	29-41	143-161	89-102
			16	19-27	98-110	32-41	20-27	118-131	77-88
			10	38-43	75-92	23-31	20-23	69-75	80-98
			0	65-78	64-73	19-23	15-20	41-57	79-96
Apple pieces	AXAZ-1	Roditis	25	Tr-26	37-66	Tr-9	16-30	188-338	4-73
			20	Tr-26	68-108	7-13	20-23	284-324	13-99
			15	28-45	87-123	11-20	18-29	221-276	3-99
			12	13-53	117-150	18-28	27-32	204-247	12-99
			9	30-50	118-154	21-25	27-29	185-259	13-57
			6	6-52	91-118	5-11	18-21	121-141	84-96
			1	9-49	77-116	2-8	Tr	58-90	48-85

Dairy and Wine Flavors

Quince pieces	AXAZ-1	Roditis	30	Tr-27	30-85	13-20	9-27	155-294	5-72
			15	Tr-28	69-105	18-30	17-25	154-225	13-67
			10	Tr-59	62-113	29-48	24-27	87-147	Tr-20
			5	9-28	19-34	5-16	6-11	35-94	21-72
			0	23-31	41-50	5-8	7-8	23	25-68
Pear pieces	AXAZ-1	Roditis	30	47-95	32-52	17-23	29-60	136-213	Tr-68
			15	53-120	45-73	42-63	36-51	150-183	Tr
			8	80-93	28-56	35-52	24-28	117-130	Tr-60
			3	65-76	8-24	11-19	9-16	36-48	Tr-116
Dried raisin berries	*Uvaferm* 299	Saint Denis	25	12-25	17-33	8-23	23-57	66-95	100-180
			15	22-24	27-44	9-22	14-25	46-58	48-75
			6	20-60	21-25	22-27	8-10	31-44	90-95

Tr: Traces.

Due to the significant effect of this compound on organoleptic characteristics, the concentration of ethyl acetate was also examined. Although in some wines relatively high ethyl acetate content was reported (up to 154 ppm), in most cases it was lower. Such ethyl acetate concentrations are usually detected in wines[19] and contribute to the aroma of the product. In addition, there was no indication of vinegar odor in the final products. On the contrary, a fruity aroma was predominant.

Acetaldehyde content in wines usually ranges between 13-40 ppm,[20] and may reach 75 ppm.[21] Acetaldehyde was detected in relatively high amounts (up to 222 ppm) in some cases, especially in fermentations of raisin extracts from the Chondrada variety. The high acetaldehyde concentrations may be attributed to coupled oxidation of ethanol and o-diphenols,[19,22] due to enhanced concentration of o-diphenols present in must.[23] However, in most cases acetaldehyde was found in normal levels.

Methanol is formed from methylated pectic substances (pectins) by the action of pectin esterase, and it usually ranges between 0.1-0.2 g/L in traditional fermentations. Methanol content was high when raisin extracts from Chondrada and Trechumena varieties were used in fermentations due to extraction.[8,9,18] Relatively high methanol concentration was also reported in wines produced by immobilized cells on dried raisin berries at 25°C (Table 1), which could be attributed to the fact that raisin berries were in contact with grape must during fermentation and therefore enhancing possible pectin extraction, as it is well known that methanol derives from methylated pectic substances by the action of pectin esterase enzyme. However, in wines produced by immobilized cells on gluten pellets, delignified cellulosic material, apple, quince and pear pieces the methanol contents were in very low levels (<102 ppm), indicating a high quality product.

With regard to the most abundant by-products in the wines, a reduction of the percentage of amyl alcohols and a relative increase of the percentage of ethyl acetate in the total volatiles were observed in the low-temperature fermentations. The reduction of concentrations of 1-propanol, isobutanol and amyl alcohols and the increase of esters content with the drop of temperature have been related to improvement of organoleptic quality.[22,24]

4.2 GC/MS Analysis

Wines produced by immobilized cells on apple and quince pieces were also analyzed by GC/MS headspace technique (Table 2). The most important compounds identified by this technique were esters, which are known to make a positive contribution to the aroma of wines, alcohols, organic acids and carbonyl compounds. Ethyl 2-methyl-butyrate identified in wine produced by immobilized cells on apple pieces is a distinctive compound occurring in apples and provides the "ripe" note to apple aroma.[23] Fatty acids identified in wine produced by immobilized cells on quince pieces are considered for their possible flavor impact as they show simultaneously low odor threshold values and enough volatility at room temperature.[22] In addition, a number of miscellaneous compounds were identified, some groups of which are known to contribute to the complexity of wine aroma, such as acetals and terpenes. Distinctive was the peak for 2-phenyl-ethanol in wine produced by apple-supported biocatalyst, which is a compound with a characteristic rose aroma.

Table 2 GC/MS Headspace analysis of volatile components of wines produced by immobilized cells S. cerevisiae AXAZ-1 on apple and quince pieces

Compound	Certainty of identification		Compound	Certainty of identification	
	apple biocatalyst	quince biocatalyst		apple biocatalyst	quince biocatalyst
esters					
Ethyl acetate	a	a	Ethyl 2-methyl-butyrate	b	nd
Ethyl butyrate	a	a	Hexyl acetate	b	nd
Ethyl decanoate	a	nd	Isobutyl acetate	a	nd
Ethyl dec-9-enoate	a	nd	n-butyl acetate	b	nd
Ethyl dodecanoate	a	nd	2-methyl-butyl acetate	a	a
Ethyl hexadec-9-enoate	a	nd	2-phenyl-ethyl acetate	b	b
Ethyl hexanoate	a	a	3-methyl-butyl acetate	a	a
Ethyl octanoate	a	a	3-methyl-butyl decanoate	b	nd
Ethyl propanoate	a	nd	3-methyl-butyl octanoate	c	nd
alcohols					
Dodecanol	nd	b	2-ethyl-1-decanol	nd	c
Ethanol	a	a	2-methyl-1-butanol	a	a
Hexadecanol	nd	b	2-methyl-1-propanol	a	a
1-butanol	a	nd	2-phenyl-ethanol	a	nd
1-hexanol	a	nd	3-methyl-1-butanol	a	a
1-propanol	a	a			
organic acids					
Acetic acid	nd	a	Nonanoic acid	nd	c
Benzenecarboxylic acid	nd	c	Undecanoic acid	nd	c
Decanoic acid	nd	c	2-ethyl-hexanoic acid	nd	b
Hexanoic acid	nd	b			
carbonyl compounds					
Acetaldehyde	a	a	2-propanone	nd	a
miscellaneous compounds					
Farnesene	a	nd	1,1-diethoxyethane	a	nd
n-pentane	nd	b	2,4,5-trimethyl-1,3-dioxalane	c	nd

nd: None detected, a: Positive identification from MS and retention times, b: Tentative identification.

4.3 Sensory Investigation

Preliminary sensory tests carried out in the laboratory have ascertained the fruity aroma, fine taste and overall improved quality of the wines produced by immobilized cells on fruit pieces at low temperatures. The produced wines were accepted by the panel (Table 3) and characterized as novel, special type wines with a distinctive aromatic potential and a pleasant, fruity taste.

Table 3 *Results of aroma and taste for wines produced by immobilized cells of S. cerevisiae AXAZ-1 on fruit pieces*

Test	Apple-supported biocatalyst	Quince-supported biocatalyst	Pear-supported biocatalyst
Aroma	7.7±0.82	7.6±1.07	6.2±0.84
Taste	7.5±0.97	6.2±0.92	5.4±0.89

0: Unacceptable, 10: Excellent.

The overall improved quality of the wines produced at low temperatures could be attributed to (i) enhanced volatiles' profile due to increased solubility,[19] (ii) increased percentage of ethyl acetate and reduction of 1-propanol, isobutanol and amyl alcohols, which are considered off flavor compounds.

4.4 Technological Consideration

Suitable supports for wine making should be cheap, of food-grade purity, abundant in nature, stable and easy to use at industrial scale. The studied supports seem to meet all the above prerequisites and could be used in low-temperature improved quality wine making. Fruit pieces appear to have an advantage over minerals (kissiris and γ-alumina). Food-grade purity of fruits is incontrovertible, they are fully compatible with wine, while the produced wines had a distinctive aroma and taste which may lead to new and different wine products. Kissiris- and γ-alumina-supported biocatalysts could be used for rapid fermentations for distillates and potable alcohol production. The experience that has been acquired enables proposition of a relatively small (5000-10000 L) "Multi Stage Fixed Bed Tower" (MFBT) bioreactor[25,26] that could be used for industrialization of immobilized cells for wine making, so as to achieve support division in at least three floors, in order to (i) reduce pressure at the bottom layers of the bioreactor which may lead to structure destruction of immobilization supports which lack the necessary hardness, such as fruits and (ii) achieve uniform distribution of the immobilized biocatalyst in the bioreactor, as fruit pieces tend to float. The low volume of the bioreactor is necessary for easier handling of the supports, while immobilization could be performed in the bioreactor.

5 CONCLUSIONS

In conclusion, kissiris, γ-alumina, gluten pellets, delignified cellulosic (DC) material and fruit pieces were found suitable yeast immobilization supports for wine fermentation. Fruit pieces seem to have a great potential since the produced wines were of improved quality, which is strengthened by their food-grade purity.

References

1. A. Margaritis and F.J.A. Merchant, *CRC Crit. Rev. Biotechnol.*, 1984, **1**, 339-393.
2. G.G. Stewart and I. Russell, *J. Inst. Brew.*, 1986, **92**, 537-558.
3. Y. Shimobayashi and K. Tominaga, *Hokaidoritsu Kogyo Shikenjo Hokoku*, 1986, **285**, 199-204.
4. M. Fumi, G. Trioli and O. Colagrande, *Biotechnol. Lett.*, 1987, **9**, 339-342.
5. K. Nakanishi and K. Yokotsuka, *Nippon Shokuhin Kogyo Gakkaishi*, 1987, **34**, 362-369.
6. S. Silva, F.R. Portugal, P. Andrade, S. Abreu, M. Texeira and P. Strehatato, *Am. J. Enol. Vitic.*, 2003, **54**, 50-55.
7. T. Argiriou, A. Kalliafas, C. Psarianos, K. Kana, M. Kanellaki and A.A. Koutinas, *Appl. Biochem. Biotechnol.*, 1992, **36**, 153-161.
8. K. Kana, M. Kanellaki, C. Psarianos and A.A. Koutinas, *J. Ferment. Bioeng.*, 1989, **68**, 144-147.
9. K. Kana, M. Kanellaki, A. Papadimitriou, C. Psarianos and A.A. Koutinas, *J. Ferment. Bioeng.*, 1989, **68**, 213-215.
10. E. Bardi, V. Bakoyianis, A.A. Koutinas and M. Kanellaki, *Process Biochem.*, 1996, **31**, 425-430.
11. E. Bardi and A.A. Koutinas, *J. Agric. Food Chem.*, 1994, **42**, 221-226.
12. Y. Kourkoutas, M. Komaitis, A.A. Koutinas, M. Kanellaki, *J. Agric. Food Chem.*, 2001, **49**, 1417-1425.
13. Y. Kourkoutas, M. Komaitis, A.A. Koutinas, A. Kaliafas, M. Kanellaki, R. Marchant, I.M. Banat, *Food Chem.*, 2003, **82**, 353-360.
14. P. Mallios, Y. Kourkoutas, M. Iconomopoulou, A.A. Koutinas, C. Psarianos, R. Marchant, I.M. Banat, *J. Sci. Food Agric.*, 2004, **84**, 1615-1623.
15. A. Tsakiris, A. Bekatorou, C. Psarianos, A.A. Koutinas, R. Marchant, I.M. Banat, *Food Chem.*, 2004, **87**, 11-15.
16. K. Kana, M. Kanellaki, J. Kouinis and A.A. Koutinas., *J. Food Sci.*, 1988, **53**, 1723-1725.
17. M.D. Cabezudo, E.F. Gorostiza, M. Herraiz, J. Fernadez-Biarange, J.A. Garcia-Dominguez, M.J. Molera, *J. Chromatogr. Sci.*, 1978, **16**, 61-67.
18. K. Kana, M. Kanellaki and A.A. Koutinas, *Food Biotechnol.*, 1992, **6**, 65-74.
19. S.R. Jackson. "Wine Science. Principles and Applications", Academic Press, 1994, pp. 184, 199, 201, 255.
20. E. Longo, J.B. Velazquez, C. Siero, C.J. Ansado, P. Calo and T.G. Villa, *World J. Microbiol. Biot.*, 1992, **8**, 539-541.
21. A.A. Koutinas and S. Pefanis. "Biotechnology of Foods and Drinks", University of Patras Ed. Patras, Greece, 1994, p. 74.
22. P.X. Etiévant. "Volatile Compounds in Foods and Beverages", Ed by Henk Maarse, 1991, Chapter 14, pp. 483-497, 507.
23. H.D. Belitz, W. Grosch. "Food Chemistry", Springer-Verlag, Berlin, Heidelberg, 1987, Chapter 18, pp. 578-621.
24. R. Vidrih and J. Hribar, *Food Chem.*, 1999, **67**, 287-294.

25. P. Loukatos, M. Kiaris, I. Ligas, G. Bourgos, M. Kanellaki, M. Komaitis and A.A. Koutinas, *Appl. Biochem. Biotechnol.*, 2000, **89**, 1-13.
26. V. Bakoyianis and A.A. Koutinas, *Biotechnol. Bioeng.*, 1996, **49**, 197-203.

CHANGES IN THE PROFILE OF MAJOR VOLATILE COMPOUNDS IN GREEK WINES STORED UNDER CELLAR TEMPERATURE CONDITIONS

Vasileios C. Siaravas, Panagiotis G. Demertzis and Konstantoula Akrida-Demertzi

Laboratory of Food Chemistry, Department of Chemistry, University of Ioannina, P.O. Box 1186, GR-45110 Ioannina, Greece

1 ABSTRACT

The objectives of this study were to analyze the major volatile constituents of four different types of Greek wines (dry white, medium dry semi-sparkling white, rosé and red wine) produced in the area of Zitsa and to determine the changes in volatile composition during storage for twelve months under cellar temperature conditions. The analysis was performed through a solvent extraction method followed by GC and GC/MS analysis. Relatively high amounts of 2-phenylethanol, especially in the monovarietal white wine samples, were found. On the other hand comparatively low concentrations of ethyl acetate, furfural, isobutanol, 1-hexanol, 1-propanol, 1-pentanol, acetaldehyde and ethyl acetate were determined. With the exception of acetaldehyde content in white wines, the observed changes in volatile composition were significant after a storage period longer than three months.

2 INTRODUCTION

Flavor is a combination of taste and aroma and is of particular importance in determining food preferences. Wine flavor depends on a number of factors, the most important of which is its chemical composition. Aroma substances are important in wine as they make a major contribution to the quality of the final product. Several hundreds of different flavor compounds such as alcohols, esters, organic acids, carbonyl compounds and monoterpenes have been found in wines. Among the carbonyl compounds of wine, aldehydes are the most significant as aroma compounds. Esters are also significant components of wines, ethyl acetate being the most important. Other major flavor compounds produced by yeast during fermentation are higher alcohols, of which amyl alcohols and isobutanol are considered to be the most important. It is the combined contribution of these compounds that forms the character of the wine. Since many viticulture and enology factors greatly influence the type and concentration of flavor compounds, the ability to determine each individual compound would provide an approach to optimize the operational conditions (i.e., canopy management of the vine, harvest parameters, juice preparation and fermentation techniques, use of yeast, lactic acid bacteria, enzymes and wine storage and aging).[1,2]

Moreover, the particular importance of each compound in the final aroma depends on the correlation between chemical composition and perception thresholds, because most of the volatile compounds are present at concentrations near or below their individual sensory thresholds.[3]

Debina is a typical white grape variety cultivated only in a continental area (Zitsa), located in Northwestern Greece. From this variety different types of white wine (dry and medium dry semi-sparkling wines) are mainly produced. Rosé wine is produced from three grape varieties: *Vitis vinifera* var. Debina, var. Vertzami (red variety cultivated on the island of Lefkada – Greece) and var. Roditis (a famous red grape cultivar highly esteemed in Greece for its potential to produce high quality dry rosé and white wines). Red wine is produced from three grape varieties: *Vitis vinifera* var. Bekari (a red grape variety cultivated in the same area –Zitsa), *Vitis vinifera* var. Roditis and *Vitis vinifera* var. Agiorgitiko (another famous red grape cultivar in Northeastern Greece). These wines are meant to be consumed within a year of their production and not to be aged. Their flavor is formed during the processing of the grapes, by chemical, enzymatic and thermal reactions in grape must and during the alcoholic fermentation.

Young white wines do not present a "bouquet" caused by chemical reactions during maturation process and should be consumed within a short time period after bottling to avoid the loss of their fresh, fruity characteristics and the formation of undesirable components that are the main causes of the deterioration. The hydrolysis of acetate esters is the most important factor that results in the loss of the fruity character of young wines. Such changes take place more rapidly when the wines are not stored under proper light and temperature conditions. Not all wines are altered during the storage time. Wines made from Riesling white variety are not adversely affected by a reasonable storage time and they are consumed 1-2 years after bottling; during this time the wines develop a characteristic "bouquet of ageing". The same is not true for white wines of aromatic varieties such as Muscat, in which oxidation of the terpenes with time results in the loss of their flowery aromas.[4-7]

If wines have been stored under proper conditions they may retain initial quality, but if they have been stored in warm or lighted areas they will have lost their best attributes. While surveying practices in California, Boulton, *et al* found cellar temperature for varietal white wines ranged from 7 °C to 24 °C averaging 16 °C in 1971. Similarly, the cellar temperature for varietal red table wine averaged 18 °C with a 10 °C to 24 °C range.[8] Nevertheless, there are no systematic studies on the effect of storage conditions on Greek wine quality.

Several analytical methods have been developed for the extraction and determination of wine flavor compounds. These include dynamic headspace analysis[9], liquid/liquid extraction,[1,3,7,10] polymeric resin/solvent extraction[1,10] and solid phase microextraction,[6] followed by chromatographic determination. Liquid/liquid extraction is a widely used sample preparation method for the determination of wine volatiles. Although it is a time-consuming technique, the extract generated contains a wide spectrum of volatile components. On the other hand, capillary gas chromatography is a technique with high selectivity and sensitivity, suitable for the determination of multiple components, at low level, in wines.[1-11]

In this work, a combined solvent extraction/GC method for the quantitative analysis of major volatiles of Greek wines produced in the area of Zitsa is reported during their storage under cellar temperature conditions for a period of 12 months. Identification of compounds was accomplished by GC-MS. Dichloromethane was the solvent used to extract the volatiles and the extracts were subjected to GC analysis without further concentration.[10]

3 METHODS AND RESULTS

3.1 Experimental

3.1.1 Wine samples. The samples of Zitsa wines (dry white, medium dry semi-sparkling white, rosé and red wines) from 1999 vintage, bottled in 2000 were supplied by the Union of Agricultural Cooperatives of Ioannina, Greece. The fermentation temperature for all types of wine was 18 °C. Samples were stored in a cellar (where temperatures ranged from 12°C to 24°C) and analyzed at specific time intervals (0, 3, 6, 9 and 12 months). Measurements were carried out in triplicate. The storage time was taken to be the time elapsed from the exact date of bottling to the date of analysis.

3.1.2 Reagents. High purity chromatographic standards and dichloromethane of analytical grade were obtained from Merck (Darmstadt, Germany) and Aldrich (Germany). Anhydrous sodium chloride was obtained by Riedel-de-Häen (Germany) and tartaric acid was obtained from Merck (Darmstadt, Germany). All reagents were used without further purification.

3.1.3 Extraction Method. A liquid/liquid extraction method was employed for the extraction of the volatile compounds from the wine samples: Wine (100 mL), CH_2Cl_2 (10 mL) as extracting agent, 1 mL of 2-octanol solution (72.04 mg L^{-1} in CH_2Cl_2) as internal standard and sodium chloride (20 g) to reduce the degree of emulsification at the wine/CH_2Cl_2 interface, were added in a 300-mL flask. The flask was cooled in melting ice and the wine/CH_2Cl_2 mixture was stirred at 200 rpm for 2 h. The wine/CH_2Cl_2 emulsion formed during stirring was separated from the aqueous layer and frozen at –20 °C. The flask was then allowed to reach room temperature, and the CH_2Cl_2 layer progressively separated from the remaining wine, was transferred without concentration into a screw-capped vial and stored at –20 °C.[10]

3.1.4 Working standard solutions. A synthetic wine solution was prepared by dissolving each of the identified flavor compounds, listed in Table 1, at 6.5 g L^{-1} tartaric acid and 11.5% (v/v) ethanol. This solution was used to evaluate the extraction recovery of the flavor compounds.[1,12] Recovery tests were performed by extracting the synthetic wine solution according to the described extraction method.

3.1.5 GC Conditions. The GC unit was a Fisons 9000 series (Fisons Instruments, Milano, Italy) gas chromatograph equipped with an FID detector. The separation column was 30 m long x 0.32 mm internal diameter fused silica capillary DB-WAX (J&W Scientific Folsom, CA, USA) with 0.25 µm film thickness. The following GC parameters were utilized: detector temperature, 240 °C; injector temperature, 200 °C; injection mode: split with split ratio 1:50; injection volume, 2 µl. Column temperature program: 40 °C (hold 7 min), program from 40 °C at the rate of 15 °C min^{-1} to 160 °C (hold 1 min) and program again from 160 °C, at the rate of 30 °C min^{-1} to 230 °C (hold 5 min). Carrier gas He: 50 kPa.

Table 1 Relative recoveries (RR) and standard deviation (s.d.) for liquid/liquid extraction of aroma compounds from synthetic wine solution.

Volatile compound	% RR	s.d.
Acetaldehyde	99.55	0.11
Ethyl acetate	90.24	2.27
1-Propanol	92.59	0.61
Isobutanol	70.39	0.49
Amyl alcohols	46.33	0.60
1-Pentanol	16.88	0.77
Ethyl lactate	36.85	0.49
1-Hexanol	30.96	0.32
Furfural	58.75	0.62
2-Phenylethanol	23.79	0.42

Calibration curves for the GC quantification were constructed by dissolving known amounts of the flavor compounds, including the internal standard (2-octanol), in aqueous ethanol solution having an alcohol content similar to that of the analyzed wines.

The CG-MS unit consisted of a Hewlett-Packard 6890 (Wilmington, DE, USA) gas chromatograph coupled to an HP-5973 mass selective detector. The GC was equipped with the same DB-WAX capillary column. The GC conditions were the same as above. The transfer line was held at 260 °C. Ionization was carried out with electron energy of 70 eV. Identification of compounds was accomplished by comparing retention times and mass spectra (SCAN mode, 28 – 550 amu) with those of reference standards using the GC-MS Willey 275.L workstation.

3.1.6 Statistical Analysis. Results are expressed as mean ± s.d. of three experiments. Normality tests were made using the Kolmogorov-Smirnov criteria. Statistical comparisons were made using the t-Student test. P-value was considered to be statistically significant when P<0.05. Calculations were performed using the MINITAB (MTBWIN) statistical package.

3.2 Results and discussion

3.2.1 Extraction, chromatographic separation and quantification. A number of volatile organic compounds were extracted, identified and analysed following the previously reported methods. The extraction procedure described previously gave satisfactory results. Previous studies[1,11,12,13] on the ability of different solvents to extract volatile compounds from alcoholic solutions showed that dichloromethane was the most efficient solvent and for this reason was used for this study. It is also important to add the correct amount of salt to improve the extraction efficiency of dichloromethane. Unfortunately, the optimal salt (NaCl) addition to wine causes phase separation difficulties, as the density of wine is becoming too close to that of dichloromethane. Several experiments were carried out to determine the best extraction conditions. The selected conditions (100 mL of wine + 10 mL CH_2Cl_2 + 20 g NaCl) have been found a reasonable compromise between extraction efficiency, concentration and extraction cleanliness. No emulsion problems were observed in the analysis of the wines examined. The relative recovery (RR) values for liquid/liquid extraction of flavor compounds from synthetic wine solution are presented in Table 1. Obtained RR values are generally comparable to those reported by other investigators[1,12,13]. As expected, the range of recoveries is wide. It should be noticed that the applied extraction method, without any concentration step, has the advantage of reducing the artefacts that are possibly formed through oxidation and preventing the loss of important volatiles due to evaporation and degradation. Therefore, the proposed method can be considered to be very suitable for analyzing the volatile constituents of wines and other alcoholic beverages.

The organic volatiles identified in the wine samples and their retention times are given in Table 2. The cited retention times are the average of at least three injections. The proposed method allows the quantification of at least 10 major aroma compounds in Zitsa wines. The average concentrations of these compounds in the wine samples are summarized in Table 3.

Table 2 *Volatile compounds identified in Zitsa wines.*

Peak no.	Volatile compound	Retention time (min)
1	Acetaldehyde	1.89
2	Ethyl acetate	2.58
3	1-Propanol	5.28
4	Isobutanol	7.11
5	Amyl alcohols	10.19
6	1-Pentanol	11.06
7	Ethyl lactate	12.26
8	1-Hexanol	12.49
I.S.[a]	2-Octanol	13.29
9	Furfural	13.67
10	2-Phenylethanol	17.98

[a], internal standard.

Table 3 Mean concentration (mg l^{-1}) of aroma compounds and standard deviation (s.d.) in Zitsa wines as a function of storage under cellar temperature conditions.

Compound	0 months Mean	s.d.	3 months Mean	s.d.	6 months Mean	s.d.	9 months Mean	s.d.	12 months Mean	s.d.
					Dry white wine					
Acetaldehyde	0.43^a	0.02	1.36^b	0.03	1.35^b	0.11	0.80^c	0.31	0.67^{ac}	0.28
Ethyl acetate	41.08^a	7.03	68.12^a	18.19	9.01^b	4.01	17.05^c	1.12	13.99^d	1.07
1-Propanol	1.31^{ab}	0.15	1.16^a	0.13	1.24^a	0.07	1.38^b	0.12	1.20^a	0.04
Isobutanol	23.76^{ab}	2.05	22.22^{ab}	1.41	21.18^a	0.46	22.90^b	0.56	20.94^a	0.27
Amyl alcohols	295.00^{ab}	15.50	294.88^a	8.87	295.01^a	2.97	296.01^a	2.64	276.37^b	2.73
1-Pentanol	0.34^a	0.07	0.32^{ab}	0.10	0.50^b	0.14	0.12^c	0.10	0.19^{ac}	0.13
Ethyl lactate	56.68^a	3.33	57.13^a	1.81	57.71^a	0.59	63.25^b	1.31	58.09^a	0.69
1-Hexanol	1.98^{ab}	0.07	2.02^a	0.04	1.89^b	0.10	1.63^c	0.01	1.60^c	0.10
Furfural	2.80^{ab}	0.33	2.70^a	0.14	3.39^c	0.08	3.58^c	0.22	3.09^b	0.05
2-Phenylethanol	155.75^a	2.08	158.48^a	1.27	169.28^b	1.06	163.45^c	0.69	156.32^a	2.11
					Medium semi-sparkling white wine					
Acetaldehyde	0.39^b	0.06	0.93^a	0.28	1.04^a	0.43	0.86^a	0.33	0.91^a	0.41
Ethyl acetate	32.10^{abc}	25.11	70.24^a	12.29	7.02^b	1.03	11.01^{bc}	6.12	17.09^c	8.09
1-Propanol	1.21^{ab}	0.19	1.13^{ac}	0.12	1.30^{abc}	0.09	1.39^{bd}	0.08	1.31^{cd}	0.08
Isobutanol	23.82^{ab}	1.96	23.11^a	0.66	22.43^a	0.65	24.90^b	0.76	23.04^a	0.55
Amyl alcohols	283.09^{ac}	2.99	278.92^{ab}	2.40	277.10^a	3.54	286.01^c	2.14	269.42^d	4.40
1-Pentanol	0.08^a	0.16	0.29^{ab}	0.20	0.71^c	0.08	0.13^a	0.15	0.46^b	0.12
Ethyl lactate	50.78^a	0.99	51.67^a	0.52	70.14^b	1.13	82.49^c	1.48	79.72^d	1.02
1-Hexanol	2.02^a	0.02	2.05^a	0.03	1.84^b	0.02	1.67^c	0.03	1.58^d	0.04
Furfural	3.09^a	0.17	3.01^a	0.08	3.77^{bc}	0.17	3.98^b	0.17	3.76^c	0.07
2-Phenylethanol	159.66^a	0.70	162.49^a	2.48	171.81^b	1.15	171.09^b	2.15	161.33^a	2.22
					Rosé wine					
Acetaldehyde	0.36^a	0.38	0.40^a	0.26	0.59^{ab}	0.27	0.81^{ab}	0.08	0.96^b	0.17
Ethyl acetate	21.61^a	4.22	19.34^a	2.37	14.91^{ab}	3.61	11.92^{bc}	1.28	7.72^c	3.42
1-Propanol	2.01^{ab}	0.24	1.93^{ab}	0.34	1.84^a	0.17	2.19^b	0.12	2.40^b	0.20
Isobutanol	18.30^a	0.65	17.97^{ab}	1.61	18.20^a	0.60	19.47^b	0.39	21.64^c	0.90
Amyl alcohols	225.28^{ab}	5.70	219.12^{ac}	5.43	216.85^a	1.40	214.53^c	1.38	230.05^b	6.38
1-Pentanol	nd		nd		nd		nd		nd	
Ethyl lactate	49.75^a	1.06	50.78^a	1.01	43.64^b	0.49	34.94^c	1.09	31.66^d	0.77
1-Hexanol	2.97^a	0.06	2.94^{ab}	0.18	2.80^b	0.05	2.57^c	0.10	2.44^c	0.04
Furfural	12.46^a	0.95	13.84^a	1.10	17.67^b	0.37	17.98^b	0.66	19.10^c	0.57
2-Phenylethanol	105.48^a	1.18	105.87^{ab}	2.64	110.18^c	2.10	106.42^a	0.73	110.35^{bc}	1.17
					Red wine					
Acetaldehyde	0.34^a	0.36	0.63^a	0.17	0.67^{ab}	0.07	1.08^b	0.28	1.96^c	0.37
Ethyl acetate	27.13^a	1.22	36.05^{ab}	10.12	32.06^{ab}	15.06	19.11^b	4.33	20.85^{ab}	15.07
1-Propanol	2.41^a	0.19	2.35^a	0.23	1.95^a	0.58	2.97^b	0.09	2.81^{ab}	0.39
Isobutanol	18.23^a	0.53	18.07^a	1.17	14.78^{ab}	3.70	19.99^b	0.23	18.98^{ab}	1.78
Amyl alcohols	216.26^a	3.35	219.77^{ab}	7.92	199.40^{ab}	31.70	230.31^b	1.75	216.20^a	7.15
1-Pentanol	nd		nd		nd		nd		nd	
Ethyl lactate	62.50^a	0.96	64.15^a	1.95	79.33^b	7.38	96.57^c	0.48	91.42^d	4.03
1-Hexanol	2.06^{ab}	0.17	2.12^a	0.17	1.93^{ab}	0.26	1.89^b	0.04	1.84^b	0.10
Furfural	30.64^a	0.19	29.94^a	1.92	31.12^a	5.39	45.93^b	1.27	51.70^b	6.22
2-Phenylethanol	105.76^a	1.70	108.05^{ab}	1.89	108.64^{ab}	5.43	113.49^c	1.74	109.48^b	0.55

$^{a-d}$ Means within rows without a common superscript are significantly different (p<0.05) according to the Student t-test

nd : not detected

3.2.2 Major volatiles in the wine samples investigated. Aroma compounds such as 2-phenylethanol, described as 'rose-like', 'sweet', and 'perfume-like'; acetates having an aromatic description of 'sweet', 'fruity', and 'banana-like' and ethyl esters, with an aromatic description of 'apple-like', 'fruity' and 'sweet' are considered to be compounds capable of exerting a positive strong influence on the wine aroma. On the contrary, the higher alcohols, which have an aromatic description of 'alcoholic', 'sweet', and 'choking' and 1-hexanol, described as 'coconut-like', 'harsh' and 'pungent', can contribute negatively to the wine aroma. Isoamyl alcohols and acetaldehyde when present in high amounts impact negative features to wine flavor. Moreover, low acetaldehyde concentrations may contribute a slight 'sherry' character[2,3,14,15].

Higher alcohols
Alcohols with more than two carbon atoms are commonly called higher alcohols or fusel oils. They may be present in healthy grapes but seldom occur at significant levels. Hexanols are the major exception to this generalization and these produce herbaceous odors in certain wines. Higher alcohols (quantitatively the most important of them are 1-propanol, isobutanol, isoamyl alcohols and 2-phenylethanol) are considered to be produced from amino acids or from hexoses through pyruvate.[16] Most of the higher alcohols found in wine occur as by-products of yeast fermentation. In table wines, the total concentration of fusel oils is reported to range between 62 and 420 mg L^{-1}.[2,5,12,13,16-19] In this range fusel oils are believed to contribute to the overall aroma complexity whereas at higher levels they may become objectionable. In dessert wines, total fusel oil concentration may range between 100 mg L^{-1} and 1,000 mg L^{-1}.[19] 2-phenylethanol has its origin in phenylalanine, an amino acid found in the must and is largely produced during yeast fermentation.[20] Glycosides of 2-phenylethanol have been found in white *V. vinifera* var. Chardonnay, Muscat Alexandria and Riesling grapes and may contribute to the total amount of 2-phenylethanol found in wines.[21] 1-hexanol and 2-pentanol are considered to be varietal compounds, characteristic of young wines.[16] Moreover, 1-hexanol can result from the physical disruption of the cell wall of the grape and it originates from unsaturated C_{18} fatty acids.[22]

In both types of white Debina wines similar **amyl alcohol** concentrations (295 mg L^{-1} for dry white and 283 mg L^{-1} for medium semi-sparkling white wine) were detected, adding positive attributes to the wine aroma. Lower concentrations of amyl alcohols were detected in rosé and red wines. The concentration of amyl alcohols (total content of 2-methyl-butanol-1 and 3-methyl-butanol-1) range between 62 and 300 mg L^{-1} for wines produced during natural fermentation.[2,5,13,16,18,23] Isoamyl alcohol (3-methyl-butanol-1) was found in French wines in concentrations ranging from 114 mg L^{-1} to 400 mg L^{-1}.[24] None of the types of Debina wines exceed the level of 300 mg L^{-1} for amyl alcohols.

The formation of higher alcohols during fermentation is markedly influenced by practices at the winery. Additional higher alcohols may occur from the metabolic action of spoilage yeasts and bacteria. T. Herraiz *et al* noted the roles of different yeast cultures in flavor production.[23]

Even lower concentrations of **isobutanol, 1-propanol** and **1-pentanol** were determined. These findings are in general agreement with those of other investigators.[2,5,12,13,18] Threshold values for 1-propanol, isobutanol and active (+) isoamyl alcohols of 500, 500 and 300 mg L^{-1}, respectively, have been reported.[24] These compounds undoubtedly have an additive effect to the wine aroma.

1-Hexanol is considered to be a varietal compound.[13,16] It was detected at concentrations of 2 mg L^{-1} in both white Debina wines. According to other investigators it was found at concentrations from 0.4 to 1.4 mg L^{-1} for white wines and from 1.4 mg L^{-1} to over 2.3 mg L^{-1} for rosé wines.[7,13,16,23] There have been also reported concentrations in the range 1.3-12.0 mg L^{-1} for Australian wines.[24]

Relatively high amounts of **2-phenylethanol**, at concentrations ranging between 154 and 160 mg L^{-1} for both white Debina wines and lower (approx. 105 mg L^{-1}) for rosé and red wines were determined. It is belived that 2-phenylethanol has a positive strong influence on the wine aroma as it can give a rose-like flavor.[25] Lower concentrations of 2-phenylethanol (7.2 - 109 mg L^{-1}) have been reported by other authors.[7,12,13,16,18,21,23,24]

Esters

The fruity character of *V. vinifera* wines is partially due to the volatile esters formed during fermentation. Lower temperatures of fermentation, removal of heavy solids before fermentation and the use of good fermentation practices (sulfur dioxide and pure yeast) seem to be the main influences on the formation and retention of esters. In addition, it was found that the higher temperatures caused earlier and increased production of some of esters but greater losses later in fermentation.[24]

An aroma compound produced during malolactic fermentation is **ethyl lactate**. The concentrations of ethyl lactate were 57 mg L^{-1} for dry white and 45 mg L^{-1} for medium semi-sparkling white wines. According to certain investigators, ethyl lactate has little odor and is of rather minor sensory consequence.[24] The amounts of this compound found in varied from a trace to 534 mg L^{-1}, for table and sparkling wines.[5,7,12,13,16,24]

Ethyl acetate concentrations have been also determined in the analyzed wine samples. They are ranging between 21 and 41 mg L^{-1}. Although it is unlikely that compounds present at sub threshold levels could exert a significant aroma influence individually, they may be representative of a group of compounds that could act synergistically[21]. At low levels (<50 mg L^{-1}) ethyl acetate may be pleasant and adds to the general fragrance complexity, while above 200 mg L^{-1} it is likely to donate a sour-vinegar off-odor[2,24]. The concentrations of this ester in various types of wine generally range between 1.5 and 175 mg L^{-1}.[2,5,13,16,17,18] Ethyl acetate is not only formed during primary alcoholic fermentation but also by a secondary acetic fermentation. Its concentration increases during aging because of the continuous oxidation of ethanol to acetic acid and esterification of this acid. In addition, ethyl acetate is also a bacterial spoilage index[16,24]. Ethyl acetate (odor threshold value: 7.5 mg L^{-1}) can seriously flaw the fragrance of the wine long before the acetic acid (odor threshold value: 200 mg L^{-1}) level reaches a concentration sufficient to make the wine undrinkable.[26]

Carbonyl Compounds

A large number of aldehydes, ketones and related compounds have been found in wines. Some of these have little sensory importance, but a few do. **Acetaldehyde** is the carbonyl compound that has received the most attention. It is undesirable as an odor in table wines but desirable in dessert wines. Acetaldehyde is the major aldehyde found in wines in concentrations usually between 1.04 and 76 mg L^{-1}.[2,5,13,14] In the present study, low acetaldehyde concentrations (approx. 0.4 mg L^{-1}) for all samples were found. This is indicative of young fresh wines. Acetaldehyde is an early alcoholic fermentation direct byproduct and its content is affected by the grape cultivar.[24] Another source of acetaldehyde in wine is the coupling of the autooxidation of o-diphenols and ethanol.[19] Acetic acid and ethanol can be formed from acetaldehyde degradation, while malolactic fermentation strains are able to degrade SO_2-bound acetaldehyde.[27] It is usually considered an odor constituent with odor threshold value at 500 µg L^{-1}.[26]

Another aldehyde occasionally having a sensory impact in wine is **furfural**, the synthesis of which involves sugar degradation. In wines the sensory threshold is about 100 mg L^{-1}.[24] Its concentration increases significantly with storage time, especially when it exceeds 2.5 years. Furthermore, in white wines very low furfural concentrations (0.02 mg L^{-1} – 6.09 mg L^{-1}) are usually observed.[7,13,14] In agreement to these studies, very low furfural concentrations (approx. 3 mg L^{-1}) were found in our white wine samples. On the contrary, higher furfural content was determined in rosé and red wine samples (approx. 12 mg L^{-1} and 31 mg L^{-1}, respectively).

3.2.3 Changes in the volatile compounds during storage. Table 3 presents the concentrations of the analyzed aroma compounds in the four wine types as a function of storage time under cellar temperature conditions. Mean values without a common superscript are significantly different (p<0.05), according to the t-Student test.

Esters

The concentrations of ethyl acetate were significantly lower in wines stored at cellar temperature for storage longer than 3 months, in white Debina wines. This could be the main cause of the loss of fresh aroma in these types of wines.[7]

On the other hand, the concentrations of ethyl lactate increased gradually with storage time in semi-sparkling and red wines (for storage period longer than 3 months). In case of white wines, a slight increase in ethyl lactate concentrations was observed after a storage period of 9 months. On the contrary, in rosé wines, there was a significant decrease in ethyl lactate concentration after 3 months storage period. These results can be explained by recognizing differences in hydrolysis-esterification equilibrium, as some other authors have reported.[7,28] Our results are also in general agreement with those of other investigators who found that the concentration of ethyl lactate in young white wines increases with storage time.[29]

Carbonyl Compounds

Furfural content increased significantly with storage particularly for storage times longer than 3 months in Debina wines and longer than 6 months in red wines. However, measured concentrations were not sufficiently high so as to adversely affect the aroma of the wines.[7]

Acetaldehyde content of Debina white wine samples significantly increased immediately after storage, within the first 3 months of the storage period. This period is normally the normal time of shipping of Debina young wines to the market after bottling. On the other hand, in rosé and red wine samples a significant increase in acetaldehyde content was observed during the second half of the 12-month storage period.

Alcohols

Observed changes in amyl alcohols, isobutanol, 1-propanol, 1-pentanol and 2-phenylethanol concentrations were significant in wines stored for periods longer than 3 months. These changes were not sufficiently high so as to adversely affect the aroma of the wines. Particularly, the increased concentrations of 2-phenylethanol can have a positive strong influence on the aroma of Debina wines.[8,25] The concentrations of amyl alcohols in all type of wines do not exceed the level of 300 mg L^{-1}.

Hexanol-1 content decreased significantly after storage time longer than 3 months in Debina (dry, semi-sparkling and rosé) wines. The mean concentration of this compound decreased also significantly in red wine samples after storage time longer than 9 months.

4 CONCLUSIONS

The combined solvent extraction/GC method without concentration allows quantification of a number of wine aroma compounds. The method can be used for the quantitative determination of these compounds at low concentrations with satisfactory precision and accuracy.

The varieties of investigated wines show high concentration levels of amyl alcohols and 2-phenylethanol. Lower levels of ethyl lactate, ethyl acetate, isobutanol and furfural, were detected. Even lower concentrations of 1-hexanol, 1-propanol, acetaldehyde and 1-pentanol were determined.

The major differences in concentration of volatile compounds between samples from Debina variety (monovarietal white wines) and multivarietal samples (rosé and red wines) are the higher 2-phenylethanol and amyl alcohols content of monovarietal wines and on their lower furfural concentrations. Within these concentration ranges, higher alcohols are believed to contribute to the overall aroma complexity of the wines and 2-phenylethanol is considered having a positive strong influence on the aroma of Debina wines.

Results showed that differences in the profile of the major volatile compounds of the wines samples were found during their 12-month storage under cellar temperature conditions. With the exception of acetaldehyde content in Debina white wines, observed changes were generally not significant during the first quarterly storage period.

Acknowledgements

The authors would like to thank the Union of Agricultural Cooperatives of Ioannina for the wine samples supplied. The authors are also grateful to the Food Certification Unit of the Network of Horizontal Laboratory Units and Centers of the University of Ioannina for the GC/MS measurements.

References

1. Y. Zhou, R. Riesen and C.S. Gilpin, *J Agric Food Chem.*, 1996, **44**, 818.
2. E. Bardi, A.A. Koutinas, C. Psarianos and M. Kanellaki, *Process Biochem.*, 1997, **32**, 579.
3. E. Falqué, E. Fernández and D. Dubourdieu, *Talanta*, 2001, **54**, 271.
4. A. Escobal, C. Iriondo and C. Laborra, *J. Chromatogr., A*, 1997, **778**, 225.

5. A. Escobal, C. Iriondo, C. Laborra, E.Eléjalde and I. Gonzalez, *J. Chromatogr. A,* 1998, **823**, 349.
6. Z. Murányi and Z. Kovács, *Microchem. J.,* 2000, **67**, 91.
7. M.S. Pérez-Coello, P.J. Martín-Álvarez and M.D. Cabezudo, *Z. Lebensm. Unters. Forsch. A,* 1999, **208**, 408.
8. R.B. Boulton, V.L. Singleton, L.F. Bisson and R.E. Kunkee, *Principles and practices of winemaking,* Chapman & Hall, New York, London, 1998, p 390.
9. M. Shimoda, T. Shibamoto and A.C. Noble, *J. Agric. Food Chem.,* 1993, **41**, 1664.
10. C. Pricer, P.X. Etiévant, S. Nicklaus and O. Brun, *J. Agric. Food Chem.,* 1997, **45**, 3511.
11. M.F. Kok, F.M. Yong and G. Lim, *J. Agric. Food Chem.,* 1987, **35**, 779.
12. V. Ferreira, A. Rapp, J.F. Cacho, H. Hastrich and I. Yavas, *J. Agric. Food Chem.,* 1993, **41**, 1413.
13. C. Ortega, R. López, J. Cacho and V. Ferreira, *J. Chromatogr. A,* 2001, **923**, 205.
14. C. García-Jares, S. García-Martin and R. Cela-Torrijos, *J. Agric. Food Chem.,* 1995, **43**, 764.
15. I. Cutzach, P. Chatonnet and D. Dubourdieu, *J. Agric. Food Chem.,* 1999, **47**, 2837.
16. S. Maicas, J.V. Gil, I. Pardo and S. Ferrer, *Food Res. Intern.,* 1999, **32**, 491.
17. S. Karagiannis, A. Economou and P. Lanaridis, *J. Agric. Food Chem.,* 2000, **48**, 5369.
18. D.H. Vila, F.J.H. Mira, R.B. Lucena and M.A.F. Recamales, *Talanta,* 1999, **50**, 413.
19. B.W. Zoecklein, K.C. Fugelsang, B.H. Gump and F.S. Nury, *Wine analysis and production,* Chapman & Hall, New York, London, 1995, p 101.
20. J.C. Mauricio, J.J. Moreno, E.M. Valero, L. Zea, M. Medina and J.M. Ortega, *J. Agric. Food Chem.,* 1993, **41**, 2086.
21. S.P. Arrhenius, L.P. McCloskey and M. Sylvan, *J. Agric. Food Chem.,* 1996, **44**, 1085.
22. V.G. Dourtoglou, N.G. Yannovits, V.G. Tychopoulos and M.M. Vamvakias, *J. Agric. Food Chem.,* 1994, **42**, 338.
23. T. Herraiz, G. Reglero, M. Herraiz, P.J. Martin-Alvarez and M.D. Cabezudo, *Am. J. Enol. Vitic.,* 1990, **41**, 313.
24. C.S. Ough and M.A. Amerine, *Methods for Analysis of Must and Wines,* 2nd edn., John Wiley & Sons, New York Chichester Brisbane Toronto Singapore, 1988, p 114.
25. J.C.R. Demyttenaere, C. Dagher, C. Sandra, S. Kallithraka, R. Verhé and N.D. Kimpe, *J. Chromatogr. A,* 2003, **985**, 233.
26. H. Guth, *J. Agric. Food Chem.,* 1997, **45**, 3027.
27. J.P. Osborne, R.M. Orduña, G.J. Pilone and S.Q. Liu, *FEMS Microbiol. Let.,* 2000, **191**, 51.
28. V. Ferreira, A. Escudero, P. Fernádez and J.F. Cacho, *Z. Lebensm. Unters. Forsch. A,* 1997, **205**, 392.
29. M.S. Pérez-Coello, M.A. González-Viñas, E. García-Romero, M.C. Díaz-Maroto and M.D. Cabezudo, *J. Chromatogr. A,* 2003, **985**, 233.

ROLE OF ANTHOCYANINS IN THE DIFFERENTIATION OF TEMPRANILLO WINES

E. Revilla, M.J. González-Reig, P. Garcinuño and E. García-Beneytez

Sección Departamental de Química Agrícola, Facultad de Ciencias, Universidad Autónoma de Madrid, 28049 Madrid, Spain.

1 ABSTRACT

The anthocyanin profile of young Tempranillo wines made with different technologies (conventional red winemaking and carbonic maceration) in different areas of Rioja (vintage 2000) has been analyzed by RP-HPLC. Data have been submitted to different multivariate statistical analysis (cluster analysis, principal components analysis, factor analysis). Results show that those wines may be classified taking into account winemaking technology and area of production. Other phenolic compounds, like flavonols, are not useful for this purpose.

2 INTRODUCTION

Phenolic compounds play a well established and relevant role in Enology, due to their direct influence on some important sensory properties of premium red wines, such as color and color stability, bitterness, and astringency.[1] Anthocyanins are the phenolic pigments responsible for color of red grapes and wines. They are accumulated in grape skins during ripening, their content increases until maturity, and they have been proposed as chemical markers to differentiate grape cultivars.[2-5] However, this analytical tool must be used carefully. Different clones have been identified in many *Vitis vinifera* cultivars, and some data suggest that the anthocyanin composition of some clones belonging to the same cultivar may be significantly different, even if they are grown in the same vineyard.[6-7] Moreover, the anthocyanin composition of grapes grown in a vineyard may change year to year as a consequence of weather conditions taking place during ripening.[7-9] During current red winemaking, contact between liquid and solid phases (fermenting must and crushed grapes) allows the extraction of polyphenols from skins to wine. Several technological procedures used in winemaking (e.g., degree of skin crushing, fermentation temperature, or length of maceration) affect the final composition of anthocyanins in red wines, and thus, their color.[1, 10-11]

Premium varietal wines play an increasingly important role in the global wine market. This has stimulated research on analytical tools to certify their authenticity and to protect consumers against fraud, especially if the grape cultivar is mentioned in bottle labels, or if legal standards prescribe the cultivar identity of certain wines.[12] Several physicochemical parameters, such as proteins, color and polyphenols, amino acids and aroma compounds, and DNA analyses have been proposed as adequate tools for this purpose. Analytical tools based on the analysis of color and phenolic compounds, and especially the analysis of anthocyanins and their ratios, seem to fit the authentification purposes for red wines made with different grape cultivars.[13-17]

The objective of our research was to classify wines made with grapes of a single cultivar, grown in different viticultural areas and using different winemaking technologies, taking into account their anthocyanin fingerprint. Other procedures proposed for the differentiation of grapes, like DNA analyses,[18-19] are invalid for this purpose. For this reason, we selected wines made with Tempranillo grapes, used to make premium red wines in several grape-growing areas of Spain, Portugal, Argentina and other countries. Among them, those produced in the Spanish region of Rioja have a long tradition of quality, especially if they are aged in oak casks. They are produced following two different technological approaches: carbonic maceration, which gives fresh, young red wines, and conventional winemaking with destemmed clusters, which usually give wines which may be submitted to aging in oak casks. Moreover, Rioja is divided in three production sub-areas (Rioja Alta, Rioja Baja and Rioja Alavesa), with distinctive agroecological characteristics. Climate in Rioja Alta presents Atlantic influence, and grapes lead to full-bodied wines with medium alcohol content and high total acidity. On the other hand, climate in Rioja Baja shows Mediterranean influence, leading to wines with high alcohol content and extract. Finally, climate in Rioja Alavesa presents Atlantic and Mediterranean influences, and wines made with grapes grown in this sub-area have average alcohol content and total acidity. Thus, a set of Tempranillo wines from Rioja, made with different technologies and grapes from different sub-areas, would meet the requirements for our study.

3 METHOD AND RESULTS

3.1 Samples

Authentic and unadulterated industrial samples of 24 Tempranillo wines produced in 2000 in several cellars of the three sub-areas (Rioja Alta, Rioja Baja and Rioja Alavesa) of "Denominación de Origen Calificada Rioja" (Spain) have been studied. Six of these wines were made by carbonic maceration, and the others by conventional procedures with destemmed grapes. Some characteristics of these wines are given in Table 1. Wines were analyzed in May and June 2001, after malolactic fermentation took place.

3.2 Reagents and Standards

Deionized water was purified with a Milli-Q water system (Millipore, Bedford, MA, USA) before using. Acetonitrile of HPLC-gradient grade was purchased from Merck (Darmstadt, Germany). Perchloric acid and trifluoracetic acid of analytical-reagent grade were obtained from Scharlau (Barcelona, Spain) and Fluka (Buch, Switzerland), respectively. All other chemicals (analytical-reagent grade) were obtained from Panreac (Mollet del Valles, Spain). Standards of several anthocyanins were prepared from red grape skins as previously described in the literature.[2]

Table 1 *Some Characteristics of Wines Studied*

Sample	Sub-area	Winemaking technique
RA-1	Rioja Alta	Conventional
RA-2	Rioja Alta	Conventional
RA-3	Rioja Alta	Conventional
RA-4	Rioja Alta	Conventional
RA-5	Rioja Alta	Conventional
RA-6	Rioja Alta	Conventional
RA-7	Rioja Alta	Conventional
RA-8	Rioja Alta	Conventional
RB-1	Rioja Baja	Conventional
RB-2	Rioja Baja	Conventional
RB-3	Rioja Baja	Conventional
RB-4	Rioja Baja	Conventional
RB-5	Rioja Baja	Conventional
AV-1	Rioja Alavesa	Conventional
AV-2	Rioja Alavesa	Conventional
AV-3	Rioja Alavesa	Conventional
AV-4	Rioja Alavesa	Conventional
AV-5	Rioja Alavesa	Conventional
MC-1	Rioja Alta	Carbonic maceration
MC-2	Rioja Alta	Carbonic maceration
MC-3	Rioja Alavesa	Carbonic maceration
MC-4	Rioja Alavesa	Carbonic maceration
MC-5	Rioja Alavesa	Carbonic maceration
MC-6	Rioja Alavesa	Carbonic maceration

3.3 Analytical and Statistical Procedures

Analyses were performed with a liquid chromatograph consisting of Waters (Milford, MA, USA) M510 and M501 pumps, a Waters 680 gradient controller, a Rheodyne 7725 injection valve furnished with a 20 µl loop, a Waters 996 photodiode array detector and a Millennium[32] workstation. The separation was carried out with a Waters Nova-Pak C_{18} steel cartridge, 3.9 x 150 mm, filled with 5 µm particles, using a Waters Sentry Nova-Pack C18 guard cartridge, 20 x 3.9 mm, both thermostated in a water bath at 32°C. The following mobile phases were used: water-acetonitrile (95:5) adjusted to pH 1.3 with perchloric acid solvent (solvent A), and water-acetonitrile (40:60) adjusted to pH 1.3 with perchloric acid (solvent B). Gradient elution was applied at 1.5 ml/min flow rate, according to the following program: linear gradient from 5% B to 10% B in 5 min, from 10% B to 20% B in 20 min, from 20% B to 25% B in 13 min, from 25% B to 35% B in 12 min, from 35% B to 100% B in 3 min, 100%B for 2 min, 100% B to 5% B in 1 min, 5% B for 10 min. Samples (20 µl) were injected. Spectra were recorded every second between 250 and 600 nm, with a band width of 1.2 nm. Samples, standard solutions and mobile phases were filtered through a 0.45 µm pore size membrane filter.

HPLC/MS analyses were carried out to confirm the identity of anthocyanins present in wines, using a Hewlett-Packard 1100 system with a PDA UV-Vis coupled to a mass spectrometer system equipped with an ESI interface. The separation was carried out using a 150 x 3.9 mm i.d., 5 μm, Waters Nova-Pack C_{18} steel column with a 20 x 3.9 mm i.d. Waters Sentry Nova-Pack C18 guard cartridge (Waters, Milford, MA, USA), both thermostated at 50 °C. The mobile phase was a linear gradient of water-acetonitrile (50:50) (solvent B), in water-acetonitrile (95:5), (solvent A) both adjusted to pH 1.3 with trifluoroacetic acid, at a flow-rate of 0.6 ml/min. The following gradient was used: 0 min, 15% B; 0-20 min, 15% to 30% B; 20-25 min, 30% to 35% B; 25-35 min, 35% to 40% B; 35-42 min, 40% B; 42-43 min, 40 to 100 % B; 43-48 min, 100% B; and 48-49 min, 100 to 15% B. *PDA.* Spectra were recorded every second between 250 and 600 nm, with a band width of 1.2 nm and chromatograms were acquired at 520 nm. MS parameters were: capillary voltage 4000 V; fragmenter ramped from 90 to 120 V; drying gas temperature 325°C; gas flow (N_2) 12 ml/min. The instrument was operated in positive ion mode scanning from *m/z* 50 to 2000 at a scan rate of 1.47 s/cycle.

Multivariate analyses to test differences among groups of wines were run using the Statgraphics 2.0 Plus statistical package.

3.4 Anthocyanin Fingerprint of Tempranillo Wines

Eight different anthocyanins, whose structures are displayed in Figure 1, have been analyzed throughout the study. Their total content, as expected, was quite different in the wines studied. Their relative contents were slightly different, depending on the sub-area of production and the technology involved in winemaking, as shown in Table 2.

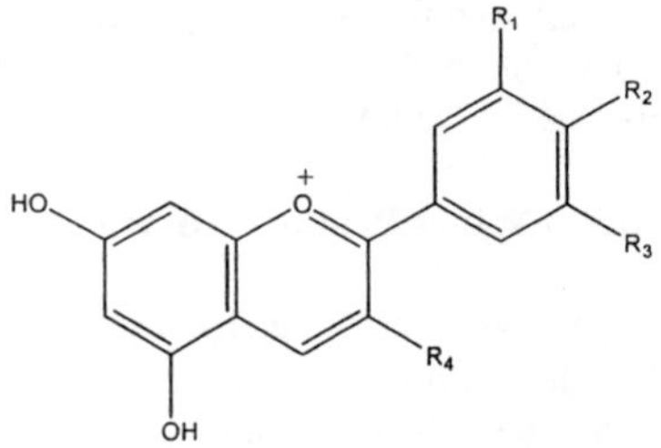

Compound	R_1	R_2	R_3	R_4
Delphinidin-3-*O*-glucoside (DfGl)	OH	OH	OH	*O*-glucose
Cyanidin-3-*O*-glucoside (CyGl)	OH	OH	H	*O*-glucose
Petunidin-3-*O*-glucoside (PtGl)	OCH_3	OH	OH	*O*-glucose
Peonidin-3-*O*-glucoside (PnGl)	OCH_3	OH	H	*O*-glucose
Malvidin-3-*O*-glucoside (MvGl)	OCH_3	OH	OCH_3	*O*-glucose
Malvidin-3-*O*-acetylglucoside (MvGlAc)	OCH_3	OH	OCH_3	*O*-acetylglucose
Peonidin-3-O-p-coumarylglucoside (PnGlCm)	OCH_3	OH	H	*O-p*-coumarylglucose
Malvidin-3-*O-p*-coumarylglucoside (MvGlCm)	OCH_3	OH	OCH_3	*O-p*-coumarylglucose

Figure 1 *Chemical structures of eight anthocyanins analyzed in Tempranillo wines.*

Wines made by carbonic maceration contain a higher proportion of MvGlCm, but a lower proportion of DfGl, CyGl and PnGl, than wines made by conventional technology using destemmed grapes. Carbonic maceration carried out in Rioja usually involves a very long contact of liquid phase with grape skins. Probably, this favors the extraction of *p*-coumaryl derivatives of anthocyanins, which is slower than the extraction of non-acylated anthocyanins when grapes are destemmed before alcoholic fermentation.[5]

On the other hand, wines made with destemmed grapes produced in Rioja Baja contain a lower proportion of DfGl and PtGl (which are considered as primitive anthocyanins[20]) and a higher proportion of PnGl and MvGl than wines made in Rioja Alta. This may be because the climate in Rioja Baja is warmer and drier than in Rioja Alta. As reported previously, the relative contents of DfGl and PtGl are lower in Tempranillo grapes grown under warm conditions than in those grown in relatively cool conditions.[9] Thus, it would be expected that wines made with Tempranillo grapes grown in a warm area (Rioja Baja) contain a lower content of DfGl and PtGl than wines made from grapes grown in a relatively cool area (Rioja Alta). The proportions of those four pigments in wines from Rioja Alavesa are intermediate among levels presented by wines from Rioja Baja and those from Rioja Alta.

Mean values of the relative content of the eight anthocyanins considered for each group of wines (RA, RB, AV and MC) are summarized in Table 3, which clearly shows the differences mentioned above. As can be noted, there are significant differences ($p<0.05$) among groups for several anthocyanins, after results obtained by one-way-ANOVA, using Fisher's least significant difference (LSD) procedure to discriminate among the means. These results suggest that the anthocyanin fingerprint of Tempranillo wines may be an analytical tool valuable for classifying them, taking into account both the technology involved in winemaking, and the sub-area of grape production for wines produced using a specified winemaking technology.

3.5 Classification of Tempranillo Wines Produced in Rioja

Different types of multivariate analyses were carried out for a better understanding of differences shown by the anthocyanin fingerprints of Tempranillo wines. To establish the relationships among the percentages of the eight anthocyanins analyzed and the samples, principal component analysis (PCA) was performed. Three principal components, with eigenvalues >1, explained 86,7% of the variability in the data (Table 4). Component 1 was mainly associated with primitive anthocyanins (DfGl, CyGl and PtGl), MvGl and its p-coumarylated derivative (MvGlCm), whereas component 2 was associated with the other three anthocyanins (PnGl, PnGlCm and MvGlAc). Furthermore, component 3 was mainly associated with MvGl and PnGl, and with their *p*-coumaryl derivatives. Thus, PCA suggests that differences observed in the anthocyanin pattern of Tempranillo wines are primarily due to the different proportions of DfGl, CyGl, PtGl, MvGl and MvGlCm, but the percentages of other three anthocyanins also contribute to differences among wines. Wines made by carbonic maceration (MC) are grouped on the left side of the plot of component 1 versus component 2 (Figure 2). On the other hand, wines made by conventional procedures with destemmed grapes (except RB-1) appear in the center and on the right side of the plot. These results suggest that technology used in winemaking affect the anthocyanin fingerprint of Tempranillo wines.

Table 2 Relative Content (%) of Eight Different Anthocyanins in Tempranillo Wines.

Sample	DfGl	CyGl	PtGl	PnGl	MvGl	MvGlAc	PnGlCm	MvGlCm
RA-1	9.44	0.55	12.26	2.50	62.70	3.44	1.04	8.07
RA-2	9.45	0.52	12.41	3.00	62.49	4.05	1.01	7.07
RA-3	10.20	0.50	12.60	2.70	61.90	3.60	0.70	7.80
RA-4	10.39	0.65	12.91	2.62	62.79	3.10	0.82	6.74
RA-5	11.74	0.73	13.63	2.99	60.24	2.73	0.79	7.16
RA-6	10.64	0.68	13.80	2.66	62.73	2.46	0.65	6.37
RA-7	11.10	0.65	13.46	2.71	61.72	2.57	0.78	7.01
RA-8	11.44	0.91	14.14	3.08	60.94	2.34	0.72	6.44
RB-1	7.46	0.75	8.99	4.29	67.99	2.53	1.01	6.99
RB-2	8.04	0.53	10.34	4.36	65.50	3.69	0.96	6.58
RB-3	9.36	0.63	10.79	3.64	64.30	3.33	0.76	7.19
RB-4	9.24	0.54	11.40	3.03	65.56	3.09	0.68	6.46
RB-5	7.03	0.45	9.85	2.86	70.41	2.52	0.68	6.18
AV-1	12.00	1.02	12.84	4.50	61.49	2.27	0.60	5.28
AV-2	8.65	0.56	11.84	2.78	67.60	2.92	0.46	5.19
AV-3	7.77	0.50	10.63	2.93	67.65	3.53	0.85	6.14
AV-4	9.73	0.78	12.80	3.08	64.16	2.77	0.73	5.95
AV-5	10.36	0.63	12.38	2.85	63.59	3.32	0.74	6.14
MC-1	5.79	0.24	9.46	1.61	68.22	2.68	0.87	11.13
MC-2	7.14	0.33	10.97	2.68	65.88	2.39	0.80	9.81
MC-3	7.02	0.25	10.20	1.83	67.23	2.61	0.72	10.14
MC-4	7.86	0.37	10.90	1.92	65.07	2.59	1.04	10.25
MC-5	6.50	0.00	10.63	2.46	67.17	2.62	0.71	9.92
MC-6	6.71	0.25	10.09	1.78	68.01	2.21	0.77	10.17

Table 3 Mean Values of the Relative Content (%) of Eight Different Anthocyanins in the Four Groups of Tempranillo Wines. The same letter indicates the means within a column were not significantly different ($p < 0.05$).

Group	DfGl	CyGl	PtGl	PnGl	MvGl	MvGlAc	PnGlCm	MvGlCm
RA	10.55 a	0.65 a	13.15 a	2.78 b	61.94 b	3.04 a	0.81 a	7.08 b
RB	8.23 b	0.58 a	10.27 c	3.64 a	66.75 a	3.03 a	0.82 a	6.68 b
AV	9.70 a	0.70 a	12.10 b	3.23 ab	64.90 a	2.96 a	0.74 a	5.74 c
MC	6.84 c	0.24 b	10.38 c	2.05 c	66.93 a	2.52 a	0.82 a	10.24 a

Cluster analysis (performed using the eight variables considered for PCA) confirms those results. The dendrogram presented in Figure 3, obtained using the nearest neighbor method with squared euclidean distances, shows that wines made by carbonic maceration (MC-1 to MC-6) grouped very close. On the other hand, wines made by conventional winemaking procedures with destemmed grape were grouped unsuccessfully if the sub-area of production (RA, RB or AV) was considered.

Table 4 *Eigenvalues, Percent of Variance, and Cumulative Percentage for the Three Principal Components after PCA*

Component number	Eigenvalue	Percent of variance	Cumulative percentage
1	4.16117	52.015	52.015
2	1.45027	18.128	70.143
3	1.32177	16.522	86.665

Discriminant analysis (DA), using a stepwise forward selection algorithm, was performed to develop a set of discriminating functions, derived from values of the eight quantitative variables (percentages of the anthocyanins considered in the study). Those functions were capable of predicting if wines were made by carbonic maceration (MC), or by conventional technology using destemmed grapes in the three sub-areas of Rioja (RA, RB and AV). Two discriminant functions, with eigenvalues higher than 1, and with P-values less than 0.05, were statistically significant at the 95% confidence level (Table 5). 100% of samples were correctly classified.

Table 5 *Eigenvalues, Relative Percentages, and P-values for Discriminant Functions after DA.*

Discriminant function	Eigenvalue	Relative percentage	P-value
1	38.1373	87.89	0.0000
2	4.91121	11.32	0.0014
3	0.341643	0.79	0.5443

Figure 4 shows the different groups of samples in the plot of discriminant functions 1 and 2. As can be noted, wines made by carbonic maceration (MC) closely grouped on the right size of the plot. On the other hand, wines made with conventional procedures using destemmed grapes appear on the right size of the plot, and grouped taking into account the sub-area of grape production. These results suggest that the anthocyanin fingerprint of

Dairy and Wine Flavors

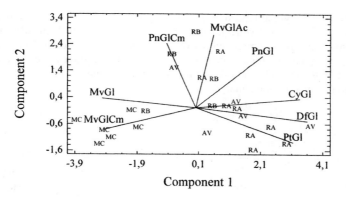

Figure 2 *Plot of the first two principal components in PCA of Tempranillo wines. Variables are percentages of eight different anthocyanins. MC: wines made by carbonic maceration. RA, RB and AV: wines made by conventional procedures with destemmed grapes from Rioja Alta, Rioja Baja, and Rioja Alavesa, respectively.*

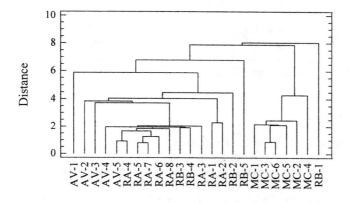

Figure 3 *Dendrogram of Tempranillo wines from Rioja. For key to samples, see Table 1.*

Tempranillo wines made by carbonic maceration is quite different from that shown by Tempranillo wines made by conventional procedures with destemmed grapes. Furthermore, despite the quite similar anthocyanin fingerprint showed by Tempranillo wines made in Rioja by conventional procedures with destemmed grapes, there are slight differences caused by the natural conditions of the sub-area of grape production than allow their differentiation by discriminant analysis.

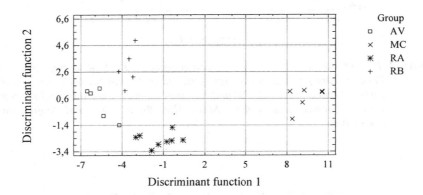

Figure 4 *Plot of discriminant functions 1 and 2 in discriminant analysis of Tempranillo wines. MC: wines made by carbonic maceration. RA, RB and AV: wines made by conventional procedures with destemmed grapes from Rioja Alta, Rioja Baja, and Rioja Alavesa, respectively.*

4 CONCLUSION

The characterization of the anthocyanin fingerprint of Tempranillo wines seems to be an analytical tool of choice for their differentiation, taking into account technological and agroecological factors, and a promising procedure for the characterization of the origin of those wines. Very probably, this analytical tool could be applied successfully to wines made with other cultivars to differentiate their technological or geographical origin.

References

1 V. Cheynier, M. Moutounet and P. Sarni-Manchado. 'Oenologie', C. Flanzy ed., Technique et Documentation Lavoisier, Paris, 1998, pp. 123-162.
2 J. Bakker and C.F. Timberlake, *J. Sci. Food Agric.*, 1985, **36**, 1315.
3 J.P. Roggero, J.L. Larice, P. Archier and S. Coen, *Rev. Fr. Oenol. (Cah. Sci.)*, 1988, **28**, 42.
4 T. Castia, M.A. Franco, F. Mattivi, G. Muggliolu, G. Sferlazzo, and G. Versini. *Sci. Aliment.*, 1992, **12**, 239.

5   E. Garcia-Beneytez, E. Revilla and F. Cabello. *Eur. Food Res. Technol.*, 2002, **215**, 32.
6   J.P. Roggero, S. Coen and B. Raggonet, *Am. J. Enol. Vitic.*, 1986, **37**, 77.
7   E. García-Beneytez, *Ph. D. Thesis*, Universidad Autónoma de Madrid, 2003.
8   M.A. Esteban, M.J. Villanueva and J.R. Lissarrague, *J. Sci. Food Agric.*, 2001, **81**, 409.
9   J.M. Ryan and E. Revilla, *J. Agric. Food Chem.*, 2003, **51**, 3372.
10 Jackson, R.S. 'Wine Science', Academic Press, San Diego 2001
11 G. Mazza. *Crit. Rev. Food Sci. Nutr.*, 1995, **35**, 341.
12 R. Wittkowski, Proceedings of the XXIV World Congress of OIV, Mainz, Germany, 1999, Vol. 2-1, pp. 118-124.
13 P. Etievant, P. Schlich, A. Bertrand, P. Symonds and J.C. Bouvier, *J. Sci. Food Agric.*, 1988, **42**, 39.
14 C. Santos–Buelga, S.S. Muñoz, Y. Gutiérrez, E. Hebrero, J.L. Vicente, P. Galindo and J.C. Rivas-Gonzalo, *J. Agric. Food Chem.*, 1991, **39**, 1086
15 I. Arozarena, A. Casp, R. Marín and M. Navarro, *Eur. Food Res. Technol.*, 2000, **212**, 108
16 B. Berente, D. De la Calle Garcia, M. Reichenbächer and K. Danzer. *J. Chromatogr. A*, 2000, **871**, 95.
17 H. Ottender, R. Marx and M. Zimmer, *Austr. J. Grape Wine Res.*, 2004, **10**, 3.
18 R. Siret, J.M. Boursiquot, M.H. Merle, J.C. Cabanis and P. This. *J. Agric. Food Chem.*, 2000, **48**, 5035.
19 E. García-Beneytez, M.V. Moreno-Arribas, J. Borrego, M.C. Polo and J. Ibañez. *J. Agric. Food Chem.*, 2002, **50**, 6090.
20 P.K. Boss, C. Davies and S.P. Robinson, *Austr. J. Grape Wine Res.*, 1996, **2**, 163.

ORGANIC ACID COMPOSITION OF *UME* LIQUEUR

Rie Kuramitsu and Shoji Furukawa

Department of Chemistry, Faculty of General Education, Akashi College of Technology Uozumi, Akashi, 674-8501, JAPAN

1 ABSTRACT

Ume liqueur (Japanese apricot liqueur) is made by soaking *ume* fruits (Japanese apricot) in shochu (distilled sake, alcohol content: 35%). Among the flavor components that influence greatly the quality of *ume* liqueur are the organic acids. During manufacturing and storage, transfer of organic acids contained in *ume* fruits into shochu and formation of ethyl esters of the organic acids were observed. Eighteen months after the start of the experiments, the content of organic acids in shochu reached a maximum level and changed only slightly thereafter. Citric and malic acids comprised approximately 90% of total organic acids, but the change of the two acids during storage of *ume* liqueur differed from each other. While the content of citric acid in *ume* liqueur did not change significantly after reaching a maximum level, the content of malic acid decreased gradually. This finding suggested that malic acid was converted into its ester more readily than citric acid.

2 INTRODUCTION

Ume liqueur is a fruit liqueur which is made by soaking *ume* fruits (Japanese apricot) together with rock sugar in shochu (which is distilled sake) for an extended time period. *Ume* liqueur has been loved by the Japanese people for ages and is manufactured in homes as well as in factories. It has a special fragrance, and the organic acids contained therein are believed to augment energy metabolism in human body. Due to the recent trend of healthy life style, consumption of *ume* liqueur has increased steadily. In the current studies, organic acids in *ume* liqueur were studied since these influence greatly the quality of *ume* liqueur. Four samples of *ume* liqueur were prepared from four sets of manufacturing conditions, with different pretreatments and different amounts of *ume* fruits. Change in the contents of organic acids during manufacturing and storage of *ume* liqueur were determined by means of high performance liquid chromatography.

3.1 Manufacturing of *Ume* Liqueur

A commercial shochu for making *ume* liqueur (a product of Hyogo Prefecture; alcohol concentration 35%), *ume* fruits (a product of Wakayama Prefecture) and rock sugar (a product of Shizuoka Prefecture) were used for experiments. Washed *ume* fruits and rock sugar were placed alternately in layers into shochu, and the containers were tightly closed and agitated from time to time. The compositions of the starting materials are shown in Table 1. One kg of *ume* fruits was the control composition (A). (B) used 1 kg of *ume* fruits whose surface was pierced with a fork to make holes. In (C) 1.5 kg of intact *ume* fruits was employed and in (D) the surface of 1.5 kg of *ume* fruits was pierced for making holes. These 4 samples were used in experiments.

3.2 Acidity and Ester Concentration

For assaying *ume* fruits, 100 g of fruits and 100 g of water were homogenized, centrifuged and the supernatant (fruit juice) was diluted to 500 ml. An aliquot of the solution was analyzed for both acidity and ester concentration by the conventional methods (both with 0.1 N NaOH/10 ml)[1], and the acidity and ester concentration in 1 kg of *ume* fruits were calculated. For assaying *ume* liqueur an aliquot was used for the estimation of the acidity and ester concentration, and from the results and the yield of *ume* liqueur (Table 2) the acidity and ester concentration of each sample were calculated.

3.3 Analysis of Organic Acids

3.3.1 Preparation of Samples for the Analysis of Organic Acids. The juice (100 ml) described in 3.2 was used as the sample for *ume* fruits. *Ume* liqueur was used in 50 ml portions for the liqueur analysis. Samples were prepared by the ion exchange resin method[2].

3.3.2 High Performance Liquid Chromatography of Organic Acids. For the analysis of organic acids a liquid chromatograph Model 633A (Hitachi Seisakusho Inc.) was employed. The chromatograph was equipped with a column packed with a cation exchange resin (Shodex Ionpack, 80 x 500 mm) and Hitachi 633 MLC Detector. The elution flow was 0.5 ml/min and pressure was 20 kg/cm^2. The effluent from the column was assayed at the wavelength of 210 nm.

4 EXPERIMENTAL RESULTS AND DISCUSSION

4.1 The Yield of *Ume* Liqueur

Four samples of *ume* liqueur were manufactured from 4 sets of starting materials shown in Table 1, and after 12 months the soaked *ume* fruits were removed. The yields of *ume* liqueur are shown in Table 2 reveals considerable difference among samples. It has been reported that 90.1% of *ume* fruits, excluding seeds, is water, 7.6% carbohydrate, and the remainder mainly organic acids[3]. In comparison to *ume* fruits before soaking in shochu, the fruits in (A) and (C) after making *ume* liqueur held many wrinkles in rind and showed low moisture contents, while the *ume* fruits which had been pierced before soaking in (B) and (D) showed high moisture contents. All of the 4 samples of soaked fruits contained sugar and alcohol. The results suggest that components of sugar-containing shochu (major components are water, sugar and alcohol) and those of *ume* fruits (major components are water and low amounts of carbohydrate and organic acids) migrated and an equilibrium was reached after eluting out of the moisture in the fruits into shochu. We expected that (C) with 1.5 kg of *ume* fruits and higher moisture content would give higher yield of *ume* liqueur than (A) with 1 kg of *ume* fruits, but the yield from (C) was only slightly higher than the yield from (A) probably due to reaching an equilibrium after elution of a rather small amount of moisture. The yields of *ume* liqueur from pierced fruits in (B) and (D) were considerably lower. The results are probably due to a lower extent of moisture elution in (B) and (D) than in (A) and (C) by the influence of component migration through holes in the rind.

Table 1 *Starting materials of Ume liqueur*

Materials / Samples		(A)	(B)	(C)	(D)
Shochu[*1]	(liter)	1.8	1.8	1.8	1.8
Ume[*2]	(kg)	1	1	1.5	1.5
Rock sugar	(kg)	1	1	1	1
Holes[*3]		No	Yes	No	Yes

[*1] Shochu or Japanese distilled liquor made from fermented corn.
[*2] *Ume* or Japanese apricot.
[*3] Surfaces of *Ume* were pierced with a fork to make holes.

Table 2 *Yields of Ume Liqueur*

Samples	(A)	(B)	(C)	(D)
Yields (ml)	2700	2510	2820	2530

4.2 The Contents of Acids, Esters and Organic Acids in *Ume* Fruits

The results of estimating the contents of acids, esters and organic acids in 1 kg of *ume* fruits before soaking in shochu are shown in Tables 3 and 4. Seven organic acids were identified and determined in *ume* fruits. Among them citric and malic acids were contained in significant amounts and sum of the two acids represented about 90% of total organic acids. The value of 55 g/kg as found here is the total content of organic acids in *ume* fruits and is close to the reported value in The Official Food Compositions of Japan[3]. The chromatographic identification and determination of organic acids are described in 4.3 and 4.4.

In comparison to the acid content, the total content of organic acids in *ume* fruits was slightly higher, and this was thought to be due to the fact that the total content of organic acids included the ester-form acids which were estimated as free acids by ion exchange resin treatment. For the sake of calculation of the extraction rate (see 4.3), the contents of acids, esters and sum of both are expressed in mol.

Table 3 *Acids and Esters in Ume Fruits*

Acids (mol/kg)	Esters (mol/kg)	Total (mol/kg)
0.758	0.062	0.820

Table 4 *Contents of Identified Organic Acids in Ume Fruits*

Organic acids (g/kg)							
Oxalic	Citric	Malic	Lactic	Formic	Acetic	Pyroglutamic	Total
1.18	24.1	25.3	3.15	0.675	0.446	0.178	55.0
							(0.799 mol/kg)

4.3 Changes of the Contents of Acids and Esters during *Ume* Liqueur Manufacturing

Changes of the acidity, ester concentration and contents of acids and esters during *ume* liqueur manufacturing and storage suggest that the organic acids in the starting *ume* fruits were extracted into the shochu, and that they are partly converted into esters by the reaction with alcohol. Then, the extraction rate is given to show what extent (%) of the sum of organic acids and esters contained in *ume* fruits extracted into shochu.

From Table 5 it is shown that the acidity in all of the 4 samples increased in the first 6 months and then gradually in the following 12 months to reach maximum levels; then the acidity decreased until 30 months from the start of experiments, time zero. On the other hand, the ester concentration reached a peak after 18 months from time zero and then increased gradually until 30 months.

The acidity in sample (A) with 1 kg of intact *ume* fruits was slightly lower than that in sample (B) with 1 kg of pierced *ume* fruits. The ester concentration was nearly identical in both samples. Since (A) gave higher yield of *ume* liqueur than (B), the change of the acids and esters as calculated by taking the yield into consideration was studied. Patterns of the change were similar to those of the change in acidity and ester concentration, but numerically (A) gave higher values than (B). The acid content showed peaks 18 months after the start of experiments, and then decreased to 30 months from time zero. The ester content continued increasing to 30 months from time zero. Changes in the combined amounts of acids and esters contents and in the extraction rates were rather small after passing 18 months when maximum levels were reached.

These results reveal that in all 4 samples the acids in *ume* fruits extracted almost completely into soaking shochu containing sugar in 6 months, and esterification of acids proceeded in parallel, and that extraction of acids continued after 6 months with some further esterification. At 18 months from time zero the acid content in *ume* liqueur reached a maximum value and the ester content also reached nearly a maximum value, thus bringing resultant *ume* liqueur to an approximate equilibrium among acids, alcohol and esters. Thereafter minimal esterification continued up to 30 months from time zero. During 18 to 30 months acidity decreased gradually while the ester content still increased gradually by contrast.

Table 5 Changes of Acids and Esters during Ume Liqueur Manufacturing

Samples	Period (1)[*1]	Acids (2)[*2]	Acids (3)[*3]	Esters (4)[*4]	Esters (5)[*5]	Total (6)[*6]	Extraction (7)[*7]
(A)	6	16.3	0.44	6.0	0.16	0.60	73.4
	18	18.2	0.49	7.0	0.19	0.68	82.9
	30	17.8	0.48	7.5	0.20	0.68	83.3
(B)	6	16.2	0.41	6.0	0.15	0.56	67.9
	18	18.7	0.47	7.1	0.18	0.65	78.9
	30	18.0	0.45	7.4	0.19	0.64	77.8
(C)	6	21.3	0.60	6.2	0.18	0.78	63.1
	18	24.1	0.68	7.7	0.22	0.90	72.9
	30	23.7	0.67	8.0	0.23	0.89	72.7
(D)	6	21.6	0.55	6.7	0.17	0.72	58.2
	18	24.9	0.63	8.1	0.21	0.84	67.9
	30	24.0	0.61	8.3	0.21	0.82	66.4

[*1] Soaking and storing period from the start of experiments (months)
[*2] Acidity (mmol/10ml)
[*3] Amount of acids (mol) = acidity x yield
[*4] Ester concentration (mmol/10 ml)
[*5] Amount of esters (mol) = ester concentration x yield
[*6] Total (mol) = combined amounts of acids and esters
[*7] Extraction rate (%) = (total amount in sample (mol)/total amount in *ume* (mol)) x 100

The extraction rates in 4 samples during the initial 18 months were as follows: (A) with 1 kg intact *ume* fruits: 82.9%, (B) with 1 kg pierced *ume* fruits: 78.9%, (C) with 1.5 kg intact *ume* fruits: 72.9%, and (D) with 1.5 kg pierced *ume* fruits: 67.9%. The extraction rates in intact fruits (A) and (C) were higher by 5 % than those in pierced fruits (B) and (D), and (A) showed 10 % higher rate than (C).

4.4 Change of Organic Acids during *Ume* Liqueur Manufacturing

An example of the chromatogram of organic acids in *ume* liqueur is shown in Figure 1. Similar chromatograms were obtained from juices made of *ume* fruits and from 12 samples of *ume* liqueur. Ten distinct peaks were found in the chromatogram. These peaks were compared with peaks for known organic acids in the chromatogram. The peaks No. 1, 5 and 9 did not correspond to peaks for any known organic acids in the chromatogram and were designated as unidentified acids. The 7 remaining peaks were identified as shown in Figure 1. The amount of each acid was calculated from its peak area in the calibration curve.

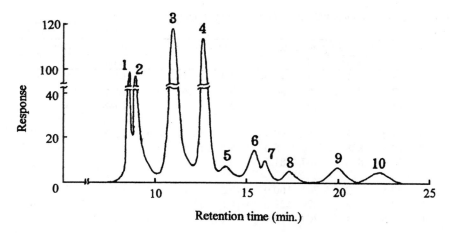

Figure 1 Chromatogram of Organic Acids in Ume Liqueur (Sample(A), 18 months)
1: unknown;   2: oxalic acid;   3: citric acid;   4: malic acid;   5: unknown;
6: lactic acid;   7: formic acid;   8: acetic acid;   9: unknown;
10: pyroglutamic acid

Table 6 shows the change of the content of organic acids during manufacturing and storage of *ume* liqueur, and the change in the extraction rate of raw materials from *ume* fruits. In all of the 4 samples the changes of the total concentration and total amount of organic acids in *ume* liqueur were close to the changes of the acidity and the acid content as described in 4.3. Thus, both total concentration and total amount of organic acids increased rapidly in the first 6 months of *ume* liqueur making, and were followed by gradual increase to reach maximum levels by 18 months from time zero, and then decreased at 30 months from time zero. Similarly, as discussed in 4.3, these results are thought to indicate that extraction of organic acids into shochu from *ume* fruits and their conversion into esters take place in the initial 18 months. In the following 12 months gradual esterification proceeds to decrease the total amount of organic acids, and the extraction rate also decreases gradually. Comparing sample (A) with 1 kg of intact *ume* fruits and sample (B) with 1 kg of pierced fruits, the total concentration of organic acids was slightly higher in (B), similarly to the changes of acidity and acid content. The total amount of organic acids and the extraction rate were higher in (A). For example, the extraction rate after 18 months in sample (A) was 73.1% which is higher than 68.2% in sample (B). Similar results were obtained with sample (C) with 1.5 kg of intact *ume* fruits and sample (D) with 1.5 kg of pierced fruits. Sample (A) showed 7% higher extraction rate of organic acids than (C).

Table 6 Changes of Organic Acids during Ume Liqueur Manufacturing

Period[1] (months)		Organic acids[2] (mg/100ml)						Amount[3] total	Extraction[4] (g)	(%)
		O	C	M	L	F	A	P		
(A)	6	23.9	607	524	119.0	18.4	15.6	5.7	1414	35.5	64.5
	18	12.8	791	569	86.4	13.3	13.8	3.2	1490	40.2	73.1
	30	11.1	775	492	78.3	12.4	16.8	2.3	1388	37.5	68.2
(B)	6	24.2	597	553	90.7	18.6	18.0	4.9	1306	32.8	59.6
	18	13.4	828	550	78.6	8.5	13.8	3.4	1496	37.5	68.2
	30	11.4	853	467	81.6	10.3	14.4	5.2	1443	36.2	65.8
(C)	6	28.5	822	741	100.0	12.5	18.1	4.7	1727	48.7	59.0
	18	12.6	1078	670	129.0	16.8	23.4	5.9	1936	54.6	66.2
	30	7.8	1071	704	73.1	9.6	17.3	3.8	1887	53.2	64.5
(D)	6	29.1	794	752	125.0	17.7	24.0	6.1	1748	44.2	53.6
	18	17.7	1114	673	142.0	22.9	25.2	7.0	2002	50.7	61.5
	30	8.1	1104	663	125.0	13.9	11.6	5.0	1931	48.9	59.3

[1] Soaking and storing period
[2] O:oxalic acid; C:citric acid; M:malic acid; L:lactic acid; F:formic acid; A:acetic acid; P:pyroglutamic acid
[3] Amount of organic acids (g) = total amount of organic acids (mg/100 ml) x yield (ml)
[4] Extraction rate (%) = (amount of organic acids in sample (g) / amount of organic acids in *ume* fruits (g)) x 100

Citric and malic acids were the major organic acids in *ume* liqueur, and the sum of the two acids was 89-92% of the total contents of organic acids after 18 months of *ume* liqueur making in all samples studied. The pattern was similar to that in *ume* fruits. The quantity of other organic acids were small. Accordingly, the change of the total amount of organic acids in *ume* liqueur during manufacturing and storage was dominated by the changes in the combined amount of citric and malic acids. Differences were found, however, between the change in the content of the two acids. In *ume* fruits the content of malic acid was higher than that of citric acid. In the first 6 months of *ume* liqueur manufacturing, citric acid was extracted from *ume* fruits into shochu slightly more than malic acid in all of the 4 samples, and considerably increased in *ume* liqueur to reach maximum values after 18 months from time zero; thereafter the composition minimally changed up to 30 months. On the other hand, the content of malic acid in *ume* liqueur was the highest after the first 18 months and then decreased slightly to 30 months from time zero. The contents of malic acid after 30 months in all of the 4 samples of *ume* liqueur were 5-15% smaller than those after 6 months. The results suggest that esterification in *ume* liqueur proceeds faster with malic acid than with citric acid. Other organic acids were contained in minor amounts and the patterns of changes during *ume* liqueur manufacturing were somewhat different among them, but the changes were thought to resemble the change of malic acid. Shinohara et al.[4] reported that lower pH accelerates the formation of esters while higher pH inhibits it, and that this effect is more distinct with dibasic acids than with monobasic acids. It is reasonably assumed that in *ume* liqueur malic acid, a dibasic acid, is more readily esterified than citric acid, a tribasic acid.

5 SUMMARY

5.1 Four samples with different pretreatment and amounts of *ume* fruits, namely, (A) 1 kg intact fruits, (B) 1 kg pierced fruits, (C) 1.5 kg intact fruits and (D) 1.5 kg pierced fruits, were used to study change of organic acids during manufacturing and storage of *ume* liqueur by means of high performance liquid chromatography.

5.2 The yields of *ume* liqueur in samples (A), (B), (C) and (D) were 2.70, 2.51, 2.82 and 2.53 liters, respectively.

5.3 The contents of acids and esters in *ume* liqueur increased in the first 6 months of *ume* liqueur manufacturing, and both reached approximately maximum levels after the following 12 months. The change in the acid content was slight thereafter while the ester content still increased gradually. The sum of the acid and ester values and the extraction rate (the sum of acid and ester contents in *ume* liqueur/that in the starting material of *ume* fruits) reached maximum levels 18 months after the start of experiments and the changes thereafter were slight.

5.4 From chromatographic data, the organic acids found present in the juice of *ume* fruits and *ume* liqueur were oxalic, citric, malic, lactic, formic, acetic and pyroglutamic and 3 unidentified acids. In both *ume* fruits and *ume* liqueur the sum of citric and malic acids were about 90% of total organic acid content.

5.5 The content of organic acids during manufacturing and storage of *ume* liqueur reached a maximum level after 18 months from the start of manufacturing and then decreased gradually. This change was dominated by the change of the sum of the amounts of citric and malic acids, but the pattern of changes of the two acids differed considerably from each other. While the content of citric acid reached a maximum after 18 months and then minimally changed up to 30 months, the content of malic acid content reached a maximum after 6 months and then decreased gradually up to 30 months from time zero. The difference was hypothesized to be due to faster esterification of malic acid than citric acid in *ume* liqueur.

From these results it was shown that in *ume* liqueur manufacturing the pretreatment and quantity of *ume* fruits greatly influenced the yield and quality of *ume* liqueur obtained.

References

1. The National Tax Administration Agency, *Brewing Analysis* 4th ed., Brewing Society of Japan, Tokyo, 1993, Chapter 10.
2. T. Sugawara and A. Maekawa, *Handbook for Food Analysis*, Kenpakusha, Tokyo, 2000, Chapter 2.

3   The Science and Technology Agency, *Standard Tables of Food Composition in Japan 5th ed.*, Kagawa Nutrition University Press, Tokyo, 2001, pp. 250-251.
4   T. Shinohara and J. Shimizu, *Nippon Nougeikagaku Kaishi*, 1981, **55**, 679-687.

Composition

THE VOLATILE COMPONENTS OF INDIAN LONG PEPPER, *PIPER LONGUM* LINN.

L. Trinnaman, N.C. Da Costa, M.L. Dewis, T.V. John

International Flavors and Fragrances Inc., Union Beach, New Jersey 07735, USA

1. ABSTRACT

As part of IFFs continuing program of looking at unusual natural products we recently analyzed *Piper longum* L. (Long pepper). This is a slender aromatic climber with perennial roots occurring in various parts of India. The dried spikes are 1–3 cm long, 4-6 mm in diameter and are grayish to dark in color. The fruits/spikes are pungent and have a pepper like flavor. Long pepper is used as a spice and in pickles and preserves. It is also used to treat bronchial trouble and as a carminative and analgesic.

Numerous terpene and sesquiterpene compounds were identified, along with a number of dienamides. Several of these amides were identified in *Piper longum* for the first time and from synthesis studies we were able to show that individual amines imparted, to a lesser or greater extent, a 'tingle effect', when tasted.

2. INTRODUCTION

Piper longum Linn. (Family *Piperaceae*) is a slender aromatic climber found in different parts of India. The erect shrub has a thick, jointed and branched root-stock. Leaves are numerous, 6 to 9 cm long, broadly ovate or oblong-oval, dark green, shiny on the top side and pale and dull underneath. Fruits are present in solitary, pedunculate, fleshy spikes 2.5 to 3.5 cm long, 5 mm thick, ovoid, oblong, erect, blackish green in color and shiny. The dried spikes are greyish to dark in color. The fruits/spikes are pungent and have a pepper-like flavor. The fruits are used in India as a spice and also in pickles and preserves.[1] The fruits of *Piper longum* (L) are frequently used to treat bronchial trouble and as a carminative and analgesic.[2] Antibacterial and fungicidal activities of *Piper longum* (L) have been reported.[3]

The present paper reports the results of a detailed study on the volatile component composition of an ethanolic extract of the dried spikes. The chemesthetic, more specifically 'tingle', properties of the plant were also investigated.

Figure 1 *Dried spikes of Piper longum.*

Chemesthetic or sensory molecules are important to the flavor industry. The word chemesthesis[4] (from Chem = chemical and esthesis = ability to perceive or feel) was adopted to describe what has been long referred to as the 'common chemical sense'.[5] The common chemical sense was coined to describe the sensory system responsible for the physical detection of irritating chemicals. The chemosensitive fibers are a subset of pain (nociceptive) and temperature (thermal) sensitive fibers. They occur throughout the body in skin and the mucous membranes as part of the somatic sensory system. The coolness of menthol, the warming of chili peppers and the tingle of carbonation are examples of the sensation of chemesthesis. Chemesthesis is a major part of oral perception.[6]

Tingle is of particular interest in the context of this paper. Tingle sensation can be described in a number of different ways. The vocabulary most commonly used to describe this sensation ranges from a numbing anesthetic effect (or tingling paresthesia) to the effect of attaching 9V battery terminals to the tongue. The feeling is generally perceived on the tongue and lips, but may also be experienced on the gums, teeth, cheeks and roof of the mouth.[6]

There are several reasons why tingle has value in flavor applications. Firstly, it synergizes with other attributes of the product, such as, carbonation and alcoholic strength. Other sensory effects, for example cooling, warming and sweetness are perceived differently, usually as being enhanced and preferred. Secondly, the effect can be perceived as refreshing or cleansing, whether on skin or in the mouth. Thirdly, antimicrobial properties of tingle chemicals could be beneficial, for example, in the mouth and underarm. Finally, the sensation can be marketed as a sensory trigger for a product claim that the consumer might miss. An example might be a numbing sensation on the scalp to demonstrate to the user the skin soothing effect of an anti-dandruff shampoo. Thus there are several obvious applications of the tingle molecules in the fragrance and flavor industry, particularly for use in gum, mouthwash, toothpaste, mints, candy and a variety of beverages.[6]

3. MATERIALS AND METHODS

3.1 Preparation of an ethanolic extract of *Piper longum* (L) spikes

The dried spikes (fruits) of *Piper longum* (L) obtained from India were used for this study. The spikes were ground (1000 g) into a coarse powder and soxhlet extracted using ethanol (2000 g) as the solvent. The extract was concentrated in a rotary evaporator under reduced pressure to 100 g. This extract was used for the GC-MS analysis.

3.2 Separation of volatile compounds

The ethanolic extract was analyzed using an HP6890 gas chromatograph with a split/splitless injector and a flame ionization detector (FID) (Hewlett Packard, Wilmington, PA). The extract was injected onto an OV-1 capillary column (50 m x 0.32 mm i.d., 0.5 μm film thickness, Restek, Bellefonte, PA) in the split (split ratio 40:1) and splitless modes. The carrier gas was helium with a flow rate of 1.0 ml/min. The injection port temperature was 250°C and the detector temperature was 320°C. The column temperature was programmed from 40°C to 270°C at a rate of 2°C/min with a holding time at 270°C of 10 minutes.

To aid in the detection of sulfur and nitrogen-containing compounds, the extract was analyzed by an HP6890 gas chromatograph equipped with an atomic emission detector (AED) (Hewlett Packard, Wilmington, PA). The column was, as above, an OV-1 capillary column and the analysis was again conducted in split (split ratio 40:1) and splitless modes. Injection and detection temperatures as well as the temperature program were as described above. All data was collected and stored using HP ChemStation software (Hewlett Packard, Wilmington, PA).

3.3 Identification of volatile compounds

Identification of components in the extract was conducted by mass spectrometry. The sample was injected onto an HP5890 GC. The chromatographic conditions for the OV-1 column were the same as those described for GC analysis. The end of the GC capillary column was inserted directly into the ion source of the mass spectrometer via a heated transfer line maintained at 280°C. The mass spectrometer was a Micromass Prospec high resolution, double-focusing, magnetic sector instrument (Walters/Micromass, Milford, MA). The mass spectrometer was operated in the electron ionization mode (EI), scanning from m/z 400 to m/z 33 @ 0.3 seconds per decade.

Spectra obtained were analyzed using the MassLib (Max Planck Institute) data system, IFF in-house libraries and the commercial Wiley 7, NIST 98 and other libraries. The identification of components was confirmed by interpretation of MS data and by GC relative retention index based on a calibration with ethyl esters, where possible.

3.4 Isolation of *N*-isobutyl-2*E*,4*E*-decadienamide (Pellitorine)

The ethanolic extract was extracted with hexane. After removal of the solvent the extract was chromatographed on a silica gel (Aldrich 70-230 mesh 60 A) column (3 cm x 60 cm) and eluted with a hexane/ethyl acetate (4:1 v/v) solvent system. Five fractions were collected based on their color. The colors of fractions 1 to 5 were yellow, green, yellow, light green and light yellow, respectively. After removal of the solvent, fraction 3 was shown to have a very intense tingling property. This fraction was subjected to NMR analysis.

3.5 Isolation of 5-(1E-dodecenyl)-1,3-benzodioxole

Spikes of the Piper Longum plant were extracted with hexane and concentrated, to give approximately 200g of concentrated hexane extract. This was distilled in a short path distillation set-up at 1.0 mm Hg and 70% of the pot charge was recovered as distillate. The benzodioxole was recovered from the distillate via crystallization. First, the distillate was blended to make a solution of 10% distillate in ethanol. This solution was put into a – 20°C freezer where it was allowed to sit for 48 hours. After 48 hours, a crop of fluffy white crystals was observed at the bottom and on the sides of the flask. The crystals were collected via vacuum filtration using a Buchner funnel and filter paper. They were re-dissolved in ethanol and put back into a -20°C freezer for 48 hours for re-crystallization. The crystals were subjected to NMR analysis.

4. RESULTS AND DISCUSSION

In total 130 compounds were identified in the ethanol extract (Table 1) some for the first time. It is interesting to note that no single class of compound dominates in either the number of components or in overall concentration. However, the chemesthetic properties are mainly concentrated in a number of amides.

The major compounds were found to be: 5-(1E-dodecenyl)-1,3-benzodioxole (27.2%), N-isobutyl-2E,4E-decadienamide (Pellitorine, 20.9%), 3-phenylpropionic acid (4.7%), beta-caryophyllene (4.7%) and beta-bisabolene (3.4%).

5-(1E-dodecenyl)-1,3-benzodioxole (**1**), the highest concentration volatile component, was positively identified by NMR analysis of the compound isolated as described in section 3.5. This is, as far as we are aware, the first time this compound has been identified in *Piper Longum* (L*)*. The identification of several other 1,3-benzodioxole compounds was postulated from mass spectral and GC data, based on the reference data obtained for this compound and some confirmed by synthesis. These compounds are also, to the best of our knowledge, new to *Piper Longum* (L). Tasting pure 5-(1E-dodecenyl)-1,3-benzodioxole showed that it had no appreciable chemesthetic properties.

(**1**)

N-isobutyl-2E,4E-decadienamide (**2**), pellitorine, was positively identified by NMR analysis of the compound isolated as described in section 3.4.

The identification of several other dienamides was postulated from mass spectral and GC data, based on the reference data obtained for compound 2 and several confirmed by synthesis. These compounds are, as far as we are aware, new to *Piper Longum* (L).

N-isobutyl-2E,4E-decadienamide was shown to have a strong tingle effect.

TABLE 1 *Identified components of Piper longum*

Peak	Component Name	Identification Method
1	ethyl acetate	MS, RI
2	acetic acid	MS, RI
3	propionic acid	MS, RI
4	acetaldehyde diethyl acetal	MS, RI
5	ethyl lactate	MS, RI
6	furfural	MS, RI
7	isovaleric acid	MS, RI
8	2-cyclopenten-1,4-dione	MS, RI
9	2-methylbutyric acid	MS, RI
10	isobutanal diethyl acetal	MS, RI
11	gamma-butyrolactone	MS, RI
12	4-ethoxypentan-2-one	MS, RI
13	benzaldehyde	MS, RI
14	5-methylfurfural	MS, RI
15	isovaleraldehyde diethyl acetal	MS, RI
16	4-hydroxyhexan-2,3,5-trione	MS
17	myrcene	MS, RI
18	valeraldehyde diethyl acetal	MS, RI
19	car-3-ene	MS, RI
20	phenylacetaldehyde	MS, RI
21	p-cymene	MS, RI
22	beta-phellandrene	MS, RI
23	limonene	MS, RI
24	Z-beta-ocimene	MS, RI
25	acetophenone	MS, RI
26	E-beta-ocimene	MS, RI
27	unknown sugar degradation compound	
28	linalool oxide I (furan)	MS, RI
29	linalool oxide II (furan)	MS, RI
30	linalool	MS, RI
31	hexanal diethyl acetal	MS, RI
32	undecane	MS, RI
33	dihydropyranone	MS
34	linalool oxide III (pyran)	MS, RI
35	linalool oxide IV (pyran)	MS, RI

36	alpha-terpineol	MS, RI
37	heptanal diethyl acetal	MS, RI
38	5-hydroxymethylfurfural	MS, RI
39	methyl 3-phenylpropionate	MS, RI
40	E-anethole	MS, RI
41	unknown BP-43	
42	undecan-2-one	MS, RI
43	1-tridecene	MS, RI
44	methyl geranate	MS, RI
45	tridecane	MS, RI
46	3-phenylpropionic acid	MS, RI
47	ethyl 3-phenylpropionate	MS, RI
48	eugenol	MS, RI
49	delta-elemene	MS, RI
50	alpha-cubebene	MS, RI
51	cyclosativene	MS, RI
52	alpha-copaene	MS, RI
53	beta-bourbonene	MS, RI
54	beta-elemene	MS, RI
55	sesquiterpene hydrocarbon BP-161-MW-204	
56	Z-alpha-bergamotene	MS
57	E-caryophyllene	MS, RI
58	sesquiterpene hydrocarbon BP-161-MW-204	
59	sativene	MS, RI
60	E-beta-farnesene	MS, RI
61	E-alpha-bergamotene	MS, RI
62	sesquiterpene hydrocarbon BP-69-MW-204	
63	alpha-humulene	MS, RI
64	2E,4E-decadienoic acid	MS, RI
65	germacrene D	MS, RI
66	benzenepropanamide	MS
67	tridecan-2-one	MS, RI
68	1-pentadecene	MS, RI
69	bicyclogermacrene	MS, RI
70	alpha-muurolene	MS, RI
71	myristicin	MS, RI
72	pentadecane	MS, RI

73	beta-bisabolene	MS, RI
74	delta-cadinene	MS, RI
75	alpha-calacorene	MS, RI
76	alpha-bisabolene	MS, RI
77	E-nerolidol	MS, RI
78	spathulenol	MS, RI
79	beta-caryophyllene oxide	MS, RI
80	globulol	MS
81	undecanal diethyl acetal	MS, RI
82	viridiflorol	MS, RI
83	humulene-1,2-epoxide	MS, RI
84	sesquiterpene compound BP-69 I	
85	isospathulenol	MS
86	delta-cadinol	MS, RI
87	sesquiterpene compound BP-69 II	
88	alpha-bisabolol	MS, RI
89	1-heptadecene	MS, RI
90	heptadecane	
91	ethyl p-methoxycinnamate	MS, RI
92	N-isobutyl decanamide	MS, NMR
93	N-isobutyl-2-decenamide	MS, NMR
94	N-isobutyl decadienamide	MS
95	nitrogen compound BP-80-MW-213	
96	5-(1E-octenyl)-1,3-benzodioxole	MS
97	1-nonadecene	MS, RI
98	N-isobutyl-2E,4E-decadienamide	MS, NMR
99	hexadecanoic acid	MS, RI
100	ethyl hexadecanoate	MS, RI
101	N-isobutyl-2E,4E-undecadienamide	MS, NMR
102	nitrogen compound BP-151-MW-237	
103	nitrogen compound BP-181-MW-253	
104	heptadecanoic acid	MS, RI
105	methyl linoleate	MS, RI
106	N-pyrrolidyl-2E,4E-decadienamide	MS, NMR
107	methyl oleate	MS, RI
108	5-(1E-decenyl)-1,3-benzodioxole	MS
109	N-piperidyl-2E,4E-decadienamide	MS, NMR

110	1-phytene	MS, RI
111	nitrogen compound BP-179-MW-251	
112	linoleic acid	MS, RI
113	9Z-octadecenoic acid	MS, RI
114	ethyl linolenate	MS, RI
115	ethyl oleate	MS, RI
116	octadecanoic acid	MS, RI
117	ethyl octadecanoate	MS, RI
118	5-(1E-undecenyl)-1,3-benzodioxole	MS
119	5-(1E-dodecenyl)-1,3-benzodioxole	MS, NMR
120	nitrogen compound BP-135-MW-275	
121	unknown BP-286	
122	5-(1E-tetradecenyl)-1,3-benzodioxole	MS
123	N-isobutyl-7-(benzodioxyl)-2-heptenamide	MS
124	N-isobutyl-5-(benzodioxyl)-2,4-pentadienamide	MS
125	nitrogen compound BP-131-MW-301	
126	nitrogen compound BP-135-MW-301	
127	N-isobutyl-11-(benzodioxyl)-undecatrienamide	MS
128	N-isobutyl-11-(benzodioxyl)-undecatrienamide	MS
129	5-hydroxy-4,7-dimethoxyflavone	MS
130	N-isobutyl-9-(benzodioxyl)-2,4-nonadienamide	MS

MS = reference mass spectrum
RI   = reference relative retention index
NMR = synthesis and nuclear magnetic resonance
Listed in order of elution from an OV1 column

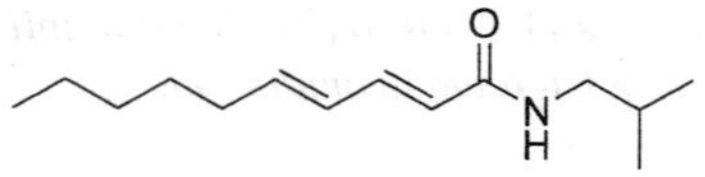

(2)

Compounds possessing tingle character are found in nature in a large number of botanical species. However, these species are concentrated into five plant families: dicotyledonous *Piperaceae, Aristolochiaceae, Rutaceae* and *Asteraceae (Compositae)*, and monocotyledonous *Poaceae*. The majority of the compounds of interest occur in the following genera of the above families: *Achilla* (Asteraceae-Anthemideae), *Acmella* (Asteraceae-Heliantheae), *Ctenium* (Poaceae), *Echinacea* (Asteraceae-Heliantheae), *Heliopsis* (Asteraceae-Heliantheae), *Piper* (Piperaceae), *Spilanthes* (Asteraceae-Heliantheae) and *Zanthoxylum* (Rutaceae).[3]

Further examples of compounds known or strongly suspected[3] of having tingle effects are:

Spilanthol/Affinin (**3**) found in the following species: *Spilanthes, Heliopsis, Acmella* and subspecies thereof and α-sanshool (**4**) found in five species of *Zanthoxylum*.

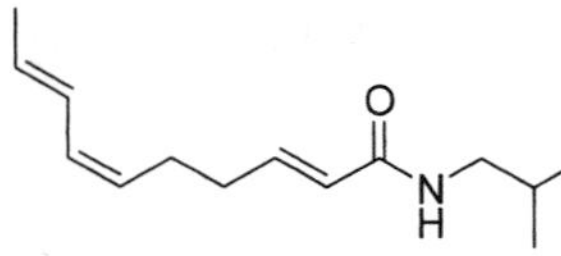

3
4 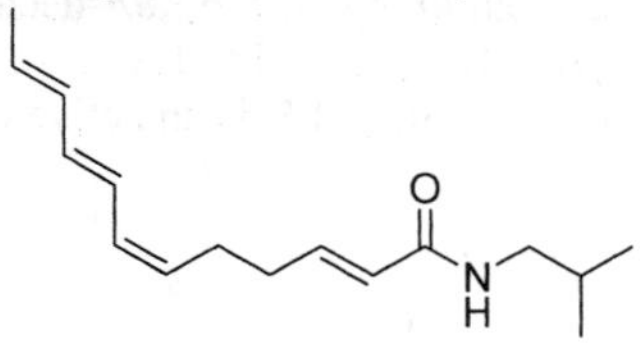

Gamma-sanshool (**5**) is found in six species of *Zanthoxylum*, hydroxy-ε-sanshool (**6**) occurs in *Zanthoxylum (Szechuan pepper)*, hydroxy-α-sanshool (**7**) is known to occur in four species of *Zanthoxylum* and two species of *Echinacea* and hydroxy-γ-sanshool (**8**) occurs in four species of *Zanthoxylum*.

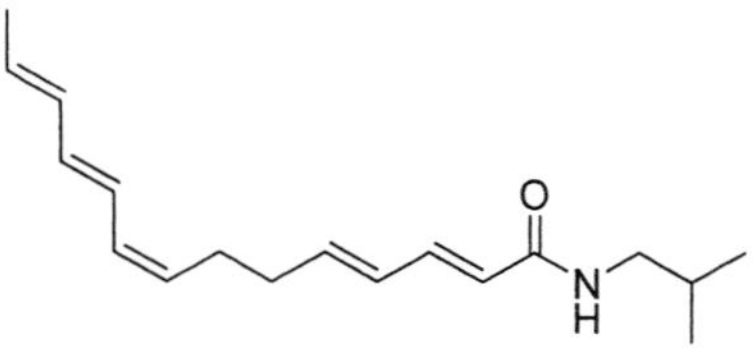

5
6 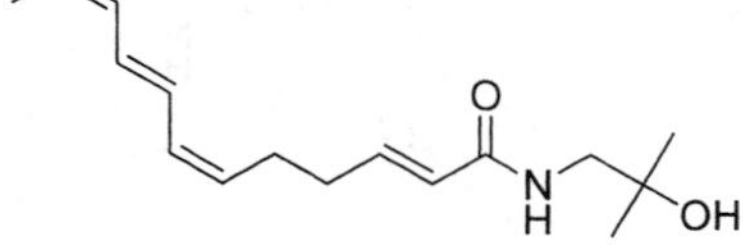
7 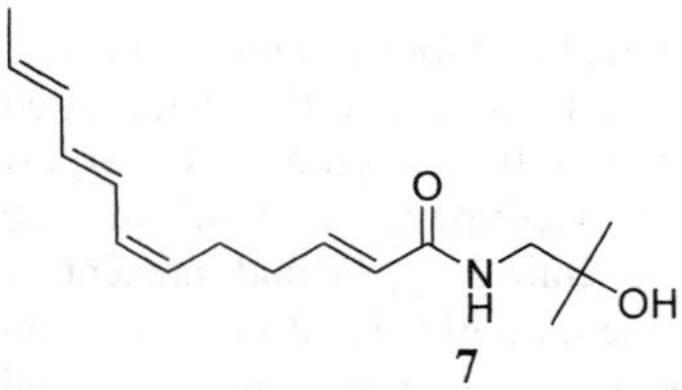
8

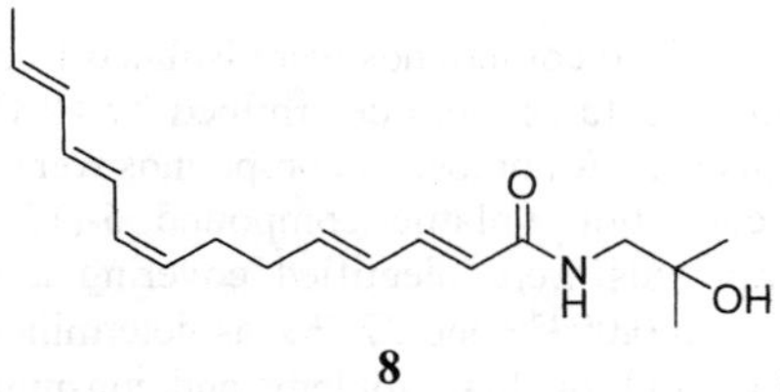

Fagaramide (**9**) found in *Zanthoxylum zanthoxyloides* and *Zanthoxylum tessmannii* and *N*-isobutyl-2*E*,4*E*,8,11-dodecatetraenamide (**10**) occur in *Ctenium aromaticum*.

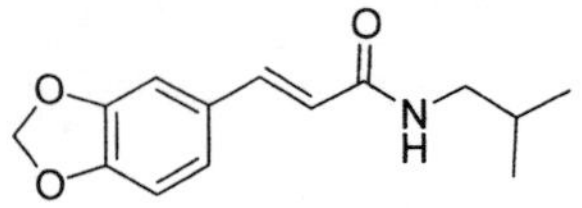

9

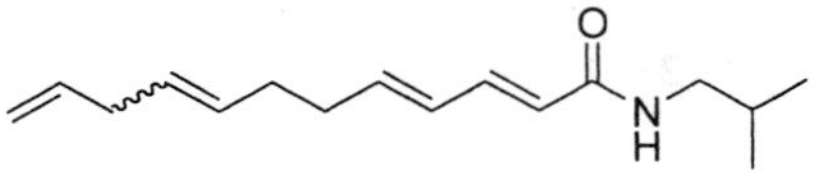

10

Isoaffinin (**11**) has been found in *Ctenium aromaticum* and *Achillea millefolium*. Pellitorine (**2**) has also been found in two species of *Zanthoxylum*, *Ctenium*, *Echinacea Pallida*, *Artemisia dracunculus* and *Piper guineese*.

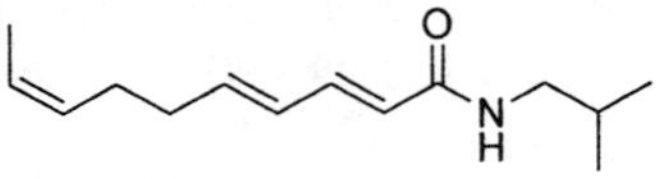

11

N-(2-methylbutyl)-2*E*,6*Z*,8*E*-decatrienamide (**12**) is found in *Spilanthes*, *Acmella* and *Heliopsis*. Bungeanool (**13**) occurs in at least two species of *Zanthoxylum*. *N*-isobutyl-2*E*,4*E*-undecenamide (**14**) is an active component in *Piper Longum* (L).

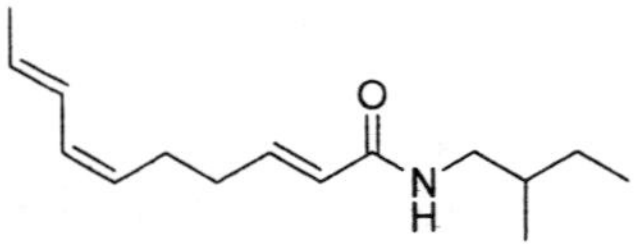

12

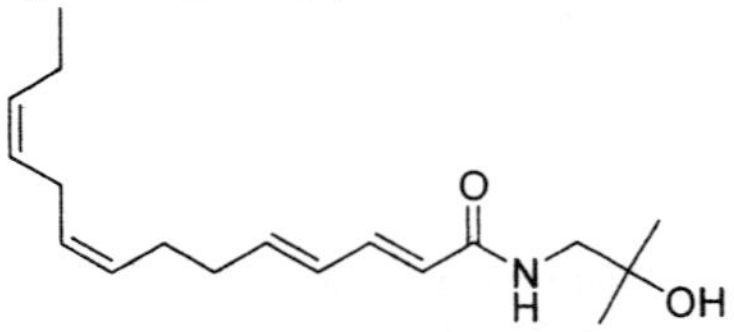

13

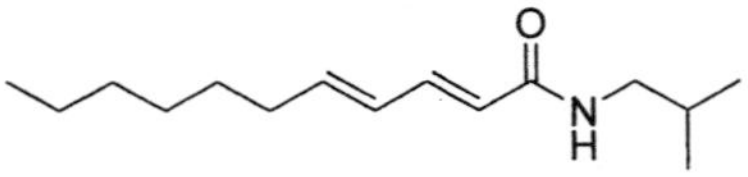

14

5. CONCLUSIONS

Two compounds were isolated from the ethanol extract of *Piper longum* (L) fruits. Their structures were determined by NMR and their chemesthetic properties determined by tasting. A number of compounds were identified for the first time, including the largest concentration volatile compound 5-(1*E*-dodecenyl)-1,3-benzodioxole. A total of 130 compounds were identified covering a wide range of chemical types and present at between 0.003% and 27.2%, as determined by GC-FID. The long history of human use to help with bronchial problems and consumption, primarily as a spice or in a pickle, as well as the presence of Pellitorine makes the fruits of *Piper longum* (L) an ideal raw material source for a potent natural tingle ingredient or molecule.

References

1. N.B. Shankaracharya, R.L. Jaganmohan, J. Puranaik and S. Nagalakssm, *J. Food Sci. Technol.*, 1997, **34**, 73 –75.
2. R.S. Thakur, H.S. Puri and A. Husain, *Major medicinal plants of India*, CIMAP, India 1989, pp. 408-410.
3. P.S. Reddy, K. Jamil, P. Madkusudhan, G. Anjani and G. B. Das. *Pharmacol. Biol.* 2001, **39**, 236 – 238.
4. B. Bryant, W.L. Silver, In *The Neurobiology of Taste and Smell*, T.E. Finger, W.L. Silver, D. Restrepo, (ed); 2nd Ed; Wiley-Liss, Inc.; New York; 2000; pp 73-100.
5. B.G. Green, J.R. Mason and M.R. Kare, *Chemical Senses, Vol. 2 Irritation*, Marcel Dekker Inc.; New York; 1990.
6. M.L. Dewis, In Chemistry and Technology of Flavours and Fragrances, D. Rowe (ed); Blackwells Publishing, Oxford; 2005; pp 199-243.

INVESTIGATION OF AROMA VOLATILES FROM FRESH FLOWERS OF SAFFRON (*CROCUS SATIVUS L.*)

M. Bergoin, C. Raynaud, G. Vilarem and T. Talou

Laboratoire de Chimie Agro-Industrielle, ENSIACET, 118 route de Narbonne, 31077 Toulouse, France

1. ABSTRACT

Saffron is a high-value spice obtained from dried *Crocus sativus* Linnaeus stigmas. It is widely used as condiment for its delicate flavour and intense color. Although the production of 1 kg of saffron requires more than 160,000 flowers, only few studies have been carried out on *Crocus sativus* L. flowers. Especially the aroma potential of the flower has not yet been subject of detailed investigation. During the saffron harvesting season in october 2002, flowers from the Quercy area (France) have been collected. After removing the stigmas, fresh flowers have been extracted by two different methods: hexane maceration was applied, leading to extracts called "concrete" and "absolute". The essential oil and the aromatic water were obtained by steam distillation. The headspace volatiles of these products were analyzed by gas chromatography, after direct injection or dynamic headspace sampling. Compounds of high olfactory impact were determined by sniffing detection GC/O and identified by mass selective detection GC/MS.

2. INTRODUCTION

Saffron, dried *Crocus sativus* Linnaeus stigmas, is a high value spice appreciated in the Mediterranean countries for its flavor, aroma and coloring properties. In 2003, the estimated global saffron market was about 170 tons. To produce 1 kg of saffron, more than 160,000 flowers are required, representing 300 kg of floral waste[1] and about 1.5 tons of leaves formerly used as cattle forage. At a global scale, *Crocus sativus* L. wastes represent a real potential.

The objective of this work was the development of new ways of adding value to saffron by-products. This approach is also known as the agroresource refining concept (ARC) for different parts of the plant. ARC has been widely used for different parts of the sunflower plant. Apart from the main product, sunflower oil, cosmetic products[2] and fuels,[3] have been produced from the oil cake. From stem and flower, essential oil was extracted and both paper pulp and low density agromaterials similar to polystyrene[4] have been developed.

The objective of this work was to apply ARC (Figure 1) concept to *Crocus sativus* L. in order to investigate its potential for obtaining aromatic and dye products from flowers, leaves and bulbs. On flowers, previous studies have been focused on characterisation of pigments present in petals as anthocyanins and flavonoids,[5,6] but to our knowledge the aromatic aspect of flower by-products has not yet been investigated. Hence, the characterisation of key odorous compounds present in saffron fresh flowers was the main goal of this study.

3. MATERIALS AND METHODS

3.1 Plant material used and agroresource refining concept (ARC)

The plant used in these experiments was *Crocus sativus* L. Flowers were harvested in Quercy area (South-West of France) and immediately shipped to us for analysis. The ARC protocol is shown in (Figure 1).

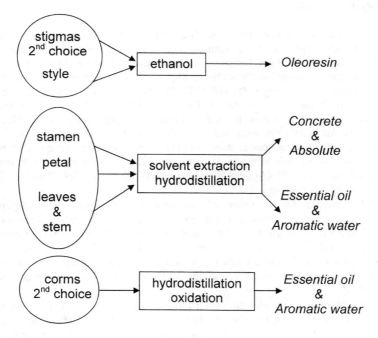

Figure 1   *Agroresource Refining Concept (ARC) applied to <u>Crocus sativus L.</u>*

3.2 Extraction

3.2.1 Hydrodistillation. Fresh flowers (800 g) were extracted by hot water (5 L) in a adapted Clevenger apparatus over eight hours. Aromatic water was recovered at the end of hydrodistilllation. Essential oil was extracted from the water using dichloromethane (3x20 mL). Solvent was then evaporated under a nitrogen flow until obtaining 1 mL of extract.

3.2.2 Maceration. Fresh flowers (70 g) were extracted by maceration in a glass vessel with two different solvents (750 mL), diethyl ether and hexane, under stirring. Times of maceration tested were: 15 min, 30 min and 60 min. After maceration, the extracts were filtered. Solvent was evaporated, under vacuum from hexane extract and under nitrogen flow from diethyl ether extract. Different concentrates, called "concretes", were obtained. From a 10 mg aliquot of the concrete, waxes were removed by using absolute ethanol (750 mg), leading to extracts called "absolutes". These extracts were filtered by a PTFE 0.45 µm filter before being analyzed.

3.3 Analytical methods

3.3.1 Aromatic Water. Dynamic Headspace[7] (DHS) technique was used to concentrate aromatic water volatiles. Aromatic water (15 g) was placed into a glass vessel during 40 min for headspace equilibration. Aroma compounds trapped on Tenax TA (130 mg) were analyzed by combined gas chromatography (GC, HP 5890, Hewlett Packard, France) and mass spectrometry (MS, HP 5971, Hewlett Packard, France) using a BPX5 column (60 m, 0.32 mm i.d., 1 µm f.t., SGE, France). Desorption of volatile compounds from Tenax (220°C, 5 min) was performed by a thermal desorption injector (CHISA, SGE device, France). An initial oven temperature of 40°C was used followed by a rate increase of 5°C.min^{-1} to 280°C. Mass spectra were obtained with 70 eV electron impact ionisation from m/z 15 to 350. Volatile compounds were identified by comparison of their spectra with those of the NIST 98 (Agilent, France) and WILEY (Hewlett Packard, France) libraries and using retention indices.[8]

3.3.2. Essential Oil. Essential oil was analyzed by a gas chromatograph (HP 5890) coupled to a mass spectrometer (HP 5971) and performed with a DB5ms column (30 m, 0.25 mm i.d., 0.25 µm f.t.). The conditions were as follows : helium carrier gas flow rate 1.3 mL.min^{-1}, temperature program; 40°C, 2°C.min^{-1}, 100°C, 4°C.min^{-1}, 250°C, 20 min; split 10 mL.min^{-1}; detector 250°C, injector 200°C. Compounds were identified by comparison of their spectra with those of the NIST 98 and WILEY libraries and using retention indices.

3.3.3 Absolute from Hexane and Diethyl ether Extracts. GC/MS-olfactometry (GC/MS-ODP) system consisted of a gas chromatograph (Agilent 6980, France) coupled to a mass spectrometer (Agilent 5973N, France) and an olfactory detector port (ODP$_2$, Gerstel GmbH, RIC, France). At the end of the DB5ms column (30 m, 0.25 mm i.d., 0.25 µm f.t., JW Agilent Technologies, USA), the effluent was split in a 1:2 ratio between mass selective detector and sniffing port. Helium was used as carried gas at a constant flow rate of 1.4 mL.min^{-1}. Oven temperature started at 90°C and was increased at a rate of 5°C.min^{-1} to 280°C over a 20 min period. The "absolute" sample (2 µL) was injected (split 10 mL.min^{-1}; 200°C injector temperature; 290°C detector temperature). An expert assessor, trained on saffron stigmas, evaluate three replicates using ODP device. Odors were described and their intensities were posteriory recorded on a scale from 1 to 5. Replicates scores were averaged to establish an aroma profile of the "absolute" samples. Additionally, an olfactory intensity device (OID) was used to mark odor-active zones in the chromatogram, in order to identify compounds by comparison of their spectra with those of the NIST 98 and WILEY libraries and using retention indices.

4. RESULTS AND DISCUSSION

The hydrodistillation of fresh saffron flowers was not enough efficient. Essential oil was extracted from aromatic water in order to be characterized. The GC/MS profile of the dichloromethane extract (Figure 2) shows that the main compound was hexadecanoic acid (20% of total area). The volatiles which contribute to the overall odor (Table 1) were: 2-ethyl-1-hexanol and phenylethyl alcohol giving both "floral" notes and phenylacetaldehyde, L-carvone and ionol giving "herbaceous" notes.[9]

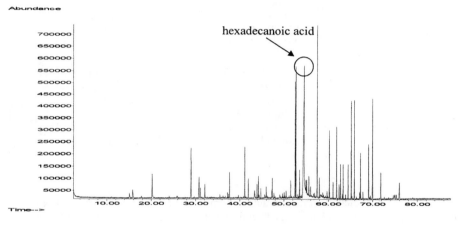

Figure 2 GC/MS chromatogram of fresh saffron flowers essential oil obtained by hydrodistillation

Headspace volatile compounds of aromatic water were extracted by DHS and identified by GC/MS. The aroma profile (Figure 3) indicated that main compounds were nonanal (50.2 %), heptanal (15.3 %) and hexanal (20.5 %). These odor active molecules give, respectively, "floral", "fatty" and "green" notes. Concerning minor compounds, some were found to be odorous and are thus susceptible of contributing to the aroma of fresh saffron flowers as benzaldehyde associated to an "almond" note, octanal to a "fatty citrus-like" note and safranal to a "saffron" note. Aromatic water contained mainly aldehydes which probably originated from oxidation of double bound compounds.

Table 1. *Identification of aroma volatile compounds in essential oil obtained from fresh saffron flowers hydrodistillation*

Idenntification[a]	Area Percent (%)	Mass spectral data [m/z (%)]	Retention Indices[b]	Descriptors given by literature[c]
2-ethyl-1-hexanol	0.5	57(100), 41(50), 55(40), 70(39), 83(38), 55(37), 84(3)	1024	Oily, sweet, slightly floral-rose odor
phenylacetaldehyde	1.5	91(100), 120(30), 92(30), 65(29), 39(10), 51(8)	1036	Powerful, penetrating pungent-green, floral and sweet odor
phenylethyl alcohol	2.7	91(100), 92(50), 122(40), 65(30)	1104	Mild, warm, rose-honey-like odor
L-carvone	3.9	82(100), 54(40), 93(58), 108(35), 150(10), 135(2)	1231	Warm-herbaceous, breadlike, penetrating and diffusive odor, somewhat spicy
ionol	1.0	121(100), 43(70), 161(69), 91(50), 105(48), 136(40), 179(30), 194(10)	1395	Sweet, oily-herbaceous, warm odor with floral-balsamic undertones

[a] Identification by comparison of their spectra with those of the NIST 98 and Wiley libraries [b] Experimental retention indices [c] S. Arctander, Perfume and Flavor Chemicals, 1994, Vol. 1-2.

The various diethyl ether extracts (15 min, 30 min and 60 min) possess four common notes among which a "burned" and two "floral honey" notes (A and B, figure 4) were very intense ("burned" note upto 3.7 units and "floral honey" notes upto 3.0 and 4.0). In addition to these notes, other odorants are present in high amounts in single sample. In the shortest time maceration a "bread" note was present, in the 30 min maceration sample it was a "pungent green" note and for the longest time maceration, an intense "peanut" note was perceived (3.7 units). It has been observed that the "burned" and "mushroom" notes were increasing with increasing maceration time whereas the "bread" one was decreasing.

A total of 14 volatile compounds possess detectable odors in hexane extracts whereas in diethyl ether extracts only eleven notes were perceived. Main notes were "burned" and "pungent green". Moreover, the 15 min maceration sample contained intense "mushroom" and "floral honey" notes (3.7 and 5.0 respectively) and the 30 min maceration one featured a second intense "burned" note (4.3) and the same "floral honey" note (4.7). The "mushroom" and "floral honey" notes were decreasing with increasing maceration time.

Composition

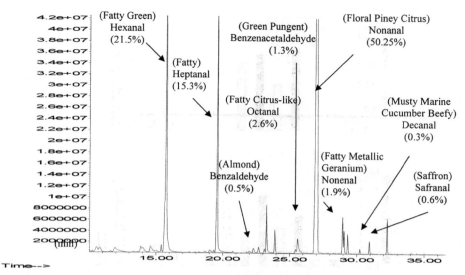

Figure 3 *Chromatogram obtained by DHS/GC-MS of aromatic water from hydrodistillation of fresh saffron flowers. Aroma descriptors used are coming from the literature[9]*

Maceration at room temperature is a softer process to obtain volatiles from fresh flowers. Extractions were made with two solvents. Firstly, diethyl ether was used for its high extraction power and secondly, hexane was used as a commonly used solvent in industrial processes. Whatever the maceration time and the solvent used, no significant variation of extraction yield was noticed (Table 2). Yellow sweet-smelling absolutes have been obtained after removing waxes. The sensory characteristics of the flower absolutes were obtained by GC sniffing. The sum of intensity values of the detected odors are visualized in the aroma profile of each sample (Figure 4 and Figure 5). A mean intensity of 0.7 was considered as noise level. The odor of active volatile compounds detected at the sniffing port could be classified in 13 categories : "green", "peanut", "mushroom", "burned", "floral honey", "bread", "floral", "pungent honey", "fruity", "fresh honey", "roasted", "pungent green" and "animal" notes.

Table 2 *Yield of extraction of fresh saffron flowers according to solvent and maceration time*

Solvent	Diethyl ether			Hexane		
Maceration time (min)	15	30	60	15	30	60
Yield of maceration (%)	0.46	0.33	0.41	0.31	0.27	0.31

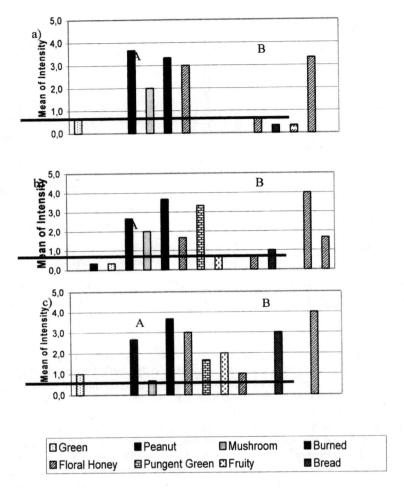

Figure 4 *Aroma profiles of fresh saffron flowers absolute corresponding to increasing duration of diethyl ether maceration 60 min a), 30 min b) and 15 min c)*
(A) : phenylethyl alcohol
(B) : 4-hydroxybenzeneethanol

Composition

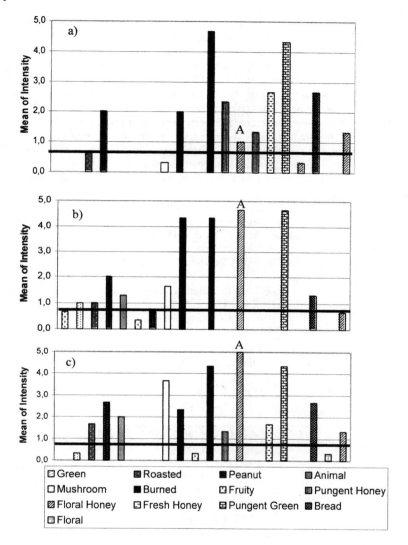

Figure 5 *Aroma profiles of fresh saffron flowers absolute corresponding to increasing duration of hexane maceration 60 min a), 30 min b) and 15 min c)*
(A): phenylethyl alcohol

In radar plots (Figures 6 and 7), the intensities of the same odor perceived at the sniffing port were summed to visualize their impact in the global aroma of the extract. The aromatic pattern is very different for diethyl ether and hexane extracts. Hexane extracts contained more odorants with "fresh honey" and "roasted" notes and also many strong odorants with pyrogenous notes like "burned", "pungent green" and "pungent honey". Diethyl ether extracts were essentially characterized by the intense, persistant and penetrating "floral honey" odor class (B).

A sensory analysis on concretes has been made by a French perfumer. Concretes from diethyl ether featured intense "honey" and "warm" notes reminding "mimosa", "broom" and "marigold" concretes. In the hexane concrete, "burned" and "green" notes were particularly present, the latter increasing with the maceration time.

Instrumental sensory results on absolutes justified the sensory evaluation made by the perfumer on concretes.

Despite their very low MS response, many compounds which contribute to absolutes aroma were identified by mass spectra and retention times of GC/MS analysis (Table 3). They are also characterized by their OID peak areas and the odors described (Figure 8). In hexane extracts the "pungent green" note was related to 2-phenylethyl ester of formic acid and the "burned" one may be undecane. The "floral honey" note (A) which is much more intense in hexane extract was identified as phenylethyl alcohol and the second one (B), which is typical of the diethyl ether extract, was identified as 4-hydroxybenzenethanol.

Table 3 *Identification of key-odor compounds of fresh saffron flowers absolute by comparison of their spectra with those of the Nist 98 and weiley libraries and literature*

Odor[a]	Identification[b]	Mass spectral data [m/z (%)]	RT(min)	RI[c]
Green	Hexanal	44(100), 56(80), 57(60), 72(35), 82(20), 67(15)	2.36	808
Animal	Heptanal	44(100), 70(70), 55(60), 57(45), 81(20), 86(10), 96(5),	2.96	906
Mushroom	1-octen-3-one	55(100), 70(80), 27(40), 83(10), 97(10), 111(2)	3.66	979
Burned	Undecane	57(100), 43(80), 71(70), 85(40) 70(5), 84(3), 99(2), 156(2)	5.30	1097
Floral honey	Phenylethyl alcohol	91(100), 44(60), 92(50), 122(35), 152(5)	5.89	1125
Pungent green	2-pheny-ethyl formate	104(100), 91(75), 51(25), 105(15), 92(10), 103(5), 122(2)	6.83	1176
Floral Honey	4-hydroxy-benzeneethanol	107(100), 138(30), 77(28) 108(4)	11.71	1367

[a] Perceived at the sniffing port [b] Identification by comparison of their spectra with those of the NIST 98 and Wiley libraries [c] RI: Retention index, Experimental retention indices

Composition

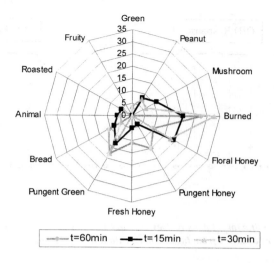

Figure 6   *Aroma profiles of notes from fresh saffron flowers absolute corresponding to increasing duration of hexane maceration*

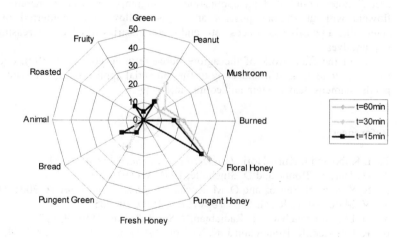

Figure 7   *Aroma profiles of notes from fresh saffron flowers absolute corresponding to increasing duration of diethyl ether maceration*

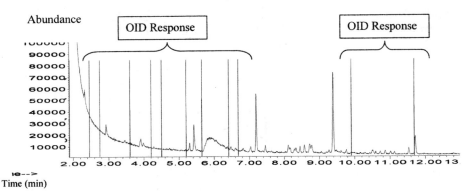

Figure 8   OID (Olfactory Intensity Device) and GC/MS responses of fresh saffron flowers absolute from hexane maceration (time=15 min)

5. CONCLUSIONS

The objective of the reported investigations was to add value to saffron flower waste material. Absolutes from hexane macerations with their intense "honey" and "pungent green" notes could find an application in fragrance industry. Concretes from saffron flowers without stigmas possess an intense yellow color conferred by carotenoïds suggesting a possible use in cosmetic and food industries which are increasingly interested in plant dyes.

In the framework of the agroresource refining concept (ARC), our research on *Crocus sativus* L. is still underway to find potential application to all plant organs (style, petals, stamens, leaves, stem and corms) into value.

References

1   I. Kubo and I. Kinst-Hori, *J. Agric. Food Chem.*, 1999, **47**, 4121.
2   M. Diot, C. Bonne and D. Sincholle, 1991, WO 91/11169.
3   S. Yorgun, S. Sensoz and O. M. Kockar, *J. Anal. Appl. Pyrolysis*, 2001, **60**, 1.
4   V. Marechal, L. Rigal, *Ind. Crops Prod.*, 1999, **10**, 185.
5   H. Ebrahimzadeh and T. Radjabian, *J. Sci. I. R. Iran*, 1998, **9**, 127.
6   R. Norbaek, K. Brandt and J. K. Nielson, *Biochem. Syst. Ecol.*, 2002, **30**, 763.
7   S. Breheret, T. Talou, S. Rapior and J. M. Bessière, *J. Agric. Food Chem.*, 1997, **45**, 831.
8   N. Kondjoyan and J. L. Berdagué, 'A Compilation of relative Retention Indices for the analysis of aromatic compounds', Edition du Laboratoire Flaveur, Saint Genes Champanelle France, 1996.
9   S. Arctander, 'Perfume and Flavor Chemicals', Allured Publishing Corporation, Carol Stream FL, 1994, **1-2**.

GC/MS ANALYSES OF THE VOLATILE COMPOUNDS OF *Tuber Melanosporum* FROM TRICASTIN AND ALPES DE HAUTE PROVENCE (FRANCE)

Gaston Vernin,[1] Cyril Párkányi[1] and Hervé Casabianca[2]

[1]Department of Chemistry and Biochemistry, Florida Atlantic University, 777 Glades Road, P.O. Box 3091, Boca Raton, FL 33431-0991, USA
[2]SCA du CNRS, Échangeur de Solaize, BP 22, F-69390 Vernaison, France

1 ABSTRACT

Black truffle (*Tuber melanosporum*) aromas from two different areas of Southern France: i) Tricastin (Richerenches) and ii) Alpes de Haute Provence (Simiane) have been studied by two different extraction methods: a) extraction using pentane and ethyl ether as the solvents, and b) solid phase micro extraction (SPME). A total of 131 compounds have been identified by GC/MS and retention indices. Chemical composition was found to be different according to areas and extraction methods. An interesting fraction of black truffles from Tricastin, obtained by SPME method, consisted of heterocyclic compounds most likely arising from the Maillard reaction between acetoin, 1,2-propanediol and hydrogen sulfide (furfural, furfuryl alcohol, 2- and 3-formylthiophenes, 2-acetylthiazole...). Some usual monoterpene hydrocarbons, citronellol, sesquiterpene hydrocarbons (β-caryophyllene, daucene) and fused aromatic compounds (methyl- and dimethylnaphthalenes), in the Tricastin sample or by solvent extraction with pentane were also identified. Several aliphatic alcohols, acids, aldehydes and sulfides were also found. It is suggested that methoxylated benzenes and toluenes (mono, di, tri, and tetra derivatives) obtained by ether extraction, as well as 1,2,5,6-tetrahydropyridine contribute to the characteristic odor and aroma of black truffles. The quantitative composition varies depending on the separation method and thus it cannot be taken into account.

2. INTRODUCTION

The remarkable organoleptic properties of truffles have been known for a long time. Several varieties of truffles exist throughout the world (see Table 1). The two varieties best appreciated by 'gourmets' are the black truffle from Perigord (France) (*Tuber melanosporum*) and the white truffle from Piedmont (Italia) (*Tuber magnatum*). Other species are of poor organoleptic quality. Among them, the Chinese truffle (*Tuber indicum*) has flooded the market since 1994-1995. These truffles are used in some food preparations by fraudulent dealers instead of the black truffle (black diamond) which is morphologically very similar. These delicious and very appreciated "black and white truffles" constitute an important market, and their prices are always increasing.

Table 1 Principal species of truffles (Tuberaceae) and their origin[a]

Species	Origin
Tuber melanosporum (or black winter truffle)	Southern France (Perigord, Tricastin, Var, Alpes de Haute Provence, Vaucluse), Spain, Italy
Tuber aestivum (summer red-grained black truffles)	Italy, France
Tuber uncinatum Chatin	France: Burgundy
Tuber magnatum (white truffles)	Italy: Piedmont, Umbria, Emilia, Romagna, Alba (San Miniato, Montferrato, around Parma, Modena, Bologna), Swiss district of Ticino, France: Rhône valley
Tuber mesentericum Vitt. (grey truffles)	North America
Tuber mesophellia glauca	Australia
Tuber indicum[b]	China

[a] Other species of truffles: *T. borchii* (white species), *T. albidum*, *T. brumale* Vitt (the winter truffle), *T. excavatum*, *T. macrosporum*, *T. nitidum*, *T. rufum*, have been studied by Talou et al.[23]

[b] The Chinese truffles, similar to black Perigord truffles, have flooded the market since 1994-1995. They are often used as such or in some food preparations. Fraudulent dealers color and sell them as authentic black truffles. A DNA test was developed at the National Institute of Agronomic Research (INRA) by G. Chevalier in order to differentiate them.[39] The price of Chinese truffles was 24 to 30 euros/kg while French truffles from Perigord ranged from 600 to 750 euros/kg (in 2002). During dry years (2003), prices are higher. Current (2004) market price of black Perigord truffles in the U.S. is $1040/kg ($500/16 oz.) or 870 euros/kg.

Numerous papers have been devoted to the chemical composition of these particular fungi. They are summarized in Table 2.

The first report on truffle aroma (*Tuber melanosporum, Tuber aestivum*) was published by Falck in 1911.[1] In 1937 Igolen reported the overall composition of truffles from Comtat Venaissin (Carpentras).[2] Between 1987-1991, Talou et al.[3-14] studied fresh canned, tinned, and imitation black truffles from Perigord. They were also studied by

Table 2 *Principal publications on various species of truffles*

Species	Author(s) (year)(s) [Ref.]
T. melanosporum, T. aestivum T. album aroma	Falck (1911) [1]
Truffles from Comtat Venaissin (Carpentras area) (overall composition)	Igolen (1937) [2]
Guide to mushrooms	Pacioni (1981) [40]
Manual on cultivation of mushrooms	Rambelli (1983) [41]
Fungi of Switzerland	Breitenbach, Kränzlin (1984) [42]
Mushrooms demystified	Arora (1985) [43]
Black truffles from Perigord	Talou et al. (1987) [3]
Artificial truffles	Patent (1986) [36]
Fresh black truffles (Perigord)	Talou et al. (1987, 1988) [3,4]
Tinned black truffles	Talou et al. (1989) [6]
Fresh canned black truffles	Talou et al. (1989, 1990) [5,8,13]
Artificial imitation of black truffle aroma	Talou et al. (1989, 1991) [10,14]
T. aestivum, T. magnatum, T. mesentericum: Role of alcohols and aldehydes	Bianco et al. (1988) [34]
Solid-phase microextraction of truffle aroma (a review)	Polesello et al. (1989) 24]
Fresh Perigord truffle aroma	Flament et al. (1991) [15]
Truffles and truffle volatiles (a review with 38 references)	Maciarello and Tucker (1994) [19]
Headspace-solid-phase microextraction (HS-SPME) of truffle aroma	Pelusio et al. (1995) [25]
Volatiles from white truffle from central Italy and their variation during deep freezing	Bellesia et al. (1996, 1997) [16,18]
Formation of sulfur compounds during storage of different species of truffles	Bellesia et al. (1996) [17]
Australian truffle-like fungus (*Mesophellia glauca*)	Leach et al. (1997) [21] Millington et al. (1998) [22]
Bacterial activity of seven species of truffles	Angelini et al. (1998) [44]
Extraction of truffle aroma using supercritical carbon dioxide	Mirande-David (1998) [37]
Tuber borchii Vitt (a review with 25 references)	Lanzotti and Iorizzi (2000) [20]
12 Different European species of truffles analyzed using GC/olfactometry and GC/MS	Talou et al. (2001) [23]
White truffle aroma obtained using HS-SPME method	Davoli et al. (2002) [26]
T. aestivum analysis using HS-SPME method	Diaz (2002) [27]
Rebuttal on truffle aroma analysis by HS-SPME method	Ibanez et al. (2002) [28]

Flament et al.[15]

The white truffle (*Tuber magnatum*) from Italy was also widely studied by Italian workers.[16-18] White truffles are easily distinguished from black truffles by the presence of a high content of *bis*-(methylthio)methane. *Tuber aestivum* is another variety of white truffle, also called summer truffle which can be harvested in Tricastin and Alpes de Haute Provence.

Reviews devoted to the volatile compounds of the truffle aroma were published by Maciarello and Tucker[19] and by Lanzotti and Iorizzi.[20]

Australian truffles (*Mesophellia glauca*) were studied by Leach et al.[21] and by Millington et al.[22] Talou et al.[23] analyzed twelve European species of truffles using GC/olfactometry and GC/MS. Several methods were used to extract truffle aroma, among them the Solid Phase Microextraction (SPME)[24] and Headspace-Solid Phase Microextraction (HS-SPME).[25-28]

The goal of this work was to study the aroma of French truffles from Tricastin and Alpes de Haute Provence, using two methods: SPME and solvent extraction with ether and pentane as the solvents.

3 EXPERIMENTAL

3.1 Samples

Truffles used in this work belong to the *Tuber melanosporum* species. They were purchased: i) on the local market of Richerenches (Tricastin), and ii) at Simiane (Alpes de Haute Provence) in the middle of December. They were kept frozen until extracted.

3.2 Extraction Procedure

3.2.1 SPME method. The two samples were subjected to Solid Phase Microextraction (SPME) using silica fiber coated with a 50 µm layer of polydimethylsiloxane according to the method reviewed by Polesello et al.[24] and modified by Diaz et al.[27] The sample from Tricastin was distilled after treatment by the SPME method, in order to obtain hydrocarbons (terpenics, aromatics and alkanes).

3.2.2 Ether extraction. The truffles from the Simiane area (20 g) were cut into thin slices and put in an Erlenmeyer flask with 100 ml ether under magnetic stirring during 12 h at room temperature. After solvent evaporation (rotary evaporator), the residue was submitted to flash chromatography on a silica gel column using pentane and ether, respectively. The pentane extract was treated using the Innoless acquisition method and selected ion monitoring (SIM). Each extract was analyzed by GC and GC/MS.

3.2.3 GC and GC/MS analyses. A Hewlett Packard gas chromatogaph equipped with an Innowax capillary column (60 m x 0.32 mm i.d., 0.5 µm film thickness) was used. The column oven temperature was programmed from 60°C (2 min) to 245°C at 2°C/min. Injector (FID) and detector temperatures were maintained at 260°C, and helium was the carrier gas (13 ml/min).

For the GC/MS analyses, the same column was used under similar conditions. Mass spectra were recorded on a Hewlett Packard gas chromatograph-mass spectrometer at 70 eV, with an ionizing temperature of 200°C.

3.2.4 Identification. Compounds were identified by comparison of their mass spectra with those reported in libraries of mass spectra (Wiley, NBS 75 KL), SPECMA data bank (Vernin et al.[29], Colon and Vernin[30]) and other databases such as those of Jennings and Shibamoto[31] (magnetic) and Adams[32] using ion trap and quadrupole mass spectrometers.

Retention indices (or Kováts indices) were calculated using linear alkanes as references and compared with those reported in the literature.[33] Reconstructed mass spectra were obtained on the basis of our SPECMA data bank.

4 RESULTS AND DISCUSSION

The two samples from Tricastin (Richerenches, A) and Alpes de Haute Provence (Simiane, B) obtained by the SPME method, and the samples obtained by solvent extraction (pentane, and ether) will be discussed separately. However, from a qualitative point of view, all identified volatile compounds (131) obtained in this work from black truffles are reported in Table 3.

4.1 Samples Obtained by the SPME Method

Gas chromatographic profiles of the two samples from Richerenches (sample A) and Simiane (sample B) are presented in Figures 1 and 2.

The two samples show significant qualitative and quantitative differences. The differences between the Tricatsin sample (A) and the Simiane sample (B) are:

i) Heavy linear aliphatic alcohols (undecanol, dodecanol, hexadecanol, heptadecanol) were found only in sample A (Richerenches). On the other hand, volatile alcohols of low molecular weight are present only in sample B (Simiane).

ii) Te linear C_4 to C_9 aliphatic acid were identified in sample A but not sample B. Acetic acid is present in both samples.

iii) The presence of aliphatic alcohols with a methylthio group at the end of the chain was confirmed in sample A. Volatile sulfur compounds, such as hydrogen sulfide, methyl mercaptan, dimethyl sulfide, and dimethyl trisulfide were found only in the sample from Simiane (B). In sample B, dimethyl sulfide was found in mixture with another product (MW = 76, base peak at m/z 44) identified using the database NBS 75 KL as ethyl mercaptan which has a similar mass spectrum. It is eluted on the Innowax column juset before isobutanal.

iv) Furfural, furfuryl alcohol, 2-formylthiopp[hene (and probably the 3 isomer), and 2-acetylthiazole, found in sample A but not sdample B, are well-know Maillard reaction products arising from the degradation of reducing sugars (acetoin – 1-hydroxy-2-propanone) alone or in the presence of hydrogen sulfide and/or ammonia. Pyrrole[34] and 2-methylpyrazine identified in fresh and canned Perigord truffle were reported in the literature.[13,15]

The presence of 4-methoxypyridine in sample A (base peak at m/z 109) indicated by the experimental data is perhaps only a tentative identification because the mass spectrum represents a mixture of two compounds. It is of interest to note the presence of γ-butyrolactone and of γ-nonalactone in sample A.

Table 3 GC/MS analysis of truffle aroma from Tricastin and Alpes de Haute Provence (France)

Compound	RI (P)[a]	MW[b]	A	B	C[c]
1. ALIPHATICS					
Aldehydes (17)					
2-Methylpropanal	750	72	-	+	-
Butanal	975	72	+	+	-
2-Methylbutanal ?	1025	86	+	+	-
Hexanal	1090	100	-	-	+
(E)-2-Methyl-2-butenal	1110	84	-	-	+
(Z)-2-Methyl-2-pentenal*	1135	98	-	-	+
(E)-2-Methyl-2-pentenal*	1150	98	-	-	+
Heptanal	1185	114	-	+	+
(E)-2-Heptenal	1280	112	-	-	+
Nonanal	1385	142	-	-	+
(E)-2-Octenal	1425	126	-	-	+
Phenylacetaldehyde	1650	120	+	+	-
(E,E)-2,4-Decadienal	1820	152	+	+	+
(E)-2-Phenyl-2-butenal*	1905	146	+	-	-
Tetradecanal*	1910	212	-	+	-
Cinnamaldehyde*	2000	132	+	-	+
(E)-5-Methyl-2-phenyl-2-hexenal*	2030	188	+	-	-
Ketones (5)					
Acetone		58			
3-Penten-2-one	1050	84	+	-	-
3-Octanone	1280	(128)	-	-	+
2-Nonanone	1385	(142)	+	-	-
(E,E)-3,5,9-Undecatrien-2-one*	2094	192	-	-	+
Esters (10)					
Ethyl 2-methylbutyrate	1050	130	-	-	+
Ethyl 3-methylbutyrate (ethyl isovalerate)	1055	130	-	-	+
Isoamyl formate	1065	(126)	-	-	+
2-Methylpropyl 2-methylbutyrate	1160	(158)	-	-	+
2-Methylbutyl 2-methylbutyrate	1250	(172)	-	-	+
Ethyl 3-hydroxybutyrate	1503	(116)	-	-	+
Ethyl hexadecanoate	2180	284	-	+	-
Ethyl oleate	2493	310	-	+	-
Ethyl linoleate	2540	308	-	+	-
Cinnamyl acetate*	2100	176	+	-	-

(Table 3 continued 1)

Alcohols (22)

Ethanol	910	46	+	+	-
2-Butanol	1000	(74)	-	+	-
1-Propanol	1005	60	-	+	-
2-Methyl-1-propanol	1070	74	-	+	-
3-Methyl-2-butanol	1117	88	-	-	+
Isoamyl alcohol	1190	(88)	-	+	+
1-Pentanol	1225	(88)	-	-	+
3-Hydroxy-2-butanone (acetoin)	1251	88	+	-	-
1-Hydroxy-2-propanone	1265	74	+	-	-
2-Methyl-3-pentanol	1310	102	+	-	-
Hexanol	1320	102	-	+	-
3-Ethyl-3-hexanol	1390	130	-	+	+
3-Ethoxy-1-propanol	1402	(104)	+	-	+
2-Butoxyethanol	1420	118	+	-	-
1-Octen-3-ol	1440	100	-	+	+
2-Ethyl-1-hexanol	1505	130	-	-	+
1-Octanol	1525	112	-	+	-
1-Undecanol	1848	(172)	+	-	-
β-Phenethyl alcohol	1860	122	+	+	-
1-Dodecanol	1936	(186)	+	-	-
1-Hexadecanol	2325	(242)	+	-	-
1-Heptadecanol	2420	(256)	+	-	-

Acids (10)

Acetic acid	1465	60	+	+	+
Propanoic acid	1540	74	-	+	+
2-Methylpropanoic acid	1580	88	-	+	+
Butyric acid	1630	88	+	-	-
2-Methylbutyric acid	1672	102	+	+	+
Hexanoic acid	1825	116	+	-	+
Heptanoic acid	1920	(130)	+	-	-
2-Ethylhexanoic acid	1960	(142)	+	-	+
Octanoic acid	2015	(144)	+	-	-
Nonanoic acid	2120	158	+	-	-

Sulfides and derivatives (10)

Hydrogen sulfide	700	34	-	+	-
Methanethiol (methyl mercaptan)	740	48	-	+	-
Dimethyl sulfide	820	62	-	+	-
Dimethyl disulfide	1065	94	-	-	+
Diisopropyl disulfide	1936	150	-	-	+
Dimethyl trisulfide	1407	126	-	+	-
2-(Methylthio)ethanol	1510	92	+	+	+
3-(Methylthio)propanol	1680	106	+	-	-

(Table 3 continued 2)

1-(Methylthio)-3-hydroxybutane ?	1700	120	+	-	-
4-(Methylthio)butanol	1780	120	+	-	-

2. AROMATICS

Hydrocarbons (3)
Toluene	1030	92	-	-	+
A xylene (*meta* ?)	1235	134	-	+	-
α,*p*-Dimethylstyrene	1280	132	+	+	-

Phenols and derivatives (21)
Anisole	1330	108	+	-	-
3-Methoxytoluene	1420	122	-	-	+
1,2-Dimethoxybenzene	1715	138	+	+	+
1,3-Dimethoxybenzene	1745	138	-	-	+
1,4-Dimethoxy-2-methylbenzene	1810	152	-	-	+
2,3-Dimethoxytoluene	1820	152	-	-	+
1,2,4-Trimethoxybenzene	2055	168	-	-	+
3-*t*-Butyl-4-methoxyphenol	2060	180	-	-	+
1,2,3-Trimethoxy-5-methylbenzene*	2075	192	-	-	+
1,2-Dimethoxy-4-(1-propenyl)-benzene (*cis*-methylisoeugenol)	2110	178	+	+	+
2-Ethylphenol	2125	122	-	-	+
Carvacrol	2170	150	-	+	-
2,4-Di-*t*-butylphenol	2355	206	-	-	+
1,2,3,4-Tetramethoxybenzene*	2385	198	-	-	+
3,4-Dimethoxybenzaldehyde	2455	166	-	-	+
β-Asarone	2485	208	-	+	-
2-Nonylphenol * ?	2690	220	+	-	-
3-Nonylphenol * ?	2710	220	+	-	-
Dibutyl phthalate	2715	278	+	-	-
4-Nonylphenol* ?	2725	220	+	-	-

3. HETEROCYCLIC COMPOUNDS (10)

2-Pentylfuran	1215	138	-	-	+
Furfural	1467	96	+	-	-
γ-Butyrolactone	1640	86	+	-	-
2-Acetylthiazole	1650	126	+	-	-
Furfuryl alcohol*	1660	98	+	-	+
2-Formylthiophene	1680	112	+	-	-
1,2,5,6-Tetrahydropyridine*	1695	83	-	-	+
γ-Nonalactone	1990	(156)	-	-	+
4-Methoxypyridine* ?	2270	109	+	-	-
2,3-Dihydro-3,5-dihydroxy-6-methyl-4-(4*H*)-pyranone*	2305	144	-	-	+

(Table 3 continued 3)

4. MISCELLANEOUS (5)

Diethyleneglycol monoethyl ether	1590	(134)	+	-	-
1,2,3,4-Tetrahydro-5,6-dimethyl-naphthalene	1750	160	-	+	-
β-Caryophyllene	1625	204	-	-	+
Daucene*	-	204	-	-	+
Citronellol*	1735	156	-	+	-

*Newly identified compounds.

[a]Retention indices were calculated on a capillary Innowax column from linear alkanes. Values reported are higher than those found on Carbowax 20M (they have been rounde off).
[b]Invisible molecular weights are given in parentheses.
[c]Samples A and B were obtained by solid phase microextraction (SPME).
The two samples A and B were purchased in winter on the local market of Richerenches near the Grignan castle (Tricastin) and Simiane (Alpes de Haute Provence), respectively. Sample C was obtained by ether extraction of sliced truffles from the Simiane area. It was divided into two fractions by flash chromatography on a silica gel column using pentane and ether as the solvents, respectively.
Sample A, obtained by the SPME, gives by distillation some monoterpene hydrocarbons (5) (α-pinene, β-pinene, δ-3-carene, limonene, p-cymene), alkanes (5) (C_{11} to C_{15}) and two alkylbenzenes
(toluene, ethylbenzene). The fused aromatic hydrocarbons were extracted with pentane from sample C, using INNOLESS acquisition method. In addition to daucene and benzaldehyde, the naphthalene derivatives - 1,2,3,4-tetrahydro-2-methylnaphthalene, 1-, and 2-methylnaphthalenes, and four dimethylnaphthalenes (with the 1,6-isomer positively identified) were identified using selected ion monitoring (SIM).

From a quantitative points of view, with the exception of the seven major constituents (2-methyl-1-propanol, isoamyl alcohol, ethanol, acetic, propionic, isobutyric and 2-methylbutyric acids; see Table 4), the two samples are quite different.
In particular, dimethyl trisulfide was not found in the Richerenches sample (A) while it accounts for 7.5% in the sample from Simiane (B). On the other hand, esters such as ethyl laurate, ethyl oleate, and ethyl linoleate are absent in the first sample (A), as well as carvacrol and β-asarone [1,2,4-trimethoxy-5(1Z)-(1-propenyl)benzene]. 1,2-Dimethoxybenzene was present in the two samples but only anisole was found in the Richerenches sample. Dibutyl phthalate is an artefact. The presence of butylated hydroxytoluene in truffle aroma has been the topic of various discussions.[26,28] It is clear that this well-known antioxidant is a contaminant arising from plastics, elastomers, solvents or food items.[26]

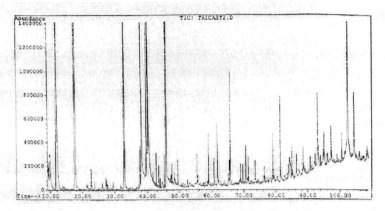

Figure 1 *GC profile of truffle aroma from Richerenches obtained from the SPME method (A)*

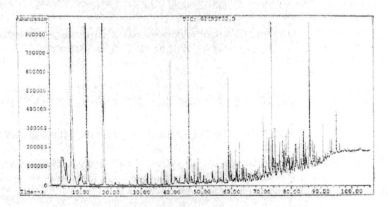

Figure 2 *GC profile of truffle aroma from Simiane obtained from the SPME method (B)*

Table 4 *Comparison of the quantitative composition of the two samples of truffles extracted by the SPME method*

Compound	Tricastin (A) (Richerenches)	Percentages[a] Alpes de Haute Provence (B) (Simiane)
Ethanol	2.5	12.8
2-Methylpropanol	10.4	8.5
Isoamyl alcohol	10.0	9.0
3-Hydroxy-2-butanone	0.6	-
Acetic acid	8.6	2.0
2-Furancarboxaldehyde	tr.	-
Propionic acid	5.6	6.0
1-Octanol	2.0	-
Isobutyric acid	10.0	3.4
2-(Methylthio)ethanol	1.0	-
Butyric acid	1.0	-
2-Methylbutyric acid	3.8	9.5
β-Phenethyl alcohol	2.4	-
γ-Butyrolactone	0.5	-
2-Acetylthiazole	tr.	-
Dimethyl trisulfide	-	7.5
Furfuryl alcohol	1.0	-
2-Thiophenecarboxaldehyde	tr.	-
Benzaldehyde	-	2.0
Hexanoic acid	0.75	-
2-Ethylhexanoic acid	0.8	-
Cinnamaldehyde	3.5	-
Cinnamyl acetate	1.0	-
2,3-Dihydro-5-methyl-4-(4H)-furanone	0.3	-
1,2-Dimethoxy-4-(1-propenyl)benzene	-	1.3
Ethyl laurate	-	1.0
Carvacrol	-	3.5
β-Asarone	-	0.8
Ethyl oleate	-	0.5
Ethyl linoleate	-	4.0

[a]The percentages of other compounds ranged from 0.05 to 0.5%. Mass spectra of numerous minor compounds have not been recorded. Polymethoxylated benzenes and toluenes have not been found in these extracts.

It must be noted that the SPME method using only polydimethylsiloxane, is said to be strongly discriminating for more polar (acids, alcohols) and very volatile compounds.[25]

Sample A from Tricastin was subjected to a post-microdistillation, after extraction according to the SPME method giving rise to a sample A' containing hydrocarbons - mainly alkanes, monoterpenes and two alkylbenzenes (see Table 5 and Figure 3).

Monoterpene derivatives such as α-pinene, camphene, limonene, camphor, α-terpineol, borneol, terpinen-4-ol, and β-ocimene were identified in Australian truffle-like fungus by Millington et al.[22] *p*-Cymene and limonene were previously reported by Flament *et al.*[15] in fresh Perigord truffle aroma.

Table 5 *GC Analysis of the hydrocarbon extract obtained by SPME acquisition method after post distillation (A')*[a]

Compounds	$t_{R(min)}$	$RI_{(p)}$	%
Alkane C_{10}	7.5	1000	-
α-Pinene	9.49	1048	10.0
Toluene	10.30	1067	16.3
Alkane C_{11}	11.69	1100	1.1
β-Pinene	12.92	1127	3.2
Ethylbenzene	14.59	1162	1.9
δ-3-Carene	14.77	1166	7.9
Alkane C_{12}	16.34	1200	1.1
Limonene	17.35	1217	8.5
Unknown (MW = 120)	18.98	1246	4.6
p-Cymene	21.10	1283	1.1
Alkane C_{13}	22.07	1300	1.7
Unknown (MW = 120)	22.28	1305	5.1
Alkane C_{14}	28.11	1400	6.0
Alkane C_{15}	34.11	1500	11.2
Unknown	34.45	1512	3.3
Miscellaneous	-	-	17.0

[a]See Figure 3.

4.2 Ether Extract of Truffles from Simiane (Alpes de Haute Provence)

The qualitative and quantitative chemical composition of this sample is totally different from that obtained by the SPME method (cf. Figure 4 and Table 6).

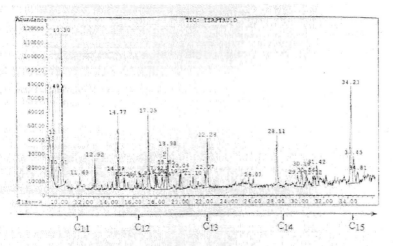

Figure 3 *GC profile of the post-distilled extract obtained by the SPME (hydrocarbons) (A')*

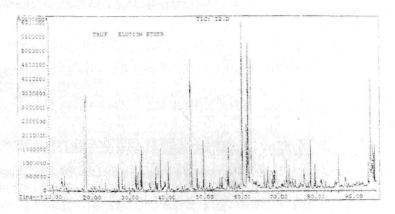

Figure 4. *GC profile of the ethyl ether extract of truffle aroma from Simiane (C).*

Table 6 *Percentages of main components of the ether extract of truffles from Alpes de HauteProvence (Simiane)*

Compound	RT^a (min)	%
β-Phenethyl alcohol	60.04	10
2-Ethylhexanoic acid ?	61.6	8
2-Methylbutyric acid	46.5	7.6
Dodecanol	62.4	7.2
An alkylphenol (nonyl ?)	94.4	6.2
2-Phenyl-2-butenal	61.2	3.7
Nonanoic acid	61.9	3.7
1,2-Dimethoxybenzene	50.0	3.0
2,4-Di-*t*-butylphenol	78.6	2.8
Benzaldehyde	38.4	2.6
3-Methoxytoluene	33.3	2.6
Hexanoic acid	56.2	2.4
1-Octen-3-ol	32.9	2.1
Isobutyric acid	40.4	1.8
1,2,3,4-Tetramethoxybenzene	79.8	1.3
1,2,5,6-Tetrahydropyridine	48.1	1.2
γ-Nonalactone	66.1	0.7
2-Ethoxy-1-propanol	28.5	0.3
Propanoic acid	39.0	tr.
Nonanal	30	tr.
Amyl alcohol	20.7	tr.
1,3-Dimethoxybenzene	51.3	tr.
2-Pentylfuran	19.9	tr.
Cinnamaldehyde	66.7	tr.

^{a} See Figure 4. Numerous compounds are present in trace amounts (< 0.5%) Total: acids: 22%, aromatics: 20%, arylaliphatics: 14%, aliphatics: 10%, heterocycles > 2%.

The major fraction includes aliphatic acids and alcohols, and aromatic compounds including benzaldehyde, 2-phenyl-2-butenal, phenols and anisole derivatives (polymethoxylated benzenes and toluenes). Also, very small quantities of aldehydes, ketones and esters are observed. Esters include ethyl 2-methylbutyrate, ethyl 3-methylbutyrate, isoamyl formate and 2-methylpropyl 2-methylbutyrate. 1-Octen-3-ol accounts for 2.1% of the mixture. β-Phenethyl alcohol is the most abundant compound (10%). The presence of 1,2,5,6-tetrahydropyridine contributes to the characteristic aroma of truffles. The parent compound, pyridine, in very dilute solution possesses this characteristic aroma which is also found in quinoline and thiazole, but to a lesser extent. Sulfides in trace amounts can also contribute to the aroma of truffles.

Dimethoxybenzenes, dimethoxytoluenes, 1,2,3- and 1,2,4-trimethoxybenzenes and 3,4,5-trimethoxytoluene were reported by Flament *et al.*[15]

4.3. GC-MS Analysis of the Pentane Extract of Black Truffle from Simiane (Alpes de Haute Provence) (SIM Technique)

Using GC-MS and the Innoless acquisition method, we found a sesquiterpene hydrocarbon, daucene, benzaldehyde, and fused aromatic hydrocarbons. The latter included 1,2,3,4-tetrahydro-2-methylnaphthalene, the two isomeric 1- and 2-methylnaphthalenes, and four (or five) dimethylnaphthalenes (one of which is the 1,6-isomer). They were not previously described in truffle aroma.

The selected ions at m/z 142 and 156 (SIM technique) (see Figure 5) show the separation of two methylnaphtalenes and dimethylnaphthalene isomers. The suggested structures of the dimethylnaphthalene isomers are reported in Scheme 1. They are classified according to their increasing retention times on a polar column, and based upon the steric hindrance of the methyl groups.

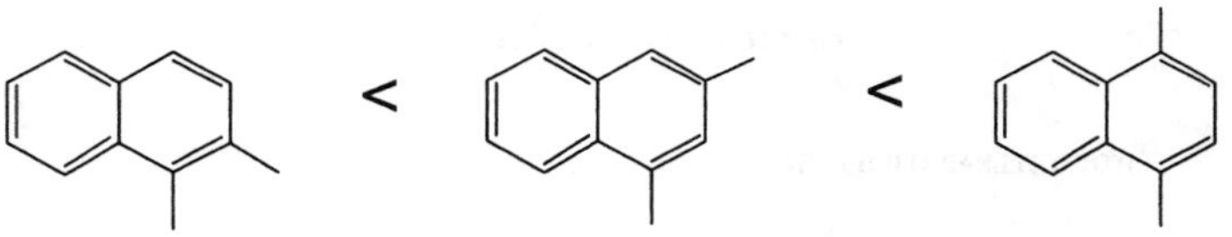

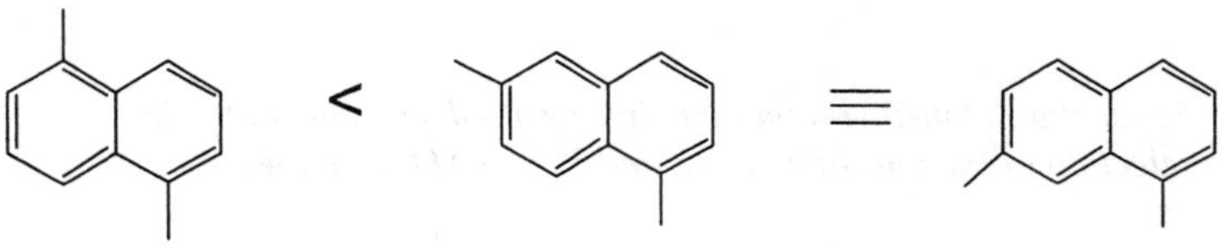

Scheme 1. *Suggested increasing elution order of dimethylnaphthalene isomers on a polar column. Only mass spectrum of the 1,6-isomer (probably in mixture with 1,7-isomer) was recovered and identified.*

4.4 Odors and Flavors of the Volatile Compounds of Truffle Aroma

Odors and flavors of the individual compounds identified in this work are reported in Table 7. Black truffle aroma from Perigord is usually described as a mixture of nuts, musk and ozone, and the odor as slightly pungent (acids), sulfurous (sulfides) and musky.[35] Leach et al.[21] found that the compounds responsible for the herbal, earthy and fungal aroma of *Mesophellia glauca* from Australia are the C_6 and C_8 alkenes and alcohols including 1-hexen-3-ol, 1-octene, 1,3-octadiene and 1-octen-3-ol.

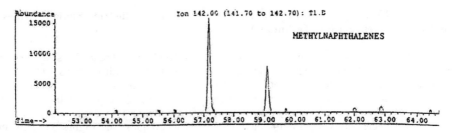

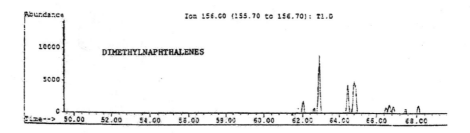

Figure 5 *Separation of methyl- and dimethylnaphthalene isomers by the selection of the ions at m/z 142 and 156 (SIM technique).*

Some authors stressed the olfactory role of : i) aldehydes and alcohols,[34] in particular 1-octen-3-ol (TV = 0.01 ppm) and the corresponding ketone, and ii) sulfur compounds (dimethyl-, mono-, di-, and trisulfides, and 1,2,4-trithiolane) in black and white truffle aroma.[25]

Pyridine and, to a lesser extent, thiazole and quinoline, possess the characteristic odor of truffle in a very diluted aqueous solution. The tetrahydropyridine found in this work should have the same odor. From Table 7 it appears that:
- herbaceous notes are due to (*E*)-2-octenal, 3-octanone;
- earthy notes are associated with 1-octen-3-ol, dodecyl alcohol, methylnaphthalenes, 1,3-dimethoxybenzenes, and carvacrol;
- burnt notes with heterocyclic compounds (2-acetylthiazole, 2-formylthiophene, furfuryl alcohol, and furfural);
- spicy notes with 1,2,4-trimethoxy-5-(1-propenyl, *cis*)-benzene;
- woody notes with β-pinene, β-caryophyllene, 2-ethylphenol, 3,4-dimethoxybenzaldehyde;
- sulfurous (onion, garlic) notes with sulfides (mono-, di-, and tri-);
- pungent notes with aliphatic acids;

Table 7   Odors/flavors of some volatile compounds of black truffles (Tuber melanosporum)

Compounds	Odors/flavors[a]
1. ALIPHATICS	
Aldehydes	
Butanal	Pungent, fruity
Hexanal	Fresh, green-like, fruity, oily
(*E*)-2-Methyl-2-butenal	Green fruit
(*E*)-2-Methyl-2-pentenal	Grassy, green, slightly fruity, powerful
(*E*)-2-Heptenal	Pungent, green
Heptanal	Harsh, nauseating, fatty
Nonanal	Fatty, floral
(*E*)-2-Octenal	Green, herbaceous, spicy
Phenylacetaldehyde	Sweet, floral, rose-like, hyacinth
(*E,E*)-2,4-Decadienal	Fatty, sweet, green, aldehydic
Tetradecanal	Sweet-fatty, waxy
Ketones	
3-Penten-2-one	Fruity
3-Octanone	Herbaceous, spicy, buttery, mushroom, fresh
2-Nonanone	Fruity, floral
Esters	
Ethyl 2-methylbutyrate	Powerful, green, fruity, pungent
Ethyl 3-methylbutyrate	Strong, fruity
Isoamyl formate	Plum
2-Methylbutyl 2-methylbutyrate	Fruity, apple
Ethyl 3-hydroxybutyrate	Sweet, fruity
Ethyl hexadecanoate	Fruity, waxy
Cinnamyl acetate	Basalmic, floral
Alcohols	
2-Butanol	Medicinal, ethereal
Propanol	Sweet
2-Methyl-1-butanol (isoamyl alcohol)	Fusel oil, whiskey
Amyl alcohol	Strong, sweet, balsamic
3-Hydroxy-2-butanone (acetoin)	Buttery
2-Methyl-3-pentanol	Fruity, green, leafy
Hexanol	Fatty, fruity
1-Octen-3-ol	Earthy, woody, mushroom, rancid, faint (TV= 0.08 ppm in soy bean)
2-Ethyl-1-hexanol	Mild, oily, sweet
1-Octanol	Sharp, fatty, waxy, citrus
2-Phenylethanol	Mild, warm, rose, honey

(Table 7 continued 1)

Dodecyl alcohol (dodecanol, lauryl alcohol)	Mild, earthy, soapy, waxy, fatty
1-Hexadecanol	Waxy, floral, mild
1-Heptadecanol	Waxy, floral, mild
Acids	
Acetic acid	Acidic, pungent, sour, vinegary
Propanoic acid	Pungent, rancid
Isobutyric acid	Penetrating, rancid butter
Butyric acid	Repulsive, sweaty, rancid butter
2-Methylbutyric acid	Fruity, sour, cheese (diluted)
Hexanoic acid (caproic acid)	Sour, rancid, musty, goat, pungent, cheesy, fatty
Heptanoic acid	Rancid, sour, sweat-like, fatty
2-Ethylhexanoic acid	Fatty, oily
Octanoic acid (caprylic acid)	Rancid, sweat, goat
Nonanoic acid	Fatty, goat
C_{10} to C_{12} acids (suggested presence)	Fatty, waxy
Sulfides	
Hydrogen sulfide	Rotten egg
Methyl and ethyl mercaptans	Strong, unpleasant
Dimethyl disulfide	Vegetable, cabbage
Diisopropyl disulfide	Alliaceous, onion, garlic
Dimethyl trisulfide	Alliaceous, diffusive, meaty, penetrating
3-(Methylthio)-1-propanol (methionol)	Sulfureous, onion-like with sweet savory, soapy-cooked vegetable nuance

2. AROMATICS

Phenols and derivatives

Anisole	Phenolic, gasoline, ethereal, anise
2-Ethylphenol	Woody
1,2-Dimethoxybenzene (veratrole)	Creamy, vanilla (diluted)
1,3-Dimethoxybenzene	Coconut, hazelnut, earthy
1,4-Dimethoxybenzene	Fennel, mild fatty

- balsamic notes with alcohol, cinnamyl acetate;
- fruity notes with aliphatic esters;
- fatty, oily and waxy notes with heavy aliphatic acids, alcohols and esters

None of them possesses the characteristic truffle aroma with the exception of 1,4,5,6-tetrahydropyridine and 1-octen-3-ol (mushroom). A synergic effect between some of these compounds cannot be excluded.

Some papers are devoted to the reconstitution of this very much studied aroma,[12,14,36] but these artificial aromas need to be improved.

5 CONCLUSIONS

In the present study new compounds have been identified, mainly aromatic derivatives, heterocyclic compounds and monoterpenes using the solvent extraction method (pentane, ethyl ether). Important differences between two samples collected in two not too distant locations (200 km) but harvested at different altitudes (200 and 600 m), respectively, in Southern France, and extracted by SPME, have been observed. Although much work has been carried out on truffle aroma, additional studies are still needed to determine the total composition of this delicious aroma and its characteristic impact compounds and their threshold values.

As far as the extraction methods are concerned, at present headspace-SPME and extraction with a supercritical fluid such as carbon dioxide[37] seem to be the most appropriate and efficient.

Acknowledgements

We are grateful to the SCA of CNRS at Vernaison (Lyon) for GC and GC/MS analyses. Thanks are also due to Dr. René Barone (FST St. Jérôme, Université d'Aix-Marseille III) for his interest in this work and to Mrs. Geneviève M.F. Vernin (FST St. Jérôme, Université d'Aix-Marseille III) for her useful collaboration.

References

1 O. Falck, *Z. Nahr. Genusm.*, 1911, **21**, 209.
2 G. Igolen, *Parfums de France*, 1937, **15**, 2.
3 T. Talou, M. Demas and A. Gaset, *J. Agric. Food Chem.*, 1987, **35**, 774.
4 T. Talou, M. Demas, and A. Gaset, *Dev. Food Sci.*, 1988, **17**, 367.
5 T. Talou, M. Demas and A. Gaset, *J. Essent. Oil Res.*, 1989, **1**, 281.
6 T. Talou, M. Demas and A. Gaset, *Flavour and Fragrance J.*, 1989, **4**, 109.
7 T. Talou, M. Demas and A. Gaset, *ACS Symp. Series*, 1989, **388**, 202.
8 T. Talou, M. Demas and A. Gaset, *ACS Symp. Series*, 1989, **409**, 346.
9 T. Talou, M. Demas and A. Gaset, *J. Sci. Food Agriculture*, 1989, **48**, 56.
10 T. Talou, L. Rigal, A. Gaset, D. Rutledge and C. Ducauze, in *11th Proc. Int. Congress on Essential Oils : Fragrances and Flavours*, 1989, No. 4, 71.
11 T. Talou, M. Demas and A. Gaset, *Perfumer and Flavorist*, 1989, **14**, 9.
12 T. Talou, M. Demas and A. Gaset, in *Thermal Generation of Aromas*, T. H. Parliment, R. J. McGorrin and C.-T. Ho, eds., American Chemical Society,

Washington, DC, 1989, p. 346.
13   T. Talou, M. Demas and A. Gaset, *Black Perigord Truffle from Aroma Analysis to AromatizerFormulation*, in *Flavors and Off-Flavors*, G. Charalambous, ed. *Proc. 6th Intl. Flavor Conf.*, Rethymnon, Crete, Greece, July 5-7, 1989, Elsevier, Amsterdam, 1990, **14**, 715.
14   T. Talou, M. Demas and A. Gaset, *Riv. Ital. EPPOS*, 1991, No. Speziale, 269.
15   I. Flament, C. Chevalier and C. Debonneville, *Riv. Ital. EPPOS*, 1994, 280.
16   F. Bellesia, A. Pinetti, A. Bianchi and B. Tirillini, *Flavour and Fragrance J.*, 1996, **11**, 239.
17   F. Bellesia, A. Pinetti, A. Bianchi and B. Tirillini, *Atti della Società dei Natur. e Matematici di Modena* (Italia), 1996, **127**, 177.
18   F. Bellesia, A. Bianchi, A. Pinetti and B. Tirillini, *Riv. Ital. EPPOS*, 1997, No. 8, 41.
19   M. J. Maciarello and A. O. Tucker, *Dev. Food Sci.*, 1994, **34**, 729.
20   V. Lanzotti and M. Iorizzi, *Proc. Phytochem. Soc. Europe*, 2000, **46**, 37.
21   D. N. Leach, S. Millington, S. G. Wylie and A. Claridge-Walker, *Abstracts*, 214th ACS Natl. Meeting, Las Vegas, NV, Sep. 7-11, 1997, AGDF-042.
22   S. Millington, D. N. Leach, S. G. Wylie and A. W. Claridge, *ACS Symp. Series*, 1998, **705**, 331.
23   T. Talou, M. Doumenc-Faure and A. Gaset, *Flavor Profile of 12 EdibleEuropean Truffles*, in *Food Flavors and Chemistry*, Special Publ., Royal Society of Chemistry, 2001, p. 274.
24   A. Polesello, L. F. Di Cesare and R. Nani, *Industrie delle Bevande* (Italia), 1989, **18**, 10.
25   F. Pelusio, T. Nilson, L. Montarella, R. Tilio, B. O. Larsen, S. Facchetti and J. Madsen, *J. Agric. Food Chem.*, 1995, **43**, 2138.
26   P. Davoli, F. Bellesia and A. Pinetti, *J. Agric. Food Chem.*, 2003, **51**, 4483.
27   P. Diaz, F. Senorans, G. Reglero and E. Ibanez, *J. Agric. Food Chem.*, 2002, **50**, 6468.
28   E. Ibanez, F. J. Senorans, P. Diaz and G. Reglero, *J. Agric. Food Chem.*, 2003, **51**, 4484.
29   G. Vernin, C. Lageot and C. Párkányi, *GC-MS (EI, PCI, NCI, SIM, ITMS) Data Bank Analysis of Flavors and Fragrances. Kováts Indices*, in *Instrumental Methods in Food and Beverages Analysis,* D. Wetzel and G. Charalambous, eds. Elsevier Science, Amsterdam, 1998, **39**, 245.
30   F. Colon and G. Vernin, *The Specma 2000 Data Bank Applied to Flavor and Fragrance Materials*, in *Instrumental Methods in Food and Beverages Analysis*, D. Wetzel and G. Charalambous, eds. Elsevier Science, Amsterdam, 1998, **39**, 489.
31   W. Jennings and T. Shibamoto, *Qualitative Analysis of Flavor andFragrance Volatiles by Glass Capillary Chromatography*. Academic Press, New York, NY, 1980.
32   R. P. Adams, *Identification of Essential Oil Components by Gas Chromatography – Ion Trap Mass Spectrometry*. Allured Publ., Carol Stream, IL , 1995. Idem, on *Quadrupole*, 2001.
33   N. W. Davies, *J. Chromatogr.*, 1990, **503**, 1.
34   L. Bianco, M. Marucchi and P. Cossa, *Industrie Alimentari* (Italia), 1988, **27**, 518.
35   M. Wexler, *International Wildlife*, 1980, **10**, 12.
36   French Pat., Publ. date 1985/1011, Appl. 19841 130; Appl. 84: 18 299.
37   A. Mirande-David, PCT A 29L001-221, French Pat. Appl. 1998-0624 FR 1341.

38   E. Kováts, *Hev. Chim. Acta*, 1958, **41**, 1915.
39   G. Chevalier, cited in the *Daily Telegraph*, July 23, 2003 (INRA, Clermont Ferrand, France).
40   G. Pacioni, *Guide to Mushrooms*, Simon and Schuster, New York, NY, 1981.
41   A. Rambelli, *Manual on Mushroom Cultivation*, Food and Agricultural Organization of the United Nations, Rome, 1983.
42   J. Breitenbach and F. Kränzlin, *Fungi of Switzerland, Vol. 1, Ascomycetes*, Verlag Mykologia, Lucerne, Switzerland, 1984.
43   D. Arora, *Mushrooms Demystified*, 2nd ed., Ten Speed Press, Berkeley, CA, 1995.
44   P. Angelini, L. Costamagna and M. Ciani, *Ann. Microbiologia*, 1998, **48**, 59.

CHARACTERIZATION OF OFF-ODOR OF LOCAL DUCK MEAT

A. Apriyantono[1], R. Hustiany[1], J. Hermanianto[1], P.S. Hardjosworo[2]

[1]Department of Food Technology and Human Nutrition, Bogor Agricultural University, Kampus IPB Darmaga, PO Box 220, Bogor 16002, Indonesia.
[2]Department of Animal Production, Bogor Agricultural University, Kampus IPB Darmaga Bogor, Indonesia

1 ABSTRACT

A major limitation for the use of local duck meat in Indonesia is its off-odor. Trained panelists described the odor of such boiled skin-on duck meat as fishy, rancid, earthy, boiled potato and bloody. Boiled duck meat had a higher off-odor intensity than the fresh one and this increased with increasing boiling time up to 40 min; no difference was observed for breast and leg meat. The volatiles identified included (*E*)-4-penten-2-ol, 1-pentanol, hexanal, (*E*)-1-octen-3-ol, nonanal, (*E*)-2-octen-1-ol, (*E*)-2-decenal, (*E*)-2-nonen-1-ol, and *trans*-2-undecenal; they were characterized as off-odor components. There were also two unidentified components with linear retention index (LRI) on DB-5 column of 1104 and 1123 and possessing the most significant off-odor of duck meat.

2 INTRODUCTION

Local duck serves as a potential source of meat in Indonesia. However, so far, local duck meat is bred mainly for egg production. A major limitation to the use of local duck meat is its off-odor. Therefore, it is very important to overcome this problem in order to fully utilize this resource as a first step to characterize the off-odor substances. This project is aimed at investigating the components responsible for off-odor and some affecting its onset.

3 MATERIALS AND METHODS

3.1 Materials

The Javanese wild duck used in this study were obtained from a local market in Bogor. They were all female, non-productive, free-range ducks with ages ranging from 1.5 to 2 years. After slaughtering, feathers were removed manually. Breast and leg parts (half with the skin-on) were deboned and stored in a refrigerator for 3 - 4 hours. The meats were prepared for sensory analysis and boiling treatment. For volatiles composition and gas chromatography-olfactometry (GC-O) analysis, some meats were stored in a freezer, thawed and ground with a meat grinder, whereas for fatty acid composition analysis, some ground meats were freeze dried and stored in a refrigerator until ready for analysis.

All chemicals used for chemical analysis were analytical grade. Fatty acid methyl ester standards were purchased from Nucheck (Denmark), alkane standards (C_7-C_{22}) were from Sigma (St Louis, MO, USA) whereas other chemicals were from Merck (Damstadt, Germany). All chemicals and flavors used in training of panelist for sensory analysis were obtained from International Flavors and Fragrances (IFF, Union City, NJ, USA), and Firmenich, Jakarta.

3.2 Methods

This study was performed in three steps. The first step was to study the difference in off-odor intensity between breast and leg of duck meat, between fresh and boiled duck meat, and the effect of boiling time on the intensity of the off-odor. The second step was to analyse volatiles and fatty acid compositions of breast and leg of duck meat. The third and last step was to analyse odor description of compounds separated by GC-O.

Sensory analysis of off-odor intensity was examined using an established method, i.e., rating test using trained panellists.[1] Extraction of fat from the meat was done using the Folsch method[2] and analysis of fatty acid composition was carried out by using IUPAC method[3] with margaric acid as the internal standard for quantification.

Isolation of volatiles was carried out using a Likens-Nickerson apparatus for 2 h with diethyl ether as the solvent. 1,4-Dichlorobenzene was used as the internal standard for quantification. Analysis was done using a GC-MS (Shimadzu QP 5000, Kyoto, Japan) with a DB-5 column (30 m x 0.25 mm, film thickness 0.25 µm), split/splitless injection technique and the temperature program as follow: initial temperature 40 °C, held for 5min and then increased to 200 °C with a ramp rate 3 °C/min, hold for 20 min at this final temperature. Mass spectra matching using Class 5000 software (Shimadzu) was used for identification of the volatiles. The identification was confirmed by matching their linear retention indices (LRI) with those reported in the literature. LRI of each volatile was calculated based on a series of alkanes run on the GC-MS with the same program as that for the sample.

Volatile extract of the duck meat was analysed using GC-O in order to obtain sensory description of each volatile component present in the extract. Analysis was done using GC (Shimadzu, type 9AM) equipped with FID and olfactometer with helium as the carrier gas. HP5 (10 m x 0.53 mm, film thickness 0.65 µm) column was used as the column. The injection technique was splitless and the temperature program was as follow: initial temperature 50 °C, held for 3 min, increased to 220 °C with a ramp rate of 8 °C/min and then held there for 3 min.

4 RESULTS AND DISCUSSION

4.1 Some Factors Affecting Off-odor Intensity

Fresh duck meat had a weak odor described as bloody, metallic and typical of that of fresh meat. It was quite surprising, since it possessed relatively low off-odor intensity. When the meat was boiled for 5 min (meat in water was cooked until boiling, after 5 min boiling the meat was removed, drained and served for sensory analysis), the intensity of its off-odor increased quite markedly (Figure 1). Trained panelists described the odor of such boiled skin-on duck meat as fishy, rancid, earthy, boiled potato and bloody. No significant difference of off-odor intensity was observed for breast and leg of either fresh, or boiled meat ($p>0.05$).

It was curious to know the effect of boiling time to off-odor intensity of duck meat, since boiling the meat increased the off-odor intensity (Figure 1). Increasing boiling time up to 40 min increased off-odor intensity (the lowest the score, the highest the off-odor intensity), but when the temperature was increased further up to 60 min resulted in slightly decreased off-odor intensity (Figure 2).

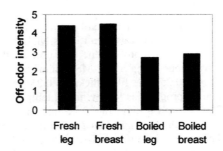

Figure 1 *Average off-odor intensity of leg and breast of fresh and boiled local duck meat with the skin-on as judged by 9 trained panelists in 3 replicates. Boiling was done by cooking meat in water until boiling and then maintained in boiling for 5 min. The intensity scale ranged from 1 to 7 where 1 is the strongest off-odor intensity and 7 is the weakest.*

4.2 Fatty Acid Composition

Eleven fatty acids were detected in duck meat, 6 of which were unsaturated fatty acids, i.e., palmitoleic, oleic, gadoleic, linoleic, linolenic and arachidonic acids (Table 1). These unsaturated fatty acids accounted for 65.2 and 66.0 % of total fatty acids present in the breast and the leg, respectively. The breast contained a higher proportion of lauric, myristic, palmitic, palmitoleic, stearic, oleic and arachidonic acids as compared to the leg meat. Similar composition for main fatty acids was also obtained for wild duck meat analysed by Cobos and his colleagues.[4] The breast of the wild duck contained a higher proportion of stearic, arachidonic and eicosaenoic acids than the leg.[4] Peking duck also had a similar composition of main fatty acids, i.e., palmitic, stearic, oleic, linoleic and arachidonic acids.[5]

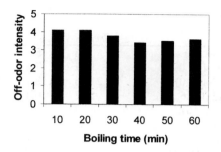

Figure 2 *Average off-odor intensity of local duck meat with the skin-on boiled in water for 10 to 60 min as judged by 10 trained panelists in 3 replicates. The intensity scale ranged from 1 to 7 where 1 is the strongest off-odor intensity and 7 is the weakest.*

4.3 Volatiles Composition

Isolation of volatile components was carried out using a Likens-Nickerson apparatus, since this method gave higher volatiles extract with higher odor intensity than volatiles extract obtained by dynamic headspace with cryogenic trapping. However, the limitation of Likens-Nickerson method is the use of heating during that may lead to additional volatile generation from reactions taking place during heating of the meat. On the other hand, heating during isolation using Likens-Nickerson method can be used as boiling process of the sample, i.e., duck meat. Therefore, in this work are isolated volatile components of the duck meat with and without boiling (for 40 min) before isolation. The isolation of the meat volatiles without boiling the meat for 40 min prior to isolation gave better results, i.e., higher concentrations of most volatile components. However, some components were present in higher concentration in the meat boiled prior to isolation, e.g., (E,E)-2,4-decadienal, 1-hexdecanol and octadecanal (Table 2). Boiling the meat for 40 min may have resulted in decreasing many volatiles formed during heating; in this case the rate of evaporation of the volatiles was higher than the rate of their formation. On the other hand, it was not the case for some compounds as mentioned before.

Table 1 *Fatty acid composition of breast and leg meat lipids of local duck meat*

Fatty Acid	Concentration			
	mg /g fat		mg/100 g fresh meat	
	Breast	Leg	Breast	Leg
Lauric acid (C-12 : 0)	9.6	5.1	90.5	62.6
Myristic acid (C-14 : 0)	10.7	7.0	101.1	85.2
Palmitic acid (C-16 : 0)	212.6	158.1	2011.0	1930.2
Palmitoleic acid (C-16 : 1)	18.2	10.4	171.9	127.0
Stearic acid (C-18 : 0)	49.7	31.5	470.4	384.7
Oleic acid (C-18 : 1)	337.9	239.5	3196.2	2924.3
Linoleic acid (C-18 : 2	159.6	133.1	1510.2	1624.7
Linolenic acid (C-18 : 3)	4.5	4.0	42.9	49.2
Arachidic acid (C-20 : 0)	2.4	2.3	22.8	28.6
Gadoleic acid (C-20 : 1)	5.3	4.5	50.3	54.8
Arachidonic acid (C-20 : 4)	9.2	4.2	87.3	51.6
Total fatty acid	819.7	599.7	7754.6	7322.9
Total saturated fatty acids	285.0	204.0	2695.8	2491.3
Total unsaturated fatty acids	534.7	395.7	5058.8	4830.9

As can be seen from Table 2, the main volatile components identified included several classes, i.e., aldehyde, alcohol, ketone, carboxylic acid, and hydrocarbon. Most of these compounds are likely lipid-derived compounds; they may be formed from degradation of lipids either during the life of the duck, or after slaughtering of the duck, especially during heating. The volatile components present in boiled duck meat analysed by Wu and Liou[6] were also mainly lipid-derived. Main volatile components present in the breast were pentanal, hexanal, (*E*)-1-octen-3-ol, nonanal and 1-hexadecanol, whereas those present in the leg were hexanal, (*E*)-1-octen-3-ol, (*E,E*)-2,4-heptadienal, nonanal, (*E,E*)-2,4-decadienal, 1-hexadecanol and octadecanal.

The breast contained a higher proportion of total unsaturated fatty acids than the leg (Table 1); this may results in higher volatile components detected in the breast (Table 2), since the volatile components are likely to be derived from degradation of lipid, especially the unsaturated fatty acids. The evidence that volatile components present in duck meat are derived from lipids comes from experiments done by Wu and Liou[6] who found that all components detected in boiled duck meat were also detected in duck fat.

Table 2 Volatiles composition of breast and leg of local duck meat with the skin-on

No	LRI_{exp}	LRI_{ref}	Volatile Compound	Fresh[a] Breast	Fresh[a] Leg	Boiled for 40 min[b] Breast	Boiled for 40 min[b] Leg
1		642[7]	3-Methylbutanal	-	-	-	94
2		697[7]	Pentanal	189	50	56	-
3	706		(E)-4-Penten-2-ol	142	20	-	-
4	745	761[7]	1-Pentanol	64	-	-	-
5	762	787[7]	Hexanal	2429	220	1057	1690
6	863	865[7]	1-Hexanol	564	-	-	-
7	881	801[7]	1-Hexanone	86	-	8	-
8	896	896[7]	Heptanal	114	6.5	22	28
9	981	982[7]	(E)-1-Octen-3-ol	1142	6	69	122
10	983		(E,E)-2,4-Heptadienal	-	-	11	133
11	984	986[7]	3-Octanone	353	43	-	-
12	1073	1067[8]	(E)-2-Octen-1-ol	2	-	-	-
13	1094		3,4-Dimethyl-2-hexanol	140	2	-	-
14	1100	1104[7]	Nonanal	597	130	102	244
15	1189		Alkane derivative	250	13	-	10
16	1251	1267[7]	(E)-2-Decenal	-	-	-	85
17	1257		Alkane derivative	6	13	-	-
18	1281	1293[9]	(E)-2-Nonen-1-ol	63	30	9	33
19	1300	1295[10]	(E,E)-2,4-Decadienal	-	1	16	609
20	1312		Alkane derivative	34	-	-	-
21	1347	1376[7]	trans-2-Undecenal	14	21	-	53
22	1353		Alkane derivative	37	-	-	-
23	1377		Alkane derivative	51	30	11	16
24	1438		Alkane derivative	42	30	14	29
25	1451		Benzene derivative	41	10	9	6
26	1457	1473[7]	2-Dodecanol	-	4	-	-
27	1468		Alkane derivative	-	9	-	-
28	1478		Alkane derivative	65	25	19	22
29	1516		Alkane derivative	-	4	-	-
30	1528		Unknown	19	-	-	-
31	1533		Alkane derivative	-	5	-	-
32	1545		Alkane derivative	-	5	-	-
33	1565		Alcohol derivative	-	1	-	-
34	1571	1568[10]	Dodecanoic acid	190	15	8	14
35	1586		Alcohol derivative	-	<1	-	-
36	1596		Unknown	-	-	-	11
37	1625		Alkane derivative	-	8	-	-
38	1643		Aldehyde derivative	-	2	-	-
39	1673		Alkane derivative	3	42	14	26
40	1677		Alkane derivative	9	12	16	34
41	1686	1676[8]	1-Tetradecanol	23	-	18	39
42	1720		Alkane derivative	-	2	-	-
43	1777		Alkane derivative	-	16	4	6
44	1795		1-Hexadecanol	463	185	1627	2352
45	1861		Aldehyde derivative	-	-	-	9
46	1898		(Z)-11-Tetradecen-1-ol-acetate	-	-	-	18
47	1906		Unknown	18	-	-	-
48	1959	1961[7]	Hexadecanoic acid	50	-	-	-
49	1980		(E)-9-Octadecenal	8	-	13	83
50	2005		Octadecanal	14	-	46	143
51	2072	2075[11]	Oleic acid	16	-	-	-

Note :
[a] = average from 2 replicates
[b] = one replicate
LRI_{exp} = LRI Experiment
LRI_{ref} = LRI Reference
- = not detected

4.4 Off-odor Components of Duck Meat

In order to obtain volatile extract with the highest off-odor intensity, volatile extracts obtained from the breast meat, the leg meat, with and without boiling for 40 min prior to isolation using Likens-Nickerson method, were compared. Volatile extract of the breast without boiling prior to isolation had the highest intensity, hence this extract was analysed by using GC-O.

As can be seen from Table 3, some compounds are likely to be responsible for off-odor of duck meat, i.e., pentanal, 1-pentanol, hexanal, (E)-1-octen-3-ol, nonanal, (E)-2-octen-1-ol, (E)-2-decenal, (E)-2-nonen-1-ol, *trans*-2-undecenal, and two unidentified compounds with LRI on DB-5 column of 1104 and 1123. The last two unidentified compounds reflected best the off-odor of the duck meat. Unfortunately the two compounds may be present in the meat in relatively low concentrations; therefore they could not be detected by GC-MS. Apart from the unidentified compounds, all other compounds are likely to be lipid-derived compounds. Sensory description of these off-odor compounds altogether were stale, woody, medicinal, grassy, beany, earthy, unpleasant, urine-like and fatty.

Free-range ducks are omnivorous animals; they eat grains, aquatic plants, aquatic animals and snail, among others. It is most probably that their off-odor may originate from either their diet, or lipid degradation taking place during handling and processing of the meat. It should be noted that boiling increased the off-odor intensity of the duck meat and increasing boiling time up to 40 min increased the intensity. This could be due to increase of compounds formed from lipid degradation, however, the increase of volatile components due to boiling was only for some components, mainly (E,E)-2,4-decadienal, 1-hexdecanol and octadecanal (Table 2). Therefore, these compounds are considered to be responsible for the increase of the off-odor intensity due to boiling, but these were not detected by GC-O analysis. This may be because (E,E)-2,4-decadienal was not detected in the unboiled breast meat volatile extract (Table 2), whereas 1-hexdecanol and octadecanal were not detected in GC-O. Alternatively, this may be because they were eluted at the end of the temperature program of the GC. Alternatively, therefore, they were not noted by the panelists.

5 CONCLUSIONS

Sensory description of boiled skin-on local duck meat was fishy, rancid, earthy, boiled potato and bloody. Boiled duck meat had a higher off-odor intensity than the fresh one and this increased with increasing boiling time up to 40 min; no difference was observed for breast and leg meat. The volatiles identified in the skin-on breast meat included pentanal, 1-pentanol, hexanal, (E)-1-octen-3-ol, nonanal, (E)-2-octen-1-ol, (E)-2-decenal, (E)-2-nonen-1-ol, and trans-2-undecenal; they were characterized as off-odor components. There were also two unidentified components with LRI of 1104 and 1123 on a DB-5 column, representing the best off-odor of duck meat.

Table 3 *Sensory description of odor compounds of breast of local duck meat with the skin-on characterized by GC-O*

No.	LRI	Compound	Odor description
1	<700	Pentanal	Woody, stale, boiled sweet potato
2	<700	-	Sour
3	<700	4-Penten-2-ol	Ethereal, green, butter
4	731	1-Pentanol	Medicinal, somewhat sweet, balsamic
5	755	Hexanal	Green, grassy
6	789	-	Board marker
7	841	1-Hexanol	Medicinal, etheral
8	878	2-Hexanone	Fried tempe, etheral
9	900	Heptanal	Fatty, nutty, oily, sweet
10	903	-	Fishy
11	954	-	Fishy, meaty
12	978	(E)-1-Octen-3-ol	Grassy, fresh, green, earthy
13	988	3-Octanone	Spicy, herbaceous
14	1021	-	Nutty, smoky, meaty
15	1027	-	Fishy, urine-like
16	1040	(E)-2-Octen-1-ol	Vegetable-like, floral
17	1049	3,4-Dimethylhexanol	Meaty, burnt
18	1063	Nonanal	Fatty, waxy, floral
19	1078	-	Putrid
20	1104	-	Stale, urine-like, unpleasant
21	1111	-	Roasted, nutty, meaty
22	1123	-	Green, beany, fishy, stale, off-odor
23	1136	Alkane derivative	Stale
24	1154	-	Fishy
25	1166	-	Rancid
26	1179	(E)-2-Decenal	Fatty, floral
27	1215	(E)-2-Nonen-1-ol	Sweet, fatty, stale
28	1263	Trans-2-Undecenal	Woody, beany, powerful
29	1315	-	Oily, brothy

Note:
Analysis was done by 4 trained panelists.
- = not detected by GC-MS

References

1. M. Meilgaard, G.V. Civille and B.T. Carr, *Sensory Evaluation Techniques*, CRC Press, Boca Raton, Florida, 1999.
2. J.M. Folsch and G.H.S. Stanly, *J. Biol. Chem.*, 1957, **226**, 497.
3. *Standard Methods for the Analysis of Oils and Derivates,* 7th edn., International Union of Pure and Applied Chemistry, Commission on Oils, Fats and Derivates, Blackwell Scientific Publications, Oxford, 1987.
4. A. Cobos. A. Veiga and O. Diaz, *Food Chem.*, 2000, **68**, 77.

5   D.P. Smith, D.L. Fletcher, R.J. Buhr and R.S. Beyer, *Poul. Sci.*, 1993, **72**, 202.
6   C-M. Wu and S-E. Liou, *J. Agric. Food Chem.*, 1992, **40**, 838.
7   R.P. Adams, *Identification of Essential Oil Components by Gas Chromatography/Mass Spectrometry*, Allured Publishing Corporation, Carol Stream, Illinois, 1995.
8   R. Triqui and G. Reineccius, *J. Agric. Food Chem.*, 1995, **43**, 1883.
9   K. Umano and T. Shibamoto, *J. Agric. Food Chem.*, 1987, **35**, 14.
10 A. Leseignur and P. Heinen, unpublished results.
11 A.D. Beal and D.S. Mottram, *J. Agric. Food Chem.*, 1994, **42**, 2880.

GLYCOSIDICALLY-BOUND AROMA COMPOUNDS PRESENT IN GREEN AND CURED VANILLA BEANS

D. Setyaningsih[1], A. Apriyantono[2] and M. T. Suhartono[2]

[1] Department of Agroindustrial Technology,
[2] Department of Food Technology and Human Nutrition
Faculty of Agricultural Technology, Bogor Agricultural University, Kampus IPB Darmaga, PO Box 220, Bogor 16002, Indonesia

1 ABSTRACT

Endogenous enzymes are believed to release aglycones during the curing process of vanilla beans. This study compared volatile aglycones composition of green and cured vanilla beans. The scalding and scratching methods of preparing vanilla beans prior to curing resulted in formation of various glycosidically-bound aroma compounds that were not detected in the green beans. Volatile aglycones of green and cured bean were dominated by aromatic derivatives and lipid-derived compounds. Vanillin content of glycoside fraction of cured beans was only 18.6% as compared to that of the green beans. Some benzene derivatives aglycones increased in the cured bean, these volatiles include guaiacol, 2-methoxy-4-methylphenol and vanillyl alcohol, whereas 4-methylphenol, 4-methoxy methylphenol, 4-ethoxymethylphenol and 3-hydroxy benzeneacetic acid decreased. The aglycon of 1-(3-methoxyphenyl)-ethanone disappeared after curing.

2 INTRODUCTION

The flavor characteristic of vanilla beans develops as a result of enzymatic, chemical, and microbiological reactions that occur during the curing process. Uncured pods are odourless. If left to ripe on the vine, the pods gradually turn from green to yellow, split open and change to dark-brown or chocolate color and develop little aroma. The aim of the curing process is to stop the vegetative development of the beans and to promote the reactions leading to the formation of the most important flavor compounds of vanilla.

It is generally accepted that phenolic glycosides are the important precursors of vanilla flavor.[1] During the curing process, various glycosides are hydrolyzed and also undergo oxidation, e.g., vanillin is assumed to be formed from glucovanillin by β-glucosidase. The other aglycones identified during the curing process include *p*-hydroxybenzaldehyde, *p*-hydroxybenzoic acid, vanillic acid, vanillyl alcohol, acetovanillin, and *p*-hydroxybenzyl alcohol.[2] Cured beans increase their vanillin content up to 14-24% upon pectinase and exogenous β-glycosidase treatment.[3] When a two step enzymatic reaction system with Viscozyme and Celluclast was used, glucovanillin extracted from green beans and its conversion to vanillin may increase 3.13 times.[4]

Several research studies on glycosidically-bound aroma compounds present in green vanilla beans have been reported. Apparently, no research work on differences in glycoside compounds of green and cured vanilla beans has been reported. Moreover, the role of endogenous β-glucosidase activity on the hydrolysis of vanilla glycosides during the curing process need better understanding. Thus, the objective of this work is to study the difference in quality and quantity of glycosidically-bound aroma compounds in green and cured vanilla beans. In addition, the effect of the use of a crude extract of endogenous enzymes from green vanilla beans, commercially available β-glucosidase from almond (emulsin) and heating on the hydrolysis of glycosides present in green and cured vanilla beans are studied.

3 METHOD AND RESULTS

3.1 Activity of β-Glucosidase during Curing

Mature green vanilla beans were divided into two lots; each was treated with a different killing procedure. Killing procedure for one lot was done by scalding in hot water at 65°C for 2-5 min (Bourbon method). The second lot was subjected to scratching i.e., three longitudinal scratches with a pin made over the full length of the bean (Guadeloupe method). These treatments were designed to study the effect of heating during killing and drying step on the extent of hydrolysis of vanilla glycosides. Our work showed that killing by scratching resulted in a lower vanillin content and lower activity of β-glucosidase during curing, and also the beans were highly susceptible to mold growth. The scalding method gave higher vanillin, an oily appearance, as well as a well developed aroma of cured beans.

The activity of β-glucosidase of crude enzymes of green vanilla beans was assayed by incubating the crude enzymes with p-nitrophenyl-β-glucopyranoside.[5] This assay only measures activity of the soluble enzymes; however, it is known that β-glucosidase is bound in the cell wall of the plant. Therefore, we investigated the possibility of bounded-β-glucosidase in the vanilla bean cell walls (insoluble enzymes) by measuring activity of β-glucosidase in the insoluble part of the beans after extraction of the crude enzymes. After scalding, β-glucosidase activity of crude enzymes (soluble enzymes) was slightly lower than the activity of the soluble enzymes in green beans, but the activity of insoluble enzymes doubled (Figure 1). This indicates the presence of heat resistant glucovanillin specific isozymes of β-glucosidase that are not detected by the standard assay. The activity of endogenous β-glucosidase was lost after scalding and the beans that were not scalded kept the activity only in the first hours during sweating.[6] However, the results of this study showed that β-glucosidase activity of the complex cell wall material was still retained during the curing process, and this activity can be considered to be responsible for the release of vanillin during curing.

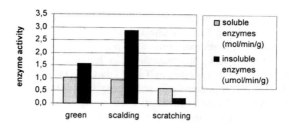

Figure 1 *β-Glucosidase activity of soluble and insoluble enzymes of green beans after scalding or scratching step of the curing process*

Crude enzymes of green vanilla beans showed a β-glucosidase specific activity of 0.79 unit/mg protein, a maximum activity at pH 5.5-6.0 and optimum temperature at 40 °C. Heating the crude enzymes at 40 °C for 5 min increased the β-glucosidase activity up to 2.9 times, and after 30 min heating the activity was still 1.7 times higher. Heating the crude enzymes at 60 °C for 5 min increased the activity 1.7 times, but at this temperature the enzyme was not stable so that the activity decreased to 0.17 times after 30 minutes. For comparison, purified vanilla β-glucosidase was found as a single protein band in native PAGE and exhibited an optimum temperature of 40°C and was fully inactivated after 30 min at 70°C in phosphate buffer.[7] Therefore, it is most probable that the main enzyme present in the extract was β-glucosidase, since the crude enzyme had similar characteristic with that of the purified β-glucosidase.

2.2 Glycosidically-bound Aroma Compounds

To study the presence of vanillin and glucovanillin in the green beans, the first step was to obtain an ethanol-water extract of vanilla beans by extracting the beans using ethanol and water (1:1).[8] Glucovanillin content was estimated by subtraction of vanillin content before and after emulsin hydrolysis of this extract. Vanillin itself was analyzed spectrophotometrically. Glucovanillin of ethanol-water extract of green beans was 304.6 μg/g dry weight, while the amount of free vanillin was negligible. The glycosides present in the extract were isolated by using Amberlite XAD-2 resin.[9] Hydrolysis of the glycosides using emulsin gave a fraction with vanillin content of 292.9 μg/g. By using crude enzyme and heating, the vanillin content was only 67.1% and 22.3% of the amount obtained by using emulsin, respectively.

Table 1 shows glycosidically-bound aroma compounds obtained by hydrolysis of the glycosides by using heating at 60°C, 124 units of emulsin and 6.64 units of crude enzyme extract. The crude enzyme having β-glucosidase activity was prepared by homogenizing green vanilla beans in bis-tris propane buffer (150 mM, pH 8.0) and dialyzing the filtered homogenate against aquadest overnight at 4°C. The volatile components were extracted from the buffer solutions containing glycosides that had been hydrolyzed by using heating or enzymes. GC-MS analysis of the aroma compounds was performed using GC-MS (GC

Shimadzu 17A, MS Shimadzu QP 5000) equipped with a fused capillary column (DB-5, 30m x 0.25mm i.d., 0.1 μm film thickness) and ethyl vanillin as the internal standard for quantification. The oven temperature was programmed 3 min isothermal at 60°C, increased at 3°C/min to 240°C, and maintained at this final temperature for 10 min.

Many of the compounds reported in Table 1 have been reported in previous studies of the free volatile flavor components of vanilla bean[1,10,11]. Glycosidically-bound volatiles of green and cured vanilla beans were dominated by lipid-derived and aromatic derivative compounds. These include aromatic aldehydes, aromatic alcohols, aliphatic esters and fatty acids which are present in relatively high levels. Other aromatic derivatives, aliphatic alcohols, alkanes, lactones, benzofuran and terpenes were present at relatively low levels. Several volatile components, mainly vanillin and vanillyl alcohol, were also found in glycoside fractions dissolved in the buffer.

Medium chain fatty acids (C_{14}-C_{18}) can be formed by hydrolysis of triglycerides or phospholipids. Aglycon γ-decalactone was detected in cured beans treated by crude enzymes. This compound may be derived from its corresponding hydroxy-carboxylic acid by releasing water molecule during curing. For comparison, many glycosidically-bound lactones have been reported in pineapple[12] and raspberry fruit.[13]

The aromatic derivative compounds isolated from vanilla beans were mainly aromatic aldehydes which consist of vanillin, 3-methoxybenzaldehyde, 4-hydroxybenzaldehyde and cinnamaldehyde. These aglycones have also been detected in green vanilla beans by other researchers [1,10,11] and also as volatile compounds of vanilla extract.[21]

The vanillin content of glycosides fractions of cured beans hydrolyzed by emulsin and crude enzymes were only 2.6% and 18.6%, respectively, as compared to that of glycosides fraction of green beans. This suggests that the hydrolysis of glucovanillin after the curing process was almost complete. The vanillin content of green bean glycoside fraction hydrolyzed by emulsin was higher than that hydrolyzed by crude enzymes. This may be due to the high activity of β-glucosidase in emulsin. However, in cured bean, addition of emulsin only produced a small amount of vanillin as compared to addition of crude enzymes. This indicates that vanillin is formed from several precursors, e.g., coniferyl alcohol[22], vanillyl alcohol[2], p-hydroxybenzaldehyde[23]. In the presence of suitable enzymes, that are available in crude enzymes these precursors were converted into vanillin.

Other aromatic compounds, such as guaiacol, 2-methoxy-4-methyl phenol and vanillyl alcohol, were present in higher amount in glycoside fraction of cured beans as compared to glycoside fraction of green beans, whereas the amount of 4-methoxy methylphenol and 3-hydroxybenzeneacetic acid were lower. Phenol, guaiacol, 4-methoxymethylphenol, and 2-methoxy-4-methylphenol were found in the scratched bean glycoside fraction in concentration more than 10 ppm each, but they were present in less than 10 ppm in scalded bean and were not present in green bean glycoside fraction at any level. The relatively high amount of phenols present in scratched beans could contribute to the strong smoky character of the beans.

Table 1 Concentration of volatile aglycones from green and cured vanilla beans (ppm)[a]

Compounds	LRI[c]	LRI ref[d]	GREEN			CURED (SCRATCHED)			CURED (SCALDED)		
			HE[b]	EM[b]	VG[b]	HE	EM	VG	HE	EM	VG
Fatty acids											
Myristic acid	1762	1759[14]					1.4				
Palmitic acid	1961	1962[14]	1.1	3.7	1.6	2.1	4.3	6.9	1.0	1.6	1.4
Linoleic acid	2132					0.9	10.8				
Oleic acid	2140	2157[15]	0.9			2.4	9.3	3.9		1.0	0.6
A fatty acid	2142						4.4			1.3	
Aliphatic alcohols											
2,3-Butanediol						3.6	3.6	10.7		0.2	
2-Ethylhexanol	1025		0.9								
13-Heptadecyn-1-ol	2129		0.7								
Esters											
Ethyl decanoate	1593		0.7						0.6		
Methyl tetradecanoate	1724	1726[16]			0.6						
Ethyl tridecanoate	1792		1.2		0.8	1.2		4.1	1.3		1.6
Ethyl pentadecanoate	1793						1.0	4.0			
Methyl palmitoleate	1901	1906[16]			0.5						
Methyl oleate	1906				0.5						
Methyl palmitate	1923	1928[17]	0.8		4.0	1.7	3.7			1.6	1.8
Ethyl palmitoleate	1970	1977[14]			0.6			2.9			0.5
Ethyl palmitate	1991	1997[14]	4.5	2.0	3.9	2.4	4.9	8.9	3.7	1.9	4.0
Methyl linoleate	2089				1.8		3.5	2.9		1.0	0.9
Methyl linolenate	2098						1.8			1.0	
Ethyl linoleate	2157		11.7	7.2	1.7	0.5	7.1	11.1		1.5	3.0
Ethyl oleate	2163	2171[14]			1.8	1.7		9.1	1.2		
Ethyl linoleolate	2163		4.1	5.1			8.1			1.5	2.1
Ethyl stearate	2192		1.1		0.7	1.0			1.7	0.2	0.4
Lactone											
gamma-Decalactone	1469	1441[12]						4.3			0.4
Alkane											
Dodecane	1201					0.8					
Tetradecane	1400		0.5			1.1			1.3		4.3
Hexadecane	1600					1.0	0.9	4.8	1.9	0.4	2.0

Compounds	LRI[c]	LRI ref[d]	GREEN			CURED (SCRATCHED)			CURED (SCALDED)		
			HE[b]	EM[b]	VG[b]	HE	EM	VG	HE	EM	VG
Heptadecane	1700									0.6	
Octadecane	1800					0.5		2.2	1.2	0.3	1.6
Aromatic alcohols											
Phenol	973	981[14]					13.9				
Benzyl alcohol	1030	1035[14]		3.5		1.8	6.0		0.9	5.7	
4-Methylphenol	1074			8.6		6.4			2.7		
Guaiacol	1082	1086[18]				5.0	45.3		0.7	1.3	
Phenylethyl alcohol	1110	1102[19]		2.5			3.7		0.4	0.8	
2-Methoxy-4-methyl-phenol	1187	1190[15]			3.2	0.5	2.5	32.4	0.3	8.5	
4-Methoxybenzyl alcohol	1282						2.5	4.2		0.6	1.1
4-Methoxymethyl-phenol	1300		4.1	148.5	25.1	0.7	11.6	5.9		9.3	4.9
1,4-Benzenediol	1341			3.4							
4-Hydroxy-benzenemethanol	1348		10.8		6.4	5.3	48.7	36.5			8.7
Eugenol	1351	1359[17]		1.6		5.5		6.0	0.4		
4-Ethoxymethylphenol	1368		4.4	86.3	11.6	1.9		9.9	0.1		
2-Hydroxy-benzenemethanol	1393				53.8					26.1	
Vanillyl alcohol	1446			65.3	13.1	14.9	44.6	106.9		5.2	8.8
Aromatic aldehydes											
4-Hydroxy-benzaldehyde	1375	1317[12]			189.5			11.8		2.3	3.1
Vanillin	1394	1392[20]	21.8	2108.6	581.1	2.6	68.1	125.2	6.1	40.9	90.9
3-Methoxy benzaldehyde	1446								1.6		3.3
Cinnamaldehyde	1643			1.6							
Aromatic ethers											
Methyl vanillylether	1392					3.8			0.1		
Ethyl vanillylether	1463	1553[14]				7.0	5.2	12.9		1.4	1.1
Aromatic ketones											
1-(3-Methoxyphenyl)-ethanone	1310			118.2						0.5	
Acetovanillone	1491	1491[14]		2.7	1.2						
Guaiacylacetone	1523					0.4					
Aromatic esters											

Compounds	LRI[c]	LRI ref[d]	GREEN			CURED (SCRATCHED)			CURED (SCALDED)		
			HE[b]	EM[b]	VG[b]	HE	EM	VG	HE	EM	VG
Methyl cinnamate	1379	1379[16]					3.9	3.0			
Methyl 4-hydroxy-benzoate	1466	1459[14]			0.6						
Methyl salicylate	1190							3.6			
Aromatic acid											
3-Hydroxy benzene acetic acid	1369						20.0	6.5		34.9	13.4
Aromatic amine											
3,4-Dimethoxy-benzeneethanamine	1446			1.8							
Aromatic alkane											
(1,1-Dimethylpropyl) benzene	1088						1.5				
(1,1-Dimethylbutyl) benzene	1169						3.0			0.2	
Miscellaneous											
2,3-Dihydro benzofuran	1220			0.8							
Caryophyllene	1419						1.2			1.3	

[a]The data shown are average of two replicates
[b]The glycosides isolated from green and cured vanilla beans were hydrolysed by HE: heat, EM: emulsin (almond glucosidase), VG : crude enzymes
[c]LRI experiment on DB-5 column
[d]LRI on DB-5 or equivalent according to 12, 14, 15, 16, 17, 18, 19, 20.

Expected oxidation products of vanillin such as vanillic acid and 4-hydroxybenzoic acid were not detected; only 3-hydroxy benzeneacetic acid was detected. This suggests that glucovanillin was not oxidized. However, free vanillin can be oxidized to the dimeric product divanillin and this does not lead to vanillic acid.[24] If vanillic acid was formed in the vanilla beans, this compound may be either oxidatively decarboxylated to 2-methoxy hydroquinone or reduced to the corresponding aldehyde[25] and alcohol.[26]

The glycoside fraction of green vanilla bean contained aromatic alcohols, such as phenylethyl alcohol, benzyl alcohol, p-cresol, p-hydroxy benzylalcohol, 2-methoxy-p-cresol, and vanillyl alcohol. Other researchers also detected these compounds in green vanilla beans.[27] The ratio of aromatic alcohols to vanillin was low in glycoside fraction of green beans, but it was significantly higher in the cured beans. Vanillyl alcohol, benzyl alcohol and 4-hydroxy benzylalcohol identified in this work could be reduction products of vanillin derivatives during curing. The glycoside of 2-methoxy-4-methylphenol may be formed during curing in the scratched bean, since 2-methoxy-4-methylphenol was not present in glycoside fraction of green beans. The high amount of the aromatic alcohols was also probably due to the low affinity of vanilla endogenous enzyme to the aromatic alcohols[6] during curing process.

Vanillin is quite toxic for white-rot fungi compared to other phenolics. Germinating wheat seed converted vanillin to vanillyl alcohol 30 times faster than to vanillic acid. Monoterpenes were also metabolized with the formation of the reduced and oxidized derivatives.[28] Various aromatic aldehydes such as vanillin, syringaldehyde and cinnamaldehyde was reduced by bacterial dehydrogenase.[29] This pathway accounted for more than 50% of the products of vanillic acid metabolism by *P. cinnabarinus*.[25] The reduction of vanillin and 4-hydroxy-benzaldehyde to vanillyl alcohol and 4-hydroxy-benzylalcohol in these experiments probably occurs through this non-specific aryl alcohol dehydrogenase.

Vanillyl-alcohol oxidase (VAO) is a flavoenzyme involved in lignin degradation by white-rot fungi *Byssochlamys fulva*,[30] *Penicillium simplicissimum*,[31] and *Lentinus edodes*.[32] VAO dehydrogenated and hydrated 4-hydroxybenzylalcohols, including vanillyl alcohol, to the corresponding aldehydes. This enzyme also converted eugenol to coniferyl alcohol and hydroxylated 4-alkylphenols to 1-(4-hydroxyphenyl) alcohols. In vanilla curing, the activity of VAO could be related to the change of 4-methylphenol, 4-methoxymethylphenol and 4-ethoxymethylphenol which was detected at high amount in glycosidic fraction of green beans and then decreased after the curing process. The glycoside of 4-methoxymethylphenol can be regarded as the precursor of aromatic aldehydes. This compound represents the physiological substrate of the VAO enzyme, involving the reaction of *p*-quinone methide product with water in the enzyme active site to form the final products 4-hydroxybenzaldehyde and methanol.[31]

The highest amounts of total aglycones were found in green bean, followed by scratched and scalded beans. Total aglycones of scratched bean were 23%, while scalded bean was 9% compared with that of green bean. It can be concluded that glycoside hydrolysis in the curing process was more complete in scalded beans than scratched beans. The curing process resulted in a variety of glycosides that were not detected in green beans; this proves that the formation of several aroma compounds also occurred when they were bound in the glycosides.

The glycoside fraction of green beans glycosides fraction hydrolyzed by emulsin resulted in 3 times higher quantity of aglycones than those hydrolyzed by crude enzyme. However, crude enzymes produced more varied compounds, mostly aliphatic esters in the green beans and aromatic alcohols in the cured beans. Similar experiments using lulo fruit glycosidase applied to lulo glycoside fraction also generated additional volatile compounds to the aglycones.[33]

Figure 2 shows the dendogram of the aromatic derivatives of the aglycones analyzed statistically by cluster analysis to classify samples according to the similarity of the composition of the aglycones. It can be seen that the similarity of green-heat, scalded-heat and scratched-heat volatiles were more than 90%. Therefore, it can be concluded that hydrolysis caused by heating resulted in similar aromatic derivative volatile aglycones between green and cured beans. The heating effect was not significant because the amount of aglycon of green and scratched beans were similar and the amount was much lower than that obtained by hydrolysis using emulsin. Hence, heating of vanilla beans during drying in an oven could be more contributed to texture, color and moisture content than for volatiles formation. On the other hand, heating also developed some specific volatiles, such as 2-ethyl hexanol, 13-heptadecyn-1-ol, 3,4-dimethoxy benzeneathanamine and 2,3-dihydrobenzofuran in the green beans glycoside fraction, as well as methyl vanillylether, guaiacylacetone and caryophyllene in cured beans. This suggests a chemical route involving heat induced oxidation of the precursor occurred during the heating stage of the curing process.

The aromatic aglycones derived from scalded cured beans and hydrolyzed by crude enzymes and emulsin were similar as were the aromatic compounds of scratched cured beans hydrolyzed by emulsin and those hydrolyzed by crude enzymes. Similar results were also found for green beans. The aromatic derivatives aglycones composition was more influenced by the type of vanilla beans (green, cured) than the type of enzyme used.

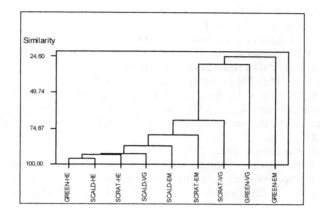

Figure 2 *Dendogram of aromatic compounds (excluding vanillin) liberated by heat, emulsin and crude enzyme from the glycoside fractions.*
aromatic derivatives aglycones of GREEN=*green bean,* SCRAT= *scratched bean,* SCALD= *scalded bean,* HE=*heat,* EM= *emulsin,* VG= *crude enzyme.*

3 CONCLUSION

The scalding and scratching methods of preparing vanilla beans prior to curing resulted in formation of various glycosidically-bound aroma compounds that were not detected in the green beans. Volatile aglycones of green and cured bean were dominated by aromatic derivatives and lipid-derived compounds. Vanillin content of glycoside fraction of cured beans was only 18.6% as compared to that of the green beans. Some benzene derivatives aglycones increased in the cured bean, these volatiles include guaiacol, 2-methoxy-4-methylphenol and vanillyl alcohol, whereas 4-methylphenol, 4-methoxy methylphenol, 4-ethoxymethylphenol and 3-hydroxy benzeneacetic acid decreased. The aglycon of 1-(3-methoxyphenyl)-ethanone disappeared after curing.

Acknowledgements

The authors thank Dr. Hokcu Suhanda, UPI, Bandung for providing GC-MS facilities and acknowledge the financial support from competitive grant X, Ministry of Education of Indonesia.

References

1 G. Leong, R.Uzio and M. Derbesy, *Flav. Fragr. J.* 1989, **4**, 163.
2 T. Kanisawa, *Kouryou*, 1993, **180**, 113
3 A.S. Ranadive, *J. Agric. Food Chem.*, 1992, **40**, 1922.
4 F. R. Teran, I. P. Amador, and A. L. Munguia, *J. Agr. Food Chem.*, 2001, **49**, 5207
5 T. J. C. Luijendijk, PhD Thesis, Leiden University, 1995.
6 M. J. W. Dignum, PhD Thesis, Leiden University, 2002.
7 E. Odoux, A. Chauwin and J. Brillouet, *J. Agr. Food Chem.*, 2003, **51**, 3168
8 A . Kaunzinger, D. Juchelka, A. Mosandl, *J. Agr. Food Chem.*, 1997, **45**, 1752.
9 Z. Y. Gunata, C. L. Bayonove, R. L. Baumes and R.E. Cordonnier, *J. Chromatogr.*, 1985, **331**, 83.
10 P. Brodelius, *Phytochem. Anal.*, 1994, **5**, 27.
11 K. Tokoro, S. Kawahara, A. Amano, T. Kanisawa and M. Indo, 'Glucosides in vanilla beans and changes of their contents during maturation' in *Flavour Science and Technology*, eds., Y. Bressiere, A. F. Thomas, John Wiley and sons, Chichester, 1990, pp. 73
12 P. Wu, M. C. Kuo, T.G. Hartman, R.T. Rosen and C.-T Ho, *J. Agric. Food Chem.* 1991, **39**, 170.
13 A. Pabst, D. Barron, P. Etievant and P. Schreier, *J. Agric. Food Chem.*, 1991, **39**, 173
14 H.J.D. Lalel, Z. Singh and S. C. Tan, *Postharvest Biol. Technol.*, 2003, **29**, 205.
15 A. Apriyantono and Indrawaty, 'Comparison of flavor characteristic of domestic chicken and broiler as affected by different processing methods' in *Food Flavors: Formation, Analysis, and Packaging Influence*, eds., E.T. Contis, C.-T. Ho, C.J. Mussinan, T.H. Parliment, F. Shahidi, A.M. Spanier, Elsevier, Amsterdam, 1998.
16 R. P. Adams, *Identification of Essential Oil Components by Gas Chromatography/ Mass Spectroscopy*, Allured Publishing Corp., 1995.
17 J. Ledauphin, H. Guichard, J. F. S. Clair, B. Picoche and D. Barillier, *J. Agric. Food Chem.*, 2003, **51**, 433.
18 T. H. Yu and C.-T. Ho, *J. Agric. Food Chem*, 1995, **43**, 1641.
19 J. Adedeji, T.G. Hartman, J. Lech, C.T. Ho, *J. Agric. Food Chem.*, 1992, **40**, 659.
20 R. Triqui and G.A. Reineccius, *J. Agric. Food Chem.*, 1995, **43**, 1883.
21 O. Negishi and T. Ozawa, *J. Chromatogr. A*, 1996, **756**, 129
22 A.S. Ranadive, K. Szkutnica, J.G. Guerrera and C. Frenkel, 'Vanillin biosynthesis in vanilla beans' in *Proc. of the Int. Congr. of Essential oils*, Singapore, 1983, pp. 147.
23 D. H. Frenkel, A. Podstolski and R. A. Dixon, *US Pat.*, 2003/0070188 A1.
24 E. Anklam, S. Gaglione and A. Muller, *Food Chem.*, 1997, **60**, 43.
25 B. Falconnier, C. Lapierre, L. Lesage-Meessen, G. Yonnet, P. Brunerie, B. Colonna-Cecaldi, G. Corrieu and M. Asther, *J. Biotechnol.*, 1994, **37**, 123.

26 K. Krisnangkura and M. H. Gold, *Phytochem.*, 1979, **18**, 2019.
27 T. Kanisawa, K. Tokoro and S. Kawahara, *Olfaction Taste XI, Proc. Int. symp.* (Eds. Kurihara, K.,N. Suzuki, H. Ogawa), 1994, Springer, Tokyo, pp. 268.
28 N. Dudai, O. Larkov, E. Putievski, H. R. Lerner, U. Ravid, E. Lewinsohn and A. M. Mayer, *Phytochem.*, 2000, **55**, 375.
29 J. Pelmont, C. Tournesac, A. Mliki, M. Barrelle and C. Beguin, *FEMS Microbiol. Lett.*, 1989, **57**, 109.
30 H. Furukawa, M. Wieser, H. Morita, T. Sugio and T. Nagasawa, *J. Biosci. Bioeng.*, 1999, **87**, 285.
31 R. H. H. Heuvel, M. W. Fraaije, A. Mattevi, C. Laane and W. J. H. van Berkel, *J. Mol. Catalysis B:Enzymatic*, 2001, **11**, 185.
32 C. Crestini and G.G. Sermanni, *J. Biotechnology*, 1995, **39**, 175.
33 C. Osorio, C. Duque and F. B. Viera, *Food Chem.*, 2003, **81**, 333.

FLAVOR STUDIES ON SOME AMAZONIAN FRUITS. 1. FREE AND BOUND
VOLATILES OF COCONA (*Solanum sessiliflorum* Dunal) PULP FRUIT

Alberto Fajardo, Alicia L. Morales[1], and Carmenza Duque[1]

[1]Department of Chemistry, Universidad Nacional de Colombia, Colombia. E-mail: almoralesp@unal.edu.co

1   ABSTRACT

The volatiles components of cocona (*Solanum sessiliflorum* Dunal) flesh obtained by liquid-liquid extraction were analyzed by HRGC, HRGC-MS, MDGC and HRGC-sniffing techniques. Results obtained show as major components, 4-hydroxy-4-methyl-2-pentanone, methyl salicylate, (E)-3-hexen-1-ol, dehydroconiferyl alcohol, benzoic, cinnamic, octanoic and decanoic acids and lactones. Glycosidically bound aroma compounds were identified by HRGC and HRGC-MS of aglycones liberated by enzymatic hydrolysis of the glycosidic extract. These aglycones mainly consisted of cinnamic acid, benzoic acid and 2-methoxy-6-vinylphenol.

2   INTRODUCTION

Cocona (*Solanum sessiliflorum* Dunal) also known as "lulo amazónico", "tupiro" or "cubiu" is a shrubby perennial, generally a meter or more tall with extremely large leaves. It bears maroon, orange-red or yellow fruits up to 10 cm in diameter with yellow flesh. The berries have a pleasant acidulous flavor, somewhat like citrus, and its aroma is a mixture of fruity, herbal, medicinal and spicy notes.[1]

The species is quite variable, particularly in size, shape and flavor of the berries. In the Colombian Amazonia, three types are distinguished[2] type I: small, round, yellow, type II: medium, pear-shaped, purple-red and Type III: large, round, resembling an apple. Cocona type II object of this study, is in greatest demand in the Colombian Amazonia.

The chemical composition of the fruit of *Solanum sessiliflorum* has been reported by Marx *et al*.[3] In this study, analyses of its main components, i. e. minerals and trace elements, carbohydrates, organic acids, fatty acids, free amino acids, biogenic amines and volatile compounds are presented. Regarding volatile compounds, the GC analysis of the water vapor distillate showed methyl salicylate (7.93%), safrol (36.65%) and palmitic acid (16.53%) as the most concentrated components. Unfortunately, authors do not report the type of cocona studied.

In a previous study, the characterization of free and glycosidically bound volatiles of lulo del Chocó or cocona type III (*S. topiro* Syn. *S. sessiliflorum* Dunal)[4] from the Pacific coast in Colombia was reported. It was found that aromatic esters (31.1 %) and carboxylic acids (31.1%) predominated in the free volatiles extract, with methyl salicylate the more abundant. Among the glycosidically bound volatiles, compounds exhibiting aromatics structures prevail followed by terpenols, aliphatic alcohols and carboxylic acids.

As a follow-up to our studies on the flavor of some fruits of the Solanaceae family,[4,5,6,7] we present the results regarding the free and glycosidically bound volatiles of cocona (*Solanum sessiliflorum*) fruit, type II, from the Colombian Amazonia.

3 MATERIALS AND METHODS

Fresh ripe fruits of cocona (*S.sessiliflorum*) type II were collected in "Granja Santo Domingo", Universidad de la Amazonia, Florencia (Caquetá) and sent by airplane to Bogotá. Fruits were selected according to table ripe parameters: reddish-yellow, 6.5 °Brix soluble solids and pulp pH 4.0. A voucher specimen was coded COL No. 385600 at Instituto de Ciencias Naturales at Universidad Nacional de Colombia. Pulp, peelings and seeds were separated. All solvents employed were analytical grade at purchase (Merck) and were redistilled before use.

3.1 Analysis of Free Volatiles

Fruit pulp (ca 1 Kg) was blended with distilled water (1.5 L) at room temperature. The juice obtained was centrifuged at 10,000 rpm, 4°C for 30 min. The clear juice was liquid-liquid extracted with pentane-dichloromethane (1:1) during 24 h.[8] The concentrated extract was subjected to capillary GC-MS (EI and PCI), multidimensional GC and capillary GC-sniffing analyses.

Liquid extract was sensorially evaluated on a trip of filter paper after solvent evaporation. The full aroma impression of the extracts was characterized by trained people of Lucta Grancolombiana S. A.

3.2 Analysis of Glycosidically Bound Volatiles

Glycosidically bound volatile compounds (GBVs) were obtained following the procedure published by Gunata.[9] The dried extract (100 mg) was subjected to enzymatic hydrolysis (overnight, pH 5.2) with a nonselective pectinase (Rohapect D5L, Rohm, Darmstadt). Phenyl β-D-glucopyranoside (1.2 mg) was used as internal standard. The liberated aglycones were extracted with diethyl ether, dried (anhydrous sodium sulphate), carefully concentrated and analysed by HRGC-MS.

3.3 Chromatographic Conditions

The analyses of free volatiles and aglycones liberated in the enzymatic hydrolysis were carried out on a Shimadzu QP 5050 using EI and PCI equipped with a J&W fused silica DB-Wax capillary column (30 m.x 0.25 mm.id., df =0.25 mm.). Split injection (1:15) was employed. Analytical conditions were: temperature program was 4 min isothermal at 50°C increased at 4°C/min to 220°C; flow rate for the carrier gas was 1.8 ml/min He. Injector and detector temperatures were kept at 220°C; electron energy 70 ev; mass range 30-300 u.

PCI-MS was performed under the following conditions: temperature source 200°C; isobutane was used as reactant gas at 2.4 x 10^{-2} Pa, as measured at the ion gauge; mass spectra were scanned in the range m/z 100-300 u.

Qualitative analysis was performed by mass spectral data studies and then verified by comparing retention indices and mass spectra with those of authentic reference substances.

3.3.1 Capillary GC-FID Odor Port Evaluation. Odor profile description of free volatiles was obtained using a sniffing port (olfactory detector, SGE) with a FID/sniffing 1:1 ratio connected to a Hewlett Packard 5890 Series II and operated under the conditions mentioned above. The effluent going to the sniffing port was heated through a glass-lined capillary (60°C) and mixed with humidified air. Each sample was sniffed three times by trained people of Lucta Grancolombiana S. A.

3.3.2 Multi Dimensional Gas Chromatography MDGC. Chiral analyses were performed using a Hewlett Packard HP 5890 Series II gas chromatograph equipped with two columns (precolumn and chiral column), two flame ionization detectors, and a SGE multidimensional capillary GC System 2000.

Preseparation of volatiles was performed on a J & W DB-Wax fused silica capillary column (30 m x 0.25 mm id, d_f= 0.25 µm) under the same conditions as described above for capillary GC analysis. A heart-cut valve device was used to perform selective transfer of sections of the chromatographed components from the pre-column onto a liquid CO_2 cold trap to be re-injected onto the second chiral column.

The chiral separation of γ- and δ-lactones enantiomers was carried out on a heptakis (3-*O*-acetyl-2,6-di-*O*-pentyl)-β-cyclodextrin (30% in OV-1701) (25 m x 0.25 mm id., d_f=0.25 µm) (Mega Capillary Laboratory Columns, Italy). The column temperature program was 40°C to 200°C at 2°C/min. Helium was used as carrier gas at 1 ml/min in both columns. Results of analyses were verified by comparison of multidimensional GC data from authentic optically pure reference substances.

4 RESULTS AND DISCUSSION

4.1 Analysis of Free Volatiles.

The aroma of the liquid extract was described as fruity, herbal, medicinal and resembled the aroma of the homogenized fresh fruit. In this extract (table 1) 51 components (93.0 % of total extract) were identified, with ethyl butanoate, methyl salicylate, (E)-2,3 butanediol and γ-octalactone being the major components.

Lactones (21.9%) and aliphatic esters (20.8%) predominate in the volatile profile followed by aliphatic alcohols (13.2%), aliphatic acids (9.3%), aromatic esters (7.3%), phenolic compounds (5.6%), aromatic acids (5.3%), carbonyl compounds (3.6%), terpenes (3.4%) and terpenols (3.5%).

Lactones are very common in fruits. Usually γ-lactones are in higher concentrations than δ-lactones[10] and frequently there is an enantiomeric excess of one of the antipodes.

The chiral analysis of some lactones (Table 1) shows a clear predominance of (S) enantiomer for γ-octalactone (ee. 36.4%) as well as for the δ-octalactone (ee. 13.0%). This predominance of the (S) enantiomer though not common in fruits, is in good agreement with that found in lulo del Chocó flesh (*Solanum topiro*)[4] and in common lulo (*Solanum vestissimum*).[11]

Another important group of compounds is the aliphatic esters. It has been suggested that they are responsible for the fruity note observed. Ethyl butanoate is the most abundant volatile. This ester is common in fruits of Solanaceae family such as lulo del Chocó (*S. topiro*)[4] and lulo (*S. vestissimun*).[5] C_6 compounds seem to be responsible for the herbal note and phenolic compounds could be responsible for the medicinal and spicy notes of the fruit aroma as in lulo del Chocó.[4]

The aroma composition of cocona tipe II (*S. sessiliflorum*) is quite different from its synonim *S. topiro* (cocona type III). Among the thirty compounds detected in type III, only 24 have been identified in cocona type II, but in quite different amounts. The main differences are the quantities of some compounds such as lactones, aromatic and aliphatic esters and the absence of terpene compounds in cocona type III. In this fruit methyl salicylate has been detected as a major component and may be responsible for the medicinal note observed.

Table 1. *Free volatile identified in cocona (Solanum sessiliflorum Dunal) pulp fruit.*

No	Compound	MW^1	Retention index[2]		%	Odour at sniffing port
			Exp.	Ref.		
1	ethyl acetate		<1000	858	2.1	
2	ethyl propanoate	102	<1000	955	2.7	
3	ethyl 2-methylpropanoate	116	<1000	956	0.3	
4	methyl butanoate	102	<1000	961	0.7	fruity
5	ethyl butanoate	116	1024	1031	10.9	sweet, red
6	ethyl 2-methylbutanoate	130	1037	1044	0.7	
7	hexanal	100	1070	1067	0.3	green
8	2-pentanol		1100	1107	0.3	
9	2-methyl-2-propanol		1115	1117	0.4	
10	3-penten-2-one[3]		1133		0.3	
11	butanol		1140	1136	0.4	
12	3-methyl-1-butanol		1217	1215	0.2	
13	(E)-2-hexenal		1227	1219	0.3	green, herbal
14	3-hydroxy-2-butanone		1269	1259	2.1	sweet, yellow
15	ethyl 2-hydroxy-2-methylbutanoate	146	1314	1303	0.4	
16	hexanol	102	1363	1353	2.1	
17	4-hydroxy-4-methyl-2-pentanone[3]	116	1371		0.6	
18	(Z)-3-hexenol	100	1394	1387	1.9	green
19	ethyl octanoate	172	1425	1423	1.2	fruity

20	octyl acetate	172	1465	1460	0.9	
21	α-copaene	204	1494	1488	0.4	
22	ethyl 3-hydroxybutanoate	132	1531	1530	0.9	sweet
23	(E)-2,3-butanediol		1546	1542	5.8	
24	unknown		1584		0.4	
25	butanoic acid		1606	1598	3.1	
26	γ-butyrolactone		1610	1599	1.5	
27	ethyl benzoate	150	1647	1647	0.8	toast
28	γ-hexalactone (racemic)	114	1672	1675	1.5	
29	α-muurolene	204	1726	1720	0.6	
30	unknown		1739		2.5	
31	δ-cadinene	204	1744	1749	1.7	
32	methyl salicylate	152	1756	1754	6.5	green, sweet,
33	δ-hexalactone	114	1771	1764	3.6	
34	trans-calamenene	202	1830	1832	0.7	
35	2-exo-hydroxy-1,8-cineol		1834	1831	1.3	
36	guaiacol	124	1840	1840	1.6	
37	γ-octalactone ee. (S) 36.4 %	142	1892	1889	5.9	sweet, caramel
38	2,6-dimethyl-3,7-octadiene-2,6-diol	170	1925	1928	1.3	
39	(E)-3-hexenoic acid	114	1928	1924	0.4	
40	2,6-dimethyl-7-octen-2,6-diol[3]	172	1931		0.9	
41	δ-octalactone ee. (S) 13.0 %	142	1942	1940	3.9	
42	octanoic acid	144	2021	2025	2.7	
43	γ-decalactone	170	2131	2125	2.2	
44	nonanoic acid	158	2142	2149	0.6	
45	eugenol	164	2164	2174	0.8	spicy, wood
46	4-vinyl guaiacol		2168	2162	0.4	
47	δ-decalactone	170	2177	2173	3.3	dairy, creamy
48	unknown		2238		0.4	
49	1,4-nonanediol	160	2251	2263	0.3	
50	decanoic acid	172	2276	2270	2.2	
51	isoeugenol	164	2270	2269	0.3	
52	unknown		2291		0.2	
53	1,2,3-propanetriol[3]		2307		1.8	
54	unknown	144	2317		1.6	
55	4-vinylphenol		2355	2372	0.3	
56	benzoic acid	122	2406	2408	3.6	
57	Vanillin		2458	2485	2.2	
58	Unknown	204	>2600	2660	0.9	
59	tetradecanoic acid	228	>2600	2695	0.3	
60	cinnamic acid	148	>2600	2832	1.7	

[1] molecular weight determinated by chemical ionisation.
[2] Retention index on DB-Wax column.
[3] Tentatively assigned from mass spectrum.

Composition 161

In contrast, cocona (type II) aroma contains a high concentration of aliphatic esters and a low amount of aromatic esters as found in lulo (*Solanum vestissimun*)[5] and naranjilla (*Solanum quitoense*).[12]

Differences in the flavor of these two types of cocona could be explained by taking into account that climatic and cultural conditions as well as genetic differences affect the flavor of fruits, as reported in tomato,[13] apple[14] and strawberry.[15]

4.2 Analysis of glycosidically bound volatiles

In the analysis of the liberated aglycones from cocona glycosidic extract 32 aglycones were identified (Table 2). They are mainly aromatic acids (60.4%), terpenols (15.3%) and phenolic compounds (12.6%). The major compounds are benzoic, cinnamic and acetic acids, methyl salicylate and benzyl alcohol.

Table 2. *Aglycones released by enzymatic hydrolysis of cocona (Solanum sessiliflorum Dunal) fruit pulp.*

No.	Compound	Retention index[1] Exp.	Ref.	Amount[3]
1	Ethanol	<1000	924	+
2	3-methyl-1-butanol	1217	1210	+
3	1-hexanol	1355	1353	+
4	(Z)-3-hexen-1-ol	1393	1387	+
5	acetic acid	1435	1438	+++
6	(E)-linalool oxide	1465	1461	+
7	butanoic acid	1602	1598	+
8	α-terpineol	1692	1699	+
9	2-butenoic acid[2]	1719		+
10	epoxylinalool[2]	1743		+
11	methyl salicylate	1776	1777	++++
12	(E)-5-hidroxy-2-methyl-1,3-dioxane	1803	1810	++
13	2-exo-hidroxy-1,8-cineol	1819	1822	++
14	dimethyloctadienediol (I)[2]	1833		+
15	Guaiacol	1865	1857	++
16	benzyl alcohol	1896	1901	+++
17	2-phenylethanol	1902	1899	++
18	Eugenol	2164	2174	+
19	4-vinylguaiacol	2185	2181	+
20	2,6-dimethyl-7-octen-1,6-diol	2229	2227	+
21	2,6-dimethyl-2(Z),7-octadien-1,6-diol	2266	2254	+
22	2,6-dimethyl-2(E),7-octadien-1,6-diol	2300	2294	++
23	4-vinylphenol	2383	2376	++
24	benzoic acid	2412	2408	++++
25	p-mentendiol[2]	2521		+++
26	phenyl acetic acid	2538	2530	++
27	Isovanillin	2547	2537	+

28	4-vinylsyringol	2568	2560	+
29	2,6-dimethyl-(2E, 6E)-octadiene-1,8-diol	>2600	2635	++
30	tetradecanoic acid	>2600	2695	+
31	3-hydroxy-5,6-epoxy-β-ionol	>2600	2776	+++
32	cinnamic acid	>2600	2835	++++

[1] Retention index on DB-Wax column.
[2] Tentatively assigned from mass spectrum.

The composition of the bound volatiles found for cocona is similar to that of the other fruits of the same family such as lulo de Chocó (*Solanum topiro*)[4], lulo (*Solanum vestissimum*)[16] and tomato (*Lycopersicum esculentum*),[17] where the aromatic structures also prevail. Methyl salicylate and 2,6-dimethyl-2(E),7-octadien-1,6-diol aglycones have been also detected in (*Solanum topiro*)[4] in free and bound form.

It is important to point out that methyl salicylate, 2-exo-hydroxy-1,8-cineol and phenolic compounds were detected as free volatiles. This fact leads us to confirm the role of the glycosidic fraction as a flavor precursor in this fruit.

Acknowledgements

Financial support by BID-COLCIENCIAS, IPICS, Uppsala University is greatly appreciated. We are particularly thankful to María Paola Castaño for the MDGC analysis and to Tatiana Osorio and Victor Armando Rincón (Lucta Grancolombiana) for their skilful support on the capillary GC-sniffing analysis. A. Fajardo thanks Universidad de la Amazonia for financial support to pursue his doctoral studies.

References

1. C. Heiser and G. Anderson. "New" Solanums. in *Perspectives on New Crops and New Uses*. J. Janick (ed), ASHS Press, Alexandria, VA. 1999, p 379-384.
2. M. S. Hernández and J. A. Barrera. 2000. SINCHI-PRONATA. Bogotá.
3. F. Marx, E. Andrade and J. Maia. *Food Res. Technol.* 1998, **206**, 364.
4. A. L. Morales, C. Duque and E. Bautista. *J. High Resol. Chromatogr.* 2000, **23**, 379.
5. M. Suárez and C. Duque. *J. Agric. Foood Chem.* 1991, **39**, 1496.
6. A. Torrado, M. Suárez, C. Duque, D. Krajewski, W. Neugebauer, and P. Schreier, *Flav. Fragr. J.* 1995, **10**, 349.
7. H. Mayorga. *Doctoral Thesis*. 2002, Universidad Nacional de Colombia. Bogotá..
8. F. Drawert and A. Rapp. *Chromatographia.* 1968, **1**, 446.
9. Y. Gunata, C. Bayonove, R. Baumes and R. Cordonier. *J. Chromatogr.* 1985, 331, 83.
10. P. Werkhoff, S. Brennecke, W. Bretschneider, M. Güntert, R. Hopp and H. Surburg. *Z. Lebensm. Unters. Forsch.* 1993, **196**, 307.
11. M. Suárez, *Doctoral Thesis*, 1992, Universidad Nacional de Colombia. Bogotá.
12. E.-J. Brunke, P. Mair and F.-J. Hammerschmidt. *J. Agric. Food Chem.* 1989, **37**, 746.
13. E. A. Baldwin, J. W. Scott, C. K. Shewmaker and W. Schuch. *Hortscience.* 2000, **35**, 1013.
14. J. K. Fellman, T. W. Miller, D. S. Mattinson and J. P. Mattheis. *Hortscience* 2000, **35**, 1026.

15. C. F. Forney, W. Kalt and M. A. Jordan. *Hortscience* 2000, **35**, 1022.
16. M. Suárez, C. Duque, H. Wintoch and P. Schreier, *J. Agric. Food Chem.* 1991, **39**, 1543.
17. C. Marlatt, C.-T.. Ho and M. Chien. *J. Agric. Food Chem.* 1992, **40**, 249.

FLAVOR STUDIES ON SOME AMAZONIAN FRUITS. 2. FREE AND BOUND VOLATILES OF ARAZÁ (*Eugenia stipitata* Mac Vaugh) PULP FRUIT

Alberto Fajardo, Jenny L. Delgado, Alicia L. Morales[1], and Carmenza Duque[1]

[1]Department of Chemistry, Universidad Nacional de Colombia, Colombia. E-mail: almoralesp@unal.edu.co

1 ABSTRACT

The Arazá (*Eugenia stipitata*), a relative of the guava family, is a large, bright yellow fruit with a sour, acidic flavor. The volatile constituents of the Arazá fruit pulp were obtained by headspace-solid phase microextraction (HS-SPME) and liquid-liquid extraction and subsequently analyzed by HRGC and HRGC-MS. Of the 66 components that were identified in total, the most abundant were *n*-hexanol, methyl 3-methythiopropanoate, ethyl hexanoate and ethyl butanoate. Glycosidically bound aroma components were identified by capillary GC and capillary GC-MS after isolation of the glycosidic fraction, obtained by Amberlite XAD-2 adsorption and methanol elution followed by hydrolysis with a commercial pectinase. In total 19 aroma aglycones were identified, consisting mainly of aromatic compounds and aliphatic acids.

2 INTRODUCTION

Arazá (*Eugenia stipitata*) is a large sized guava relative, with bright yellow fruit that has an excellent sour-acid flavor. Fruits are very juicy, up to five inches wide and possess a strong fruity aroma. The fruit is very attractive to consumers because of its delicious tart flavor. The fruits are used to make drinks, popsicles, ice cream and arazá liquor.

Eugenia stipitata includes two subspecies known in Colombia as Brazilian and Peruvian ecotypes, respectively. Peruvian ecotype is pear-shaped; its flavor resembles passion fruit and peach and due to its superb organoleptic characteristics is enjoyed by Colombian consumers.[1] Despite the pleasant flavor of arazá only one study[2] of its volatile constituents has been reported. In this study, volatiles from headspace of frozen fruits of araça-boi from Brazil were characterized. A complex pattern of sesquiterpenes with germacrene D as the major compound was found. Recently, chemical composition of araça-boi has been reported.[3]

Based on our knowledge, studies on the free and glycosidically bound volatiles of the Peruvian ecotype have not been published. This research is a contribution to the knowledge of the Amazonian flora and is a continuation of our studies about the aroma of tropical fruits, studying free and glycosidically bound volatiles.

3. MATERIALS AND METHODS

Fresh ripe fruits (*Eugenia stipitata*) were collected in "Granja Santo Domingo" Universidad de la Amazonia, Florencia (Caquetá) and sent by airplane to Bogotá. Fruits were selected according to table ripe parameters: bright yellow, 4.1°Brix soluble solids and pulp pH 2.8. A voucher specimen was coded as COL No. 494092 at Instituto de Ciencias Naturales at Universidad Nacional de Colombia. Pulp, peelings and seeds were separated. All solvents employed were analytical grade at purchase (Merck) and were redistilled before use.

3.1 Analysis of Free Volatiles

3.1.1 Headspace-solid phase microextraction (HS-SPME)
According to the methodology described by Azodanlou,[4] the following conditions for HS-SPME were established. The sample (30 g of arazá pulp) was placed in 50 ml vials, and allowed to equilibrate for 40 minutes. Sampling was carried out in the headspace using DVB/Carboxen/PDMS, 50/30 fiber during a period of 15 minutes; volatile compounds were desorbed for 3 minutes in the splitless injection port of a gas chromatograph HP 5890 with a HP 5970 mass selective detector.

3.1.2 Liquid-liquid Extraction
Fruit pulp (ca 1 Kg.) was blended with distilled water (1.5 L) at room temperature. The juice obtained was centrifuged at 10000 rpm, 4 °C for 30 min. The clear juice was liquid-liquid extracted with pentane-dichloromethane (1:1) for 24 h.[5] The concentrated extract was subjected to capillary GC-MS (EI and PCI), multidimensional GC and capillary GC-sniffing analyses.
Liquid extract was sensorially evaluated on a strip of filter paper after solvent evaporation. The full aroma impression of the extracts was characterized by trained people of Lucta Grancolombiana S. A.

3.2 Analysis of Glycosidically Bound Volatiles

Glycosidically bound volatile compounds (GBVs) were obtained following the procedure published by Gunata[6]. The dried extract (100 mg) was subjected to enzymatic hydrolysis (overnight, pH 5.2) with a nonselective pectinase (Rohapect D5L, Rohm, Darmstadt). Phenyl β-D-glucopyranoside (1.2 mg) was used as internal standard. The liberated aglycones were extracted with diethyl ether, dried (anhydrous sodium sulphate), carefully concentrated and analysed by HRGC-MS.

3.3 Chromatographic Conditions

The analyses of free volatiles and aglycones liberated in the enzymatic hydrolysis were carried out on a Shimadzu QP 5050 using EI and PCI equipped with a J&W fused silica DB-Wax capillary column (30 m x 0.25 mm id., df =0.25 mm.). Split injection (1:15) was employed. Analytical conditions were: temperature program 4 min isothermal at 50 °C increased at 4 °C/min to 220 °C, flow rate for the carrier gas was 1.8 ml/min He. Injector and detector temperatures were kept at 220 °C; electron energy 70 eV; mass range 30-300 u.

PCI-MS was performed under the following conditions: temperature source 200 °C; isobutane was used as reactant gas at 2.4 x 10^{-2} Pa, as measured at the ion gauge; mass spectra were scanned in the range m/z 100-300 u.

Qualitative analysis was performed by mass spectral data studies and then verified by comparing retention indices and mass spectra with those of authentic reference substances.

Capillary GC-FID Odor Port Evaluation
Odor profile description of free volatiles was obtained using a sniffing port (olfactory detector, SGE) with a FID/sniffing 1:1 ratio connected to a Hewlett Packard 5890 Series II gas chromatograph and operated under the conditions mentioned above. The effluent going to the sniffing port was heated through a glass-line capillary (60 °C) and mixed with humidified air. Each sample was sniffed three times by trained people of Lucta Grancolombiana S. A.

3.3.2 Multi Dimensional Gas Chromatography MDGC.
Chiral analyses were performed using a Hewlett Packard HP 5890 Series II gas chromatograph equipped with two columns (precolumn and chiral column), two flame ionization detectors, and a SGE multidimensional capillary GC System 2000.

Preseparation of volatiles was performed on a J & W DB-Wax fused silica capillary column (30 m x 0.25 mm id, d_f= 0.25 µm) under the same conditions as described above for capillary GC analysis. A heart-cut valve device was used to perform selective transfer of sections of the chromatographed components from the pre-column onto a liquid CO_2 cold trap to be reinjected onto the second chiral column.

The chiral separation of γ- and δ-lactones enantiomers was carried out on a heptakis (3-*O*-acetyl-2,6-di-*O*-pentyl)-β-cyclodextrin (30% in OV-1701) (25 m x 0.25 mm id., df=0.25 µm) (Mega Capillary Laboratory Columns, Italy). The column temperature program was 40°C to 200 °C at 2 °C/min. Helium was used as carrier gas at 1 ml/min in both columns. Results of analyses were verified by comparison of multidimensional GC data from authentic optically pure reference substances.

4. RESULTS AND DISCUSSION

4.1 Analysis of Free Volatiles

In order to achieve a representative information on free volatiles present in arazá, two techniques of extraction were used, HS-SPME and the usual liquid-liquid extraction.[5] Table 1 lists the compounds identified by both methods, with their GC peak area percents and their odor at sniffing port.

Through HS-SPME 35 compounds were identified, with ethyl hexanoate, ethyl octanoate and ethyl acetate being the major constituents.

The mixture of volatile compounds obtained by liquid extraction of arazá pulp shows a characteristic aroma of this fruit, with sweet, fruity, aqueous and fragant notes. In this extract, a total of 56 volatile compounds (88% of the total extract) were identified with ethyl esters of C_6, C_8 and C_2 acids as major components, as was found by SPME.

With both methods, aliphatic esters (66.6%, 50.6%) predominate in the arazá volatile profile, followed by alcohols (14.3%, 13.2%), carbonyl compounds (7.1%, 7.7%) and carboxylic acids (5.7%, 2.3%). It is important to point out that terpenes (6.2%) were detected only by SPME and lactones (11.7%) only in the liquid extract.

These results are quite different from those reported for Brazilian ecotype[2] where the terpenes (74.6%) are the predominant group with Germacrene D as major component, and the amount of esters is only 1.7%. This flavor variation could be explained not only by the methodology used, but also by climatic and cultural differences between these cultivars.

The olfactometric analysis performed in this study did not reveal any character impact components, although the panelists gave the description "like arazá" to some chromatographic zones corresponding to esters such as methyl 3-methyl thiopropanoate, ethyl 3-methyl thiopropanoate and ethyl 3-hydroxybutanoate. These results show the important contribution of thioesters and hydroxyesters to the exotic flavor of this fruit.

Other important compounds in the flavor of arazá are (Z)-2-hexenal, 1-hexanol and 3-hexenol due to their sweet, green, fruity notes and δ-hexalactone and δ-octalactone with their sweet notes.

The enantiomeric composition of identified lactones (Table 1) shows a clear predominance of the S-enantiomer in γ-lactones whereas in δ-octalactone the R-antipode dominates, as has been found in other tropical fruit grown in Colombia.[7,8]

Table 1. *Volatile Compounds Identified in Arazá (Eugenia stipitata Mac Vaugh) Pulp Fruit.*

No.	Compound	MW^1	Ri^2	Area % L-L	Area % SPME	Odor at sniffing port
1	ethyl acetate		860	nd	11.3	
2	ethanol		920	nd	12.9	
3	ethyl propanoate	102	957	3.7	0.3	
4	ethyl 2-methylpropanoate	116	958	2.5	0.5	
5	propyl acetate	102	965	1.9	nd	
6	ethyl butanoate	116	1026	4.9	1.2	
7	ethyl 2-methylbutanoate	130	1053	3.3	1.7	sweet
8	butyl acetate	116	1064	0.4	0.4	
9	hexanal	100	1074	1.2	0.2	fruity, apple green
10	2-pentanol		1109	0.4	nd	
11	3-methylbutyl acetate		1115	nd	0.4	
12	ethyl pentanoate	130	1138	0.4	0.5	
13	ethyl 2-butenoate	114	1167	0.4	nd	

14	methyl hexanoate	130	1178	0.1	5.1	
15	(Z)-2-hexenal		1208	0.4	nd	green
16	ethyl hexanoate	144	1233	6.9	13.9	fruity, green
17	(Z)-β-ocimene	136	1242	nd	6.2	
18	3-hydroxy-2-butanone		1269	3.6	0.3	sweet
19	n-hexyl acetate	144	1278	4.3	5.9	
20	(Z)-3-hexenyl acetate	142	1310	2.1	1.2	pear flesh
21	2-heptanol	116	1320	0.4	nd	fruity, sweet, pear
22	ethyl heptanoate	158	1326	nd	0.4	
23	hexyl 2-methylpropanoate	172	1338	nd	0.4	
24	ethyl 2-hydroxypropanoate	118	1341	1.1	nd	
25	1-hexanol	102	1353	4.9	1.2	fruity, red
26	(Z)-3-hexenol	100	1379	2.9	0.2	green, sweet
27	methyl octanoate	158	1384	nd	0.4	
28	hexyl butanoate	172	1414	nd	0.4	
29	hexyl 2-methyl butanoate	186	1425	0.5	1.8	wood, moist, rust
30	ethyl octanoate	172	1434	nd	12.4	fruity, sweet
31	methyl 3-methylthiopropanoate	134	1436	7.9	2.7	fruity, like arazá
32	(Z)-3-hexenyl pentanoate	184	1466	nd	0.3	
33	methyl 3-hydroxybutanoate	118	1477	0.4	nd	
34	benzaldehyde	106	1499	0.4	nd	
35	ethyl 3-hydroxybutanoate	132	1509	3.2	nd	fruity, like arazá
36	ethyl nonanoate	186	1526	nd	0.4	
37	(E)-2,3-butanediol		1533	1.7	nd	
38	ethyl 2-octenoate	170	1543	0.4	0.2	
39	ethyl 3-methylthiopropanoate	148	1554	2.6	0.4	fruity, like arazá
40	(Z)-2,3-butanediol		1572	1.1	nd	
41	unknown	158	1600	0.6	nd	
42	hexyl hexanoate	200	1607	nd	0.2	
43	ethyl decanoate	228	1630	nd	2.6	red, sweet
44	ethyl (Z)-4-decenoate	198	1638	0.4	1.6	
45	ethyl 3-hydroxyhexanoate	160	1659	0.4	nd	fruity, green
46	γ-butyrolactone		1670	1.3	nd	
47	γ-hexalactone e.e (S) 53.4%	114	1739	1.3	nd	
48	(Z)-3,7-dimethyl 2,6-octadien-1-ol,	154	1772	1.0	nd	
49	δ-hexalactone	114	1795	0.9	nd	fruity, red, honey
50	unknown	168	1808	7.5	2.6	dairy
51	(E)-3,7-dimehyl 2,6-octadien-1-ol	154	1812	0.4	nd	
52	hexanoic acid	116	1828	1.6	5.7	
53	ethyl 3-hydroxyoctanoate	188	1855	0.8	nd	
54	phenethyl alcohol	122	1892	0.4	nd	
55	unknown		1930	0.3	nd	
56	(Z)-3-hexenoic acid	114	1934	0.3	nd	
57	γ-octalactone e.e (S) 36.6%	142	1944	2.2	nd	
58	δ-octalactone e.e (R) 46.8%	142	1990	3.3	nd	red, sweet

No	Compound				
59	δ-nonalactone	188	2016	0.1	nd
60	Furaneol	128	2036	2.1	nd	fruity, sweet
61	unknown	196	2041	3.6	4.0
62	ethyl tetradecanoate	196	2055	2.0	nd
63	γ-decalactone	202	2155	0.4	nd
64	2,6-dimethyl-5(E),7-octadien-2,3-diol	170	2156	0.5	nd
65	δ-decalactone	202	2220	1.5	nd
66	decanoic acid	172	2247	0.4	nd
67	γ-dodecalactone	198	2375	0.7	nd
68	4-oxo-α-ionol	208	>2600	0.4	nd
69	5,8-epoxy-6-megastimen-3,9-diol	226	>2600	1.5	nd

[1] Molecular weight determined by chemical ionization
[2] Retention index on DB-Wax column

Esters similar to those found in this study have been identified as important contributors of tropical fruit flavors belonging to Myrtacea family such as guava,[9,10] strawberry guava[11] and champa.[12]

4.2 Analysis of Glycosidically Bound Volatiles

The liberated aglycons profile from arazá glycosidic extract is showed in Table 2. In total, 21 substances were identified for the first time as bound aroma constituents in this fruit. The identified aglycons mainly consisted of aromatic structures (52.0%) and these were followed by aliphatic acids (21.0%), aliphatic alcohols (12.0%) and C_{13}-norisoprenoids (11.0%). As can be seen in table 2, the profile of bound volatiles is dominated by cinnamic acid, that represents nearly 32% of the total amount of aglycons.

Table 2. *Volatile Identified After Enzymatic Hydrolysis of a Glycosidic Extract of Arazá (Eugenia stipitata) Fruit Pulp.*

No	Compound	Ri^1	Amount[2]
1	2-methyl-1-propanol	1100	+
2	2-pentanol	1109	
3	3-methyl-1-butanol	1207	+
4	1-hexanol	1353	++
5	(Z)-3-hexen-1-ol	1379	++
6	acetic acid	1454	++
7	ethyl 3-hydroxybutanoate	1509	+
8	3-methylbutanoic acid	1671	++
9	hexanoic acid	1828	++
10	benzyl alcohol	1897	+
11	phenethyl alcohol	1902	+
12	Furaneol	2036	++

13	coniferyl alcohol	2120	+
14	4-vinylguaiacol	2190	+
15	benzoic acid	2418	+++
16	4-oxo-α-ionol	>2600	+
17	3-hydroxy-β-ionone	>2600	+
18	3-hydroxy-5,6-epoxy-β-ionone	>2600	++
19	3-hydroxy-5,6-epoxy-β-ionol	>2600	+
20	cinnamic acid	>2600	++++
21	vomifoliol	>2600	++

[1] Retention index on DB-Wax column
[2] Range of concentrations μg/Kg pulp: +<500; ++ = 500-3000; +++ = 3000-5000; ++++> 5000 μg/Kg pulp.

Glycoconjugates of cinnamic acid have been identified as natural precursors of cinnamic acid derived volatiles in tropical fruits such as *Physalis peruviana* and *Psidium guajava*[13].

Among bound volatiles found in arazá, 2-pentanol, 1-hexanol, *cis*-3-hexenol, ethyl 3-hydroxybutanoate, Furaneol, hexanoic acid, phenethyl alcohol and 3-oxo-α-ionol were also identified as free aroma constituents of the fruit.

Acknowledgements

Financial support by BID-COLCIENCIAS, IPICS, Uppsala University is greatly appreciated. We are particularly thankful to María Paola Castaño for the MDGC analysis and to Tatiana Osorio and Victor Armando Rincón (Lucta Grancolombiana) for their skilful support in the capillary GC-sniffing analysis. A. Fajardo thanks Universidad de la Amazonia for financial support to pursue his doctoral studies.

References

1. M. S. Hernández, J. A. Barrera, D. Paez, E. Oviedo, H. Rubio. Aspectos biológicos y conservación de frutas promisorias de la Amazonia colombiana. Instituto Amazónico de Investigaciones Científicas-SINCHI. Universidad de la Amazonia. Ed. Produmedios. Bogotá D.C. 2004. pp 41-58.
2. H. Rogez, R. Buxant, E. Mignolet, J. N. S. Souza, E. M. Silva, I.Larondelle. *Eur. Food Res. Technol.* 2004, **218**, 380.
3. M. R. B. Franco, T. Shibamoto. *J. Agric. Food Chem.* 2000, **48**, 1263.
4. R. Azodanlou, C. Darbellay, J. L. Luisier, J. C. Villettaz, and R. Amado, *Z. Lebensm Unters Forsch B,* 1999, **208**, 254.
5. F. Drawert, A. Rapp. *Chromatographia.* 1968, **1**, 446.
6. Y. Gunata, C. Bayonove, R. Baumes, R. Cordonier. *J. Chromatogr.* 1985, **331**, 83.
7. L. Morales, C. Duque, E. Bautista. *J. High Resol. Chromatogr.* 2000, **23**, 379.
8. M. Suárez, *Doctoral Thesis,* 1992, Universidad Nacional de Colombia. Bogotá.
9. J. A. Pino, A. Ortega, A. Rosado, *J. Essent. Oil Res.* 1999, **11**, 623.
10. J. A. Pino, R. Marbot, C. Vásquez. *J. Agric. Food Chem.* 2002, **50**, 6023.
11. J. A. Pino, R. Marbot, C. Vásquez. *J. Agric. Food Chem.* 2001, **49**, 5883.

12. C. Osorio. Personal communication. 2004.
13. S. Latza, D. Gansser, R. G. Berger. *Phytochem.*. 1996, **188**, 122.

Formation of Flavors

GENERATION OF POTENTIALLY NEW FLAVORING STRUCTURES FROM THIAMINE BY A NEW COMBINATORIAL CHEMISTRY PROGRAM

René M. Barone[1], Michel C. Chanon[1], Gaston A. Vernin[2] and Cyril Párkányi[2]

[1]Laboratoire AM3, UMR CNRS 6009, Faculté des Sciences de St Jérôme, Case 561, F-13397 Marseille Cédex 20, France
[2]Department of Chemistry and Biochemistry, Florida Atlantic University, 777 Glades Road, P.O. Box 3091, Boca Raton, Fl. 33431-0991. USA

1 ABSTRACT

This study is devoted to the theoretical investigation of the thermal degradation of thiamine. Degradation of this important vitamin leads to its loss during cooking, processing, and storage of foods. The first part of this contribution describes the program which consists of three modules: (i) the drawing of starting materials, (ii) the representation of reactions, and (iii) the combinatorial chemistry program which generates the products which could be theoretically formed. The final file contains 3166 structures, 1113 of which are linear and 2053 cyclic, respectively. In the second part of the study, some of these structures are reported. They have been divided in five- and six-membered rings, each including alicyclic and heterocyclic compounds. Among alicyclic compounds, cyclopentanone and cyclohexanone derivatives were found. The heterocycles include furans, thiophenes, 1,2-dithioanes, thiazoles, pyrans, 1,3-dioxanes, 1,2-dithianes, pyrimidines, and their derivatives. Selected references on the synthesis of the basic skeletons are also presented.

2 INTRODUCTION

This study represents a continuation of our research on the computer-aided organic synthesis as applied to studies of formation of aroma compounds in the Maillard reaction and model systems [1-6] and, in this case, investigates the theoretical thermal degradation of thiamine (vitamin B_1).

Thermal destruction and loss of thiamine during cooking, processing, and storage of foods (particularly meats) has been known for a long time (1945). Nevertheless, the first paper on the reaction products of thiamine degradation was reported only in 1969 by Arnold et al.[7] They found 2-methylfuran, 2-methylthiophene, and their 4,5-dihydro derivatives in heated systems of thiamine in phosphate buffer (pH 6.7). Three years later Dwivedi and Arnold[8] identified 4-methyl-5-(β-hydroxyethyl)thiazole as a major degradation product in sealed tubes held in a boiling water bath for 30 min., in tubes open to the air on a boiling water bath for 30 min., and in autoclaved sealed tubes heated to 121 °C for 30 min.

Güntert et al.[9,10] heated aqueous solutions of thiamine hydrochloride at different erated, respectively (Figure 1). concentrations and at different pH values in an autoclave for various times. They identified 96 products by GC/MS, mostly heterocyclic compounds (furans, thiophenes, thiazoles, 1,3-dithiolanes, 1,3,5-dithiazines) and some aliphatic precursors such as α-diketones, α- and β-ketonethiols, 3-mercaptopropanal, formaldehyde, acetaldehyde, their corresponding acids, ammonia and hydrogen sulfide.

Some of the products identified are identical to those obtained from the reaction of acetoin, aldehydes and ammonium sulfide in model systems related to Maillard reaction.[11,12] They possess a pleasant aroma of cooked and roasted meat.

The development of microcomputers allowed us to develop a new combinatorial chemistry program. This program was then applied to this important reaction generating flavours from known precursors. Several hundred compounds were postulated and the results are reported here.

The use of computers in organic synthesis has been reviewed by Barone and Chanon.[13,14] More recently, a different approach, iterated reaction graphs (IRG) that simulates complex chemical reaction systems, such as the Maillard reaction, was described by Patel et al.[15] Our previous results led us to apply our computer program to the degradation products of thiamine from known precursors. Several hundred suggested compounds were found and some of them are reported here, in order to propose new aroma compounds.

3 PROGRAM

The computer program was written in Visual Basic, using a PC, Windows, Pentium III (750 MHz) and a 800x600 screen. It resembles a reaction vessel with a soup of molecules (SM) inside. When the program starts, it tries to transform each structure of the SM using known reactions. For example, from structure **1**, structures such as **2** and **3** (by nucleophilic addition), **4 – 8** (by nucleophilic substitution), and **9** and **10** (by elimination) are generated, respectively (Figure 1).

The program consists of three modules: (i) drawing of starting materials: the user draws the different molecules and saves them in a file, (ii) drawing of reactions which can take place during the process of generation, and (iii) the combinatorial chemistry program which generates the products that can be theoretically formed. The aim of this program is not to predict the most probable products, but to generate as many products as possible in order to identify and/or suggest the presence of new flavouring compounds in the mixtures resulting from the thermal degradation of thiamine.

It is possible to have several files of reactions. The user can manage them as he wants. The user can save all the reactions in one file, or can save each reaction in separate files: for example, a file for nucleophilic substitution, another for elimination, etc.

Formation of Flavors

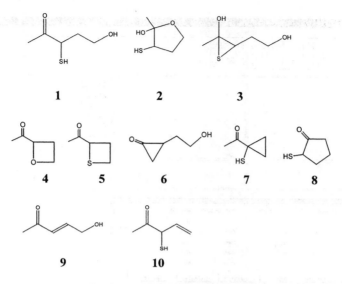

Figure 1 *Examples of structures generated from structure 1 by nucleophilic addition, nucleophilic substitution, and elimination, respectively.*

A view of the screen during the input of an elimination reaction is shown next (Figure 2).

The window is divided into four parts: The two upper left parts concern the description of the reaction. For an elimination reaction (HO–C–CH → C=C), in the upper left part the user draws the substructure to find in a structure (here HO–C–CH for elimination of H_2O) and in the right part the product of the reaction (C=C) (carbons are represented by dots).

For each reaction it is possible to code tests to discard chemically unsound solutions. For example, if phenol is submitted to the elimination reaction, the program should find the corresponding substructure (HO–C–CH) and should propose a solution for Scheme 1. The window is divided into four parts: The two upper left parts concern the description of the reaction. For an elimination reaction (OH-C-CH -> C=C), in the upper left part the user draws the substructure to find in a structure (here OH-C-CH for elimination of H_2O) and in the right part the product of the reaction (C=C) (carbons are represented by dots).

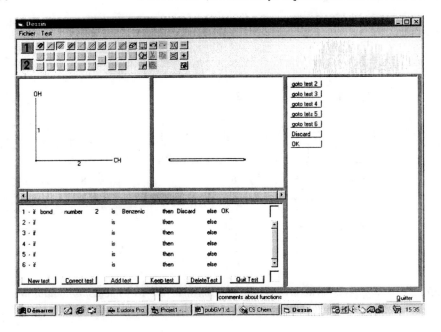

Figure 2. *Input of an elimination reaction.*

For each reaction it is possible to code tests to discard chemically unsound solutions. For example, if phenol is submitted to the elimination reaction, the program should find the corresponding substructure (HO-C-CH) and should propose a solution for Scheme 1:

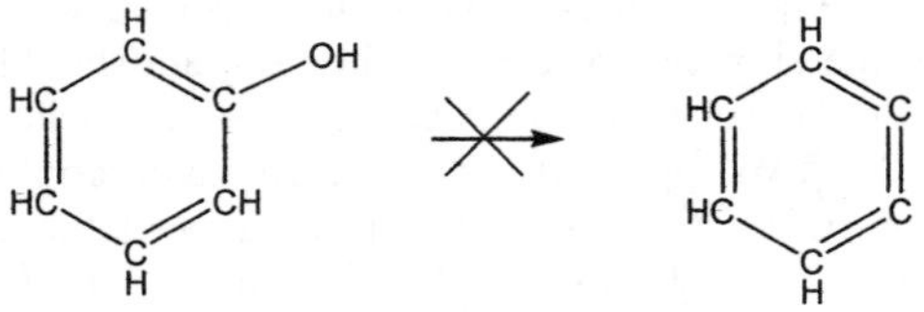

Scheme 1. *Example of a solution discarded by the program.*

The test displayed in Fig. 2 (discard the solution if the bond is in a benzene ring) allows the program to reject such a solution. The tests have the following form:

IF <entity> IS <function> THEN <action1> ELSE <action2>

Formation of Flavors

<Entity> is an atom or a bond in the structure; <function> is a chemical element, a substructure: depending on whether it is present or not action1 or action2 is performed. Action1 and Action2 are: go to another test, discard the solution, or the solution is OK.

In order to reduce the number of solutions several constraints can be applied by another subroutine: it is possible to discard solutions which contain rings of size 3, 4, 7, 8 and > 8; bridged rings; bridgehead double bond, linear or cyclic structures. The user can select the tests that he wants before running the program.

The progress of the program is as follows:

Selection of a reaction file, then selection of compounds files: it is possible to select one file or two. If the user selects one file, the flowchart of the program can be summarized by the following instructions:

FOR I = 1 TO NBMOL   (NBMOL = number of structures in the file)
    load structure I
FOR J= 1 to NBRXN   (NBRXN = number of reactions in the file)
    load reaction J
    search if reaction J is possible on structure I
    if yes the program generates the corresponding products
        (storage option)
NEXT J
NEXT I

This option is a kind of intramolecular mode.
If the user selects two files of structures, the flowchart is the following:

FOR I = 1 TO NBMOL1   (NBMOL1 = number of structures in file 1)
    load structure I
FOR J = I to NBMOL2   (NBMOL2 = number of structures in file 2)
    load structure J
FOR K= 1 to NBRXN   (NBRXN = number of reactions in the reaction file)
search if reaction K is possible between structures I and J,
    if yes the program generates the corresponding products
        (storage option)
NEXT K
NEXT J
NEXT I

In the case when there is a combination of the two compounds, it is an intermolecular option. The storage option in the flowchart indicates that the program can work automatically (all generated structures are saved) or interactively (when a solution is displayed the user can save the proposed structures or delete them). A view of the screen during the execution of the program is shown next (Figure 3):

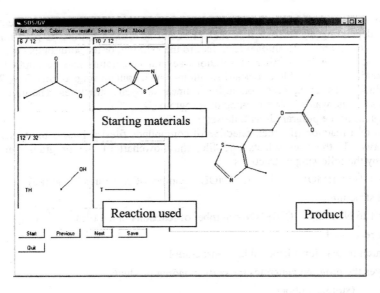

Figure 3. *View of the screen during the generation of products.*

The program, called GRAAL (**G**énérateu**R** d'**A**rômes **AL**imentaires) is written in Visual Basic and runs on Windows compatible PC.

4 DATA FOR THE PROGRAM

The results of the theoretical degradation of thiamine, which was experimentally studied by Güntert *et al.*[9,10], are described in Scheme 2.

The structures shown in Scheme 2 were coded and saved in the file F1.

In the reaction files the basic mechanisms, i.e. nucleophilic substitution, nucleophilic addition, elimination, reduction, and oxidation, were also coded.

The program was executed several times. In the first run each reaction was applied to each structure of file F1 (intramolecular option) and the program generated another file, called F2, which contained 29 compounds. In the entire study we discarded structures containing rings of size <5 and size >6, and/or bridges. For example, from structure **1** (Figure 1), from nucleophilic addition and nucleophilic substitution only structures **2** and **8** were generated; structures **3 – 7** were discarded. In the interactive mode, the user saves (or not) the structures which are presented, one by one.

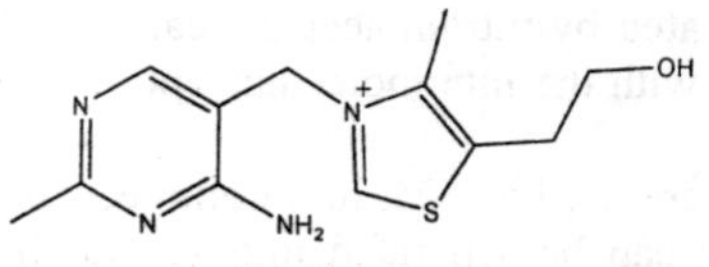

Formation of Flavors

Thiamine

H₂S, NH₃, HCOOH, CH₃COOH, CH₂O, CH₃CHO, HSCH₂CHO, HSCH₂CH₂CH₂OH

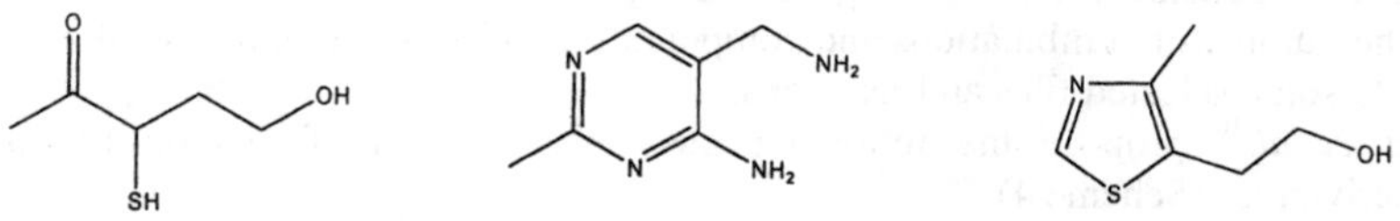

Scheme 2. *Basic structures generated by the thermal degradation of thiamine that have been taken into account in the SM*

When the program finishes, it calculates the mass of each compound and ranks them in increasing order of mass. The different combinations tried in this study are depicted in Scheme 3.

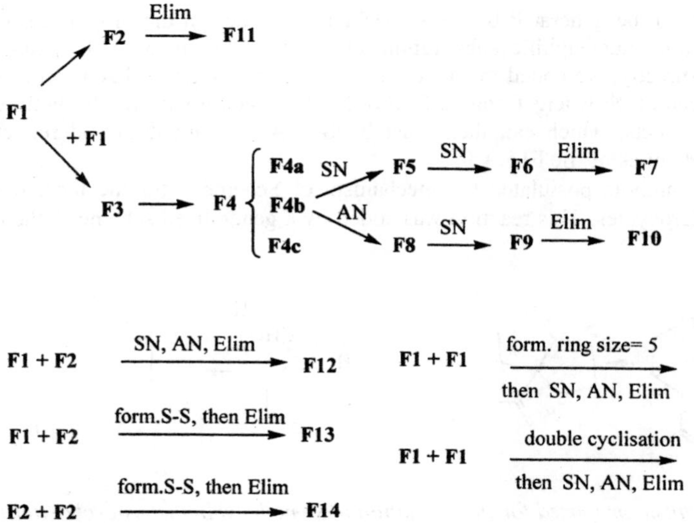

```
              Elim
        F2 ─────→ F11
       ↗
  F1
       ↘  + F1              SN        Elim
              ↘     F4a ─SN→ F5 ─────→ F6 ─────→ F7
         F3 → F4 { F4b ↘AN
                     F4c ──→ F8 ─SN→ F9 ─Elim→ F10
```

F1 + F2 —SN, AN, Elim→ F12	F1 + F1 —form. ring size=5, then SN, AN, Elim→ F15
F1 + F2 —form.S-S, then Elim→ F13	F1 + F1 —double cyclisation, then SN, AN, Elim→ F16
F2 + F2 —form.S-S, then Elim→ F14	

Scheme 3. *The different combinations for this work. F1, F2, Fi ... are the names of the files generated. Elim, SN, indicates the reactions used. When all the reactions have been tried there is no indication above the arrow (example: F1 -> F2). File F3 was generated by combining F1 with itself.*

Most of the files were generated by intramolecular reactions. File F3 was generated by combining file F1 with itself, with the intermolecular option. In this case the program generated 283 structures.

When structures of file F3 were used as SM, in intramolecular mode, 1509 structures were generated. These structures can be separated into several files: linear compounds (File F4a: 836 compounds), compounds with one ring (size = 5 or 6) (file F4b: 634 structures) and 39 structures with two rings (File F4c).

Since the number of combinations and compounds could be quite immense, the study was limited to some selected files and reactions.

Güntert et al.[10] propose the following mechanism for the formation of some thiophene derivatives (Scheme 4):

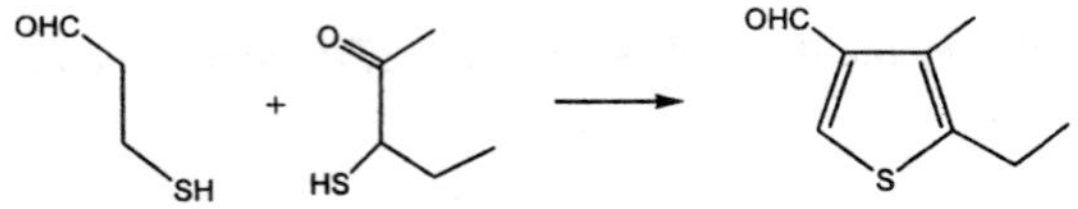

Scheme 4. *Formation of 2-ethyl-3-methyl-4-carboxaldehyde thiophene.*

This product can be generated by basic mechanisms coded in the program, such as nucleophilic addition, nucleophilic substitution, elimination, but in order to obtain the desired products directly, we coded this reaction in one step in a generalized form, using superatom T instead of S, where T can be O, S or N. We tested it with file F1 with itself, generating six products, which can then react by SN, AN, elimination and reduction, leading to new structures of file F15.

Similarly the authors postulated the mechanism of Scheme 5 for the formation of some bicyclic heterocycles. This reaction was coded by a generalized scheme (Scheme 5, right).

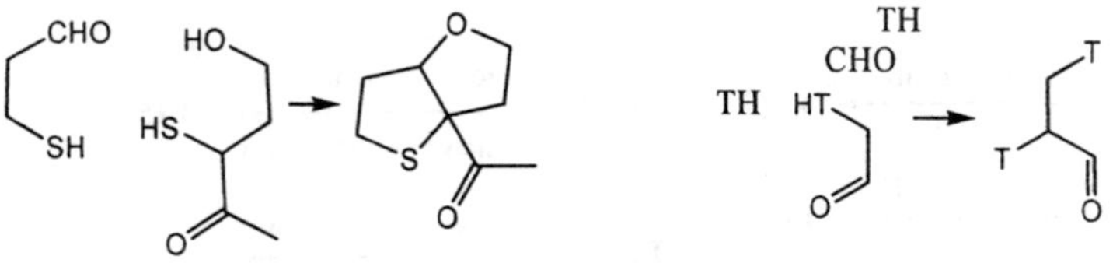

Scheme 5. *Mechanism proposed for the formation of some bicyclic heterocycles (left) and the general scheme coded in the program for this reaction (right). The symbol T means an heteroatom: O, N or S.*

This reaction was tested again with file F1, leading to file F16.

Since the same product can be found in the different files which have been generated, an option was included in the program to merge two files, which, simultaneously, eliminates identical structures. So it was possible to merge all the files generated, step by step, into one file which does not contain identical structures. The different files generated have been merged into one final file containing 3166 structures.

Another option allows us to separate linear structures and cyclic structures. The 3166 structures have been divided into two files: the first one contains 1113 linear structures and the second one 2053 cyclic structures. This file can also be sorted by the number of rings: there are 939 structures with one ring, 936 with two rings, and 178 with three rings. Of course several structures are only intermediates, such as those in Scheme 6.

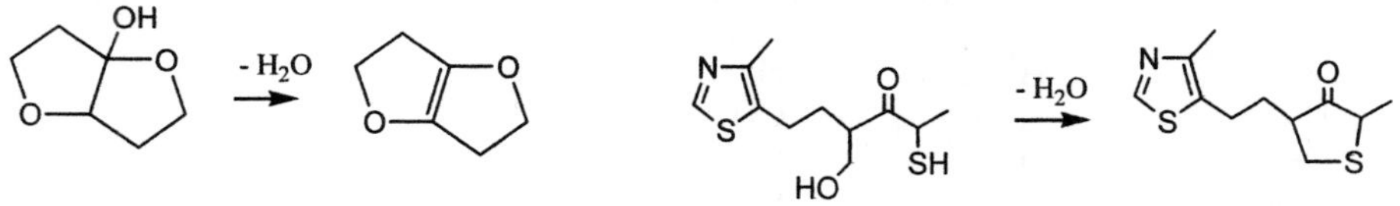

Scheme 6. *The files can contain structures which are intermediates.*

The last option allows the user to print the results.

5 RESULTS AND DISCUSSION

5.1 Five-membered rings

Alicyclic compounds with a cyclopentane ring are reported in Scheme 7 and Tables 1 and 2.

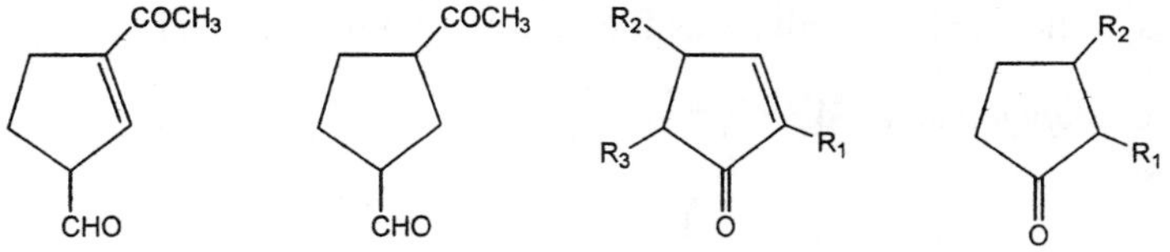

1 (MW : 138)   2 (MW : 140)   3 (Table 1)   4 (Table 2)

Scheme 7. *Alicyclic compounds (five-membered rings)*

Table 1. Cyclopent-2-en-1-ones **3**

R_1	R_2	R_3	MW
H	H	H	82
H	H	CHO	110
CHO	H	H	110
$COCH_3$	H	H	124
H	H	$COCH_3$	124
CH_3	CHO	CH_3	138
$(CH_2)_2CHO$	H	H	138
H	H	$(CH_2)_2CHO$	138
$SCH_2CH=CH_2$	H	H	156
H	$S(CH_2)_2CHO$	H	170

None of 2-cyclopentenones **3** reported in Table 1 have been found in food aromas. However, mass spectra of the 2-alkyl derivatives (**3**; $R_2 = R_3 = H$; $R_1 = CH_3$, C_2H_5, C_3H_7, C_4H_9, C_5H_{11} have been described by Buchbauer et al.[16] and Jennings and Shibamoto.[17] The butyl derivative was found in a santolina essential oil (R_1 = Bu).[18]

Table 2. Cyclopentanones **4***

R_1	R_2	MW	R_1	R_2	MW
H	H	84	OC_3H_7	H	142
CHO	H	112	$SCH=CH_2$	H	142
$(=CH_2)$	$(=O)$	124	$S\ C_2H_5$	H	144
$COCH_3$	H	126	SCHO	H	144
CH_2CHO	H	126	S-SH	H	148
SCH_3	H	130	$SCH_2CH=CH_2$	H	156
$CH=CH_2$	CHO	138	SC_3H_7	H	158
C_2H_5	CHO	140	$SCOCH_3$	H	158
$(CH_2)_2CHO$	H	140	$S(CH_2)_2CHO$	H	172
$OCH_2CH=CH_2$	H	140	H	$S(CH_2)_2CHO$	172

Formation of Flavors

* Five additional compounds have also been found (4a-4e) :

4a (MW=110) 4b (MW=124) 4c (MW=126) 4d (MW=126) 4e (MW=140)

Among the cyclopentanones **4** (Table 2), the following have been described in the literature: 2-ethyl-3-carboxaldehyde **4**; R_1 = C_2H_5, R_2 = CHO (Otera et al.[19]), 2-methylthio **4**; R_1 = SCH_3, R_2 = H (Seebach et al.[20,21]; Iriuchijima et al.[22]; Gregoire et al.[23]), 2-ethylthio **4**; R_1 = SC_2H_5, R_2 = H (Wladislaw et al.[24]) and a dimethyl derivative by thermal degradation of fruc-Valine Amadori intermediate (Vernin et al.[25]).

The parent compound and the 2,3-dihydro derivative were found in a curuba fruit (Froehlich et al.[26]), in roasted coffee and model system (Baltes et al.[27,28]), respectively. 2-Methylene-1,3-cyclopentanedione **4a** (Chantarasini et al.[29]) and 2- (**4b, 4c**) and 2,4-disubstituted-1,3-cyclopentanediones **4d** (Cant and Vandewalle[30]; Demir and Enders[31]) have also been described (See Table 2).

The formulas of heterocyclic compounds are given in Scheme 8.

Among furans **5a** reported in Table 3, 2-methylfuran has been identified in coffee aroma by Baltes and Bochman[27,28] and 2-methyl-3-furanylcyclopentanone (**5b**: MW = 164) by Royen et al.[32]

Among 4,5-dihydrofurans (see Table 4), the 2-methyl, 2-methyl-3-formyl, and 3-acetyl derivatives have been found after photolysis of thiamine by Van Dort et al.[33]

From the *p*-tolylsulfonyl hydrazones of the readily available 4,5-dihydro-3(2*H*)-furanones, Gelin et al.[34] obtained 4,5-dihydrofurans (**6**, R_1 = CH_3, R_2 = H; Table 4). The compound was found among the products resulting from autoclave degradation of thiamine in aqueous solution with different concentrations and different pH values for various times (Güntert et al.[10]).

2-Methyl-2,5-dihydrofuran (**7**, R = H; Table 5) was found by Gianturco et al.[35] in coffee aroma. Its synthesis was reported by isomerization of α-allenic alcohols in acidic media by Gelin et al.[34]

Among tetrahydrofurans **8** listed in Table 6, the natural compounds are the 2-methyl-3-hydroxy- and the (*Z*)- and (*E*)-3-mercapto-2-methyl derivatives identified in beef aroma by Van den Ouweland and Peer.[36,37]

3(2*H*)-Dihydrofuranones **9** (Table 7) are widely distributed in food flavors. Synthesis of these compounds from 2-(dimethylamino)-4-methylene-1,3-dioxolanes was first described by Carpenter et al.[38] They are easily converted to muscarine derivatives values by perfumers due to their pleasant and varied odors (Noyori et al.[39,40]).

186 Food Flavor and Chemistry: Explorations into the 21st Century

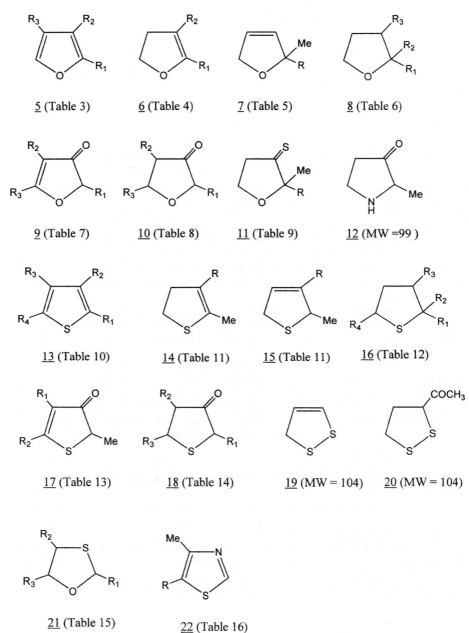

Scheme 8. *Heterocyclic compounds (five-membered rings)*

Table 3. Furans 5a

R_1	R_2	R_3	MW
CH_3	H	H	82
$CH=CH_2$	CH_3	$COCH=CH_2$	162
CH_3	$CH=CH_2$	$COCH=CH_2$	162
$CH=CH_2$	CH_3	COC_2H_5	164
C_2H_5	CH_3	$COCH=CH_2$	164
C_2H_5	CH_3	COC_2H_5	166

5b   5c

Table 4. 4,5-Dihydrofurans 6

R_1	R_2	MW
CH_3	H	84
CH_3	CHO	112
H	$COCH_3$	112
CH_3	$(CH_2)_2CHO$	126
CH_3	$COCH_3$	126
$CH=CH2$	$COCH_3$	138
CH_3	SCHO	144
CH_3	$S(C=O)CH_3$	156
(Table 4 continued)		
$CH_2CH=S$	$COCH_3$	170
CH_3	$S(CH_2)_2CHO$	172

Table 5. 2,5-Dihydrofurans 7

R	MW
H	84
OCH_3	114
$(CH_2)_2CHO$	126
OC_2H_5	128
$SCH_2CH=CH_2$	156
$CH(C=S)CHO$	170
$S(CH_2)_2CHO$	172

Table 6. Tetrahydrofurans $\underline{8}$

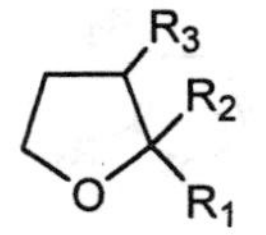

$\underline{8}$

R_1	R_2	R_3	MW
H	H	$COCH_3$	114
CH_3	OCH_3	H	116
(=CH_2)		$COCH_3$	126
CH_3	CH_2CHO	H	128
CH_3	H	$COCH_3$	128
CH_3	H	CH_3CHO	128
CH_3	OC_2H_5	H	130
CH_3	$COCH=CH_2$	H	140
$CH=CH_2$	$COCH_3$	H	140
C_2H_5	$COCH_3$	H	142
CH_3	COC_2H_5	H	142
CH_3	H	SCHO	146
CH_3	$SCH_2CH=CH_2$	H	158
CH_3	H	$SCOCH_3$	160
$CH_2CH=S$	H	$COCH_3$	172
CH_3	$S(CH_2)_2CHO$	H	174

Only two compounds (MW = 198) : R_1 : $-CH=CH_2$, R_2 : $COCH_3$ and R_1 : CH_3, R_2 : $=CH_2$ have an acetyl group in the 5 position.

Table 7. 3(2H)furanones $\underline{9}$

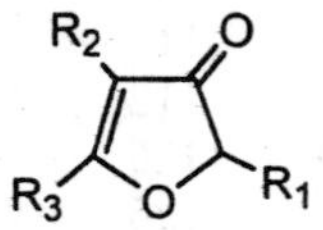

$\underline{9}$

R_1	R_2	R_3	MW
CH_3	CH_3	H	112
CH_3	C_2H_5	H	126
CH_3	CH_3	CH_3	126

Table 8. *4,5-Dihydro-3(2H)-furanones* 10

R_1	R_2	R_3	R_4	MW
H	CH_3	H	H	100
H	$CH=CH_2$	H	H	112
H	CH_3	CH_3	H	114
H	C_2H_5	H	H	114
CH_3	CH_3	H	H	114
CH_3	$CH=CH_2$	H	H	126
H	CH_3	CH_3	CH_3	128
H	CH_3	C_2H_5	H	128
H	C_2H_5	H	CH_3	128
CH_3	C_2H_5	H	H	128
H	CH_3	CHO	H	128
CH_3	CHO	H	H	128
CH_3	$CH_2CH=CH_2$	H	H	140
H	CH_3	$CH_2CH=CH_2$	H	140
CH_3	$COCH_3$	H	H	142
H	CH_3	$COCH_3$	H	142
H	CH_3	$(CH_2)_2CHO$	H	156
CH_3	$(CH_2)_2CHO$	H	H	156

Among dihydro-3(2*H*)-furanones **10** (see Table 8) the 2-methyl derivative (**10**; R_1 = CH_3, R_2 = R_3 = R_4 = H) has been found in guava aroma (Idstein *et al.*[41]), in green and ripened pineapple (Umano *et al.*[42]), in roasted almonds (Takei[43]), and in several model systems including serine, threonine and sucrose (Baltes *et al.*[27,28]), hydrolyzed gelatin and glucose (Roedel and Habisch[44]), and valine-glucose (Vernin *et al.*[45]). The odor is sweet, roasted and solvent-like with a brown, rummy and nut-like note while its taste is crusty, caramel, nutty and astringent with light creamy, almond note, bread, buttery top-note.

4,5-Dihydrofuran-3(2*H*)-thiones **11** (Tables 9) are only rarely found in food flavors.

Table 9. *4,5-Dihydro-3(2H)furanthiones* __11__

[Structure: 11a (4,5-dihydro-3(2H)furanthione with Me and R substituents) ⇌ 11b (3-mercaptofuran with SH and Me), R = H]

    __11a__   __11b__

R	MW
H	126
OCH2CH=CH2	172
OC$_3$H$_7$	174
SCH$_2$CH=CH$_2$	188
SC$_3$H$_7$	190

The only compounds of this class at aromas and model systems such as cysteine/reducing sugars are the following unsaturated 3-mercaptofurans: 2-methyl- (Tressl et al.[46]), 2,5-dimethyl- (Silvar et al.[47]) and 4- (or 5)-methyl- (Van den Ouweland and Peer[37]; Farmer and Mottram[48]). They can exist in tautomeric forms as their saturated derivatives.

[Structure: tautomeric equilibrium of substituted furanthione (R$_1$, R$_2$) with 3-mercaptofuran form, R$_1$ = H]

They possess beef and meat flavors similarly as their thiophene homologs.

Güntert et al.[9] also identified two thiadioxabicyclooctanes from the photolysis of thiamine :

[Structure: thiadioxabicyclooctane with R substituent, R = H, Me]

The only thiophene among those listed in Table 10 was the 2-ethyl-4-formyl-3-methyl derivative (__13__; R$_1$ = C$_2$H$_5$, R$_2$ = CH$_3$, R$_3$ = CHO) identified by Güntert et al.[9] from thermal degradation of thiamine.

Table 10. Thiophenes 13

R_1	R_2	R_3	R_4	MW
$CH=CH_2$	CH_3	CHO	H	152
CH_3	$CH=CH_2$	CHO	H	152
CH_3	OH	CH_3	$CH=CH_2$	154
CH_3	C_2H_5	CHO	H	154
C_2H_5	CH_3	CHO	H	154*
$CH=CH_2$	CH_3	$COCH=CH_2$	H	178
CH_3	$CH=CH_2$	$COCH=CH_2$	H	178
CH_3	C_2H_5	COC_2H_5	H	182
C_2H_5	CH_3	COC_2H_5	H	182
CH_3	OH	H	cyclopentanone	196
CH_3	OH	H	dihydrofuran-methyl	196
CH_3	OH	tetrahydrofuran-methyl	H	198
CH_3	OH	tetrahydrothiophene-methyl	H	214

Dihydrothiophenes **14** and **15** (Scheme 8) have not been found in food flavors (Table 11).

Table 11. *Some 2,3- and 2,5-dihydrothiophene derivatives*

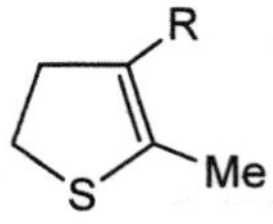

14	15
14a : R = CH_2CHO	**15a** : R = CH_2CHO
14b : R = $S(CH_2)_2CHO$	**15b** : R = $S(CH_2)_2CHO$

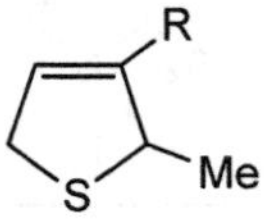

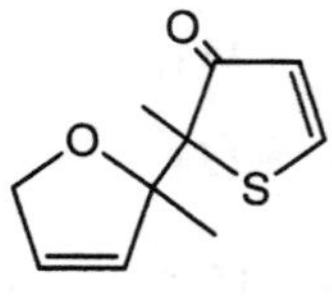

MW = 196 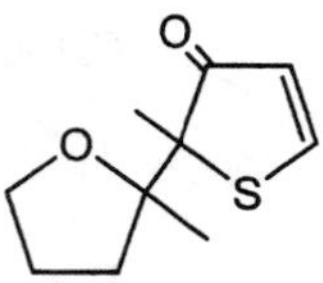   MW = 198

Tetrahydrothiophenes **16** are listed in Table 12 :

Table 12. *Tetrahydrothiophenes* **16**

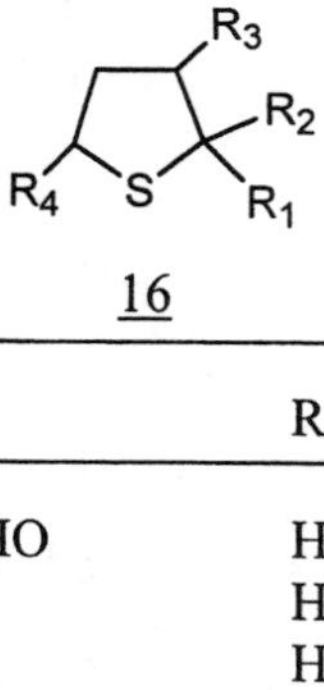

R_1	R_2	R_3	R_4	MW
CH_3	H	CH_2CHO	H	144
CH_3	$COCH=CH_2$	H	H	156
$CH=CH_2$	$COCH_3$	H	H	156
C_2H_5	$COCH_3$	H	H	158
CH_3	COC_2H_5	H	H	158
CH_3	$(CH_2)_2CHO$	H	H	172
CH_3	H	$S(CH_2)_2CHO$	H	190
CH_3	$COCH=CH_2$	$COCH_3$	H	198
$CH=CH_2$	$COCH_3$	$COCH_3$	H	198
$CH=CH_2$	$COCH_3$	H	$COCH_3$	198
CH_3	$COCH=CH_2$	H	$COCH_3$	198
CH_3	COC_2H_5	$COCH_3$	H	200
C_2H_5	$COCH_3$	$COCH_3$	H	200
CH_3	COC_2H_5	$COCH_3$	H	200

The following **16a**, **16b**, **16c** identified by Güntert et al.[9] have not been found by the program :

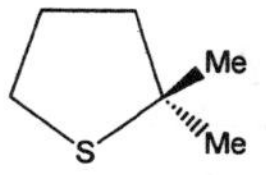

16a 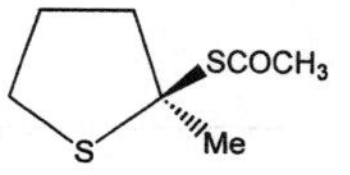   **16b** 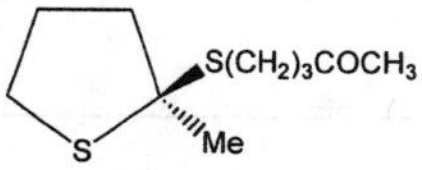   **16c**

Another group of aroma compounds are 3(2H)-thiophenones 17 (See Table 13).

Table 13. *3(2H)-Thiophenones* **17**

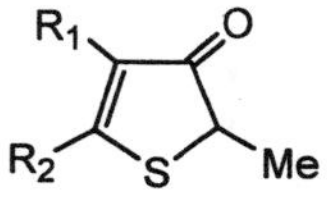

R_1	R_2	MW
H	H	114
CH_3	H	128
CH_3	CH_3	142
C_2H_5	H	142
CH_3	$CH_2CH=S$	186

Among 4,5-dihydro-3(2H)-thiophenones **18a** (Table 14), the parent compound (**18a**; $R_1 = R_2 = R_3 = R_4 = H$) was identified in coffee aroma for the first time by Stoll et al.[49,50]. The 2-methyl derivative (**18** ; $R_1 = CH_3$, $R_2 = R_3 = R_4 = H$) was also reported in coffee aroma and from thermal degradation of thiamine by Dwivedi and Arnold.[8] The 2,5-dimethyl derivative (**18**; $R_1 = R_4 = CH_3$, $R_2 = R_3 =H$) was found in cysteine/ribose/phospholipids model system by Farmer et al.[51] and results from the thermal degradation of thiamine (Güntert et al.[10]) (Table 14).

Table 14. *4,5-Dihydro-3(2H)-thiophenones* **18**

18

R_1	R_2	R_3	R_4	MW
CH_3	H	H	H	116
CH_3	H	CH_3	H	130
CH_3	CH_3	H	H	130
C_2H_5	H	H	H	130
CH_3	H	$CH=CH_2$	H	142
CH_3	$CH=CH_2$	H	H	142
CH_3	H	CHO	H	144
CH_3	H	C_2H_5	H	144
C_2H_5	H	CH_3	H	144
CH_3	H	CH_3	CH_3	144
CH_3	C_2H_5	H	H	144
$CH=CH_2$	H	$CH=CH_2$	H	154
CH_3	H	$CH_2CH=CH_2$	H	156
C_2H_5	H	H	$CH=CH_2$	156
$CH=CH_2$	H	H	C_2H_5	156
CH_3	H	CH_3	$CH=CH_2$	156
CH_3	$CH2CH=CH_2$	H	H	156
CH_3	C_3H_7	H	H	158
C_2H_5	H	H	C_2H_5	158
CH_3	H	CH_3	C_2H_5	158
CH_3	H	C_3H_7	H	158
CH_3	$COCH_3$	H	H	158
CH_3	H	H	CH_2CHO	158
CH_3	H	CHO	CH_3	158
CH_2CHO	H	H	CH_3	158
CH_3	CH_2CHO	H	H	158
CH_3	H	$COCH_3$	H	158
CH_3	H	CHO	$CH=CH_2$	170
CH_3	H	H	$OCH_2CH=CH_2$	172
CH_3	H	CHO	C_2H_5	172
CH_3	H	CH_2CH_2CHO	H	172
CH_3	CH_2CH_2CHO	H	H	172
CH_3	H	OC_3H_7	H	174
$CH=CH_2$	H	H	$CH_2CH=S$	186
CH_3	H	H	$SCH_2CH=CH_2$	188
CH_3	H	CH_3	$CH_2CH=S$	188
C_2H_5	H	$CH_2CH=S$	H	188
CH_3	H	H	SC_3H_7	190

1,2-Dithiolanes and 1,3-oxathiolanes are reported in Table 15.

Thiazoles **22** are widely distributed in food flavors. Among 4-methyl-5-substituted derivatives **22**, the 5-vinyl derivative (**22** ; R = CH_2=CH, Table 16) was found by Kinlin et al.[52] in roasted filberts and by Dwivedi and Arnold[8] from thermal degradation of thiamine. It possesses nutty and cocoa odor. The 5-ethyl derivative (**22** ; R = C_2H_5) was reported by Buttery et al.[53], Pittet and Hruza[54], and Tabacchi[55]. It was also found in a β-mercapto-acetaldehyde-ammonium sulfide model system by Hwang et al.[56].

Table 15. *Five-membered ring heterocyclic compounds containing two heteroatoms* 19-21

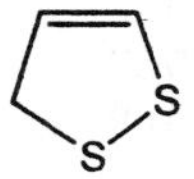

19 (MW = 104)
1,2-dithiolane

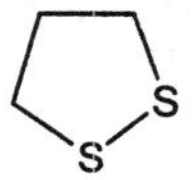

20a (MW = 106)

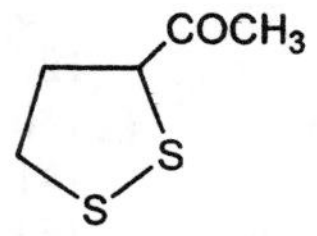

20b (MW = 148)

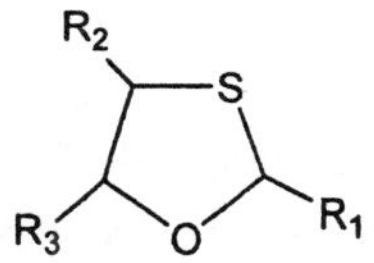

21
1,3-oxathiolanes

	R1	R2	R3	MW
(a)	H	CH=CH$_2$	CH$_3$	130
(b)	H	CH$_3$	CH=CH$_2$	130
(c)	CH$_3$	CH$_3$	CH=CH$_2$	144
(d)	CH$_3$	CH=CH$_2$	CH$_3$	144

Table 16. *Thiazoles* 22

22

R	MW	R	MW
CH=CH$_2$	125	(CH$_2$)$_2$OCH$_3$	157
C$_2$H$_5$	127	CH$_2$CH=S	157
CH$_2$CH=NH	140	(CH$_2$)$_3$CHO	169
CH$_2$CHO	141	(CH$_2$)$_2$OCH=CH$_2$	169
(CH$_2$)$_2$CH=CH$_2$	153	(CH$_2$)$_2$OC$_2$H$_5$	171
(CH$_2$)$_3$CH$_3$	155	(CH$_2$)$_2$OCHO	171

5.2 Six-membered rings

Among alicyclic six-membered rings (see Scheme 9) only 4-alkyl-1,3-cyclohexanediones have been reported by Vandewalle et al.[57]. Heterocyclic compounds are reported in Scheme 10.

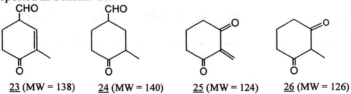

23 (MW = 138)   24 (MW = 140)   25 (MW = 124)   26 (MW = 126)

Scheme 9. *Six-membered ring alicyclic compounds*

Formation of Flavors

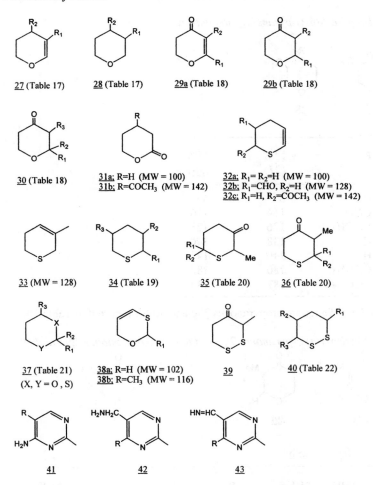

Scheme 10. *Six-membered ring heterocyclic compounds containing one or two different O and/or S atoms in their rings*

Tetrahydropyran (**28**; $R_1 = R_2 = H$, Scheme 10, Table 17) was earlier reported by Smakman and de Boer[58].

Table 17. *4,5-Dihydro-5(6H)-pyrans* 27 *and tetrahydropyrans* 28

R₁	R₂	MW (27)	MW (28)
H	H	84	86
H	CHO	114	116
H	COCH₃	126	128
CHCH=CH₂	H	138	140
COC₂H₅	H	140	142
COCH=CH₂	COCH₃	180	182
COC₂H₅	COCH₃	182	184

5,6 Dihydro-4(4*H*)- 29a tetrahydropyranones 29b and 30 are reported in Table 18.

Table 18. *5,6-Dihydro-4(4H)-pyranones* 29a *and tetrahydropyranones* 29b *and* 30

	29a	29b			30
R₁	MW	MW	R₁	R₂	MW
H	112	114	H	H	114
CH₃	126	128	H	CH₃	128
CH=CH₂	138	140	H	CH=CH₂	140
C₂H₅	140	142	H	C₂H₅	142
			CH₃	CH=CH₂	154
			CH₃	COCH=CH₂	182
			CH=CH₂	COCH₃	182
			CH₃	COC₂H₅	184
			C₂H₅	COCH₃	184

δ-Valerolactone (**31a**; R = H, Scheme 10) was found in curuba fruit (*Passiflora moltissima*) by Fröhlich et al.[26] Thiapyran **34** and thiapyranone **35**, **36** derivatives are reported in Tables 19 and 20, and 1,3-dioxanes **37a** and 1,3-oxathianes **37b** in Table 21.

Table 19. *2,3,4,5-Tetrahydro-(6H)-thiapyranes* **34**

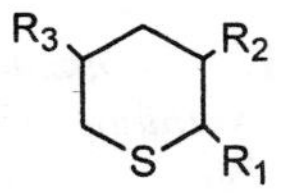

34

R_1	R_2	R_3	MW
H	H	H	102
H	CHO	H	130
$COCH_3$	H	H	144
H	$COCH=CH_2$	H	156
H	COC_2H_5	H	158
H	CHO	$COCH_3$	172
$COCH_3$	H	$COCH=CH_2$	198
$COCH_3$	H	COC_2H_5	200

Table 20. *2,4,5,6-Tetrahydro-3(3H)-thiapyranones* **35**

and *2,3,5,6-tetrahydro-4(4H)-thiapyranones* **36**

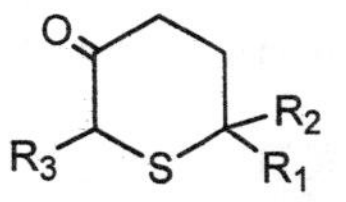

35 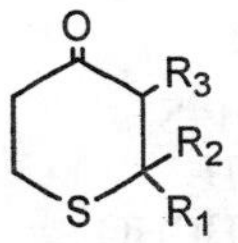   **36**

R_1	R_2	R_3	MW
H	H	H	116
H	CHO	H	144
$COCH_3$	H	H	158
H	$COCH=CH_2$	H	170
H	COC_2H_5	H	172
H	CHO	$COCH_3$	186
$COCH_3$	H	$COCH=CH_2$	216
$COCH_3$	H	COC_2H_5	218

Among 1,3-oxathianes (**37b**; X = S, Table 21) the 2-methyl derivative (R_1 = CH_3, R_2 = R_3 = H) was found in *Vitis vinifera* by Krammer et al.[59], and the 6-propyl and higher homologs in yellow passion fruit by Winter et al.[60].

Table 21. *1,3-Dioxanes* <u>37a</u> (X = Y = O) and *1,3-oxathianes* <u>37b</u> (X = O, Y = S)

R_1	R_2	R_3	MW 37a	37b
H	H	H	88	104
(=CH_2)		H	100	116
H	CH_3	H	102	118
H	OH	H	104	120
H	H	CH=CH_2	116	130
H	CH=CH_2	H	116	130
H	C_2H_5	H	118	132
H	H	C_2H_5	118	132
H	CH_3	CH=CH_2	128	144
H	H	$COCH_3$	130	146
H	CH_3	C_2H_5	130	146
CH=CH_2	CH=CH_2	H	140	156
CH_3	CH=CHCH_3	H	142	158
CH_3	C_3H_7	H	144	160
C_2H_5	C_2H_5	H	144	160
H	CH_3	$COCH_3$	144	160
H	CH_2CH=S	H	146	162
H	CH=CH_2	$COCH_3$	156	172
CH_3	CH=CHCHO	H	156	172
CH=CH_2	CH_2CHO	H	156	172
H	CH_2CH=CH_2	H	156	172
H	CH=CH$COCH_3$	H	156	172
CH_3	CH_2CH_2CHO	H	158	174
C_2H_5	CH_2CHO	H	158	174
H	$CH_2COC_2H_5$	H	158	174
H	(CH_2)$_2COCH_3$	H	158	174
H	C_2H_5	$COCH_3$	158	174
CH=CH_2	C(=S)CH_3	H	172	188
CH_3	C(=S)C_2H_5	H	174	190
C_2H_5	C(=S)CH_3	H	174	190

Formation of Flavors

The parent compound 1,2-dithiane (**40**; $R_1 = R_2 = H$, Table 22) was formed by pyrolysis of cysteine with xylose (Ledl and Severin[61]).

Table 22. *1,2-Dithianes* **40**

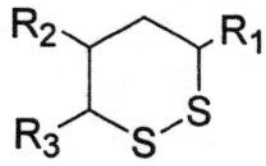

40

R_1	R_2	R_3	MW
CH_3	$COCH_3$	H	176
$CH=CH_2$	$COCH_3$	H	188
CH_3	$COCH=CH_2$	H	188
CH_3	COC_2H_5	H	190
CH_3	$COCH=CH_2$	$COCH_3$	230
$CH=CH_2$	$COCH_3$	$COCH_3$	230
CH_3	COC_2H_5	$COCH_3$	232
C_2H_5	$COCH_3$	$COCH_3$	232

Pyrimidines such as **41a, 41b, 41c**, have not been found in food flavors (See Scheme 11).

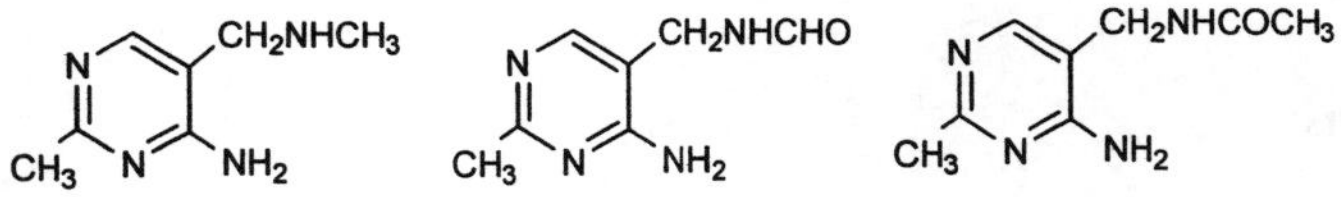

41a (MW = 152)   **41b** (MW = 166)   **41c** (MW = 180)

Scheme 11. *Some pyrimidine derivatives from 41 found by the program*

As it can be shown, very few compounds found by the program have been reported in food flavors and the Maillard reaction (or model systems). Examples of substituents are:

$(S)_n$ CH (CH = CH_2) CO CH_3 (n = 1,2)
S CH ($COCH_3$) CH_2 CHO
S CH (CH_3) CO CH_2 CHO
S S CH_2 CH = CH_2
S S CH (CH_3) CO C_2 H_5
S S CH (CH_2CHO) $COCH_3$
S S CH (CH_3) CO CH_2 CHO
O $(CH_2)_2$ COC (= S) CH_3
O $(CH_2)_2$ C (= S) CO CH_3
and
CH_2NHR (R=CHO, $COCH_3$, $(CH_2)_2CH_3$, $(CH_2)_2CHO$, $(CH_2)_2CHS$, $CH(C_2H_5)COCH_3$, etc.

NHCHO, $NHCOCH_3$, $NH(CH_2)_2CH_3$, $NHCH(CH_3)COC_2H_5$, etc. for 5 substituted-2-methyl-4-aminopyrimidines are unusual in the aroma of roasted products. Furthermore, β-carbonyl compounds can react as in the Maillard reaction and model systems to afford polyheterocyclic compounds.

A summary of the numerous papers devoted to the natural occurrence and synthesis of alicyclic and heterocyclic derivatives found by the program is given in Table 23.

6 CONCLUSION

The above results demonstrate that the use of a computer and appropriate programs to find all the products possibly formed from well known starting materials is of interest. However, the selectivity of the all possible reactions giving expected products is poor and has to be improved.

Acknowledgments

Thanks are due to Mrs R.M. Zamkotsian and G.M.F. Vernin for their assistance in this work.

Table 23. *Synthesis or Natural Occurrence of Alicyclic or Heterocyclic Derivatives Predicted by the Program*

Compounds	Origin and References
1. Alicyclic compounds	
3-Acetyl cyclopentanone-1-carboxaldehyde (Scheme 7, 2)	Synthesis (62, 63)
2-Cyclopentanone (Table 1, 3: $R_1 = R_2 = R_3$)	Roast coffee (27,28) Curuba fruit (26)
2-Cyclopentenone-2-carboxaldehyde (Table 1, 3: R_1 = CHO, $R_2 = R_3$ = H)	Synthesis (64)
2-Acetylcyclopent-2-enone (Table 1, 3, R_1 = COCH$_3$, $R_2 = R_3$ = H)	Synthesis (65)
Cyclopentanone (Table 2, 4, $R_1 = R_2$ = H)	Roast coffee (27,28) Curuba fruit (63)
Cyclopentanone-1-carboxaldehyde (Table 2, 4: R_1 = CHO, R_2 = H)	Synthesis (66-68)
2-Acetylcyclopentanone (Table 2, 4: R_1 = COCH$_3$, R_2 = H)	Synthesis (69-73)
2-Oxocyclopentyl acetaldehyde (Table 2, 4: R_1 = CH$_2$CHO, R_2 = H)	Synthesis (74) Synthesis (75)
2-Methylthiocyclopentanone (Table 2, 4: R_1 = SCH$_3$, R_2 = H)	Synthesis (23,76)
2-Ethyl-3-oxocyclopentane-1-carboxaldehyde (Table 2, 4: R_1 = C$_2$H$_5$, R_2 = CHO)	Synthesis (29)
3-(2-Oxocyclopentyl) propionaldehyde (Table 2, 4: R_1 = CH$_2$CHO, R_2 = H)	Synthesis (77,78)
2-Allyloxycyclopentanone (Table 2, 4: R_1 = OCH$_2$CH = CH$_2$, R_2 = H)	Synthesis (79)
2-Ethylthio-cyclopentanone (Table 2, 4: R_1 = SC$_2$H$_5$, R_2 = H)	Synthesis (24)
2-Propylthio-cyclopentanone (Table 2; 4: R_1 = SC$_3$H$_7$, R_2 = H)	Synthesis (80,81)
2-Methylene-1,3-cyclopentadione (Table 2, 4a)	Synthesis (29, 82)
4-Ethyl-1,3-cyclopentanedione (Table 2, 4c)	Synthesis (MS) (30,31)
2,4-Dimethyl-1,3-cyclopentanedione (bench 2, 4e)	Synthesis (MS) (30,31)

(Table 23 continued 1)

2. Furans and derivatives

2-Methylfuran	Synthesis (83)
(Table 3, <u>5a</u>: R_1 = CH_3, R_2 = R_3 = H)	Roast coffee (27,28)
2-(2-Methyl-1-cyclopentyl)furan	Synthesis (32)
(Table 3, <u>5b</u>)	
5-Methyl-2,3-dihydrofuran	Synthesis (83)
(Table 4, <u>6</u> : R_1 = CH_3, R_2 = H)	Thiamine degradation (10)
5-Methyl-4-formyl-2,3-dihydrofuran	Photolyzed thiamine (33)
(Table 4, <u>6</u> : R_1 = CH_3, R_2 = CHO)	
4-Acetyl-2,3-dihydrofuran	Photolyzed thiamine (33)
(Table 4, <u>6</u> :R_1 = H, R_2 = $COCH_3$)	
2-(4,5)-dihydro-2-methyl-3-furylthio)-2-methyl-tetrahydrofuran	Synthesis (84)
(Table 4, <u>6</u> : R_1 = 3-(2-methyl-2-tetrahydrofurylthio), R_2 = H)	
2,5-Dihydro-2-methylfuran	Synthesis (35)
(Table 5, <u>7</u> : R = H)	Synthesis (34)
2-Methoxy-2-methyltetrahydrofuran	Synthesis (83)
(Table 6 , <u>8</u> : R_1 = CH_3, R_2 = OCH_3)	
2-Ethoxy-2-methyltetrahydrofuran	Synthesis (MS) (85)
(Table 6, <u>8</u> : R_1 = CH_3, R_2 = OC_2H_5, R_3 = H)	
2-Methyl-3-(thioacetic acid) tetrahydrofuran	Synthesis (86)
(Table 6 , <u>8</u> : R_1 = CH_3, R_2 = H, R_3 = $SCOCH_3$)	
2,4-Dimethyl-3(2H)-furanone	Synthesis (38,40)
(Table 7, <u>9</u>: R_1 = R_2 = CH_3)	
2-Methyl-4,5-dihydro-3(2H)-furanone	Guava (41)
(Table 8, <u>10</u>: R_1 = CH_3, R_2 = R_3 = R_4 = H)	Model systems (27,28,44,45)
	Pineapple (42)
2-Ethyl-4,5-dihydro-3(2H)-furanone	Model systems (27,28)
(Table 8, <u>10</u>: R_1 = C_2H_5, R_2 = R_3 = R_4 = H)	
2,2-Dimethyl-4,5-dihydro-3(2H)-furanone	Synthesis (87)
(Table 8, <u>10</u>: R_1 = R_2 = CH_3, R_3 = R_4 = H)	Synthesis (88)
2-Ethyl-2-methyl-4,5-dihydro-3(2H)-furanone	Synthesis (89)
(Table 8, <u>10</u>: R_1 = CH_3, R_2 = C_2H_5, R_3 = R_4 = H)	Synthesis (90)
2,5-Dimethyl-2-ethyl-4,5-dihydro-3(2H)-furanone	Synthesis (89)
(Table 8, <u>10</u>: R_1 = R_4 = CH_3, R_2 = C_2H_5, R_3 =H)	

Formation of Flavors

(Table 23 continued 2)

---

3. Thiophenes and derivatives

2-Methyl-3-(2*H*)-thiophenone (Table 13, <u>17</u>: $R_1 = R_2 = H$)	Synthesis (91)
2-Methyl-4,5-dihydro-3(2*H*)-thiophenone (Table 14, <u>18</u> : $R_1 = CH_3$, $R_2 = R_3 = R_4 = H$)	Synthesis (92,93) Meat flavor (9, 94) Roasted sesame seeds (9 (95) Model systems(51,96,97) Stereoisomers (98,99) White wine (100)
2,2-Dimethyl-4,5-dihydro-3(2*H*)-thiophenone (Table 14, <u>18</u> : $R_1 = CH_3$, $R_3 = R_4 = H$)	Photochemical(101) Synthesis (102)
2-Ethyl-4,5-dihydro-3(2*H*)-thiophenone (Table 14 , <u>18</u> : $R_1 = C_2H_5$, $R_2 = R_3 = R_4 = H$)	Synthesis (87)
3*H*-(1,2)-Dithiole (Table 15, <u>19</u>)	Synthesis (103, 104)
1,2-Dithiolane (Table 15, <u>20a</u>)	Synthesis (105, 106, 107)
3-Acetyl-1,2-dithiolane (Table 15, <u>20b</u>)	Synthesis (108)

4. Thiazoles and derivatives

4-Methyl-5-vinylthiazole (Table 16, <u>22</u> : $R = CH=CH_2$)	Filberts (52)
5-Ethyl-4-methylthiazole (Table 16, <u>22</u> : $R = C_2H_5$)	Coffee aroma flavor (54, 109) (MS) (110)
4-Methyl-5-(2-formyloxyethyl)thiazole (Table 16, <u>22</u> : $R = CH_2CH_2OCHO$)	Synthesis (111)
5-(2-Ethoxyethyl)-4-methylthiazole (Table 16, <u>22</u> : $R = C_2H_4OC_2H_5$)	Synthesis (112)
4-Methyl-5-(2-acetyloxyethyl)thiazole (Table 16, <u>22</u> : $R = C_2H_4COOCH_3$)	Synthesis (10, 113, 114)
4-Methyl-5-(2-propanoyloxy-ethyl)thiazole (Table 16, <u>22</u> : $R = C_2H_4COOC_2H_5$)	Synthesis (115)

5. Alicyclic compounds containing six-membered rings

2-Methyl-1,3-cyclohexanedione (Scheme 9, <u>26</u>)	(MS) (57, 116, 117)

(Tabl 23 continued 3)

6. Pyrans	
3,4-Dihydro-(2*H*)-pyran (Table 17, <u>27</u>: $R_1 = R_2 = H$)	Synthesis (118, 119, 120)
Tetrahydro-(2*H*)-pyran (Table 17, <u>28</u>: $R_1 = R_2 = H$)	(MS) (58, 121, 122)
Tetrahydro-(2*H*)-pyran-4-carboxaldehyde (Table 17, <u>28</u> : $R_1 = H$, $R_2 = CHO$)	Synthesis (123)
4-Acetyltetrahydro-(2*H*)-pyran (Table 18, <u>28</u> : $R_1 = H$, $R_2 = COCH_3$)	(124)
2,3-Dihydro-5-methyl-4-(4*H*)-pyranone (Table 18, <u>29</u> : $R_1 = H$, $R_2 = CH_3$)	Synthesis (125, 126)
3-Methyl tetrahydro-4(4*H*)-pyranone (Table 18, <u>30</u> : $R_1 = R_2 = H$, $R_3 = CH_3$)	Synthesis (127, 128)
7. Thiopyrans	
Tetrahydro-(2*H*)-thiopyran (Table 20, <u>35</u> : $R_1 = R_2$ (Table 19, <u>34</u> : $R_1 = R_2 = R_3 = H$)	Synthesis (129) (MS) (130)
Tetrahydro-3-(3*H*)-thiopyranone (Table 20, <u>35</u>: $R_1 = R_2 = R_3 = H$)	Model systems (malt) (94) Thiamine (thermal degradation (10)
8. 1,3-Dioxanes	
1,3-Dioxane (Table 21, <u>37</u> : $X = O$, $R_1 = R_2 = H$)	(MS) (57)
2-Methylene-1,3-dioxane (Table 21, <u>37</u> : $X = O$, $R_1 = R_2 = (= CH_2)$)	Synthesis (131, 132)
2-Methyl-1,3-dioxane (Table 21, <u>37a</u> : $R_1 = CH_3$, $R_2 = R_3 = H$)	Synthesis (133, 134)
4-Vinyl-1,3-dioxane (Table 21, <u>37a</u> : $R_1 = R_2 = H$, $R_3 = CH=CH_2$)	Synthesis (135, 136)
2-Vinyl-1,3-dioxane (Table 21, <u>37a</u> : $R_1 = CH=CH_2$, $R_2 = R_3 = H$)	Synthesis (137)
2-Ethyl-1,3-dioxane (Table 21, <u>37a</u> : $R_1 = C_2H_5$, $R_2 = R_3 = H$)	Synthesis (MS) (57) Synthesis (133)
4-Ethyl-1,3-dioxane (Table 21, <u>37a</u>: $R_1 = R_2 = H$, $R_3 = CH_3$)	1H-NMR (138)

(Table 23 continued 4)

9. *1,3-Oxathianes*	
1,3-Oxathiane	1H-NMR (139)
(Table 21, <u>37b</u> : $R_1 = R_2 = R_3 = H$)	
2-Methyl-1,3-oxathiane	(MS) (140)
(Table 21, <u>37b</u> : $R_1 = CH_3$, $R_2 = R_3 = H$)	1H-NMR conformation
	(139, 141)
2,2-Dimethyl-1,3-oxathiane	13C-NMR conformation
(Table 21, <u>37b</u>: $R_1 = R_2 = CH_3$, $R_3 = H$)	(141, 142)
2-Ethyl-1,3-oxathiane	(MS) (140)
(Table 21, <u>37b</u>: $R_1 = C_2H_5$, $R_2 = R_3 = H$)	
2-Substituted-1,3-oxathianes	Synthesis((134)
(Table 21, <u>37b</u>: $R_1 = CH_3$ (or C_2H_5), $R_2 = R_3 = H$)	
2-Alkyl-2-methyl-1,3-oxathianes	1H-NMR (143)
(Table 21, <u>37b</u>: $R_1 = CH_3$, $R_2 = C_2H_5$, $R_3 = H$)	
10. *Pyrimidines*	
4-Amino-5-methylaminomethyl)-2-methylpyrimidine	Synthesis (144)
(Scheme 11, <u>41a</u>)	
4-Amino-5-(formamidomethyl)-2-methylpyrimidine	Reactivity studies (145)
(Scheme 11, <u>41b</u>)	
4-Amino-5-(acetamidomethyl)-2-methylpyrimidine	Synthesis (145, 146)
(Scheme 11, <u>41c</u>)	

References

1 R. Barone, M. Chanon, G. Vernin and J. Metzger, *Riv. Ital. EPPOS,* 1980, **3**, 136.
2 R. Barone, M. Chanon, G, Vernin, M. Petitjean and J. Metzger, *Parfums, Cosmétiques et Arômes,* 1981, **38**, 71.
3 R. Barone, M. Chanon, G. Vernin and J. Metzger, *Proceedings VIIIth International Congress Essential Oils,* Cannes-Grasse, Prodarom Publ. 7, rue Gazan, Grasse, 1981, p. 608.
4 M. Petitjean, G. Vernin, J. Metzger, R. Barone and M. Chanon, in *The Quality of Foods and Beverages. Chem. Technol.* G. Charalambous, G. E. Inglett, eds., Academic Press, New York, 1981, p. 253.
5 G. Vernin, C. Párkányi, R. Barone, M. Chanon and J. Metzger, *J. Agric. Food Chem.,* 1987, **35**, 761.
6 G. Vernin, J. Metzger, P. Azario, R. Barone, M. Arbelot and M. Chanon, in *Food Science and Human Nutrition,* G. Charalambous, ed. Elsevier, Amsterdam, The Netherlands, 1992, **29**, 75.
7 R.G. Arnold, L.M. Libbey and R.C. Lindsay, *J. Agric. Food Chem.*, 1969, **17**, 390.
8 B.S. Dwivedi and R.G. Arnold, *J. Food Sci.,* 1972, **37**, 689.

9   M. Güntert, J. Brüning, R. Emberger, M. Köpsel, W. Kuhn, T. Thielman and P. Werkhoff, *J. Agric. Food Chem.*, 1990, **38**, 2027.
10 M. Güntert, J. Brüning, R. Emberger, R. Hopp, M. Köpsel, H. Surburg and P. Werkhoff, in '*Flavor Precursors*', R. Teranishi, G. R. Takeoka and M. Güntert, eds,. Academic Press, New York, 1980.
11 G. Vernin, J. Metzger, A.M. Sultan, A.K. El-Shaffei and C. Párkányi, in *Food Flavor and Safety. Molecular Analysis and Design, ACS Symposium Series*, 1993, **528**, 36.
12 M. Güntert, H.J. Bertram, R. Emberger, R. Hopp, H. Sommer and P. Werkhoff, in *Sulfur Compounds in Foods, ACS Symposium Series*, 1994, **564**, 199.
13 R. Barone and M. Chanon, in *Computer Aids to Chemistry*, G. Vernin, and M. Chanon, eds., 1986, Chapter 1, p. 19.
14 R. Barone and M. Chanon, in *The Encyclopedia of Computational Chemistry*, P.V.R. Schleyer, N. L. Allinger, T. Clark, J. Gasteiger, P. A. Kollman, H. F. Schaefer and P. R. Schreiner, eds., John Wiley and Sons, Chichester, 1998, p. 2931.
15 S. Patel, J. Rabone, S. Russell, J. Tissen and W. Klaffke, *J. Chem. Inf. Comput. Sci.*, 2001, **41**, 926.
16 G. Buchbauer, H. Kalchhauser, P. Wolschann, M. Yahiaoui and D. Zakarya, *Monatsh. Chem.*, 1994, **125**, 1091.
17 W. Jennings and T. Shibamoto, *Agric. Biol. Chem.*, 1976, **40**, 1031.
18 G. Vernin, *J. Essent. Oil Research*, 1991, **3**, 49.
19 J. Otera and Y. Niibo, *J. Org. Chem.*, 1989, **54**, 5003.
20 D. Seebach and M. Teschner, *Tetrahedron Letters*, 1973, **51**, 5113.
21 D. Seebach and M. Teschner, *Chem. Ber.*, 1976, **109**, 1601.
22 S. Iriuchijima, K. Tanokuchi, K. Takodoro and G. Tsuchihashi, *Agric. Biol. Chem.*, 1976, **40**, 1031.
23 B. Gregoire, M. C. Carre, and P. Caubere, *J. Org. Chem.*, 1986, **51**, 1419.
24 B. Wladislaw, H. Viertler, P.R. Olivato, I.C. Calegao, V. Pardini and I.C. Rittner, *J. Chem. Soc. Perkin Trans. 2*, 1980, **3**, 453.
25 G. Vernin, C. Boniface, J. Metzger, T. Obretenov, J. Kantasubrata, A.M. Siouffi, J.L. Larice and D. Fraisse, *Bull. Soc. Chim. Fr.*, 1987, 681.
26 O. Froehlich, C. Duque and P. Schreier, *J. Agric. Food Chem.* 1989, **37**, 421.
27 W. Baltes and G.Z. Bochman, *Lebensm.-Unters. Forsch.*,1987, **184**, 179.
28 W. Baltes and G.Z. Bochman, *Lebensm.-Unters. Forsch.*,1987, **185**, 5.
29 N. Chantarasiri, P. Dinjraseri, C. Threbtaranonth, Y. Threbtaranonth, and C. Yenjai, *J. Chem. Soc., Chem. Comm.*, 1990, **4**, 286.
30 E. Cant and M. Vandewalle, *Org. Mass Spectrom.*,1971, **5**, 1197.
31 A.S. Demir and D. Enders, *Tetrahedron Letters*, 1989, **30**, 1705.
32 L.A. Van Royen, R. Mijingheer and P.J. De Clercq, *Bull. Soc. Chim. Belges*, 1984, **93**, 1019.
33 H.M. Van Dort, L.M. Van der Linde and D. Dejke, *J. Agric. Food Chem.*, 1984, **32**, 454.
34 R. Gelin, S. Gelin and M. Albrand, *Bull. Soc. Chim. Fr.*, 1972, 720.
35 M.A. Gianturco, P. Friedel and V. Flanagan, *Tetrahedron Letters*, 1965, **23**, 1847.
36 G.A.M. Van den Ouweland and H.G. Peer, Ger.Öffen. 1,932,800 (C107d,A 23p), 08 Jan. 1970.

37 G.A.M. Van den Ouweland and H.G. Peer, *J. Agric. Food Chem.*, 1975, **23**, 501.
38 B.K. Carpenter, K.E. Elemens, E.A. Schmidt and H.M.R. Hoffmann, *J. Amer. Chem. Soc.*, 1972, **94**, 6213.
39 R. Noyori, Y. Hayakawa, S. Makino, N. Hayakawa and H. Takaya, *J. Amer. Chem. Soc.*, 1973, **95**, 4103.
40 R. Noyori, *Japan Kokai*, 7589,360 (Cl.C07D, A61K, B01J) 17 Jul. 1975, Appl. 73 139,926,13, Dec. 1973
41 H. Idstein and P. Schreier, *J. Agric. Food Chem*, 1985, **33**, 138.
42 K. Umano, Y. Hagi, K. Nakahara, A. Shoji and T. Shibamoto, *J. Agric. Food Chem.*, 1992, **40**, 599.
43 Y. Takei and T. Yamanishi, *Agric. Biol. Chem.*, 1974, **38**, 2329.
44 W. Roedel and D. Habish, *Nahrung*, 1989, **33**, 449.
45 G. Vernin, J. Metzger, T. Obretenov, K.N. Suon and D. Fraisse, *Developments in Food Science, Flavors and Fragrances*, 1988, **18**, 999.
46 R. Tressl, E. Kersten, C. Nittka and D. Rewicki, in *Sulfur Compounds in Foods*, *ACS Symposium Series*, 1994, **564**, 224.
47 R.Z. Silvar, *Lebensm.-Unters. Forsch.*, 1992, **195**, 112.
48 L.J. Farmer, D.S. Mottram and F.B. Whitfield, *J. Sci. Food Agric.*, 1989, **49**, 347.
49 M. Stoll, M. Winter, F. Gautschi, I. Flament and B. Willhalm, *Helv. Chim. Acta*, 1967, **50**, 628.
50 M. Stoll, M. Winter, F. Gautschi, I. Flament and B. Willhalm, *Helv. Chim. Acta*, 1967, **50**, 2065.
51 L.J. Farmer and D.S. Mottram, *J. Sci. Food Agric.*, 1990, **53**, 505.
52 T.E. Kinlin, R. Muralidhara, A.O. Pittet, A. Sanderson and J P. Walradt, *J. Agric. Food Chem.*, 1972, **20**, 1021.
53 R.G. Buttery, L.C. Ling and R.E. Lundin, *J. Agric. Food Chem.*, 1973, **21**, 488
54 A.O. Pittet and D.E. Hruza, *J. Agric. Food Chem.*, 1974, **22**, 264.
55 R. Tabacchi, *Helv. Chim. Acta*, 1974, **57**, 324.
56 S.S. Hwang, J.T. Carlin, Y. Bao, G.J. Hartmann and C.T. Ho, *J. Agric. Food Chem.*, 1986, **34**, 538.
57 M. Vandewalle, N. Schamp and H. De Wilde, *Bull. Soc. Chim. Belges*, 1967, **76**, 111.
58 R. Smakman and T.J. De Boer, *Org. Mass Spectrom.*, 1968, **1**, 403.
59 G.E. Krammer, M. Güntert, S. Lambrecht, H. Summer, P. Werkhoff, J. Kaulen and A. Rapp, in *Chemistry of Wine Flavor, ACS Symposium Series*, 1998, **714**, 53.
60 M. Winter, A. Furrer, B. Willhalm and W. Thomman, *Helv. Chim. Acta*, 1976, **59**, 1613.
61 F. Ledl and T. Severin, *Chem. Mikrobiol. Technol. Lebensm.*, 1973, **2**, 155.
62 B.M. Trost and K. Hiroi, *J. Amer. Chem.Soc.*, 1975, **97**, 6911.
63 C.W. Jefford and C.G. Rimbault, *J. Amer. Chem. Soc.* 1978, **100**, 295.
64 G.A. Kraus and Z. Wan, *Syn. Letters*, 1997, **11**, 1259.
65 D. Liotta, M. Saindane, C. Barnum, H. Ensley and P. Balakrishnan, P. *Tetrahedron Letters*, 1981, **22**, 3043.
66 G.E. Gream, D. Wege and M. Mular, *Aust. J. Chem.*, 1974, **27**, 567.
67 B.P. Chandrasekhar, U. Schmid and R. Schmid, *Helv. Chim. Acta*, 1975, **58**, 1191.
68 J. Nakayama, *J. Chem. Soc. Perkin Trans 1*, 1976, **5**, 540.
69 P.A. Magriotis, W.V. Murray and F. Johnson, *Tetrahedron Letters*, 1982, **23**, 1993.

70 S. Tsuboi, K. Arisawa, A. Takeda, S. Sato and C. Tamura, *Tetrahedron Letters,* 1983, **24**, 2393.
71 R. M. Coates and S. J. Hobbs, *J. Org. Chem.,* 1984, **49**, 140.
72 J.B. Paine, J.R. Brough, K. Buller and E.E. Erikson, *J. Org. Chem,.* 1987, **52**, 3986.
73 M. Sawamura, H. Nagata, H. Sakamoto and Y. Ito, *J. Amer. Chem. Soc.,* 1992, **114**, 2586.
74 S. Tsunoi, I. Ryu, S. Yamasaki, H. Fukushima, M. Tanaka, M. Komatsu and N. Sonoda, *J. Amer. Chem.Soc.,* 1996, **118**, 10670.
75 N. Palani, T. Rajamannar and K.K. Balasubramanian, *Synlett,* 1997, **1**, 59.
76 M. Matsugi, K. Gotanda, K. Murata and Y. Kita, *Chem. Commun.* 1997, **15**, 1387.
77 G.A. Molander and P.R. Eastwood, *J. Org. Chem.,* 1995, **60**, 4559.
78 J.M. Aurrecoechea, R. Fananas, M. Arrate, J.M. Gorgojo and N. Aurrekoetxea. N. *J. Org. Chem.,* 1999, **64**, 1893.
79 D. Desmaele and N. Champion, *Tetrahedron Letters,* 1992, **33**, 4447.
80 D. Scholz, *Liebigs Ann. Chem.,* 1984, **2**, 259.
81 D. Scholz, *Monatsh Chem.,* 1984, **115**, 655.
82 P.E. Eaton, W.H. Bunnelle and P. Engel, *Can. J. Chem.,* 1984, **62**, 2612.
83 R. Paul and S. Tchelitcheff, *Bull. Soc. Chim. Fr.,* 1950, 520.
84 K.B. De Roos, G. Sipma, S. Van den Bosch, K. Dirk and J. Stoffelsma, *Ger. Öffen.,* 2,458,609/Cl.Co7D, A33L, 19 Jun 1975.
85 J.R. Dias and C. Djerassi, *Org. Mass Spectrom.,* 1972, **6**, 385.
86 A. Goeke, *Phosphorus, Sulfur Silicon Relat. Elem.,* 1999, **153**, 303.
87 J.A. Durden and M.H.J. Weiden, *J. Agric. Food Chem.,* 1974, **22**, 396.

88 M. Bertrand, J. P. Dulcere, G. Gil, J. Grimaldi and P. Sylvestre-Panthet, *Tetrahedron Letters,* 1976, **16**, 1507.
89 V.M. Vlasov, T.A. Favorskaya, A.S. Medvedeva and L.P. Safronova, *Zh. Org. Khim.,* 1968, **4**, 365.
90 E. Anklam, R. Ghaffari-Tabrizi, H. Hombrecher and S. Lau, *Helv. Chim. Acta,* 1984, **67**, 1402.
91 A. Manzara and P. Kovacic, *J. Org. Chem.,* 1974, **39**, 504.
92 G.A. Hunter and H. McNab, *J. Chem. Soc. Chem. Comm.,* 1990, **5**, 375.
93 G.A. Hunter and H. McNab, *J. Chem. Soc. Perkin Trans 1,* 1995, **10**, 1209.
94 P. Werkhoff, J. Bruening, R. Emberger, M. Güntert, M. Koepsel, W. Kuhn and H. Surburg, *J. Agric. Food Chem.,* 1990, **38**, 777.
95 O. Nishimura, H. Masuda and S. Mihara, *Koryo,* 1990, **165**, 91.
96 Y. Zhang and C.T. Ho, *J. Agric. Food Chem.,* 1991, **39**, 777.
97 Y. Chen, J. Xing, C.K. Chin and C.T. Ho, *J. Agric. Food Chem.,* 2000, **48**, 3512.
98 A. Mosandl, U. Hener and H.D. Fenske, *Liebigs Ann. Chem.,* 1989, **9**, 859.
99 B. Unterhalt, L. Kerckhoff and M. Moellers, *Sci. Pharm.,* 2000, **1**, 101.
100 S. Karajiannis and P. Lanaridis, *Vitis,* 2000, **39**, 71.
101 P.Y. Johnson and C.E. Hatch, *J. Org. Chem.* 1975, **40**, 3502.
102 A.G. Schultz and T.H. Fedynshyn, *Tetrahedron,* 1982, **38**, *1761.*
103 C.W. Chen and C. T. Ho, *J. Agric. Food Chem.,* 1998, **46**, 220.
104 C.W. Chen, R.T. Rosen and C.T. Ho, in *Flavor Analysis, ACS Symposium Series,* 1998, **705**, 152.

105   R. Singh and G. Whitesides, *J. Am. Chem. Soc.*, 1990, **112**, 6304.
106   A.S. Kiselyov, L. Strekowski and V.V. Semenov, *Tetrahedron*, 1993, **49**, 2151.
107   M. Hamaguchi, T. Misumi and T. Oshima, *Tetrahedron Letters*, 1998, **39**, 7113.
108   U. Schmidt, *Liebigs Ann. Chem.*, 1963, **670**, 157.
109   O.G. Vitzthum and P. Werkhoff, *J. Food Sci.*, 1974, **39**, 1210.
110   A. Haag and P. Werkhoff, *Org. Mass Spectrom.*, 1976, **11**, 511.
111   R.F. Brown, M.D. Kinnick, J.M. Morin, R.T. Vassileff, F.T. Counter, E.O. Davidson, P.W. Ensminger, J.A. Eudaly and J.S. Kasher, *J. Med. Chem.*, 1990, **33**, 2114.
112   Y.T. Chen and F. Jordan, *J. Org. Chem.*, 1991, **56**, 5029.
113   M.M. Litvak, *Khim.-Farm.Zh.*, 1991, **25**, 64.
114   F. Leeper and D. H. C. Smith, *J. Chem. Soc. Perkin Trans 1*, 1995, **7**, 861.
115   H. Bretschneider and K. Biemann, *Monatsh.*, 1950, **81**, 647.
116   O.H. Mattson, *Acta Chem. Scand.*, 1968, **22**, 2479.
117   T. Yanami, M. Kato, M. Miyashita, A. Yoshikoshi, Y. Itagaki and K. Matsuura, *J. Org. Chem.*, 1977, **42**, 2779.
118   T.C. Snapp and A. E. Blood, US Pat. 3,766,179 (Cl. 260 244R, CO7D), 16 Oct. 1973, Appl. 787,205, 26 Dec. 1968.
119   T.C. Snapp, *Ger. Öffen.* 2,346,943 (Cl. CO7D), 4 Apr. 1974, US Appl. 290,294, 19 Sept. 1972.
120   J. Delaunay, *C.R. Hebd. Seances Acad. Sci. Ser. C*, 1976, **282**, 391.
121   J.E. Collin, *Int. J. Mass Spectrom. Ion Phys.*, 1968, **1**, 213.
122   N.I. Shuikin, B.L. Lebedev and I.P. Yakovlev, *Izv. Akad. Nauk SSSR, Ser. Khim.*, 1967, **3**, 644.
123   H. Klein and W. Grimme, *Angew. Chem.*, 1974, **86**, 742.
124   A. Prelog, W. Bauer, G.H. Cookson and G. Westoo, *Helv. Chim. Acta*, 1951, **34**, 736
125   R. Garry, L. Nyffenegger and R. Vessiere, *Bull. Soc. Chim. Fr.*, 1974, 933.
126   C. Eskenazi, G. Chommet and M. G. Richaud, *J. Heterocycl. Chem.*, 1976, **13**, 253.
127   D. Crich and S. M. Fortt, *Tetrahedron Letters*, 1988, **29**, 2585.
128   M. Chini, P. Crotil, C. Gardelli and F. Macchia, *Tetrahedron*, 1994, **50**, 1261.
129   H. Konda and A. Negishi, *Japan.* 7247,035 (Cl. C 07D. B01J), 27 Nov. 1972, Appl. 70,112,996, 18 Dec. 1970.
130   Q. Porter and J. Baldas, *Mass Spectrometry of Heterocyclic Compounds*, Wiley Interscience, New York, N.Y., 1971.
131   B.G. Yasnitskii, E.G. Ivanycek and R. Sarkisyants, *Referat. Zh. Khim.* 1971, Abstr. N° 22h 288.
132   Z. Wu, R.R. Stanley and C.U. Pittman, *J. Org. Chem.* 1999, **64**, 8386.
133   D.L. Rakhmankulov, S.S. Zlostskii, U.N. Uzikova and Y. M. Paushkin, *Dokl. Akad. Nauk SSSR*, 1974, **218**, 156.
134   K. Fuji, M. Ueda, K. Sumi, K. Kajiwara, E. Fujita, T. Iwashita and I. Micera, *J. Org. Chem.*, 1985, **50**, 657.
135   J. Thivolle-Cazat and I. Tkatchenko, *J. Chem. Soc. Chem. Comm.*, 1982, **19**, 1128.
136   A. Yanagisawa, S. Haboue and H. Yamamoto, *J. Amer. Chem. Soc.*, 1989, **111**, 366.
137   M.L. Peterson, *Ger. Öffen.* 2,524,040 (Cl. C07D) 19 Feb. 1976, US Appl. 495,510,07, Aug. 1974.
138   D. Tavernier and M. Anteunis, *J. Magn. Resonance*, 1974, **13**, 181.

139 P. Pasanen, *Suom Kemistilehti*, 1972, **45**, 363.
140 K. Pihlaja and P. Pasanen, *Org. Mass Spectrom.*, 1971, **5**, 763.
141 K. Pihlaja, J. Jokila and U. Heinonen, *Finn. Chem. Lett.*, 1974, **8**, 275.
142 K. Pihlaja and P. Pasanen, *Suom. Kemistilehti*, 1973, **40**, 273.
143 P. Pasanen, *Finn. Chem. Lett.*, 1974, **1**, 49.
144 H. Hirano, *J. Pharm. Soc. Jap.*, 1955, **75**, 249.
145 R.F. Evans and A.V. Robertson, *Austral. J. Chem.*, 1973, **26**, 1599.
146 H. Morimoto, N. Hayashi, T. Naka and S. Kato, *Chem. Ber.*, 1973, **106**, 893.

GENERATION OF ALDEHYDES FROM MAILLARD REACTION OF GLUCOSE AND AMINO ACIDS

Jiangang Li and Chi-Tang Ho

Department of Food Science, Rutgers University, 65 Dudley Road, New Brunswick, NJ 08901-8520

1 ABSTRACT

Maillard reaction was recognized as the major source for the generation of flavor and color in thermally processed foods. Many steps in Maillard reaction can produce carbonyl compounds. The most important of them is Strecker degradation, during which amino acids react with dicarbonyl compounds, followed by decarboxylation and oxidation to produce Strecker aldehydes. Strecker aldehydes are important flavor compounds and active intermediates for other reactions. The reactivity and volatility of those carbonyl compounds make them difficult to analyze using traditional methods. Cystamine easily reacts with volatile carbonyl compounds to quantitatively produce thiazolidines, which are much more stable than their corresponding carbonyl compounds. In this study, we have employed the cystamine derivatization reaction to study the generation of aldehydes from reaction of glucose with different amino acids.

2 INTRODUCTION

Maillard reaction was recognized as the major source for the generation of flavor and color in thermally processed foods. It is a multiple step reaction with many intermediates and products. The analytical challenge resulting from the volatility and reactivity of its many intermediate products is partially responsible for the difficulty in the thorough understanding of the Maillard Reaction. Many steps in Maillard reaction can produce carbonyl compounds. The most important of them is Strecker degradation, during which amino acids react with dicarbonyl compounds, followed by decarboxylation and oxidation to produce Strecker aldehydes.[1] Strecker aldehydes are important flavor compounds and active intermediates for the other reactions. Some Strecker aldehydes from amino acids are listed in Table 1.

The reactivity and volatility of those carbonyl compounds make them difficult to analyze using traditional methods, such as headspace techniques, purge and trap, liquid-liquid extraction and simultaneous distillation and extraction.[2] Cystamine easily reacts with volatile carbonyl compounds to quantitatively produce thiazolidines, which are much more stable than their corresponding carbonyl compounds. The reaction mechanism of cystamine and aldehydes were studied and proposed in different ways.[3-4] This reaction has been used to analyze low molecular weight carbonyl compounds in foods.[5-6] The employment of cystamine derivatization reaction could be an analytical break through for the analysis of carbonyl compounds in Maillard model reactions.

Table 1 *Strecker aldehydes from their respective amino acids*

Strecker aldehydes	Amino acids
Formaldehyde	Glycine
Acetaldehyde	Alanine
Isobutyraldehyde	Valine
Isovaleraldehyde	Leucine
2-Methylbutyraldehyde	Isoleucine
Methional	Methionine
Phenylacetaldehyde	Phenylalanine
α-Mercaptoacetaldehyde	Cysteine
α-Hydroxyacetaldehyde	Serine
2-Hydroxypropanal	Threonine

3 MATERIALS AND METHODS

3.1 Preparation of Thiazolidine Derivatives from Acetaldehyde, Isobutyraldehyde, Isovaleraldehyde, 2-Methylbutyraldehyde

Cystamine chloride (1.5 g) was dissolved in 50 mL distilled water. Phosphate buffer (5 ml 0.2 M) were added to the solution and the pH was adjusted to 7.0. Carbonyl compound (0.5 g) was dissolved in the solution, which was then kept at room temperature for 3 hours under a nitrogen atmosphere. The reaction solution was extracted twice with methylene chloride (20 ml) to obtain acidic extract. The water phase was adjusted to pH 8.0 with 4% aqueous sodium hydroxide solution, and then extracted with methylene chloride (30 ml for 2 times) to obtain basic extract. The basic extract was washed with water and dried by anhydrous magnesium sulfate. The solvent was evaporated to obtain the reaction product.

3.2 Preparation of Thiazolidine Derivatives from Methional and Phenylacetaldehyde

Cystamine (1.5 g) was dissolved in 20 ml methanol and 30 ml distilled water. Phosphate buffer (5 ml 0.2 M)were added to the solution and the pH adjusted to 7.0. Aldehyde (0.5 g) was dissolved in the solution, which was then stirred at room temperature for 3 hours under nitrogen gas. Methanol was removed by vacuum. The reaction solution was extracted with methylene chloride (20 ml, twice) to obtain acidic extract. The aqueous phase was adjusted to pH 8.0 with 4% aqueous sodium hydroxide (NaOH), and extracted twice with methylene chloride (30 mL) to obtain basic extract. The basic extract was washed with water and dried over anhydrous magnesium sulfate. The solvent was evaporated to obtain the reaction product.

3.3 Study of Maillard Reaction in Model System of Glucose and Amino Acids in Aqueous Solutions

Glucose was heated for 5 hours in weak alkaline solution to facilitate the Maillard reaction with seven different amino acids, namely L-methionine, L-phenylalanine, L-glycine, L-alanine, L-valine, L-leucine and L-isoleucine. Reactions of glucose-methionine and glucose-phenylalanine were studied for time dependency in both weak alkaline and weak acid solutions.

The Maillard model reaction product was then derivatized using cystamine reaction and then applied to GC and GC-MS for analysis of the carbonyl compounds from Maillard reactions.

3.4 Reaction of Amino Acid and Glucose in Weak Alkaline Solution

Amino acid (10 mmol) and 1.80 g glucose (10 mmol) were dissolved in 100 mL distilled water and the pH adjusted to 8.0 with 4% aqueous NaOH. The reaction solution was heated in a closed bottle at $100°C$ for 0.5, 1, 1.5, 2, 2.5, 3, 4, and 5 hours.

3.5 Reaction of Amino Acid and Glucose in Weak Acid Solution

The reaction procedure was same as that described under 2.4, except that the solution was adjusted to alkaline pH of 5.5 with phosphoric acid instead of pH of 8.0 in the above reaction.

3.6 Cystamine Reaction with Model Reaction Products

After cooling the model reaction solution to room temperature, 10 mL 0.2 M phosphate buffer (pH 7.2) and 1.14 g (10 mmol) cystamine chloride were dissolved in the reaction solution. The solution pH was adjusted to 7.0 as needed using phosphoric acid or NaOH solution. The solution headspace was filled with nitrogen gas. The reaction solution was then kept at room temperature for 3 hours. The reaction solution was readjusted to pH 8.0 with 4% aqueous NaOH and then twice extracted with 25 ml methylene chloride. The combined methylene chloride extract was dried over anhydrous $MgSO_4$. Tridecane in methylene chloride (5.0 µL, 20.0% (v/v)) was mixed into the extract as an internal standard. The extract was concentrated to 0.5 mL at a temperature below $0°C$ by nitrogen gas. The concentrated solution was subjected to GC and GC-MS analyses.

4 RESULTS AND DISCUSSION

4.1 Synthesis of Thiazolidine Derivatives by Cystamine and Aldehydes

Aldehydes of low molecular weight were soluble in water. They reacted readily with cysteamine aqueous solution to produce corresponding thiazolidine derivatives. Aldehydes of lower molecular weight have high solubility in water, therefore the reaction with cystamine can occur in an aqueous solution. Aldehydes of higher molecular weight are not readily dissolved in water, a methanol-water system was employed for cystamine reaction. The reaction occurs at room temperature in phosphorus buffered neutral solution (Huang et al. 1998). The nitrogen gas was effective to prevent the oxidation of aldehydes and thiazolidines.

After the reaction, the solution pH changed to between 6.5 and 7. At this time, extraction with methylene chloride was employed to remove any unreacted aldehyde and other possible non-alkaline side products. The thiazolidine molecules have an alkaline amino group, which have strong polarity at acidic and neutral phases. Therefore, most of the thiazolidine stays in the water phase during the acidic extraction. Addition of NaOH water solution into the water phase resulted in a turbid suspension, which was then extracted to obtain a clear sticky liquid. The solubility of thiazolidine in water decreased with the increase of solution pH. The chemical properties of thiazolidines are similar to those of Aldol condensation products. They are stable in neutral and alkaline conditions, but decompose at low pH. Acidic groups in silica gel decompose thiazolidines at room temperature, therefore those compounds could not be applied to normal phase column chromatography for the purpose of purification. After purification the reaction product purity was over 99% as tested by GC.

4.2 The Mass Spectra and ^{1}HNMR Spectra of 2-substituted Thiazolidine Derivatives

Thiazolidine, colorless liquid, MS: (m/Z), 89 ([M]$^+$, 100%), 91 ([M+2]$^+$, 4.7%), 88 ([M-1]$^+$, 38.2%), 86 ([M-1-2H]$^+$, 1.2%), 61 (11.7%), 60 (14.5%), 59 (18.7%), 56 ([M-1-S]$^+$, 3.1%), 43 (57.3%), 42 (31.0%). ^{1}H NMR: δ 1.72 (1H, s, H-3), 2.82 (2H, t, H-5), 3.15 (2H, t, H-4), 4.13 (2H, s, H-2).

2-Methylthiazolidine, clear liquid, MS: (m/Z), 103 ([M]$^+$, 95.6%), 105 ([M+2]$^+$, 4.5%), 102 ([M-1]$^+$, 11.3%), 88 ([M-CH$_3$]$^+$, 100%), 70 ([M-1-S]$^+$, 5.5%), 61 (16.7%), 59 (22.4%), 56 ([M-CH$_3$-S]$^+$, 80.1%), 44 (47.7%), 42 (21.7%), 41 (11.0%). ^{1}H NMR: δ 1.55 (3H, d, H-6), 1.65 (1H, s, H-3), 2.93-3.03 (3H, m, H-5 and H-4a), 3.54 (1H, m, H-4b), 4.55 (1H, q, H-2).

2-Isopropylthiazolidine, clear liquid, MS: (m/Z), 131 ([M]$^+$, 6.7%), 133 ([M+2]$^+$, 0.3%), 116 ([M-CH$_3$]$^+$, 0.4%), 102 ([M+1-CH$_3$-CH$_3$]$^+$, 0.7%), 90 ([M+2-C$_3$H$_7$]$^+$, 5.3%), 88 ([M-C$_3$H$_7$]$^+$, 100%), 84 ([M-CH$_3$-S]$^+$, 6.5%), 70 ([M+1-CH$_3$-CH$_3$-S]$^+$, 11.3%), 61 (10.2%), 56 ([M-C$_3$H$_7$-S]$^+$, 4.4%), 44 (5.1%), 42 ([C$_3$H$_7$-1]$^+$, 3.0%). ^{1}H NMR: δ 1.01 (3H, d, H-7), δ 1.02 (3H, d, H-8), 1.66 (1H, s, H-3), 1.86 (1H, m, H-6), 2.69-2.93 (3H, m, H-5 and H-4a), 3.45 (1H, m, H-4b), 4.24 (1H, d, H-2).

2-Isobutylthiazolidine, clear liquid, MS: (m/Z), 145 ([M]$^+$, 17.1%), 147 ([M+2]$^+$, 0.8%), 130 ([M-CH$_3$]$^+$, 9.2%), 102 ([M-C$_3$H$_7$]$^+$, 2.2%), 98 ([M-CH$_3$-S]$^+$, 10.1%), 90 ([M+2-C$_4$H$_9$]$^+$, 5.0%), 88 ([M-C$_4$H$_9$]$^+$, 100%), 70 ([M-C$_3$H$_7$-S]$^+$, 7.6%), 61 (8.3%), 56 ([M-C$_4$H$_9$-S]$^+$, 19.0%), 44 (8.2%), 41 (7.6%). ^{1}H NMR: δ 0.94 (6H, d, H-8 and H-9), 1.57-1.86 (4H, m, H-3, H-6 and H-7), 2.82-2.99 (3H, m, H-5 and H-4a), 3.55 (1H, m, H-4b), 4.51 (1H, t, H-2).

2-(1-methylpropyl)thiazolidine, clear liquid, MS: (m/Z), 145 ([M]$^+$, 4.2%), 147 ([M+2]$^+$, 0.2%), 130 ([M-CH$_3$]$^+$, 0.6%), 116 ([M-C$_2$H$_5$]$^+$, 0.8%), 102 ([M-C$_3$H$_7$]$^+$, 0.3%), 98 ([M-CH$_3$-S]$^+$, 3.6%), 90 ([M+2-C$_4$H$_9$]$^+$, 5.4%), 88 ([M-C$_4$H$_9$]$^+$, 100%), 84 ([M-C$_2$H$_5$-S]$^+$, 4.1%), 70 ([M-C$_3$H$_7$-S]$^+$, 9.2%), 61 (8.0%), 56 (2.7%), 44 (4.3%), 41 (5.7%). ^{1}H NMR: δ

0.96 (3H, t, H-8), 1.05 (3H, d, H-9), 1.28 (2H, m, H-7), 1.60-1.80 (2H, m, H-3 and H-6), 2.73-2.99 (3H, m, H-5 and H-4a), 3.55 (1H, m, H-4b), 4.37 (1H, dd, H-2).

2-(2-methylthioethyl)thiazolidine, clear liquid, MS: (m/Z), 163 ([M]$^+$, 16.5%), 165 ([M+2]$^+$, 1.6%), 162 ([M-1]$^+$, 2.6%), 148 ([M-CH$_3$]$^+$, 3.8%), 135 ([M-C$_2$H$_4$]$^+$, 13.2%), 116 ([M-CH$_3$-S]$^+$, 49.0%), 102 ([M-CH$_3$-S-CH$_2$]$^+$, 3.8%), 90 ([M+2-CH$_3$-S-C$_2$H$_4$]$^+$, 6.6%), 88 ([M-CH$_3$-S-C$_2$H$_4$]$^+$, 100%), 75 (3.6%), 70 (7.5%), 61 (42.0%), 56 ([M-CH$_3$-S-C$_2$H$_4$-S]$^+$, 39.3%). ^{1}H NMR: δ 1.70 (1H, s, H-3), 2.01 (2H, m, H-6), 2.10 (3H, s, H-9), 2.63 (2H, t, H-7), 2.80-3.10 (3H, m, H-5 and H-4a), 3.41 (1H, m, H-4b), 4.59 (1H, t, H-2).

2-Phenylmethylthiazolidine, clear liquid, MS: (m/Z), 179 ([M]$^+$, 0.8%), 181 ([M+2]$^+$, 0.1%), 178 ([M-1]$^+$, 1.1%), 135 (2.9%), 132 (13.4%), 103 ([M+1-C$_6$H$_5$]$^+$, 6.3%), 102 ([M-C$_6$H$_5$]$^+$, 1.7%), 91 ([C$_7$H$_7$]$^+$, 34.6%), 90 ([M+2-C$_7$H$_9$]$^+$, 12.1%), 88 ([M-C$_7$H$_9$]$^+$, 100%), 77 (C$_6$H$_5$, 7.8%), 65 ([C$_5$H$_5$]$^+$, 11.5%), 61 (17.6%), 56 (1.2%), 51 (5.4%), 44 (6.2%), 39 (4.2%). ^{1}H NMR: δ 1.71 (1H, s, H-3), 2.85-3.04 (4H, m, H-5, H-4a and H$_{6a}$), 3.20 (1H, q, H$_{6b}$), 3.50 (1H, m, H-4b), 4.75 (1H, t, H-2), 7.29 (5H, m, phenyl).

4.3 Reaction of Amino Acids and Glucose at Weak Aqueous Alkaline Solution

Glucose was reacted with seven different amino acids for 5 hours in weak alkaline aqueous solution. Thiazolidine derivatives from aldehydes are summarized in Table 2.

Table 2 *Amounts of thiazolidine dereivatives produced from selected Strecker degradation reactions (mg/kg amino acid)*

	Gly	Ala	Val	Leu	Ilu	Met	Phe
Thiazolidine	1387	1756	1409	262	4171	90	396
2-Methylthiazolidine	711	28630	795	146	640	208	240
Thiazolidine from Strecker aldehyde	1387	28630	33688	40524	39971	1920	5407

From the above Table, it is obvious that Strecker aldehydes are the major aldehyde products of the Maillard Model reaction at the selected reaction condition. Different amino acids have different reactivity.

4.4 Reaction Time Dependency of Glucose-Methionine and Glucose-Phenylalanine Reaction in Both Alkaline and Acid Solution

The reaction of glucose-methionine and glucose-phenylalanine in both weak alkaline and weak acid solutions was studied for 0.5, 1, 1.5, 2, 2.5, 3, 4, and 5 hours. The results from the cystamine derivatization method are listed in Tables 3 and 4.

Table 3 *Aldehydes from glucose-phenylalanine model reaction (mg/kg phenylalanine)*

Reaction Time (hr)		0.5	1.0	1.5	2.0	2.5	3.0	4.0	5.0
	pH								
Formaldehyde	5.5	15	43	129	185	226	271	277	317
	8.0	36	113	130	205	295	335	374	396
Acetaldehyde	5.5	4	11	40	47	51	62	66	85
	8.0	55	73	84	106	169	200	225	240
2-Phenylacetaldehyde	5.5	772	741	1259	1799	2563	2917	3296	4290
	8.0	970	1150	1490	2210	2420	2838	3548	5407

Table 4 *Aldehydes from glucose-methionine model reaction (mg/kg methionine)*

Reaction Time (hr)		0.5	1.0	1.5	2.0	2.5	3.0	4.0
	pH							
formaldehyde	5.5	5	8	7	9	8	12	10
	8.0	16	11	14	11	37	41	81
acetaldehyde	5.5	85	123	122	153	171	153	181
	8.0	33	57	81	85	162	189	199
methional	5.5	45	119	220	400	393	618	618
	8.0	140	684	1069	1117	1321	1248	1249

The effect of pH is very strong in Maillard model reactions of both. Glucose-methionine and glucose-phenylalanine. Weak alkaline solution favors the Maillard reaction and production of corresponding Strecker aldehydes. Both Strecker aldehydes kept increasing during the 5 hour reaction period. Phenylalanine is more reactive than methionine. The production of formaldehyde and acetaldehyde increases with reaction period in both cases. The pH effects on the variation patterns of different aldehyde are different.

References

1 J.E. Hodge, *J. Agric. Food Chem.* 1953, **1**, 928-943.
2 E. Ibanez, S. Lopez-Sebastian, E. Ramos, Tabera, G., *Food Chem.* 1998, **63**, 281-286.
3 H.C.H. Yeo, T. Shibamoto, *J. Agric. Food Chem.* 1991, **39**, 370-373.
4 T.C. Huang, L.Z. Huang, C.-T. Ho, *J. Agric. Food Chem.* 1998, **46**, 224-227.
5 A. Yasuhara, T. Shibamoto, *J. Food Sci.* 1989, **54**, 1471-1472.
6 A. Yasuhara, T. Shibamoto, *J. Chromatogr.* 1991, **547**, 291-298.

THE BIOSYNTHESIS OF FURANONES IN STRAWBERRY: ARE THE PLANT CELLS ALL ALONE?

A. Kyriacou[1] and I. Zabetakis[2]

[1]Department of Dietetics and Nutritional Science, Harokopio University of Athens E-mail: mkyriacou@hua.gr
[2]Laboratory of Food Chemistry, Department of Chemistry, University of Athens, Athens 157-71, Greece, E-mail: izabet@chem.uoa.gr

1 ABSTRACT

The facultative methylotrophic bacterium *Methylobacterium extorquens* has been isolated from strawberry callus cultures and its role in the biosynthesis of strawberry flavor (2,5-dimethyl-4-hydroxy-2H-furan-3-one or furaneol®) has been studied. The microorganism has been cultivated with 1,2-propanediol - a substrate that exists in strawberries – and methanol and 7 days later 2-hydroxypropanal (lactaldehyde) was detected as the DNP-derivative. This result in conjunction with the observation that strawberry cells can biosynthesize furaneol from lactaldehyde is discussed. We propose a symbiotic relationship between the bacteria and the strawberry cells: the strawberry cells provide a niche for the bacteria whereas the bacteria oxidize 1,2-propanediol to lactaldehyde; the latter is utilized by the plant cells in flavor biosynthesis.

2 INTRODUCTION

One of the most important flavor components in strawberry is 2,5-dimethyl-4-hydroxy-2H-furan-3-one (DMHF, Fraision®, Furaneol®).[1] Because of the great importance of DMHF as a potent flavoring, several studies of the biosynthesis of this molecule in either strawberry fruits or strawberry callus cultures have been carried out and recently reviewed.[2] The precursors of DMHF are still not fully identified: in one study in strawberries, it was found that all carbons of DMHF originate from D-fructose[3] whereas the yeast *Zygosaccharomyces rouxii* utilises D-fructose 1,6-biphosphate to produce DMHF.[4]

Two possible earlier precursors of DMHF are 1,2-propanediol and 2-hydroxypropanal (lactaldehyde). 1,2-Propanediol has been found as a natural component in strawberries.[5] The bacterium *Methylobacterium extorquens* that grows on strawberry plants, using them as a niche habitat, has been shown to be capable of oxidizing 1,2-propanediol to lactaldehyde.[6] Strawberry calli, when provided with lactaldehyde, biosynthesized high levels of DMHF (Figure 1).[7]

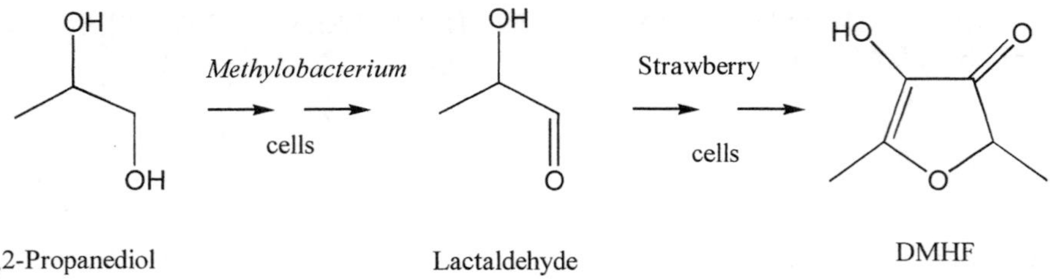

1,2-Propanediol → (Methylobacterium cells) → **Lactaldehyde** → (Strawberry cells) → **DMHF**

Figure 1 *The bacterial bioconversion of 1,2-propanediol to lactaldehyde and the linked plant bioconversion of lactaldehyde to DMHF in strawberry.*

Lactaldehyde may be a key precursor in the biosynthesis of DMHF and chemically it can be used in the synthesis of a wide range of flavor molecules (e.g. methyl-butanal) and compounds of high biological interest (e.g. 6-deoxyhexoses).[8] The bacterial bioconversion of the diol to lactaldehyde has also been studied in other bacterial species such as the Flavobacterium[9] and the bacterial strain SA-1 isolated from soil.[10] The characterization of the bacterial methanol dehydrogenase from *Methylocystis* sp. GB25 has also been reported.[11] In this paper, we report our latest results on the bioconversion of 1,2-propanediol to lactaldehyde in suspension cultures of *M. extorquens*.

3 METHODS AND RESULTS

3.1 HPLC Analysis of Lactaldehyde and the Culture of *Methylobacterium extorquens*

The HPLC analysis of lactaldehyde was performed as described before, using a Techopak 10-ODS (250 x 3.9 mm, HPLC Technology Ltd, UK) coupled to a M-Bondapak guard column.[6] Quantitative HPLC determinations were conducted using a Waters model 600E pump and Waters model 996 Photodiode Array Detector. The injection volume was 20 μl and the mobile phase was 60% H_2O – 40% CH_3CN (isocratic; flow rate 1.5 ml/min; detection at 365 nm).

Lactaldehyde was synthesized following the method of Zabetakis *et al.*[7] Linear responses from HPLC-UV (r=0.9997) were obtained for the aldehyde in the concentration range 0.5-50 μg/ml in acetonitrile, and a standard curve was constructed. In all runs, the aldehyde was analyzed in the supernatant of the cultures.[6] All analyses were carried out in triplicate.

The bacterial strain of *Methylobacterium extorquens* IMI 366607 CABI Bioscience (UK) used in this study, has been isolated from strawberry plants.[6] The culture was maintained at 4°C for up to one month on slants of a medium containing glycerol (1% w/v), peptone (1% w/v) and agar (1.5% w/v).[6]

The shake-flask experiments were carried out in 500ml conical flasks containing 100 ml medium. 1,2-Propanediol was added in different concentrations (0.1, 0.5, 0.75, 1.0% (v/v). Methanol was also added when it is mentioned, in a concentration of 0.25% (v/v). The nitrogen source was 1% peptone for the standard run. Yeast extract 1% (w/v), NH_4Cl

0.2% (w/v) supplemented with yeast extract 0.5% (w/v) and $NaNO_3$ 0.4% (w/v) supplemented with yeast extract 0.5% (w/v) was also examined. The total nitrogen was estimated to be approximately 8% in the nitrogen source (w/v). 1,2-Propanediol was separately filter sterilized and added to the medium.

A 72 h culture was used as inoculum (1% v/v). The inoculum size was approximately 10^7 cfu/ml. Cultures were grown at 30° C in a shaker, at an agitation rate of 180 rpm. Samples of 1 ml were taken to measure the absorbance at 550 nm. Sterilized growth medium was used as the blank. The specific growth rate was estimated by using the cell numbers during the exponential growth phase.

A calibration curve was plotted by measuring the absorbance at 550 nm of individual dilutions of the cell suspension against the blank. The equation $Y = 7.29 \times 10^8 X + 144659$, where X represents the absorbance and Y represents the number of cells, was obtained; $R^2 = 0.9920$. The cell counts of *M. extorquens* cultures were calculated using the above equation after absorbance measurements.

3.2 The Growth Curve of *M. extorquens* and the Bioformation of Lactaldehyde

The growth of *M. extorquens* in a medium containing 0.75% (v/v) 1,2-propanediol, 0.25% (v/v) methanol and peptone 1% (w/v) as a nitrogen source (this mixture is defined as standard) and the bioformation of lactaldehyde were screened over a course of 10 days (Figure 2). The bacteria follow a rather typical exponential growth phase followed by a stationary one. The exponential growth from day 2 to day 4 is rather sharp. The maximum levels of lactaldehyde (3517 ppm) were obtained when the bacterial growth has reached the stationary phase (day 7). After this day, the levels of lactaldehyde are depleting (2912 ppm on day 10). It was thus decided, in the experiments reported here, to harvest the bacterial cells and determine the levels of lactaldehyde on day 7.

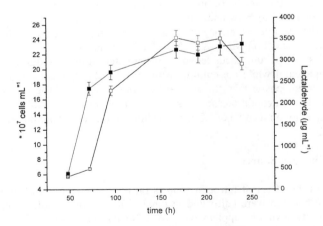

Figure 2 *The growth curve of the bacteria (■) and the levels of the produced lactaldehyde (□).*

3.3 Effect of Propanediol in the Presence or Absence of Methanol

In this series of experiments, the effect of different levels of the diol on the formation of lactaldehyde was studied while keeping the levels of methanol constant in all experiments (0.25%, v/v). Because of the sharp exponential growth that has been observed in figure 2, the bacterial growth in this series of experiments was studied for a time course of 96 hours. In addition to the standard run, three different levels of the diol were used: 0.1%, 0.5% and 1% (v/v). On day 7, the cells were centrifuged, and the levels of lactaldehyde were determined in the supernatant (Table 1). In control runs, where 1% methanol but no 1,2-propanediol was used, lactaldehyde was not detected as previously reported.[6] The results for the standard run were very close to the ones observed in the previous series (about 3500 ppm). The highest levels of the aldehyde (77% higher than the standard run) were found when 0.5% (v/v) 1,2-propanediol was used followed by the run using 1% (v/v) 1,2-propanediol (17% higher than the standard run).

Table 1 *Levels of lactaldehyde at day 7 when Methylobacterium extorquens was grown with various levels of 1,2-propanediol either with or without methanol*

Levels of 1,2-propanediol	With methanol (0.25% v/v)	Without methanol
0.1 % (v/v)	1310 ± 103	n.d.
0.5 % (v/v)	6200 ± 345	n.d.
0.75 % (v/v) (standard)	3500 ± 145	3400 ± 128
1 % (v/v)	4097 ± 189	6900 ± 295

n.d. : not detected, values are presented as the average ± S.D. of triplicate runs

Given the facultative methylotrophy of *M. extorquens*[6], we examined whether the bacteria needed methanol to grow and produce the aldehyde or if the diol could suffice as the necessary carbon source. *M. extorquens* was grown using three different levels of 1,2-propanediol (0.1, 0.5 and 1%) whereas 0.75% of 1,2-propanediol and 0.25% of methanol were used in the standard run. Lactaldehyde was not detected in the cultures with 0.1 and 0.5% of 1,2-propanediol while the cultures with 1% of 1,2-propanediol produced 103% higher levels of lactaldehyde than the standard cultures (Table 1). These results suggest that the optimum growth conditions are obtained when the methyl providing compound(s) (either 1,2-propanediol on its own or 1,2-propanediol and methanol) are present in a (total) ratio of about 1%.

3.4 Effect of Nitrogen Source

In order to assess the effect of nitrogen sources on the bioconversion of propanediol to lactaldehyde, the bacteria were cultivated in three different sources of nitrogen, namely yeast extract on its own or supplemented with $NaNO_3$ or NH_4Cl. The results were compared to the standard run where peptone was used as nitrogen source.

The microorganism does not grow when NH_4Cl or $NaNO_3$ was used as a sole nitrogen source. The levels of lactaldehyde determined in the different cultures were 3350 ± 150 ppm (control), 1328 ± 59 ppm (when yeast extract was used), 531 ± 42 ppm (when yeast extract and $NaNO_3$ were used), 300 ± 22 ppm (when yeast extract and NH_4Cl were used). The levels of lactaldehyde in these cultures were dramatically lower than in the standard run.

4 CONCLUSION

With this work, we further reinforced the role of *M. extorquens* on the bioconversion of 1,2-propanediol to lactaldehyde in strawberry. We investigated the effects of carbon and nitrogen sources and the substrate availability on this bioconversion. Currently we are working on the identification and complete characterization of the enzymes involved in this bioconversion with the aim to fully understand the role of *M. extorquens* in the biosynthesis of furanones in strawberry.

References

1   I. Zabetakis, J.W. Gramshaw and D.S. Robinson, *Food Chem.*, 1999, **65**, 139.
2   K.G. Bood and I. Zabetakis, *J. Food Sci.*, 2002, **67**, 2.
3   M. Wein, E. Lewinsohn and W. Schwab, *J. Agric. Food Chem.*, 2001, **49**, 2427.
4   L, Hecquet, M. Sancelme, J. Bolte and C. Demuynck, *J. Agric. Food Chem.*, 1996, **44**, 1357.
5   I. Zabetakis and J.W. Gramshaw, *Food Chem.*, 1998, **61**, 351.
6   I. Zabetakis, *Plant Cell Tiss. Org. Cult.*, 1997, **50**, 179.
7   I. Zabetakis, P. Moutevelis-Minakakis and J.W. Gramshaw, *Food Chem.*, 1999, **64**, 311.
8   C-H. Wong, F.P. Mazenod and G.M. Whitesides, *J. Org. Chem.*, 1983, **48**, 3493.
9   A. Willets, *Biochim. Biophys. Acta*, 1979, **588**, 302.
10 Y. Tanaka, K. Fujii, A. Tanaka and S. Fukui, *Hakko Kog. Zasshi*, 1975, **53**, 354.
11 S. Grosse, K.D. Wendlandt and H.P. Kleber, *J. Basic Microb.*, 1997, **37**, 269.

EVOLUTION OF VOLATILE COMPOUNDS AND SENSORY RANCIDITY IN PURIFIED OLIVE OIL DURING STORAGE UNDER NORMAL AND ACCELERATED CONDITIONS (25-75 °C)

M.D. Salvador, S. Gómez-Alonso, V. Mancebo-Campos and G. Fregapane.

Departamento de Química Analítica y Tecnología de Alimentos,
Universidad de Castilla - La Mancha, Ciudad Real, España E-13071

1 ABSTRACT

In the course of the lipid autoxidation reaction in foods, a series of compounds are formed, causing off-flavors and rancidity, loss of nutritional value and ultimately consumer rejection of the product. Autoxidation is therefore the main cause of olive oil quality deterioration and its reaction rate determines the shelf-life of the oil. This study describes the evolution of volatile compounds and sensory rancidity in purified olive oil (POO) stored in darkness at temperatures ranging from 25 to 75 °C. The results show that there was a high linear correlation ($r^2 = 0.999$) between the time taken to reach the rancidity threshold and the induction period of the reaction causing formation of 2,4-decadienal. This could therefore be a useful instrumental means of measuring sensory recognition of rancidity, at least in POO.

2 INTRODUCTION

In the course of the autoxidation reaction in foods, a series of compounds are formed, causing off-flavors and rancidity, loss of nutritional value and ultimately consumer rejection of the product. Autoxidation is therefore the main cause of olive oil quality deterioration and its reaction rate determines the shelf-life of this typically Mediterranean product.

Lipid oxidation occurs fairly slowly at room temperature, and hence accelerated methods should be employed to estimate the oxidative and flavor stability of the product in a relatively short period of time. Temperature is often the parameter used to increase the oxidation rate in accelerated shelf life testing (ASLT); however, there are limitations since several studies have shown that the reaction mechanism is significantly different at temperatures above 60 °C.[1]

The decomposition of hydroperoxides, which form during the first stage of triacylglycerol and free fatty acid autoxidation and photooxidation, eventually leads to a complex mixture of volatile compounds known as secondary oxidation products. Many of these have a very low sensory recognition threshold and as such are responsible for rancidity and the consequent loss of olive oil organoleptic quality. For this reason, the determination of volatile compounds, together with sensory evaluation, is one of the best means of verifying the oxidation state of oils and fats.[2] In the case of olive oil, it is known that the volatile fraction contains mainly aldehydes (e.g., hexanal, nonanal, 2-decenal, pentanal), alcohols (e.g., 2-penten-1-ol and 1-octen-3-ol) and aliphatic acids (hexanoic, heptanoic, octanoic and nonanoic); some sensory thresholds and descriptors of these compounds have been studied.[3]

Sensory analysis is a very useful tool for obtaining information on consumer acceptance or rejection of a food product. Moreover, comparative studies of organoleptic and instrumental methods are very useful for establishing relationships on the basis of which instrumental results can be extrapolated for assessment of the sensory quality of a food product.[3,4]

This study describes the generation of volatile compounds and sensory rancidity in purified olive oil (POO) stored in darkness at temperatures ranging from 25 to 75 °C. Before conducting research into virgin olive oil, basic research on the oxidation process in POO (in absence of pro- and antioxidants to avoid confusing effects) is needed to fully understand the complex influence of accelerated test conditions on the decomposition rate of the oxidized olive oil triacylglycerol matrix. The time taken to reach the limits established by EU Regulations and the sensory rancidity threshold was also assessed in order to address the effect on shelf-life and flavor stability.

3 METHODS

3.1 Purified Olive Oil (POO) Preparation

Cornicabra virgin olive oil was stripped of pro- and antioxidants and trace metals by adsorption chromatography.[5] 100 g of virgin olive oil in 1000 ml distilled hexane were passed through a column (i.d. 2 cm) filled with 70 g alumina (type 507c neutral, Fluka Buchs, Switzerland) activated for 4 h at 180 °C, and collected in darkness. The absence of antioxidants in the POO was verified using the AOCS official method (method number Ce 8-89) for α-tocopherol and the method previously described by those authors for phenolic compounds.[6]

3.2 Oxidation Experiments

Twelve 36.6 g (40 ml) samples of POO were stored in darkness at different temperatures (25, 40, 50, 60 and 75 °C) in 125 ml open amber glass bottles (i.d.: 4.2 cm; surface area exposed to the atmosphere: 13.85 cm^2). One bottle was taken from the incubator for analysis at scheduled times.

3.3 Volatile Compounds (Adapted from Jelen et al.[7])

Three grams of oil sample were placed in a 10 ml headspace vial and kept at 28 °C for one hour. The Teflon–lined septum covering the vial was pierced with a SPME (Solid Phase Micro-Extraction) needle and a fiber (100 µm divinylbenzene/carboxene/ poly(dimethylsiloxane) (DVB/CAR/PDMS) (Supelco Inc., Bellefonte. PA) was exposed to the oil headspace for 20 min. The fiber was then retracted into the needle and immediately transferred and desorbed for 5 min in the gas chromatograph injection port. A gas chromatograph equipped with a FID detector was used. Compounds were resolved on a HP-5 fused silica column (30 m x 0.32 mm x 0.25 µm, Agilent Technologies, U.S.A.) under the following conditions: injection port temperature 240 °C; helium flow 2 ml/min; oven temperature ramp 35 °C for 5 min, 4 °C/min up to 100 °C and then 15 °C/min up to 220 °C (final hold 5 min). Volatile compounds were identified by comparison with standard substances. The following reference compounds were used: hexanal from Sigma Chemical Co. (St. Louis. MO); heptanal, octanal, nonanal, *t*-2-hexenal, *t*-2-heptenal, *t*-2-octenal and *t,t*-2.4-decadienal from Aldrich Chemie (Steinheim. Germany); and *t*-2-decenal from Fluka Chemie (Switzerland).

3.4 Sensory Analysis

POO samples were assessed for aroma changes only, at twelve different stages of the oxidation process by a Sensory Panel of eight assessors from the University of Castilla-La Mancha and the Protected Designation of Origin "Montes de Toledo" (Toledo. Spain). The purpose of the sensory analysis was to determine the recognition threshold of the rancid defect and to correlate it with the chemical composition of the oil at that stage of the oxidation process. The recognition threshold is defined as the level of a stimulus at which the specific stimulus can be recognized and identified. A rank probability plot is a useful tool for testing whether a set of individual thresholds are normally distributed; this graph also serves to locate the group threshold of the stimulus (75% of the correct answers by the panel). Assessors were therefore asked to evaluate differences between the aroma of the fresh POO (reference oil) and the twelve samples of POO removed from the incubator at different times. Fifteen ml of the oil were poured into a standard olive oil tasting glass (COI/T.20/Doc.no.5. 1987; corresponding to UNE 87021:1992). and the panelists marked the oil samples in which they could recognize the defect on an appropriate form (Method of Investigating Sensitivity of Taste; ISO 3972:1996, corresponding to UNE 87003-2000).

All experiments and analytical determinations were carried out at least in duplicate.

3.5 Statistical Analysis.

Statistical analyses were performed using SPSS 11.5 statistical software (SPSS Inc. Chicago. IL).

4 RESULTS

Results showed that after the induction period (IP), the decrease in the unsaturated fatty acids followed a pseudo-zero-order kinetic, whereas the formation of secondary oxidation products followed a pseudo-first-order kinetic whose rate reaction constant also increased exponentially with temperature in the range studied, from 25 up to 75 °C.[8]

4.1 Volatile Compounds

Many volatile compounds have a very low sensory recognition threshold and are therefore responsible for rancidity and the consequent loss of olive oil organoleptic quality. For this reason, determination of volatile compounds, together with sensory evaluation, is one of the best means of verifying the oxidation state of fats and oils.

The evolution of the total content in volatile compounds during storage of the purified olive oil (POO) at different temperatures (25-75 °C) is depicted in Figure 1. As expected, the concentration of volatile compounds greatly increased with the storage temperature due to the known effect of temperature on the decomposition rate of triacylglyceride hydroperoxides.[1,2,9]

The major volatile compounds found in the oxidized POO were hexanal, 2-heptenal and 2-octenal, decomposition products of linoleate hydroperoxides; the least abundant was 2,4-decadienal, also formed from linoleic acid. Other aldehydes found in significant amounts, such as heptanal, octanal, nonanal and 2-decenal, were formed by the cleavage of oleate hydroperoxides. Finally, 2-hexenal was formed through decomposition of linolenate hydroperoxides.[9] The relative oxidation rate of the three main olive oil unsaturated fatty acids (oleate, linoleate and linolenate) was approximately 1:12:22,[8] which is close to that observed by Lea (1:12:25).[10] The absolute decreases in these fatty acids were 1.2%, 0.9% and 0.2% for oleic, linoleic and linolenic acids respectively. The principal oxidized fatty acid, then, was oleate, concentrations of which are much higher in the olive oil triacylglycerol matrix (80.2%, 4.8% and 0.5% for the three fatty acids respectively).

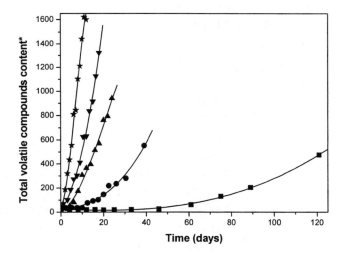

Figure 1 *Evolution of total volatile compounds* during storage at different temperatures.* ■, 25; ●, 40; ▲, 50; ▼, 60; ★, 75 °C. *expressed as the sum of the main components found: hexanal, 2-hexenal, heptanal, 2-heptanal, octanal, 2-octenal, nonanal, 2-decenal and 2,4-decadienal.

4.2 Sensory Rancidity

The purpose of the sensory analysis was to determine the sensory group's recognition threshold of rancidity and to correlate it with the chemical composition of the oil at that stage in the oxidation process. POO samples were assessed at twelve different stages in the oxidation process (for aroma changes only) by a Sensory Panel of eight assessors from the University of Castilla-La Mancha and the Protected Designation of Origin "Montes de Toledo" (Toledo, Spain).

Table 1 reports the values of the data for the oxidation indexes studied, corresponding to the group rancidity threshold determined by sensory analysis (75% correct answers in the recognition threshold test) in the course of oxidation in POO at the different experimental temperatures. As expected, the time taken to reach the rancidity threshold decreased considerably with rises in temperatures, from 63 days at 25 °C down to 4 days at 75 °C.

Table 1 *Oxidation index values at the sensory rancidity threshold in POO*

	Temperature (°C) and time (day)				
	25 (62.8 d)	40 (14.1 d)	50 (10.2 d)	60 (8.3 d)	75°C (4.2 d)
PV (meq/kg)	49.96 ± 1.98^b	41.86 ± 0.70^a	60.60 ± 0.78^e	68.06 ± 0.46^d	55.88 ± 1.40^c
K232	3.36 ± 0.06^b	2.84 ± 0.02^a	3.66 ± 0.03^c	3.96 ± 0.03^d	3.53 ± 0.53^{c,d}
K270	0.060 ± 0.000^b	0.046 ± 0.003^a	0.070 ± 0.003^c	0.100 ± 0.002^e	0.074 ± 0.021^d
AnV	3.15 ± 0.27^a	3.64 ± 0.06^b	5.79 ± 0.07^c	9.17 ± 0.04^d	9.58 ± 0.21^e
oxTG (%)	2.23 ± 0.06^b	1.95 ± 0.12^a	3.21 ± 0.03^c	3.08 ± 0.09^c	3.02 ± 0.27^c
DTG + PTG (%)	0.10 ± 0.01^b	0.05 ± 0.00^a	0.16 ± 0.02^c	0.20 ± 0.02^d	0.25 ± 0.03^e
C18:1 (%)*	0.33 ± 0.09^b	0.27 ± 0.09^b	0.38 ± 0.03^b	0.74 ± 0.02^c	0.11 ± 0.05^a
C18:2 (%)*	6.01 ± 0.20^b	4.39 ± 0.45^a	6.83 ± 0.02^b	8.09 ± 0.03^c	6.33 ± 0.74^b
C18:3 (%)*	11.52 ± 0.52^b	8.92 ± 0.55^a	12.70 ± 0.11^b	14.84 ± 0.11^c	11.94 ± 0.84^b
Hexanal**	34.8 ± 9.7^a	34.6 ± 8.2^a	162.7 ± 47.7^b	211.6 ± 8.4^{b,c}	244.3 ± 53.3^c
2-Hexenal**	4.19 ± 0.41^a	4.04 ± 0.44^a	12.03 ± 0.66^b	19.21 ± 1.02^c	21.55 ± 1.45^d
Heptanal**	1.63 ± 0.51^a	2.89 ± 0.29^a	10.15 ± 1.82^b	11.32 ± 1.06^b	12.17 ± 2.20^b
2-Heptenal**	22.0 ± 2.42^a	34.8 ± 3.6^b	104.2 ± 6.5^d	94.6 ± 1.3^c	123.9 ± 9.7^e
Octanal**	2.17 ± 1.23^a	2.64 ± 0.30^a	9.50 ± 2.21^b	10.00 ± 1.12^b	14.92 ± 3.15^c
2-Octenal**	1.97 ± 0.23^a	2.11 ± 0.26^a	8.46 ± 0.40^b	9.97 ± 0.73^c	14.96 ± 0.85^d
Nonanal**	1.44 ± 0.32^a	3.40 ± 0.66^a	7.81 ± 1.84^b	8.58 ± 0.37^b	13.16 ± 3.33^c
2-Decenal**	1.40 ± 1.05^a	2.53 ± 0.22^b	2.49 ± 0.18^b	5.17 ± 0.24^c	9.23 ± 0.26^d
2,4-Decadienal**	0.07 ± 0.01^a	0.25 ± 0.04^c	0.20 ± 0.02^b	0.58 ± 0.02^d	0.75 ± 0.03^e
Total volatiles**	69.7 ± 15.1^a	90.5 ± 3.7^a	314.5 ± 62.3^b	373.5 ± 9.1^b	454.9 ± 71.0^c

PV, peroxide value; K232 and K270, UV absorption characteristics at 232 and 270 nm; AnV, anisidine value; oxTG, oxidized triacylglycerols; DTG+PTG, dimmers and polymers of TGs.
* Decrease with respect to the initial content.
** Concentration expressed as arbitrary SPME-GC peak area unit.
Mean values in the same row with different letters are statistically different ($p \leq 0.05$).

In addition, depending on the storage temperature used, many oxidation indexes showed statistically significant different values at the time taken to reach the sensory rancidity recognition threshold. It is especially interesting to note that the total volatiles content, corresponding to the time in which the rancidity recognition threshold in POO was reached, increased with storage temperature. The threshold did not therefore reflect the total or individual content of any one volatile compound in the oxidized oil, and may rather have reflected the ratio of concentrations of different substances.

Further analysis of the relationship between the rancidity threshold and SPME-GC volatile compound concentrations suggested that 2,4-decadienal was a key compound in the relationship between these two methods. This has suggested by other authors in regards to seed oils.[9] Indeed, according to Figure 2 the rancidity threshold coincided with the induction period (IP) for the kinetics of 2,4-decadienal formation. Figure 2 reports this behaviour at 50 °C, but the same effect was observed at all of the other four temperatures studied (25-75 °C). Such instrumental measurement of the IP of 2,4-decadienal could therefore provide a key to the relationship between sensory perception of oxidized olive oil and analytical chemistry methods.

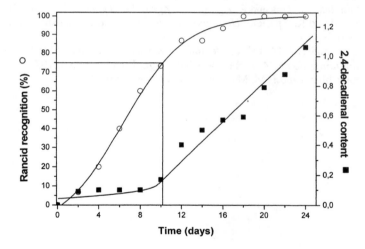

Figure 2 *Changes in 2,4-decadienal content and rancidity recognition threshold in the course of oxidation at 50 °C.* ○, correct answers in rancidity recognition; ■, 2,4-decadienal content (arbitrary GC peak area unit).

In a previous paper,[8] it was suggested that the time required to reach the rancidity threshold was related to the IP of the 2,4-decadienal formation reaction. Analysis of the results in the temperature range studied (25-75°C) showed a high linear correlation ($r^2 = 0.999$) between these two parameters. This instrumental determination may therefore be useful in measuring sensory recognition of rancidity, at least in POO. The value of the slope of the linear regression (0.97) was close to 1, and therefore the time taken to reach the rancidity threshold is approximately the same as the IP for formation of 2,4-decadienal.

Acknowledgements

This research project was supported by the CICYT (AGL 2001-0906).

References

1. E.N. Frankel, *Trends Food Sci. Technol.*, 1993, **4**, 220.
2. E.N. Frankel, *Lipid Oxidation*. The Oily Press Ltd, Dundee, 1998.
3. S. Gómez-Alonso, M.D. Salvador and G. Fregapane, *J. Agric. Food Chem.*, 2002, **50**, 6812.
4. S. Gómez-Alonso, M.D. Salvador and G. Fregapane, *J. Am. Oil Chem. Soc.*, 2004, **81**, 177.
5. H.H. Jelen, M. Obuchowska, R. Zawirska-Wojtasiak and E. Wąsowicz, *J. Agric. Food Chem.*, 2000, **48**, 2360.
6. C.H. Lea, *J. Sci. Food Agric*, 1952, **3**, 586.
7. M.T. Morales, J.J. Ríos and R. Aparicio, *J. Agric. Food Chem.*, 1997, **45**, 2666.
8. J. Harwood and R. Aparicio, *Handbook of Olive Oil*, Aspen Publishers, Gaithersburg, MD, 2000, p. 459.
9. K. Warner and N.A.M. Eskin, *Methods to Assess Quality and Stability of Oils and Fat-Containing Foods*, AOCS Press, Champaign, Illinois, 1995, p.107.
10. N.V. Yanishlieva and E.M. Marinova, *Food Chem.*, 1995, **54**, 377.

DEWATERING-IMPREGNATION-SOAKING IN NONCONVENTIONAL SOLUTIONS AS SOURCES OF NATURAL FLAVORANTS FOR COLOMBIAN AZÚCAR VARIETIES OF MANGO (*Mangifera indica*) AND PEROLERA PINEAPPLE (*Ananas comosus*).

A. L. Morales[1], M. P. Castaño, D. C. Sinuco, G. Camacho[2] and C. Duque[1].

[1]Department of Chemistry, Universidad Nacional de Colombia, Colombia.
[2]ICTA, Universidad Nacional de Colombia., Colombia.

1 ABSTRACT

Dewatering-impregnation-soaking (DIS) is a useful technique for fruit preservation. A process for DIS of pineapple var. Perolera and mango var. Azúcar using sucrose, sorbitol and glycerol is described. In this study we approached DIS from the perspective of transfer of fruit flavor into the dehydrating agents used. For the purposes of this study we analysed free volatiles by headspace-SPME-HRGC-MS and glycosidic aroma precursors through the formation of TFA derivates of glycosidic extract by HRGC-MS (NCI) of fruits and products of DIS process. Free volatiles and some glucosides were transferred in DIS process. These results suggest that dehydrating agents could be utilized as flavoring agents in other foods.

2 INTRODUCTION

Osmotic dehydration, better defined as dewatering-impregnation-soaking (DIS) in a concentrated solution, is a technique that increases solid concentration in foods. The process is carried out by immersing whole fruits or pieces in concentrated solutions of substances compatible with the material to be treated. Once contact takes place, there are three spontaneous fluxes of mass transfer. Two fluxes occur from the interior of the food into the solution, a major flux of water and a minor one composed of the solutes capable of crossing the semi permeable membranes of the food. In a reverse process, the third flux, solutes move from the solution into the food; this flux, though, is always less significant than water transport.[1]

Compounds or solutions, used as osmotic dehydration agents are selected based on three factors: sensory characteristics of the food to be processed, molecular masses of the solutions and costs.[2] Sucrose and corn syrup are the most frequently used agents as these are highly compatible with fruit. However, these agents are susceptible to microbial degradation and fermentation and are discarded after osmotic dehydration, increasing process costs.

In order to achieve a maximum benefit of osmotic dehydration products, we examined on the use of other substances as alternatives to the traditionally used agents. Sorbitol and glycerol were selected due to their capacity to decrease water activity and retarding crystallization conditions,[3] and were compared with sucrose syrup, a traditional dehydrating agent.

Sorbitol and glycerol have been used in some studies on osmodehydrated vegetables with good results in terms of the sensory qualities and mass transfer of the processed products.[4,5] Bolin et al.,[4] found that osmotic syrups can be reconcentrated and reused for osmotical water removal through at least five complete cycles without adversely affecting the fruit being concentrated, even though the properties of the osmotic medium do change. Shipman et al[5] developed a method, by soaking celery in glycerol, to improve the texture in the reconstituted dehydrated product.

Previous studies of DIS have been focused on two aspects: achieving a major efficiency in the process[6,7] and obtaining foods with good sensory and nutritional characteristics.[8,9,10,11] Since aroma is one of the best indicators of fruit quality, its retention during DIS of muskmelon[12,13] and kiwi[14] has been studied. However, to the best of our knowledge, studies on flavor enrichment of the osmotic solutions with that of fruits has not been reported.

The fruits selected for this study, mango (*Mangifera indica*) var. "Azúcar" or "Vallenato" and pineapple (*Ananas comosus*) var. "Perolera" were of Colombian varieties highly appreciated because of their superb sensory qualities such as aroma and taste. The process for osmotic dehydration was studied in these fruits with two purposes, the first was to obtain high quality fruits for direct consumption, and the second was to take advantage of osmotic solutions enriched with the fruit aroma in order to use them as flavoring or aromatic agents.

3 MATERIALS AND METHODS

Mangoes of the "Azúcar" variety were collected in Anolaima (Cundinamarca, Colombia) and selected according to table ripe parameters: yellow and red rind, 24° Brix soluble solids and pulp pH of 4.0. Pineapples of the "Perolera" variety were collected in Lebrija (Santander, Colombia) and selected according to established ripeness parameters: yellowish-orange rind, easily detachable crown leaves with reddish tips, pulp pH of 3.5 and 12.5° Brix soluble solids.

3.1 Osmotic Dehydration

Fruits were washed, peeled and chopped. Pineapples were cut in triangular shapes of 2x2x1 cm and mangoes were cut in 1x1 cm cubes. Osmotic agents used were 70% sorbitol, 70% sucrose syrup and 65% glycerol.

Fruit pieces (ca 200 g) were introduced in polyethylene bags and the different osmotic agents were added to the bags in a ratio 1:3 (w/w). Fruits were allowed to dehydrate in the osmotic agent whit stirring at 80 rpm, at 30°C, over 4 h. After this process, fruit pieces were separated from the osmotic agent, rinsed with potable water and pat dried with paper towels. Each treatment was performed in triplicate with different fruit at a time.

Fresh and dehydrated fruit pieces were analyzed to determine free and glycosidically bound volatiles, moisture content, soluble solids, water activity value (Aw), color and sensory qualities. Physicochemical parameters were evaluated using the usual methodology.[15] Fruit color was measured with a Chromameter CR-300 and expressed in luminosity (L), red (a) and yellow (b) units. Osmotic solutions (OS) were analyzed to determine water activity (Aw) and soluble solids (° Brix). Additionally, the visible spectra (350-650 nm) of the osmotic solutions were recorded using a UV-VIS Biorad Smart-SPEC 3000 spectrophotometer. Each experiment was performed in triplicate.

3.2 Analysis of Free and Bound Volatiles (GBVs)

Free volatiles in fresh and osmodehydrated fruits (OF) and in osmotic solutions (OS) were analyzed by headspace solid phase microextraction (HS-SPME) and GC-MS using the work conditions previously established.[16,17]

Glycosidically bound volatile compounds (GBVs) were obtained following the procedure published by Gunata[18], the extract was analysed as TFA derivates[19] by GC-MS under EI and NCI conditions.[20]

3.3 Sensory Evaluation

A trained panel of eight members (Lucta Grancolombiana S.A) evaluated sensory qualities of osmodehydrated fruits. Sensory qualities in the used OS were analyzed after dilution by a multiple comparison method. The sensory qualities considered were: aroma, taste and color intensities of the different dehydrating agents after the processes. Furthermore, the aroma of OS was described with the same characteristic descriptors as fruits aroma.

4 RESULTS AND DISCUSSION

All resulting OS acquired a characteristic mango or pineapple aroma respectively, being polyol the most intense. The aroma profile of the OS showed a similar profile for the three mango and pineapple syrups, verifying the fruit aroma enrichment of these OS. However, in the sucrose syrup were found stronger ripe, overripe and fermented notes, showing susceptibility to fermentation and non-desirable aroma formation in this product.

Fruit taste in the sucrose syrup was described as the most intense, followed by glycerol and sorbitol syrups, due to their aftertaste masking the fruit flavour.

4.1 Evaluation of recycled osmotic solutions

For monitoring aroma changes during DIS process, we evaluated total aroma[21] of recycled OS, showing increasing values after each cycle, sorbitol readings showed the highest change.

4.1.1 Free volatiles. Free volatiles identified in mango and in the solutions obtained after osmodehydration as well as their relative abundance, are listed in Table 1. In the HS-SPME of the fruit and in the three osmotic solutions reused, we found a high percentage of terpenic components, mainly monoterpes and sesquiterpenes being the last present in minor proportion. Most of these compounds show smell notes that contribute to the total aroma of the mango variety considered in this study. For example, α-pinene, β-pinene, 3-carene, β-mircene, α-terpinene, limonene among other monoterpenes, contribute with the green rind and citric notes; while sesquiterpenes like α-gurjunene, β-cariofilene, α-humulene and β-selinene contribute to the fruity, pulp and ripe odors.[16]

Additionally, furfural was found in the sucrose syrup. Furfural and its derivates are formed as breakdown products of carbohydrate, whose formation is catalized by the acids leached from the fruit.[4] Hence, sucrose syrup could favor off-flavors of the fruit under the osmodehydration conditions used.

Table 1. *Detected compounds in mango fruit and in osmotic solutions by SPME-GC-MS*

N°	COMPOUND	RI (DB-Wax)	Fresh fruit	Osmotic solutions		
				Sucrose	Sorbitol	Glycerol
1	ethyl acetate	872	+	-	-	-
2	Ethanol	900	++	+++	++	++
3	ethyl butanoate	1020	+	-	-	-
4	α-pinene	1021	++++	++	+++	+++
5	β-pinene	1105	+	+	++	++
6	ethyl 2-butenoate	1114	+	+	+	+
7	3-carene	1148	++++++	+++	+++	+++
8	β-mircene	1154	++++	+	++	+
9	α-terpinene	1175	+	+	+	+
10	Limonene	1190	+++	+	+	+
11	β-phellandrene	1199	+++	+	+	+
12	γ-terpinene	1240	+	-	+	+
13	α-terpinolene	1271	+++	+	++	+
14	Furfural	1489	-	+	-	-
15	α-gurjunene	1515	+	-	+	+
16	β-cariofilene	1582	+++	+	++	++
17	α-humulene	1660	++	+	+	+
18	β-selinene	1706	+++	+	+	++
19	3,5-dihydroxy-6-methyl-2,3-dihydro-4H-pyran-4-ona	2325	-	++	-	-
20	Benzoic acid	2453	+	-	-	-
21	5-hydroxy-methyl-furfural	2598	-	+++	-	-
22	4-hydroxy-benzoic acid	2700	+	-	-	-

+= 0-1%, ++ >1-2%, +++ >2-5 %, ++++>5-10%, +++++>10-50%, ++++++>50%, - no detected

For pineapple, as shown in Table 2, during the OD process, a transference of methyl and ethyl esters of butanoic and hexanoic acid through sucrose, sorbitol and glycerol syrups occurs. These compounds and furaneol have been reported as potent odor-active components responsibles for the aroma of pineapple var. Perolera.[17]

In the HS-SPME analysis of the osmodehydrated fruits with the different agents, some monoterpenes and some esters in mango and pineapple respectively, were detected in lower concentrations, compared with those found in the fresh fruit. Another important difference occurs in the osmodehydrated fruits with sucrose, where we found a major amount of ethanol compared with the fresh fruit, as well as 3-hidroxy-methyl-furfural, furfural and 2,3-dihidro-3,5-dihidroxy-6-methyl-4H-piran-4-one.

The increase of ethanol level suggests the occurrence of fermentation process. Regarding the sugar degradation compounds, their presence is presumably due to the increase in the sugar level that could contribute to the higher levels of Maillard/sugar derived products. Hence, the use of sorbitol and glycerol agents represents an advantage for the DIS process because aroma enriched solutions of better quality were obtained with these agents.

Table 2. *Detected compounds in pineapple fruit and in osmotic solutions by SPME-GC-MS*

N°	COMPOUND	RI (DB- Wax)	Fresh fruit	Osmotic solutions		
				Sucrose	Sorbitol	Glycerol
1	Ethyl acetate	900	+++	+++++	++++	+++++
3	Ethanol	920	++	+++++	+++++	+++
4	Methyl 2-methylbutanoate	930	++++	++++	+++++	+++++
5	Methyl butanoate	984	++	+	+	+
6	Ethyl butanoate	1002	+	+++	+	+++
7	Ethyl 2-methylbutanoate	1023	+++	+++	+	++
8	3-methylbuthyl acetate	1120	+	+++++	+++	+++
9	Methyl hexanoate	1189	++++	++++	+++++	+++++
10	Ethyl hexanoate	1236	++++	+++++	++++	+++
11	Methyl-E-3-hexenoate	1248	+++	+	+	+
13	Ethyl-E-2-hexenoate	1295	+++	+	+	+
14	Methyl octanoate	1388	+++++	-	+	+
15	Ethyl octanoate	1434	+++++	-	-	-
16	Methyl-E-4-octenoate	1438	++	-	-	-
18	Methyl-E-3-octenoate	1463	+	-	-	-
21	Methyl-3-acetoxy-2-methylbutanoate	1507	+++	-	-	-
22	Methyl 2-acetoxybutanoate	1528	+++	-	-	-
23	Methyl 3-methylthiopropanoate	1534	++++	-	++++	+++
24	Methyl 2-acetoxybutanoate	1545	++	-	-	-
25	Methyl-3-acetoxy-2-methylbutanoate	1564	+	-	-	-
26	Methyl decanoate	1604	++	-	-	-
27	Methyl benzoate	1624	+	-	-	-
28	Ethyl benzoate	1678	+	-	-	-
31	Methyl 5-acetoxyhexanoate	1781	+	-	-	-
32	Ethyl 5-acetoxyoctanoate	1962	+			
34	Furaneol	2022	+++	+	+	+
36	5-hidroxymethylfurfural	2507	-	++++	-	-

+= 0-1%, ++ >1-2%, +++ >2-5 %, ++++>5-10%, +++++>10-50%, ++++++>50%, - no detected

4.1.2 Glycosidically bound volatiles (GBVs). In the osmotic process, free volatile compounds flow into the osmotic solution. This fact suggests that transference of aroma precursors may be feasible too during the process. Taking this into account we proceeded to analyse the glycosidically bound volatiles present in the osmotic agents.

Table 3 shows the names and concentration of the glycosides identified in mango fresh fruit and in the osmotic solutions resulting in the DIS process. Sorbitol is the best extracting agent for GBVs followed by glycerol. On the other hand only two GBVs were detected when sucrose was used. In the case of pineapple, only two glucosides of furaneol could be identified in the samples.

GBVs can be used as fragrant materials with long-lasting efficacy[22] and osmotic solutions can be reused many times.[23] Hence, osmotic solutions have high potential as natural flavouring agents to be used in food, cosmetic or pharmaceutical industry because free and bound aroma content can be increased each time as it was demonstrated here.

Tabla 3. *TFA derivates of glycosides detected for HRGC-MS in mango pulp and in reused solutions of DIS process*

Compound	RI (DB-5)	Fresh pulp	Osmotic solutions		
			Sorbitol	Glycerol	Sucrose
3,7-dimethyl-2-octen-1,8-diol of β-D-glucopyranoside	1508	+	+	-	-
Furaneyl-β-D-glucopyranoside Isomer I	1724	+++	+++	++	+
Furaneyl-β-D-glucopyranoside Isomer II	1729	+++	+++	++	+
Benzyl-β-D-glucopyranoside	1756	+++	+++	+	-
2-phenyl-ethyl-β-D-glucopyranoside	1836	++	++	+	-
β-D-glucopyranoside of 3-phenyl propanoate	2136	+++	+	-	-
Coniferyl-β-D-glucopyranoside	2167	+	-	-	-
2-phenyl-ethyl-rutinoside	2207	+++	-	-	-
(6S, 9R)-vomifolyl-β-D-glucopyranoside	2322	+	+	-	-

+ 0 – 0.1 mg/kg, ++ 0.1-0.5 mg/kg, +++ 0.5- 2 mg/kg, ++++>2 mg/kg of pulp or of osmotic solutions

4.1.3 Color evaluation. Color transfer from the fruit to the osmotic agents was measured through the VIS spectrum of OS obtained after the DIS process. The absorption spectra of sucrose, sorbitol and glycerol syrups show an absorption maximum at 406 nm (0.533, 0.873 and 0.606 respectively) for mango and (0.465, 0.270 and 0.269 respectively) for pineapple. As can be noted during the process, a significant flow of pigments from the fruit to the osmodehydrating agents occurs, being greater with sorbitol for mango and with sucrose for pineapple.

Results here reported suggest the DIS process could be applied to develop flavoring agents where color would play an important role.

4.2 Evaluation of Osmodehydrated Fruit

As shown in Table 4, and considering changes in water activity, sorbitol is the strongest dehydrating agent followed by, glycerol and sucrose syrup. Sucrose syrup is a commonly used osmotic dehydrating agent. Although, results show that sucrose syrup has the least dehydrating strength, it is close to that of glycerol and sorbitol syrup. Hence these agents can be used as a sucrose replacement since they are less prone to bacterial decay. Solid gain was greater in mango dehydrated in sorbitol syrup, closely followed by sucrose syrup. Total solid gain shows a direct influence on the flavor of the dehydrated fruit. On the other hand, mango pieces dehydrated in glycerol were perceived as the least sweet.

Table 4. *Mass transfer parameters of mango in DIS*

Parameter	Initial Conditions				Products DIS					
	Fruit	Suc	Sorb	Glyc	Suc		Sorb		Glyc	
					OF	OS	OF	OS	OF	OS
%H	72				62.8		54.0		57.3	
Aw	0.971	0.798	0.170	0.171	0.77	0.939	0.959	0.891	0.968	0.919
°Brix	24	70.0	67.2	56.5	31.0	48.6	32.0	45.5	31.0	44.2
SG					3.2		4.7		2.3	
WL					19.5		36.2		31.2	

OF: Osmodehydrated Fruit; OS: Osmotic solution obtained after second DIS; Suc: Sucrose; Sorb: Sorbitol; Glyc: Glycerol.

As shown in Table 5, mass transference between the different osmotic agents and fresh pineapple was similar to that of mango. The greatest solid gain took place in pineapple dehydrated with sorbitol. Although this gain is quite high it imparts an unpleasant aftertaste in the dehydrated fruit.

Table 5. *Mass transfer parameters of pineapple in DIS*

Parameter	Initial Conditions				Products DIS					
	Fruit	Suc	Sorb	Glyc	Suc		Sorb		Glyc	
					OF	OS	OF	OS	OF	OS
%H	84.1				72.9		65.1		69.3	
Aw	0.997	0.798	0.170	0.171	0.987	0.962	0.967	0.918	0.979	0.940
°Brix	11.4	70.0	67.2	56.5	19.0	42.0	32.0	47.6	23.0	39.2
SG					7.8		7.8		3.8	
WL					25.9		38.5		30.1	

OF: Osmodehydrated Fruit; OS: Osmotic solution obtained after second DIS; Suc: Sucrose; Sorb: Sorbitol; Glyc: Glycerol.

Color measurements performed both in fresh and osmodehydrated fruits (Table 6) report an increase in luminosity (L) and yellow tones (b), and loss of red (a) tones in the osmodehydrated fruits. These results show that during DIS process, sucrose and sorbitol extract the highest amount of pigments from mango and pineapple respectively, without darkening of fruits.

Table 6 *Color measurement in DIS process of mango and pineapple*

Fruit	Mango				Pineapple			
	L	a	b	ΔE	L	a	b	ΔE
Fresh	62.22	20.71	59.50		69.99	-3.53	38.18	
Sorbitol	63.11	9.61	63.67	11.89	68.95	-5.89	46.43	8.64
Glycerol	67.22	12.05	69.19	13.92	69.89	-5.83	40.51	3.27
Sucrose	69.76	17.48	75.57	18.04	69.98	-5.99	42.01	4.55

$\Delta E = (\Delta L^2 + \Delta a^2 + \Delta b^2)^{1/2}$

Sensory evaluation. Results of sensory analyses of fresh and dehydrated mango with different agents show significant differences in all evaluated parameters. The aroma was significantly greater in fresh mango than in osmodehydrated mango. Sweetness increased significantly in mango dehydrated with sucrose syrup, whereas with sorbitol and glycerol the increase was not significant compared to fresh mango. Total flavor was compared in all dehydrated mango products and there were no significant differences among them. Brightness of the osmodehydrated fruit was enhanced with all agents compared to fresh fruit pulp.

Pineapple dehydrated with sorbitol, glycerol and sucrose syrup had better color, brightness and sweetness than fresh fruit. Significant taste differences between osmodehydrated fruits and fresh pineapple were found. Additionally, as was expected fruit aroma was significantly reduced with the treatment.

5 CONCLUSIONS

Results here obtained in DIS process of mango and pineapple confirm the transfer of important volatile compounds, flavor precursors and pigments from the fruit to the OS, suggesting that these syrups can be used as flavorings or aromatic agents in the food, cosmetic or pharmaceutical industry.

Potential applications of these dehydrating agents in order to obtain final products with better sensory qualities are under research.

Acknowledgements

Financial support by Colciencias-BID Colombia, IPICS - Uppsala University, Sweden is greatly appreciated. We are particularly thankful to professor Patricia Restrepo and to Lucta Gran Colombiana S.A for their skilful support in the sensory analysis.

References

1. D. Torregiani and G. Bertolo,. *J. Food Eng.*, 2001, **49**, 247-253.
2. L. Raoult-Wack, *Trends Food Sci. Technol.*, 1994, **5**, 255-260.
3. S. Guilbert, "Aditivos y agentes depresores de la actividad de agua". In: Acribia, *Aditivos y auxiliares de fabricación en las industrias agroalimentarias.* p. 195-220. México: Acribia edit 1990.
4. H. R. Bolin, C. C. Huxsoll, R. Jackson and K. C. Ng, *J. Food Sci.*, 1983, **48**, 202-205.
5. J. W. Shipman, A. R. Rahman, R. A. Segars, J. G. Kapsalis, and D. E. Westcott, *J. Food Sci.*, 1972, **37**, 568-571.
6. M. Dalla Rosa, F. Giroux, *J. Food Eng*, 2001, **49**, 223-236.
7. G. Giraldo, P. Talens, P. Fito, A. Chiralt. *J. Food Eng.* , 2003, **58**, 33-43.
8. H. H. Nijhuis, H. M. Torringa, S. Muresan, D. Yuksel, C. Leguijt, and W. Kloek, *Food Sci. Technol.*, 1998, **9**, 13-20.
9. H. Mujica-Paz, A. Valdez-fragoso, A. López-Malo, E. Paolou, and J. Weltichanes, *J. Food Eng.*, 2003, **56**, 307-314.
10. R. Moreira and A. M. Sereno, *J. Food Eng.*, 2003, **57**, 25-31.
11. T. Valencia Rodriguez, A. M. Rojas, C. A Campos and L. N, Gershenson. *Lebensm-wiss. U.-Technol*, 2003, **36**,415-422.
12. R. Lo Scalzo, C. Papadimitriu, G. Bertolo, A. Maestrelli and D. Torreggiani, *J. Food Eng.*, 2001, **49**, 261-264.
13. Maestrelli, R. Lo Scalzo, D. Lupi, G. Bertolo and D. Torreggian, *J. Food Eng.*, 2001, **49**, 255-260.
14. P. Talens, I. Escriche, N. Martinez-Navarrete and A. Chiralt, *Food Res. Int.*, 2003, **36**, 635-642.
15. AOAC. "Official Methods of Analysis". Washington, D.C. Association of Official Analytical Chemists 1990.
16. M. P. Castaño, A. L. Morales and C. Duque, *Submitted Food Chem.* 2004.
17. D. C. Sinuco, A. L. Morales and C. Duque, *Submitted Rev. Col. Quim.* 2004.
18. Y. Z. Gunata, C. L. Bayonove, R. L Baumes and R. E. Cordonier, *J. Chromatogr.*, 1985, **331**, 83-90.
19. S. G. Voiring, R. L Baumes, S. M. Bitteur, Z. Y. Günata and C. L. Bayonove, *J. Agric. Food Chem.*, 1990, **38**, 1373-1378.
20. A. L. Morales, C. Duque and E. Bautista, *J. High Resol. Chromatogr.*, 2000, **23**, (5), 379-38.
21. R. Azodanlou, C. Darbellay, J. L. Luisier, J. C. Villettaz, and R. Amado, *Z. Lebensm Unters Forsch B*, 1999, **208**, 254-258.
22. Ikemoto, Mimura and Kitahara, *Flav. Frag. J.*, 2003, **18**, 45-47 2003.
23. E. Garcia Martínez, J. Martínez Monzó, M. M. Camacho and Martínez-Navarrete, *Food Res. Int.*, 2002, **35**, 307-313.

Analysis

Analysis

APPLICATION OF GCxGC (COMPREHENSIVE 2DGC) IN FLAVOR ANALYSES

Hajime Komura[1] and Mineko Kawamura[2]

[1] Suntory Institute for Bioorganic Research (SUNBOR),
Wakayamadai 1-1-1, Shimamoto, Mishimagun, Osaka 618-8503, Japan
[2] Suntory Ltd., Process Development Department,
Yamazaki 5-2-5, Shimamoto, Mishimagun, Osaka 618-0001, Japan

1 ABSTRACT

Fifteen lemon-flavored soft drinks, one lemon-flavored low-alcoholic ready-to-drink beverage, and fresh lemon juice were analyzed on GCxGC. The majority of these commercial products showed very similar spot pattern, but slightly different from the fresh lemon juice, which contained much less spots corresponding to monoterpene alcohols. In contrast, unique spots were observed for some samples, but these could not be identified. Of the ready-to-drink low-alcoholic beverage, heat-challenge test was performed and analyzed by GCxGC. Decrease of limonene and linalool, while increase of deterioration compounds including *p*-cymen-8-ol were observed by heating. Thus it was shown that GCxGC is a powerful tool to distinguish the differences/changes of the components between samples.

2 INTRODUCTION

When analysing food and beverage flavor components using GC especially in regards to their flavor characteristics, the target compounds playing an important role in total flavor profile are often hidden under the peaks of co-existing components in larger quantities. In such cases, re-separation of the region where the target compound should appear, on a column with different separation characteristics, can frequently resolve the components co-migrated in the first column. This can be performed on a GC with a column-switching device.[1] However, conventional column switching on a two dimensional GC allows us to analyse only a few regions in one injection. Therefore when one has several regions of interest, one must repeat the injections until all the regions of interest are analyzed. In contrast to the conventional two dimensional GC, a new two dimensional GC method, GCxGC or comprehensive two dimensional GC enables us to perform two dimensional analysis throughout the entire chromatogram region to give the result as a two dimensional map, with one injection, within the first chromatograph time domain. This technique was first proposed by Liu and Phillips[2] and is commercially available from few sources.[3] After extensive modifications, primarily on the modulation method, detector types, and data processing[4,5] satisfactory results are obtained. The heart of GCxGC, the modulator, is a system that can trap the effluents from the first column and spontaneously re-inject them as pulses to the second column for the ultra fast GC separation.[2]

The major benefit of the GCxGC is not limited to its high peak resolving power furnished by GC multiplied by GC, but also to the polarity information of the components. It is possible to estimate the polarity of each components from the position on the two dimensional map when first non-polar then polar second column combination is applied.[6] In addition, the chromatogram as a two dimensional map allows us to perform pattern recognition much more readily than in the case of one dimensional chromatogram, and comparison of samples for their similarity or difference become much more obvious. Utilizing the GCxGC, commercially available lemon-flavored beverages were analyzed and compared to each other as well as with natural lemon juice. In addition, the change in the volatile components upon heat challenge was monitored with an alcoholic ready-to-drink (RTD) product.

3 METHOD AND RESULTS

3.1 Sample Preparation

Fifteen commercially available lemon-flavored soft drinks in glass or PET bottles, or in paper cartons, were purchased at shops in Osaka, and stored in a refrigerator until sample preparation. Natural lemon juice was obtained by squeezing fresh lemon fruits purchased in Osaka. A ready-to-drink (RTD) low-alcoholic beverage (alcohol content: 5%) in 350 ml aluminium can was purchased in Osaka, and the heat-challenge experiment was performed at 50 °C for 1, 3, 6, 9, 12, and 18 days. Heat challenged samples were stored in a refrigerator with unheated sample until sample analysis.

All the samples (200 ml) were distilled under reduced pressure at *ca.* 2.0 kPa (*ca.* 15 mmHg), on a water-bath heated to *ca.* 40 °C, and a finger-type cold trap placed between the distillate receiver and vacuum pump was cooled with dry-ice/ethanol mixture, while a condenser and a receiver were cooled with coolant maintained at *ca.* −5 °C. Distillation was finished after the residue level reached *ca.* 50 ml, and the distillates in the receiver and in the cold trap were combined for extraction. Combined distillate (*ca.* 150 ml) was extracted 3 times with 150 ml each of methylene chloride, and the extracts were combined and dried over anhydrous sodium sulphate. Extract was concentrated to 1 ml under atmospheric pressure, and the concentrate thus obtained was stored in a refrigerator until GCxGC analysis.

3.2 GCxGC Apparatus and the Analysis Condition

The GCxGC apparatus, a modified version of an Agilent Technologies' GC 6890 model equipped with an FID, was purchased from Zoex Corporation (Lincoln, NE, U.S.A.). The modulator was replaced with a KT2003 Loop Modulator[7] with an E50 electric valve controller from Beljer Electronics AB (Sweden). The hot jet pulse for the modulator with the duration of 150 msec at every 7.5 sec was generated by heating the pulsed gas through the hot jet tip maintained at as high a temperature as possible with the set point 350 °C. A continuous cold nitrogen flow was obtained by cooling the pressurized nitrogen gas through heat exchanger cooled with liquid nitrogen. Flow rates of the nitrogen gas for the hot jet pulse and the continuous cold stream were *ca.* 5 L/min for both at 500 and 300 kPa, respectively.

A set of columns, a methylsilicon SPB-1 (15 m x 0.25 mm id x 1.0 μm df) for the first dimension and a Supelcowax-10 (0.7 m x 0.1 mm id x 0.1 μm df) for the second dimension connected with the capillary modulation column without phase materials (1.5 m x 0.1 mm id) was used throughout the experiment. For the analysis, helium gas with constant pressure at 137.9 kPa (20.0 psi) was applied to the SPB-1 column head, 1.0 μL sample was injected manually with 1:2 split ratio while the oven was held at 50 °C. After injection, the oven temperature was held at 50 °C for 5 min before employing a temperature gradient to 220 °C at the rate of 4 °C /min, and held for 50 min at 220 °C.

For obtaining the analysis data, the FID signal was collected at the rate of 100 Hz using Agilent Technologies' ChemStation software, and exported as the 'csv' file format to be processed with the visualizing software GC-Image supplied from Zoex Corporation to generate two dimensional maps.

The analytical conditions, especially the column head pressure and the oven temperature condition were obtained by "trial and error" method, changing the conditions to get reasonably orthogonal spots of hydrocarbon mixtures.

3.3 Identification of Components

After the analysis of fresh lemon juice (Figure 1), authentic reference samples of the components reported in lemon juice[8] or deteriorated lemon juice were analyzed under the same conditions. Two-dimensional maps of the samples with those of authentic chemicals were superimposed to identify spots for each component. Unless the position on the two-dimensional map agreed exactly with those of the authentic components, the spots were left unidentified, and no further attempts were made to identify spot components. This is because without further evidence such as mass spectral data or other structural information, it is almost impossible to identify compounds. Particularly in the case of soft drinks and RTD's, these may contain unexpected flavor components foreign to natural lemon juice, hence identification of the components would be much more difficult than in the case of natural lemon juice.

3.4 Components in Soft Drinks

The majority of the lemon-flavored soft drinks gave similar spot patterns. Typical example is shown in Figure 2. Seven samples among 15 were in this group. However, as shown in Figures 3 to 5, some samples contain unique components not present in others. Two more samples out of 15 were similar to Sample A, two more to Sample B, and one more to Sample C. Judging from their longer retention times on the second dimension relative to major components observable on the fresh juice analysis map, they are likely more polar components than monoterpene mono-alcohols. The specific component in the Sample C (Figure 5) can be quite polar component, because the component gave a long streak along the second dimension familiar to fatty acids. Specific spot observed in Sample A (Figure 3) and Sample B (Figure 4) could not be elucidated either, but they are rather sharp thus not as polar as the specific component in the Sample C.

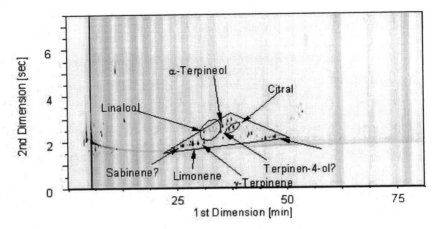

Figure 1 GCxGC map of fresh lemon juice. Regions specified with circles are of the monoterpene alcohols and of citral. Spots were assigned by superimposing the two dimensional GC maps of the fresh lemon juice with those of each authentic compound. For GC conditions, see text.

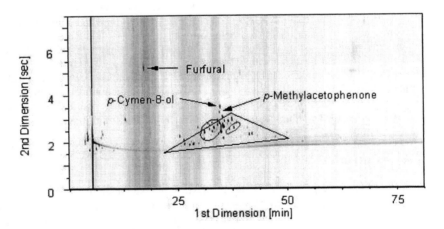

Figure 2 GCxGC map of a typical lemon-flavored soft drink. Spots for furfural, p-cymen-8-ol, and p-methylacetophenone were assigned by comparing with those of the authentic compounds.

Analysis

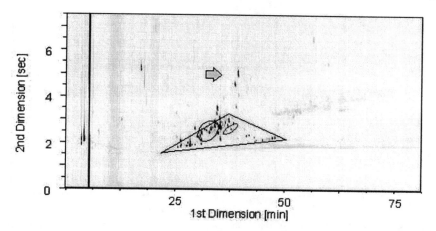

Figure 3 GCxGC map of the lemon-flavored soft drink Sample A. The arrow represents the specific spot observed only for this type.

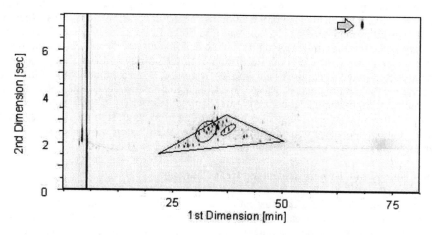

Figure 4 GCxGC map of the lemon-flavored soft drink Sample B. The arrow represents the specific spot observed only for this type.

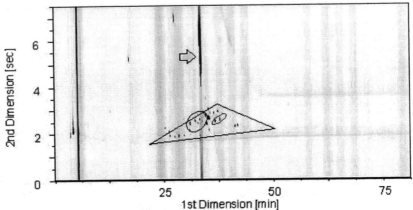

Figure 5 GCxGC map of the lemon-flavored soft drink Sample C. The arrow represents the specific spot observed only for this type.

The region where major components such as monoterpenes should appear is enlarged as Figure 6. The spots in the Region 1 present in fresh juice, expected to be some monoterpene hydrocarbons from their position, are missing in all soft drink samples. Among the samples, difference in the amounts of sesquiterpene hydrocarbons (Region 2), citral, as well as monoterpene alcohols was also obvious. Presence of alcohol components at higher concentration might partly due to the deterioration of terpenic hydrocarbons and citral under aqueous acidic condition common to lemon-flavored beverages, as furfural was already formed in almost of all samples examined. Therefore, it would be interesting to correlate the pattern changes caused by the change of components with deterioration of the samples. An example can be seen in the Sample B (inset D) in contrast to the Sample C(inset C); weaker spots, *i.e.*, smaller amount of citral in contrast to the darker spots, *i.e.*, larger amount of polar components including the deterioration products of citral,[9-12] *p*-cymen-8-ol and *p*-methylacetophenone.

3.5 Changes in Components During Heat Challenge

Lemon flavored beverages are susceptible to deterioration even during storage at room temperature. In order to visualize the changes in the components, a heat challenge test of a RTD low-alcoholic beverage sample was performed and the samples were analyzed. The two dimensional maps of the heated samples were superimposed according to the length of the heated period, and the changes in the components were examined. The maps for the unheated sample (Day 0) and Day 12 sample are shown in the Figures 7 and 8, respectively. Three different trends are apparent during heating, *i.e.*, A) constant decrease (shown with wedges in Figure 7), B) increase once then decreasing afterwards, and C) constant increase (shown with wedges in Figure 8). These are easily noted as the spot pattern changes. Among the spots identified, limonene and linalool decreased, while furfural, 5-hydroxymethylfurfural (HMF), *p*-cymen-8-ol, and 1,8-terpin hydrate increased. However, in this particular RTD case, increase of *p*-methylacetophenone, a known deterioration compound of citral, was not observed. This should be coincided with the fact that no citral was observed even in the non-heated sample. Among the spots that showed changes in their intensities upon heat challenge, only few could be identified because of the lack of their mass spectral data.

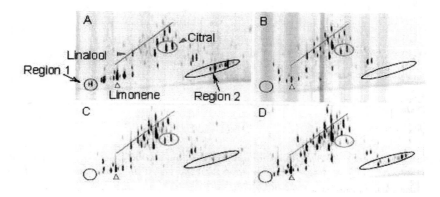

Figure 6 *Comparison at the region for the major components: (A) of fresh lemon juice; (B) of soft drink Sample C; (C) of soft drink Sample B; (D) of soft drink Sample A. Polar compounds not present in fresh juice above the line are commonly observed in soft drinks while spots in region 1 are missing.*

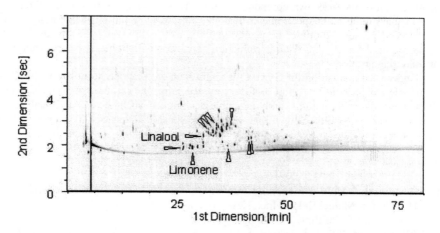

Figure 7 *GCxGC two-dimensional map of a low-alcoholic RTD without heat challenge. Wedges represent the spots that diminished upon heating. Unassigned spots were not identified.*

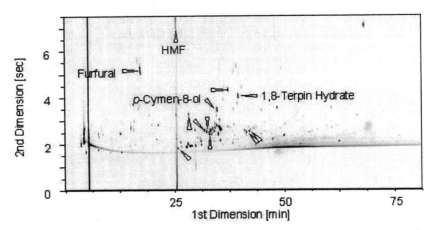

Figure 8 *GCxGC two-dimensional map of a low-alcoholic RTD after heated for 12 days. Wedges represent the spots that increased their intensities upon heating. Unassigned spots were not identified.*

4 CONCLUSION

As discussed above, GCxGC has proven to be a potent analytical tool particularly when determining the changes of, or differences between, samples in question. Unfortunately, identification of the unknown components on the two dimensional map is not facile, as a regular quadrupole mass spectrometer available for conventional GC/MS cannot be used for GCxGC because sampling frequency of the quadrupole mass spectrometer is not as fast as required for GCxGC data. Introduction of a time-of-flight mass spectrometer can so far solve this problem.

Under such a situation for GCxGC, it might be more practical to utilize GCxGC as it is, in order to pick up the point to focus, by accepting the limitations. Then one could supplement these limitations with available techniques such as conventional column switching style two dimensional GC with mass spectral detection. In such case, even element-specific detectors can be applicable to support identifying target components.

References

1. W. Bertsch, *"Multidimensional Chromatography, Techniques and Applications"*, ed., H.J. Cortes, Marcel Dekker Inc., New York and Basel, 1990, p. 74.
2. Z. Liu and J.B. Phillips, *J. Chromatogr. Sci.*, 1991, **29**, 227.
3. P.J. Marriott, P.D. Morrison, R.A. Shellie, M.S. Dunn, E. Sari, and D. Ryan, *LC•GC Europe*, December 2003, 2.
4. P.J. Marriott, *"Multidimensional Chromatography"*, eds., L. Mondello, A.C. Lewis, and K.D. Bartle, John Wiley & Sons Ltd., 2002, p. 77.
5. J.V. Hinshaw, *LC•GC Europe*, February 2004, 2.
6. J. Beens, R. Tijssen, and J. Blomberg, *J. Chromatogr., A*, 1998, **822**, 233.

7   E.B. Ledford, Jr., J.R. TerMaat, and C.A. Billesbach, *Zoex Corporation, Technical Note*, 2003, KT030606-1.
8   *"Volatile Compounds in Food, Qualitative and Quantitative Data"*, 7th edn., eds., L.M. Nijssen, C.A. Visscher, H. Maarse, L.C. Willemsens, and M.H. Boelens, TNO Nutrition and Food Research Institute, Zeist, 1996, Chapter 5E, p. 5.22.
9   K. Kimura, H. Nishimura, I. Iwata, and J. Mizutani, *J. Agric. Food Chem.*, 1983, **31**, 801.
10 P. Schieberle, H. Ehrmeier, and W. Grosch, *Z. Lebensm. Unters. Forsch.*, 1988, **187**, 35.
11 P. Schieberle and W. Grosch, *J. Agric. Food Chem.*, 1988, **36**, 797.
12 B.C. Clark, Jr. and T.S. Chamblee, *"Off-flavors in Foods and Beverages"*, ed., G. Charalambous, Elsevier Science Publisher, Amsterdam, 1992, p. 229.

AROMA ACTIVE NORISOPRENOIDS IN ORANGE AND GRAPEFRUIT JUICES

Kanjana Mahattanatawee[1,4], Russell Rouseff[1], Kevin Goodner[2] and Michael Naim[3]

[1]University of Florida, Institute of Food and Agricultural Sciences, Citrus Research and Education Center, Lake Alfred, FL USA
[2]USDA, ARS, Citrus and Subtropical Products Laboratory, Winter Haven, FL USA
[3]The Hebrew University of Jerusalem, Institute of Biochemistry, Food Science & Nutrition, Faculty of Agricultural Food and Environmental Quality Sci., Rehovot, Israel
[4]Siam University, Department of Food Technology, Faculty of Science, 235 Petkasem Rd., Phasicharoen, Bangkok 10163, Thailand

1 INTRODUCTION

Norisoprenoids are C9-C13 fragments from carotenoid plant pigments that are extremely potent odorants. Norisoprenoids have been identified as aroma impact compounds in a wide range of beverages, foods and spices, including, tea, tomato, and saffron.[1-3] Orange juice aroma is the result of a complex mixture of volatiles which lacks a specific character impact compound. Numerous studies have identified and quantified the major volatiles in orange juice in an effort to duplicate this aroma.[4-6] However, when combined, the volatiles present at the highest concentrations did not duplicate orange juice aroma, suggesting that essential aroma components were missing. Citrus carotenoids have been studied primarily for their role as pigments and have generally been ignored as a source of aroma compounds. Known carotenoid norisprenoid precursors such as β-carotene, α-carotene, neoxanthin, β-cryptoxanthin, lutein, violaxanthin and canthaxanthine have been identified in orange juice using HPLC.[7-11] Lycopene and β-carotene are the major carotenoids in Ruby red grapefruit juice.[12,13] Since orange and grapefruit juice contain carotenoids with the necessary structural features which could act as norisoprenoid precursors, we hypothesize that there may be additional norisoprenoids present in these juices. Grapefruit juice should have fewer norisoprenoids than orange juice because fewer carotenoids are present.

2 METHOD AND RESULTS

2.1 Chemical and Juice Samples

Most aroma standards were purchased from Aldrich (Milwaukee, WI). Octanal, limonene, linalool, nonanal, hexanal, decanal, dodecanal, 1,8 cineole, citronella, terpinen-4-ol, β-sinensal, β-myrcene, nootkatone, ethyl butyrate, acetaldehyde, geraniol, and nerol were obtained as gifts from SunPure (Lakeland, FL). α-Ionone was obtained as a gift from Danisco (Lakeland, FL). β-Damascenone and p-1-menthen-8-thiol were obtained from Givaudan (Lakeland, FL). Both 4-mercapto-4-methyl-2-pentanone and 4-mercapto-4-

methyl-2-pentanol were synthesized in our laboratory and 3-mercaptohexan-1-ol was bought from Interchim (Montlucon, France). Additional experimental details can be found in earlier publications.[14]

Late-season Valencia oranges (from Haines City Citrus Growers Association, Haines City Florida) were juiced using an FMC juice extractor at the Citrus Research and Education Center (CREC), Lake Alfred, Florida. The oranges were juiced using a commercial FMC juice extractor model 291 with standard juice settings and passed through a FMC model 035 juice finisher with a 0.02 inch screen (FMC Corp., Lakeland, FL). The freshly squeezed juice was immediately chilled and NaCl (36 g/100 mL of juice) added to inhibit enzymatic reactions. The pasteurised Ruby red grapefruit juice was a commercial sample purchased from a local supermarket.

2.2 Sampling of Headspace Volatiles

SPME fiber types were evaluated for their ability to concentrate norisoprenoids in the headspace of citrus juices. Headspace sampling was employed to identify and characterize citrus norisoprenoids as it involved minimal sample manipulation and would not co-extract carotenoids which might produce norisoprenoid artefacts in the 220 °C GC injector. Orange juice was fortified with four norisoprenoids: β-cyclocitral, α-ionone, β-damascenone and β-ionone and analyzed in the usual manner. Ten mL of juice was added to a 40 mL screw cap glass vial containing a micro stirring bar and sealed with a Teflon coated septa. The bottle and contents were placed in a combination water bath and stirring plate set at 40 °C and equilibrated for 30 min. Three SPME fiber types: 50/30 μm DVB/Carboxen/PDMS on a 1 cm and 2 cm StableFlex fiber, carbowax and PDMS (Supelco, Bellefonte, PA) were compared. Each fiber was inserted into the headspace of the fortified juice and exposed for 30 min. Subsequently, the fiber was thermally desorbed in the GC injector port for 5 min (220 °C). As shown in Figure 1, the most effective norisoprenoid fiber was 50/30 μm DVB/Carboxen/PDMS on 2 cm StableFlex fiber and it was used for the study of norisoprenoids in both orange and grapefruit juices with the exception that exposure time was extended to 45 min.

2.3 GC with FID and Olfactory Detection

Separation was accomplished with a HP-5890 GC (Palo Alto, CA) using either a DB-wax column (30 m x 0.32 mm. i.d. x 0.5 μm, J&W Scientific; Folsom, CA) or Zebron ZB-5 column (30 m x 0.32 mm. i.d. x 0.5 μm, Phenomenex, Torrance, CA). Column oven temperature (for DB-wax) was programmed from 40 to 240 °C (but 40 to 265 °C for ZB-5) at 7 °C/min with a 5 min hold. Additional experimental details can be found in recent publications.[15] Aroma intensities reported are the average from four GC-O trials. Identification of aroma-active components was based on the combination of sensory descriptors, standardized retention indices and mass spectra.

Fifty-nine aroma active components were detected in freshly squeezed orange juice and 40 were detected in commercial Ruby red grapefruit juice using SPME headspace sampling. Since the primary goal of this study was to determine if additional aroma active norisoprenoids were present, the region where norisoprenoids standards were known to elute was given the most scrutiny. Using standards of β-cyclocitral, α-ionone, β-ionone and β-damascenone, the target retention time region was established to be between 12 and 20 min. The inverted aromagram and concurrent chromatogram for this region are shown in Figure 2 for the case of orange juice. The identical plot from grapefruit juice is shown in Figure 3. Four norisoprenoids; β-cyclocitral, β-damascenone, α-ionone and β-ionone are detected in orange juice and only β-damascenone and β-ionone are detected in ruby red grapefruit juice.

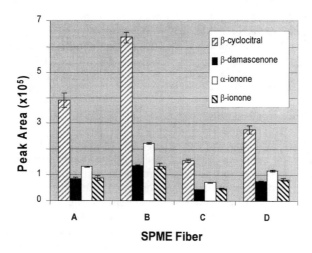

Figure 1 *Comparison of SPME fiber types to concentrate norisoprenoids in an orange juice fortified with each of the four norisoprenoids after 30 min. exposure at 40 °C. A = 50/30 mm DVB/Carboxen/PDMS, B = same as A, but 2 cm, C = Carbowax and D = PDMS, DVB = divinyl benzene, PDMS = polydimethysiloxane. Error bars represent one standard deviation from triplicate analyses.*

2.4 GC-MS Confirmations

GC-MS was employed to confirm the identities of the aroma active volatiles identified in the GC-O experiments. SPME volatiles were separated and analyzed using a Finnigan GCQ ion trap mass spectrometer (Finnigan, Palo Alto, CA) equipped with a DB-5, 60 m x 0.25 mm i.d., capillary column (J&W Scientific, Folsom, CA). The injector temperature and transfer line temperature were 200 and 250 °C, respectively. Helium was used as the carrier gas at 1 mL/min. The oven temperature program consisted of a single thermal gradient from 40 to 275 °C at 7 °C/min. The MS was set to scan from mass 40 to 300 at 2.0 scans/s in the positive ion electron impact mode. The ionization energy was set at 70 eV. Chromatographic peaks were identified using NIST 98 and Wiley, 6th Edition, database. Only those compounds with spectral fit value equal to or greater than 800 were considered positive identifications. Final identification was based on the combination of spectral matches and alkane linear retention index values (Kovat's Index) and aroma characteristics. Standards were used to confirmation identification, by comparing the resulting fragmentation pattern, retention index value and aroma descriptor. [16]

Analysis

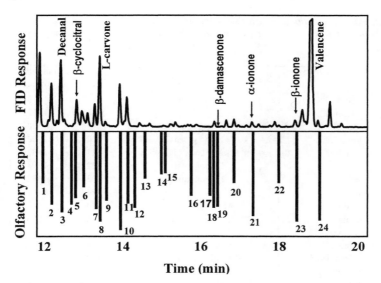

Figure 2 *GC-FID (top) and average time-intensity of four GC-O runs by two panellists (inverted, bottom) of fresh orange juice on ZB-5 column. Peaks 5, 19, 21 and 23 correspond to norisoprenoids identified in Table 1. (taken from[15])*

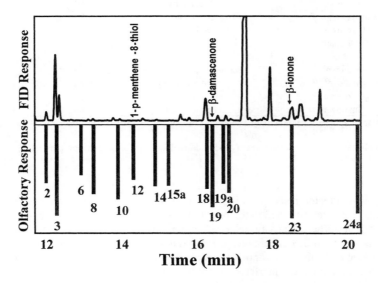

Figure 3 *GC-FID (top) and average time-intensity of four GC-O runs by two panellists (inverted, bottom) of Ruby red grapefruit juice on ZB-5 column. Peak numbers correspond to the compounds in Table 1.*

Table 1 *Identification, retention characteristics and aroma descriptors of aroma active compounds in fresh orange juice observed in the general region where norisoprenoids elute.*

No.	Compound	Aroma descriptor	Linear retention index	
			ZB-5	DB-wax
1	Terpinen-4-ol[b]	Metallic, musty	1175	1619
2	(Z)-4-decenal[a]	Green, metallic, soapy	1188	1542
3	Decanal[b]	Green, soapy	1198	1508
4	(E,E)-2,4-nonadienal[a]	Fatty, green	1209	1702
5	β-cyclocitral[b]	Mild floral, sweet, hay-like	1214	1632
6	Nerol[a]	Lemongrass	1222	1798
7	Neral[b]	Lemongrass	1236	1692
8	L-carvone[b]	Minty	1242	1747
9	Unknown	Metallic/ woody	1247	
10	Geraniol[a]	Citrus, geranium	1265	1853
11	Unknown	Soapy, almond	1274	
12	1-p-menthene-8-thiol[a]	Grapefruit	1281	1619
13	(E,Z)-2,4-decadienal[a]	Metallic, geranium	1293	1759
14	Geranial[a]	Green, minty	1310	1742
15	(E,E)-2,4-decadienal[a]	Fatty, green	1314	1819
15a	Unknown	Metallic	1340	
16	α-terpinyl-acetate[a]	Sweet	1349	1663
17	4,5-epoxy-(E)-2-decenal[a]	Metallic, fatty	1375	2010
18	Unknown	Sweet nutty	1380	
19	β-damascenone[b]	Tobacco, apple, floral	1383	1829
19a	Unknown	Solventy	1400	
20	Dodecanal[a]	Soapy	1403	1722
21	α-ionone[b]	Floral	1426	1863
22	Unknown[a]	Fermented, rancid butter	1459	
23	β-ionone[b]	Floral, raspberry	1484	1951
24	Unknown	Nutty	1510	
24a	Unknown	Passion fruit	1586	

[a] Identified by matching linear retention index values on ZB-5 and/or DB-wax, and aroma descriptors with standards.

[b] Identified using linear retention index matches (ZB-5 and/or DB-wax), aroma descriptor plus MS spectrum with standards

To achieve greater selectivity for the norisoprenoids of interest, selected ion chromatograms were reconstructed in the retention region where norisoprenoids standards were found to elute. The selectivity achieved is demonstrated in Figure 4. Several m/z values from the spectrum of standards were evaluated to find that which provided the best signal to noise ratio. In this way optimal signal from the norisoprenoid of interest was observed while minimizing interference from non-norisoprenoid components. For example, the selected ion chromatogram using m/z 137 provided a stronger signal for β-cyclocitral than that from m/z = 152 but was not as selective. The following ions were monitored for the specific norisoprenoids: β-cyclocitral, m/z = 137 and 152; β-damascenone, m/z = 175 and 190; α-ionone, m/z = 177 and 192; β-ionone, m/z = 177 and 192. Although only a single ion has been shown for each norisoprenoid in Figure 4, two or more selective ions were employed to confirm the presence of specific norisoprenoids

Analysis

in cases where the signal was too weak to provide a useful mass spectrum. The selected ions of m/z = 177 and 192 were employed for the selective detection of α-ionone and β-ionone. Selected ion chromatograms at m/z 177 provided excellent signal strength and selectivity for β-ionone but little signal for α–ionone which shares this common ion. This latter norisoprenoid was apparently present at much lower concentrations than β-ionone and needed a more sensitive means of detection which was achieved using m/z = 192 (Figure 4). β-Damascenone could not be detected using ion trap MS. This norisoprenoid has a major peak at m/z = 121, but this ion is shared with many terpenes which are ubiquitous in citrus samples. β-Damascenone was detected by GC-O but not by FID nor ion-trap MS demonstrating that human nose is more sensitive for β-damascenone (odour threshold 0.002 µg/Kg) than the instruments used in this study. However, by employing quadrupole mass spectrometer in the single ion monitoring mode at m/z values 175 and 190, the selectivity and sensitivity for β-damascenone peak was greatly improved. A peak corresponding to β-damascenone could be detected at the expected retention time. The ion trap mass spectral data for β-cyclocitral, α-ionone and β-ionone were: β-cyclocitral; 41(100), 137(55), 79(45), 83(30), 109(28), 67(25), 123(21), 81(16), 94(15), 119(14), 152 M+(7), α-ionone; 121(100), 93(90), 177(67), 91(64), 192 M+(29), 77(24), 109(22), 136(21), 92(19), 159(12), β-ionone; 177(100), 105(20), 133(19), 91(14), 178(14), 161(10), 107(9), 119(9), 192 M+(0.32).

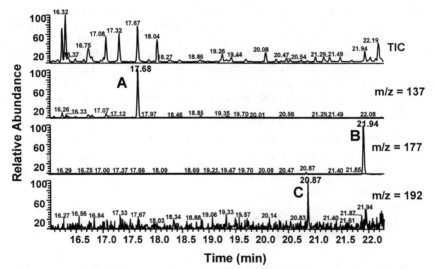

Figure 4 Comparison between total ion chromatogram and selected ion chromatograms (SIC) A: β-cyclocitral (m/z = 137), B: β–ionone (m/z = 177), C: α–ionone (m/z = 192). (taken from[15])

2.5 Relative Aroma Impact of Norisoprenoids

Many of the 59 aroma active compounds detected in fresh orange juice had similar aroma qualities. Therefore, they were categorized into seven groups based upon similarity of aroma descriptors. These seven groups were: 1. **citrusy/minty** - 1,8 cineole, nonanal, 3-mercapto-hexan-1-ol, citronellal, nerol, neral, L-carvone, geraniol, 1-p-menthene-8-thiol, geranial, nootkatone and unknown at LRI 963 (ZB-5), 2. **metallic/mushroom/geranium** - 1-octen-3-one, β-myrcene, octanal, terpinolene, (Z)-2-nonenal, terpinen-4-ol, (E,Z)-2,4-decadienal, 4,5-epoxy-(E)-2-decenal, β-sinensal and unknowns at LRI 1247 and 1589 (ZB-5), 3. **roasted/cooked/meaty/spice** - methional, 2-acetyl-2-thiazoline, unknowns at LRI 1380, 1459, 1510 and 1718 (ZB-5), 4. **fatty/soapy/green** - hexanal, (E)-2-hexenal, 3-(Z)-hexen-1-ol, 1-octanol, E-2-nonenal, (Z)-4-decenal, decanal, (E,E)-2,4-nonadienal, (E,E)-2,4-decadienal, dodecanal, (E,Z)-2,6-nonadienal and unknown at LRI 1274 (ZB-5), 5. **sulfury/solventy/medicine** - acetaldehyde, carbon disulfide, dimethyl sulfide, dimethyl disulfide, 2-methyl-3-furanthiol, 4-mercapto-4-methyl-2-pentanone, dimethyl trisulfide, 4-mercapto-4-methyl-2-pentanol and an unknown at LRI 818 (ZB-5), 6. **floral** - linalool, β-cyclocitral, β-damascenone, α-ionone, and β-ionone and 7. **sweet fruity** - ethyl-2-methylpropanoate, ethyl butyrate, ethyl-2-methylbutyrate, ethyl hexanoate and α-terpinyl acetate. All norisoprenoids in this study belong to group 6 because they possess a floral note juice concentrations. Linalool is also in this category. The relative intensities of the seven aroma groups with similar GC-O descriptors are summarized in table 2. Based on peak intensities from GC-O experiments, the entire floral group contributes approximately 10 % of the total orange juice aroma. Linalool contributes 2.2 %, α-ionone contributes 2.1 %, β-ionone contributes 2.2 %, β-cyclocitral contributes 1.7 % and β-damascenone contributes 1.9 %. Therefore, the four norisoprenoids contribute approximately 8 % of the total orange juice aroma and 78 % of the total floral group intensity and are the primary contributors to the floral note in orange juice. In grapefruit juice, the entire floral group contributes approximately 9 % of the total grapefruit juice aroma. Linalool contributes 3 %, β-ionone contributes 3 % and β-damascenone contributes 3 %. Therefore, the two norisoprenoids contribute approximately 6 % of the total grapefruit juice aroma and 66 % of the total floral group intensity and are the primary contributors to the floral note in grapefruit juice.

Table 2 *Comparison of aroma group intensities from orange and grapefruit juice GC-O data.*

	Orange juice	Grapefruit Juice
1. Citrusy/minty	25	18
2. Metallic/mushroom/geranium	17	22
3. Roasted/cooked/meaty/spice	13	13
4. Fatty/soapy/green	17	16
5. Sulfury/solventy/medicine	10	15
6. Floral	10	9
7. Sweet fruity	8	7
Total	100	100

3 CONCLUSION

Four norisoprenoids, were identified in orange juice and were also detected in Ruby red grapefruit juice using a combination of GC-O and GC-MS. β-Cyclocitral and α-ionone are reported for the first time in any citrus juices. β-Damascenone and β-ionone were observed in both orange and grapefruit juice. Ruby red grapefruit juice had fewer norisoprenoids which could serve as carotenoid precursors than orange juice.

Norisoprenoids in orange juice contribute approximately 8 % of total aroma intensity as determined from combined GC-O aromagram peak height and 78 % of the total floral category. Norisoprenoids in grapefruit juice contribute approximately 6 % of total aroma intensity as determined from combined GC-O aromagram peak height and 66 % of the total floral category.

References

1. Sanderson, G. W.; Co, H.; Gonzalez, J. G. *J. Food Sci.* 1971, **36**, 231.
2. Ishida, B. K.; Mahoney, N. E.; Ling, L. C. *J. Agric. Food Chem.* 1998, **46**, 4577.
3. Tarantilis, P. A.; Polissiou, M. G. *J. Agric. Food Chem.* 1997, **45**, 459.
4. Moshonas, M. G.; Shaw, P. E. *J. Agric. Food Chem.* 1994, **42**, 1525.
5. Shaw, P. E.; Moshonas, M. G. *Food Science & Technology (London)* 1997, **30**, 497.
6. Nisperos-Carriedo, M. O.; Shaw, P. E. *J. Agric. Food Chem.* 1990, **38**, 1048.
7. Isoe, S.; Hyeon, S. B.; Sakan, T. *Tetrahedron Lett.* 1969, 279.
8. Demole, E.; Enggist, P.; Winter, M.; Furrer, A.; Schulte-Elte, K. H.; Egger, B.; Ohloff, G. *Helv. Chim. Acta* 1979, **62**, 67.
9. Strauss, C. R.; Wilson, B.; Anderson, R.; Williams, P. J. *Am. J. Enol. Vitic.* 1987, **38**, 23.
10. Williams, P. J.; Sefton, M. A.; Francis, I. L. Glycosidic precursors of varietal grape and wine flavor. in *Flavor precursors: Thermal and enzymatic conversions*; R. Teranishi; G. R. Takeoka and M. Guntert, eds.; American Chemical Society: Washington, DC, 1992; pp 74.
11. Enzell, C. R. Influence of curing on the formation of tobacco flavor. in *Flavour '81*; P. Schreier, Ed.; de Gruyter: Berlin, New York, 1981; pp 449.
12. Lee, H. S. *J. Agric. Food Chem.* 2000, **48**, 1507.
13. Rouseff, R. L.; Sadler, G. D.; Putnam, T. J.; Davis, J. E. *J. Agric. Food Chem.* 1992, **40**, 47.
14. Lin, J.; Rouseff, R. L.; Barros, S.; Naim, M. *J. Agric. Food Chem.* 2002, **50**, 813.
15. Mahattanatawee, K.; Rouseff, R. L.; Valim, F. M.; Naim, M. *J. Agric. Food Chem.* 2004, **In Press**.
16. Valim, M. F.; Rouseff, R. L.; Lin, J. *J. Agric. Food Chem.* 2003, **51**, 1010.

BLACK TRUFFLE FLAVOR: INVESTIGATION INTO THE IMPACT OF HIGH-BOILING-POINT VOLATILES BY GC-OLFACTOMETRY

O. Jansen, T. Talou, C. Raynaud and A. Gaset

Laboratoire de Chimie Agro-industrielle, INPT-ENSIACET, 118 Route de Narbonne, 31077 Toulouse, France

1 ABSTRACT

Black Perigord Truffles (*Tuber melanosporum* Vitt.) are hypogeous mushrooms that are highly appreciated for their intense and typical flavor. Although more than 100 major and minor compounds have been reported as fresh truffle headspace constituents, little is known about the contribution of minor, high-boiling-point compounds to the overall flavor. Throughout the 2002 and 2003 truffle harvesting seasons, 123 truffle specimens from all major production regions have been analyzed by gas chromatography. Two complementary headspace sampling techniques have been used, dynamic headspace trapping on TENAX, and SPME extraction. Both olfactometric (GC-O) and instrumental detection (GC-MS) were applied and helped to identify the major black truffle odorants. Twelve of them were found to correspond to minor, high-boiling-point compounds. Among these potent odorants, aldehydes and sulphur compounds are particularly well represented.

2 INTRODUCTION

Black truffles (*Tuber melanosporum* Vitt.) are hypogeous mushrooms found in southern European countries. They grow in symbiosis with certain host trees, especially oaks, and are harvested in the winter months. Black truffles are highly appreciated by gourmets; their retail price can reach up to 1200 euros/kg. This extraordinary commercial value is almost exclusively due to the mushroom's unique aroma. However, the volatile compounds released by black truffles have not been studied until the early 1980's.[1] Later on, several scientists[2,3,4] applied dynamic headspace (DH) trapping to truffle effluents, using versatile sorptive materials like Tenax. Gas chromatographic separation and identification by mass selective detection allowed determination of a large number of headspace constituents. Altogether, more than 100 compounds have been reported.

In the 1990's, a novel headspace sampling technique emerged; solid phase microextraction (SPME).[5] The fiber coatings of relatively low polarity proved to be highly efficient on truffles, extracting up to 73 compounds.[6] The two methods seem to cover a different range of potential aroma constituents and can thus be considered as complementary.

The present study's objective was to elucidate which of the black truffle's headspace constituents really contribute to its so much appreciated aroma. This has been achieved by GC-Olfactometry (GC-O) experiments with detection frequency evaluation.[7] As far as we know, GC-O has never before been applied to black truffles on a large scale. To obtain a complete picture of potential odorants, both sampling techniques (DH and SPME) were used. As the low-boiling-point odor active compounds of the mushroom are already well known,[8] focus was given to the medium- and high-boiling-point odorants. On a non-polar chromatographic column, the latter can be defined as those eluting after the last major volatile compound, 2-methylbutanol.

Literature study of truffle volatiles[2,3,4,6] and their retention indices was completed by instrumental headspace analysis with mass spectrometric (MS) detection, aiming at the identification of the whole range of previously sniff-detected odor-active compounds. Experimental conditions were similar to those used in GC-O studies. Again, both DH and SPME were used on a large number of fresh truffle specimens.

3   MATERIALS AND METHODS

3.1   GC-Olfactometry based on dynamic headspace trapping

The olfactometric panel consisted of 12 assessors. In accordance with the detection frequency approach,[7] panelists were not subjected to any selection procedure or specific training.

GC-O runs were carried out on 12 fresh truffle specimens which had been selected for their flawless organoleptic quality. Each of these truffles was used for six individual GC-O runs. For the first run, the specimen was brushed, dried with tissue paper, and cut into quarters. For subsequent runs, fine slices were removed from the inner surfaces to reconstitute a fresh cut prior to each headspace sampling. Six additional truffle specimens were analyzed by one panelist each. Hence, the number of GC-O experiments totalled 78.

For headspace sampling, the truffle was placed in a 300 ml hermetic vessel. Equilibration time was 5 min. At ambient temperature, volatiles were then purged over a period of 5 min onto a Tenax TA trap by the means of a 100 ml/min flow of helium. Desorption was achieved by heating the trap to 240°C during 5 min, using the previously described[2] DCI device.

The analytes were flushed by helium carrier gas onto a BPX-5 column (60 m, i.d. 0.32 mm, film thickness 1 µm) from SGE (Austin, TX) installed on a DELSI DN 200 chromatograph (DELSI instruments, Suresnes, France). Initial temperature was 45°C for 4 min, followed by an increase of 2°C/min up to 65°C, then 8°C/min up to 250°C. Final temperature was held for 5 min.

The GC effluent was split in a ratio of 70:30 between sniffing port and FID. Panelists signalled beginning and end of each olfactory detection by the means of a dedicated software.[9] They were also asked to assign a descriptive attribute from a list of nine odor classes which had been defined in consensus after a series of preliminary test runs.

Detection frequencies were calculated odor zone by odor zone. They are stated as percentages relative to the total of 78 GC-O runs.

3.2 GC-Olfactometry based on SPME extraction

Eight GC-O runs were carried out on 4 fresh truffles by two expert assessors. Each truffle was used for two GC-O runs. The sample was placed in a 300 mL headspace vessel. Volatiles were extracted over a period of 60 min by an SPME fiber (SUPELCO, Bellefonte, PA). Two types of coatings were used: PDMS (100 μm) and DVB-Carboxen-PDMS (50/30 μm). Desorption was achieved by placing the fiber over a 3 min period in the injector of the gas chromatograph (see below) at 250°C.

The analytes were flushed by helium carrier gas onto a DB-5MS column (30 m, i.d. 0.25 mm, film thickness 0.25 μm) from J&W Scientific (Folsom, CA) installed on an AGILENT 6890 series chromatograph (AGILENT Technologies, Palo Alto, CA). Initial temperature was 35°C for 10 min, followed by an increase of 10°C/min up to 280°C. The GC effluent was split in a ratio of 70:30 between sniffing port (Odo II, SGE, Austin, TX) and mass selective detector (see 3.4). Detection frequencies were calculated odor zone by odor zone. They are stated as percentages relative to the total of 8 GC-O runs.

3.3 GC-MS based on dynamic headspace trapping

Seventy fresh truffles were subjected to dynamic headspace trapping followed by GC-MS analysis. The sample was placed in a 200 ml headspace vessel. Volatiles were purged over a period of 60 min onto a Tenax TA trap by means of a 100 ml/min flow of helium. Desorption was achieved by placing the trap for a 5 min period in the injector of the gas chromatograph (see below) at 210°C.

The analytes were flushed by helium carrier gas onto a BPX-5 column (50 m, i.d. 0.32 mm, film thickness 1 μm) from SGE (Austin, TX) installed on a HP 5890 chromatograph (HEWLETT-PACKARD, Palo Alto, CA). During desorption, volatiles were cryofocalized on-column by liquid nitrogen cooling. Initial oven temperature was 45°C for 4 min, followed by an increase of 2°C/min up to 65°C, then 8°C/min up to 250°C. Final temperature was held for 10 min.

To avoid saturation of the mass selective detector (HP 5971, Hewlett-Packard, Palo Alto, CA), it was activated after elution of the major black truffle volatiles, at a retention time of 20.50 min, corresponding to a linear retention index (LRI) of ca 800 (BPX -5 column). Mass spectra were obtained with 70 eV electron impact ionization.

3.4 GC-MS based on SPME extraction

Thirty-one fresh truffles were subjected to dynamic headspace trapping followed by GC-MS analysis. The sample was placed in a 200 ml headspace vessel. Volatiles were extracted over a period of 60 min by an SPME fiber (SUPELCO, Bellefonte, PA). Three types of coatings were used: PDMS (100 μm), DVB-PDMS (65 μm) and Carboxen-PDMS (75 μm). Desorption was achieved by placing the fiber over a 3 min period in the injector of the chromatograph (see below) at 250°C.

The analytes were flushed by helium carrier gas onto a DB-5MS column (30 m, i.d. 0.25 mm, film thickness 0.25 μm) from J&W scientific (Folsom, CA) installed on an AGILENT 6890 series chromatograph (Agilent Technologies, Palo Alto, CA). Initial temperature was 35°C for 10 min, followed by an increase of 10°C/min. up to 280°C. Mass selective detection was achieved by an Agilent 5973 instrument. Spectra were obtained with 70 eV electron impact ionization.

4 RESULTS AND DISCUSSION

4.1 GC-Olfactometry based on dynamic headspace trapping

Thirty-nine odor zones were detected, of which 22 were situated in the retention time zone corresponding to medium- and high-boiling-point compounds. A minimum detection frequency of 25% was fixed as criterion for a "major odorant". Table 1 shows the 8 major odorants in the range of medium- and high-boiling-point headspace constituents.

Table 1 *Major Odor zones Detected in GC-O on Black Truffles After Dynamic Headspace Trapping. Only Detections Corresponding to Medium- and High-boiling-point Headspace Constituents Are Shown*

detection frequency [%]	retention index (BPX-5)	descriptive odor class
88	849	fruity, floral
85	984	raw mushroom
78	761	fruity, floral
63	974	cooked vegetable
36	785	plastic, burnt
32	1093	musty
31	807	green
28	775	fruity, floral

4.2 GC-Olfactometry based on SPME extraction

Seventeen odor zones were detected of which all were situated in the retention time zone corresponding to medium- and high-boiling-point compounds. Like in 4.1, a minimum detection frequency of 25% was fixed as criterion for a "major odorant". Table 2 shows the 17 major odorants.

The profile of sniff-detected aroma constituents varies considerably with the employed trapping/extraction technique. As both methods equally bear the risk of either exaggeration of an odorant's impact due to specific affinity to the sorptive material, or of its discrimination by the latter, mean values of detection frequency were calculated, thus according the same weight to both sets of results. Detection frequencies below the threshold of 25% with one of the two techniques were also into taken account when averaging the data. Table 3 shows the 12 major odorants that attain a mean detection frequency of 25%.

Table 2 *Major Odor Zones Detected in GC-O on Black Truffles after SPME Extraction*

detection frequency [%]	retention index (BPX-5)	descriptive odor class
100	1007	raw mushroom
100	1060	musty
88	914	cooked vegetable
88	984	raw mushroom
75	1160	buttery, lacteous
63	761	fruity, floral
63	974	cooked vegetable
63	1093	musty
50	849	fruity, floral
38	866	fruity, floral
38	874	plastic, burnt
38	1037	plastic, burnt
38	1265	musty
25	807	green
25	955	fruity, floral
25	1108	musty
25	1119	musty

Table 3 *Major Odor Zones Detected in GC-O on Black Truffles with Both DH Trapping /SPME Extraction Methods*

HD detection frequency [%]	SPME detection frequency [%]	mean detection frequency [%]	retention index (BPX-5)	descriptive odor class
85	88	87	984	raw mushroom
78	63	70	761	fruity, floral
88	50	69	849	fruity, floral
63	63	63	974	cooked vegetable
19	100	60	1007	raw mushroom
10	100	55	1060	musty
32	63	48	1093	musty
0	88	44	914	cooked vegetable
0	75	38	1160	buttery, lacteous
31	25	28	807	green
15	38	27	866	fruity, floral
13	38	26	1037	plastic, burnt

4.3 GC-MS based on dynamic headspace trapping

Forty-two compounds have been detected with a detection frequency exceeding 5% on the total of 70 analyzed fresh truffle specimens. Figure 1 shows an example of a gas chromatogram of headspace volatiles.

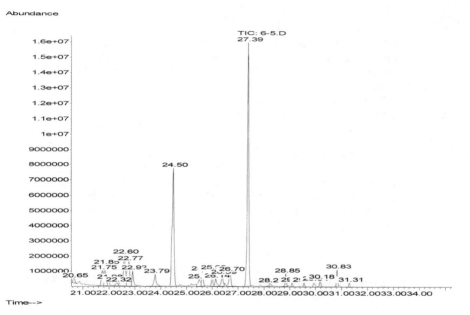

Figure 1 *Example of a chromatogram obtained in GC-MS analyses of a fresh truffle after dynamic headspace trapping. Detection (MS) was not activated until 20.50 min (LRI 800) to avoid detector saturation by major low-boiling-point volatiles.*

Analyte identifications according to the "Wiley275" database (Wiley, Chichester, UK) were confirmed by calculation of linear retention indices (LRI, see Table 6). Peaks appearing in the chromatogram shown in Figure 1 are listed in Table 4.

When averaging the quantitative results of the 70 analysis on fresh truffles, aromatic ethers (mostly methoxybenzenes) account for 59.4% of the peak area, relative to the range of medium- and high-boiling-point headspace constituents.

4.4 GC-MS based on SPME extraction

Seventy-one compounds have been detected with a detection frequency exceeding 5% on the total of 31 analyzed fresh truffle specimens. Figure 2 shows an example of a gas chromatogram obtained with sampling technique.

Analyte identifications according to the "NIST 98" (NIST, Gaithersburg, MD) database. Peaks appearing in the example chromatogram (Figure 2) are listed in Table 5. On the average of the 31 analysis, aromatic ethers accounted for 56.1% of the peak area.

Table 4 *Example of Compounds Identified in a Fresh Truffle's Headspace after Dynamic Headspace Trapping on Tenax TA*

Retention time [min]	compound	retention time [min]	compound
20.65	N.N-dimethyl formamide	26.14	3-Octanone
21.75	Ethyl-2-methylbutanoate	26.39	Isobutyl 2-methylbutanoate
21.82	Ethyl-3-methylbutanoate	26.48	Dimethyl trisulphide
21.98	unknown	26.70	2-Methylbutyl isobutanoat
22.32	unknown	27.39	3-Methyl anisole
22.60	2-(Methylthio)ethanol	28.23	1-Octanol
22.77	2-Methylbutyl acetate	28.85	2-Methylbutyl 2-methylbutanoate
22.92	1-Hexanol	29.09	Nonanal
23.79	Isobutyl isobutanoate	29.56	3-Ethyl anisole
24.50	Anisole	29.94	2-Phenyl ethanol
24.74	Propyl 2-methylbutanoate	30.18	1,2-Dimethoxy benzene
25.50	unknown nitrogen compound	30.83	1,4-Dimethoxy benzene
25.62	3,5-Dimethyl-4-heptanol	30.94	1,3-Dimethoxy benzene
25.99	1-Octene-3-ol	31.31	Decanal

Table 5 *Example of Compounds Identified in a Fresh Truffle's Headspace after SPME Extraction*

retention time [min]	compound	retention time [min]	compound
12.72	Anisole	17.78	2-Phenyl ethanol
13.70	Ethyl tiglate	18.33	1,2-Dimethoxy benzene
14.95	1-Octene-3-ol	18.67	1,4-Dimethoxy benzene
15.06	3-Octanone	18.72	1,3-Dimethoxy benzene
15.51	Octanal	19.17	unknown
15.85	3-Methyl anisole	19.25	2-Phenylethyl formate
16.05	Limonène	19.37	Decanal
17.38	1-Octanol	19.83	3,4-Dimethoxy toluene
17.60	2-Methylbutyle 2-methylbutanaote	20.02	2,5-Dimethoxy toluene
17.64	Nonanal	20.29	3,5-Dimethoxy toluene
17.71	3-Ethyl anisole	21.02	Ethyl 2,5-dimethoxy benzene

Analysis

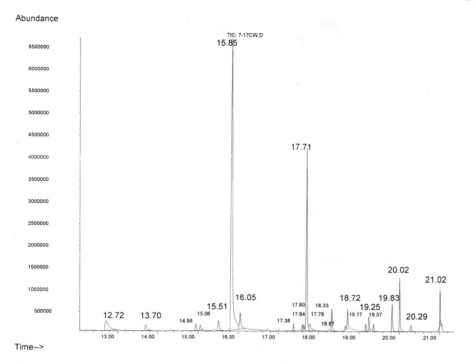

Figure 2 *Example of a chromatogram obtained in GC-MS analyses of a fresh truffle after SPME extraction. The same retention index (LRI $_{BPX-5\ column}$) region as in figure 1 is shown (LRI 800 – LRI 1400).*

In conclusion, the spectrum of compounds detected in fresh truffles tends to vary depending on the trapping/extraction method, as previously observes in GC-O experiments. SPME favors the detection of very high-boiling-point volatiles whereas "light" and polar headspace constituents seem to be discriminated. However, 35 medium- and high-boiling-point compounds were common to both techniques. They are listed in Table 6 with their detection frequencies [%] on the total of analyzed specimens per trapping/extraction technique.

Table 6 *Fresh Truffle Headspace Constituents of Which the Detection is Common to Both Trapping/Extraction Methods*

linear retention index	compounds	detection frequency DH (n=70)	detection frequency SPME (n=31)	reported in literature
866	1-Hexanol	87	9	yes
880	2-Methylbutyl acetate	91	6	(yes)[a]
905	Heptanal	7	24	yes
918	Anisole	99	94	yes
921	Isobutyl isobutanoate	74	6	yes
960	3,5-Diméthyl-4-heptanol	69	21	no
966	unknown nitrogen compound	67	15	no
967	Benzaldehyde	43	39	yes
969	Dimethyl trisulfide	27	9	yes
984	1-Octen-3-ol	99	82	yes
990	3-Octanone	90	79	yes
999	3-Octanol	43	55	yes
1004	Isobutyl 2-methylbutanoate	79	24	(yes)[a]
1007	Octanal	59	73	yes
1009	2-Methylbutyl isobutanoate	66	12	yes
1020	3-Methyl anisole	99	100	yes
1030	Limonène	10	73	yes
1037	3-(Methylthio)propanoic acid	34	15	no
1049	1-Methyl-2-pyrrolidone	16	12	no
1075	1-Octanol	71	64	yes
1103	2-Methylbutyl 2-methylbutanoate	99	100	yes
1106	Nonanal	81	97	yes
1109	3-Ethyl anisole	84	97	yes
1114	2-Phenyl ethanol	83	79	yes
1146	1,2-Dimethoxy benzene	99	100	yes
1166	1,4-Dimethoxy benzene	86	79	yes
1169	1,3-Dimethoxy benzene	79	85	yes
1188	Naphthalene	11	21	yes
1200	Phenylethyl formate	43	73	no
1207	Decanal	91	100	no
1238	3,4-Dimethoxy toluene	31	82	yes
1250	2,5-Dimethoxy toluene	59	94	yes
1268	3,5-Dimethoxy toluene	7	36	yes
1318	Ethyl 2,5-dimethoxy benzene	26	76	no
1370	1,2,4-Trimethoxy benzene	16	79	yes

[a] literature reports detection of an isomer

4.5 Odor identification

On the basis of the instrumental data presented in sections 4.3 and 4.4 as well as reviewing the literature available on black truffle volatiles,[2,3,4,6] we established associations between chemical compounds and the major odor active retention time zones presented in Table 3 (see above). Retention index matches were confirmed by reference data on odor notes[10,11] and olfactory detection thresholds.[11,12] Odorant identification results are shown in Table 7.

Table 7 *Identification of Fresh Truffle Major Odorants*

mean detection frequency [%]	retention index (BPX-5)	descriptive odor class	identification
87	984	raw mushroom	1-Octen-3-ol
70	761	fruity, floral	Ethyl 2-methylpropanoate[a]
69	849	fruity, floral	Ethyl 2-methylbutanoate
63	974	cooked vegetable	Dimethyl trisulfide
60	1007	raw mushroom	Octanal
55	1060	Musty	(E)-2-Octenal
48	1093	Musty	2-Isopropyl-3-methoxy-pyrazine[a]
44	914	cooked vegetable	3-(Methylthio)-propanal
38	1160	buttery, lacteous	(E)-2-Nonenal
28	807	Green	Hexanal
27	866	fruity, floral	1-Hexanol
26	1037	plastic, burnt	3-(Methylthio)-propanoic acid

[a] no instrumental detection in this study, identification based on retention index, descriptive note and olfactory detection threshold only

5 CONCLUSIONS

Headspace-GC-MS and headspace-GC-Olfactometry were carried out on a large number of fresh truffle samples. In order to overcome selectivity problems at the volatiles extraction step, two different methods were used: dynamic headspace with trapping on Tenax TA and pseudo-static headspace with solid phase microextraction (SPME). Focus was given to the detection medium- and high-boiling-point minor headspace constituents. In both olfactory and instrumental detection, SPME efficiency tended to increase with the boiling point of the analytes, whereas with Tenax trapping, the opposite trend was observed.

According to the GC-O results obtained by the two extraction methods, 12 odor zones attained an average detection frequency of 25% and were qualified as black truffle major odorants. In GC-MS, 35 compounds were detected after application of both dynamic headspace trapping and SPME extraction. Their role as truffle headspace constituents was thus confirmed independent of the selectivity concerns. In the range of medium- and high-boiling-point compounds, aromatic ethers accounted for more than 50% of the total peak area.

When matching GC-O results with both our instrumental data and literature information on black truffle headspace constituents, all twelve 12 major odorants could be identified. Most of them proved belong to the following chemical classes: aldehydes, esters, alcohols, and sulphur compounds. Aromatic ethers, on the contrary, did not seem to contribute in a considerable manner to black truffle aroma.

Acknowledgements

The authors would like to thank to European Commission for financial support of this research project in the framework on the Marie Curie Industry Host Fellowship HPMI-CT-1999-00060 and PEBEYRE S.A. (Cahors, France) for supplying truffle samples.

References

1. K.H. Ney and G. Freitag, *Gordian*, 1980, **9**, 214
2. T. Talou, M. Delmas and A. Gaset, *J. Agric. Food Chem.*, 1987, **35**, 774
3. I. Flament, C. Chevallier and C. Debonneville, *Proceedings "9èmes Journées Internationales des Huiles Essentielles"*, 30 Août - 1 Sept 1990, Digne les Bains, France, 280
4. F. Pelusio, T. Nilsson, L. Montanarella, R. Tilio, B. Larsen, S. Facchetti and J.Ö. Madsen, *J. Agric. Food Chem.*, 1995, **43**, 2138
5. Z. Zhang and J. Pawliszyn, *Anal. Chem.*, 1993, **65**, 1843
6. P. Diaz, E. Ibanez, F.J. Senorans and G. Reglero, *J. Chromatography A*, 2003, **1017**, 207
7. J.P.H. Linssen, J.L.G Janssens, J.P. Roozen and M.A. Posthumus, *Food Chem.*, 1993, **46**, 367
8. M. Delmas, A. Gaset, C. Montant, P.J. Pébeyre and T. Talou, *Black truffle aroma*, American patent US 4,906,487 (1987), European patent EP 0 257 666 (1990)
9. A. Janssens and J.P Roozen, *SNIF software*, 1991, Dept. ATV, Wageningen University, The Netherlands
10. S. Arctander, *Perfume and Flavor Chemicals (Aroma Chemicals)*, 1969, Montclair, NJ, USA,
11. M. Rychlik, P. Schieberle and W. Grosch, *Compilation of Odor Thresholds, Odor Qualities and Retention Indices of Key Food Odorants*, 1998, Deutsche Forschungsanstalt fuer Lebensmittelchemie, Technische Universitaet Muenchen, ISBN 3-9803426-5-4
12. http://www.odour.org.uk

Analysis

EVIDENCE OF THE PRESENCE OF (*S*)-LINALOOL AND OF (*S*)-LINALOOL SYNTHASE ACTIVITY IN *VITIS VINIFERA* L., CV. MUSCAT DE FRONTIGNAN

G. M. de Billerbeck[1], F. Cozzolino[2] and C. Ambid[1]

[1]INP-ENSAT, Equipe *Arômes et Biotechnologies*, 27 chemin de Borderouge, 31326 Castanet-Tolosan Cedex, France. E-mail: debiller@ensat.fr
[2]DEGUSSA Food Ingredients, Z.I. du Plan, BP 82067, 06131 Grasse Cedex, France

1 ABSTRACT

Among the monoterpenic alcohols identified as responsible for the particular flavor of *Vitis vinifera* L., cv. Muscat de Frontignan grape berries, linalool is known to be a key component. To date, the enantiomeric form of free linalool and the presence of a linalool synthase (LIS) activity responsible of its synthesis in this plant have not yet been determined. In the present study, the main results demonstrate for the first time that S-linalool is the principal enantiomeric form biosynthesized in the berry. A good relationship between free and bound S-linalool bioproduction and S-LIS activity time evolution is observed during berry ripening. Not detectable in cells cultivated in vitro, this activity prevails in the berry skin and in adult leaves. The whole of these experimental data is discussed and integrated in the general phenomenon of maturation.

2 INTRODUCTION

Linalool (3,7-dimethyl-1,6-octadien-3-ol), a tertiary chiral acyclic monoterpene alcohol, is known to be a major component of the floral scent or of the aroma and flavor of numerous plant species.[1,2] (*R*)-linalool is found in many flowers, such as bitter orange and lavender[3] and in basil leaves.[2] This compound has a fine fresh floral odour-character in contrast to the (*S*)-enantiomer which presents a more herbal, sweet odour-note.

(*S*)-linalool accumulates mainly in coriander fruits, apricot ($S/R = 85/15$)[4] and orange peel. In grapevine, linalool has been shown to contribute to the characteristic varietal aroma of cultivars Muscat.[5] In Muscat d'Alexandrie berries, more than 95% of the glycosidically bound linalool is present as the (*S*)-enantiomer.[4] In Muscat d'Ottonel, free (*S*)-linalool is also found at high enantiomeric purity ($S/R = 98/2$ in berries exocarp and mesocarp and $S/R = 65/35$ in leaves).[6] In *Clarkia breweri* flowers, Pichersky et al.[1] found that the biosynthesis of this (*S*)-monoterpenic alcohol was realized from geranyl pyrophosphate (GPP) in one step by a (*S*)-linalool synthase (LIS) activity (Figure 1). More recently, a cDNA coding a (*R*)-linalool synthase was isolated from *Mentha citrata* leaves[7] and *Artemisia annua* aerial organs.[8]

To date, the stereochemistry of free linalool and the presence of a linalool synthase activity responsible of its biosynthesis in *Vitis vinifera* L., cv. Muscat de Frontignan have not yet been determined. In this work, the absolute configuration and enantiomeric ratio of free linalool was determined in must of this cultivar. The existence of the corresponding enzyme was also demonstrated and analyzed in different organs and during berry development. The whole of these experimental data is discussed and integrated in the general phenomenon of maturation.

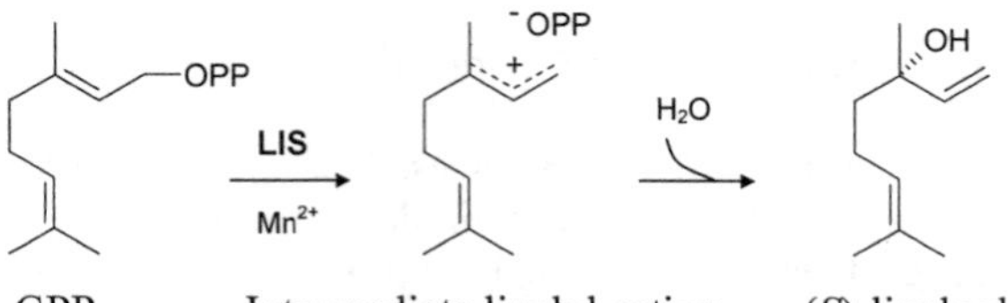

Figure 1 *Scheme of the formation of (S)-linalool from GPP by the action of (S)-linalool synthase (LIS).*

3 MATERIALS AND METHODS

3.1 Plant Material

Grape berries of *Vitis vinifera* L. cv. Muscat de Frontignan were collected weekly from vines (clone n° 454 grafted on 3309) growing in the vineyard of Seignous (Rabastens, France) throughout the 2000 and 2001 seasons. All samples were of sound fruit with no detectable symptoms of disease or damage. Whole berries or berry skins were separately frozen in liquid nitrogen and stored at –80 °C until required.

3.2 Berry Measurements

Must from squeezed berries were used to measure total reducing sugars by separate determination of glucose and fructose concentrations using hexokinase, glucose-6-phosphate dehydrogenase and phosphoglucose isomerase according to the supplier's protocol (Boehringer Mannheim, France). Total acidity was evaluated by acid-base titration.

3.3 Enzyme Extraction

All steps were performed at 4 °C. Aliquots of frozen ground tissue (1 g each) were extracted for 15 min under agitation in 5 ml 50 mM Tris HCl pH 7.0 buffer, containing 10 mM DL-dithiothreitol, 1% (w/v) Polyvinylpyrrolidone (av. mol. wt. 40,000), 25 mM KCl and 10 mM L-ascorbic acid. The homogenate was then sonicated and centrifuged at 17,000 g for 10 min. 1 ml of the supernatant was used as crude extract for enzyme assay.

3.4 Enzyme Assay

The specific activity of linalool synthase was determined according to the method of Pichersky *et al.*[1] with modifications. 250 mM Na_2MoO_4 was added to the incubation medium in order to reduce hydrolysis of [1-^{3}H]-GPP by endogenous phosphohydrolase activities.[9] After incubation at 30 °C during 2 h with [1-^{3}H]-GPP (5 mM, 20 µCi mmol^{-1}) as substrate, monoterpene alcohols were extracted with pentane, separated by thin layer chromatography and revealed by autoradiography. Radioactivity of the [^{3}H]-linalool spot was determined by liquid-scintillation counting. The LIS specific activity was calculated in relation to the total protein content of the sample as measured by the Bio-Rad protein assay based on the method described by Bradford.[10] No radiolabeled products were detected when cell-free extracts were previously boiled for 10 min (control).

3.5 Analysis of Free and Bound Linalool Accumulated in Grape Berries

Frozen grape berries were deseeded by breaking them carefully in a mortar containing liquid nitrogen. The material was grounded with a Dangoumau blender previously cooled in liquid nitrogen. 10 g of the sample were introduced into 500 ml hermetic vessels containing 10 mL of buffer (300 mM Tris HCl, pH 8), 10 µl of octan-1-ol as internal standard (0.1% v/v in ethanol 95% v/v). Headspace volatile compounds were extracted by solid phase microextraction (SPME) and subjected to GC analysis under conditions described previously.[11] The concentration of glycosidically bound linalool was measured according to the method of Günata et al.[5]

The enantioselective analysis of linalool was performed with an enantio-MDGC/MS system consisting of a Carlo Erba 8600 GC coupled with a PROSPEC magnetic mass spectrometer. The precolumn of the MDGC was a non-polar HP1 column (Agilent, 50 m × 0.2 mm i.d., 0.5 µm film thickness) and the main column was a MEGA chiral column (25 m × 0.25 mm i.d., 0.25 µm film thickness, phase DMePe BETACDX). Identification of (R) and (S) enantiomeric forms of linalool in the sample was performed by comparison of retention time and mass spectrum to that of the corresponding, optically pure standards.

All the experiments described were duplicated using different batches of samples and similar results were obtained.

4 RESULTS AND DISCUSSION

4.1 Evidence of the Presence of a Free (S)-Linalool Enantiomer in Berries of *Vitis vinifera* L., cv. Muscat de Frontignan

The free linalool fraction was analyzed in Muscat de Frontignan must using the enantio-MDGC/MS system described under Materials and Methods. A typical main chiral column chromatogram is shown in Figure 2A. As calculated from the enlargement in Figure 2B, the enantiomeric purity of free linalool was largely in favor of the (S)-configured isomer with a ratio S/R = 99/1. Comparable enantiomeric ratios were found previously concerning free linalool in Muscat d'Ottonel berry exocarp and mesocarp (S/R = 98/2)[6] and bound linalool in Muscat d'Alexandrie berries (S/R = 95/5).[4] Although the absolute configuration of linalool has not yet been determined in musts of all Muscat cultivars, it can be inferred from these results that the linalool stereoisomer responsible of the typical Muscat varietal aroma is the (S)-enantiomer.

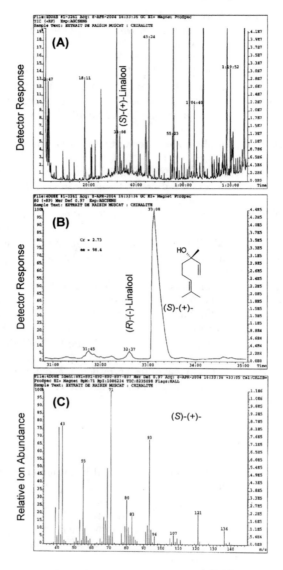

Figure 2 *Stereochemical analysis of free linalool in must of* V. vinifera *L., cv. Muscat de Frontignan. (A) Capillary MDGC/MS analysis (total ion chromatogram) demonstrating the presence of (S)-linalool (R_t = 33.08 min). (B) Enlargement of the previous chromatogram and determination of the presence of the (S)-antipode (99%) and (R)-antipode (1%) by coincidence of retention time with that of the corresponding, optically active pure standard. (C) Mass spectrum of the compound identified as (S)-linalool by chiral chromatography corresponds to that of authentic linalool.*

Analysis 275

4.2 Evolution of Free and Bound (S)-Linalool during the Maturation of Muscat Grape

To study the accumulation dynamics of free and bound (S)-linalool during the Muscat grape berry development, the time evolution of total acidity and of reducing sugars (glucose + fructose) was measured. As described in previous studies, little glucose and fructose while substantial acids accumulated during the phase of development before véraison. As shown in Figure 3, this phenomenon occurred 8 weeks postflowering.

Free and bound (S)-linalool accumulation in berries started 10 weeks postflowering (Figure 3). Free (S)-linalool concentration reached a maximum of 750 µg kg^{-1} FW 12 weeks postflowering, then declined but remained relatively high (400 µg kg^{-1} FW). In the same time, glycosidically bound (S)-linalool followed a different kinetics, primarily at the end of maturation. It is interesting to note that in contrast to other Muscat cultivars and as shown previously by other authors,[5] the concentration of bound (S)-linalool was significantly lower than that of the free form during the grape berry ripening.

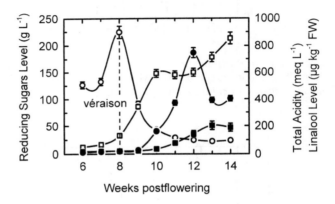

Figure 3 *Reducing sugars (□), total acidity (○), free (●) and bound (S)-linalool (■) evolution during grape berry ripening. The dashed line indicates the time of véraison (8 weeks postflowering).*

4.3 Localization of (S)-Linalool Synthase Activity

(S)-linalool synthase activity was determined in different organogenic and cellular structures (Figure 4). Only cell suspensions cultivated *in vitro* did not present this activity at detectable levels. LIS activity was greater in adult leaves than in young ones. In the berry, the exocarp presented higher LIS activity than the mesocarp.

4.4 (S)-Linalool Synthase Activity during the Ripening of the Berry

The LIS time evolution during berry development has been analyzed in the exocarp where its activity was found higher than in the mesocarp or in the whole berry. Figure 5 shows that LIS activity, almost undetectable before véraison (3 × 10^{-3} fkat mg^{-1} prot.), increased significantly one week later, reached a maximum 12 weeks postflowering (70×10^{-3} fkat mg^{-1} prot.) and remained constant. A good relationship between bound (S)-linalool accumulation in the berry and LIS activity time evolution was observed during berry ripening. The depletion of free (S)-linalool 12 weeks postflowering cannot be explained by the development of the enzyme activity.

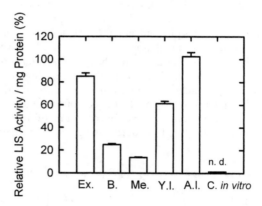

Figure 4 *Determination of (S)-linalool synthase activity in berry exocarp (Ex.), whole berry (B.), berry mesocarp (Me.), young leaves (Y.l.), adult leaves (A.l.) and in cell suspensions cultivated* in vitro *(C. in vitro).*

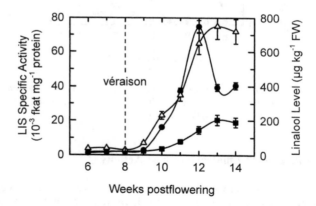

Figure 5 *Time evolution of LIS specific activity in berries exocarp ($\triangle$) and of free ($\bullet$) and bound (S)-linalool ($\blacksquare$) in berry during ripening phenomenon. The dashed line indicates the time of véraison (8 weeks postflowering).*

5 CONCLUSION

For the first time, the absolute configuration and enantiomeric ratio of free linalool was determined in must of *V. vinifera* L., cv. Muscat de Frontignan. The enantiomeric purity of this monoterpenic alcohol was largely in favor of the (S)-stereoisomer with a ratio $S/R = 99/1$. The enzymatic activity responsible for its biosynthesis has been analyzed in different organs and throughout the development of the berry. A good relationship between free and bound (S)-linalool accumulation and LIS activity time evolution was observed during berry ripening. This activity was not detectable in cells cultivated *in vitro*, but prevailed in berry skin and in adult leaves. These results indicate that LIS activity is highly regulated during the berry development in the field or by the *in vitro* conditions of culture.

References

1. E. Pichersky, R.A. Raguso, E. Lewinsohn and R. Croteau, *Plant Physiol.*, 1994, **106**, 1533.
2. A. Akguel, *Nahrung*, 1989, **33**, 87.
3. P. Kreis and A. Mosandl, *Flavour Fragance J.*, 1992, **7**, 187.
4. C. Salles, PhD Thesis, Université de Montpellier II, France, 1989.
5. Y.Z. Günata, C. Bayonove, R.L. Baumes and R.E. Cordonnier, *J. Sci. Food Agric.*, 1985, **36**, 857.
6. F. Luan and M. Wüst, *Phytochemistry*, 2002, **60**, 451.
7. A.L. Crowell, D.C. Williams, E.M. Davis, M.R. Wildung and R. Croteau, *Arch. Biochem. Biophys.*, 2002, **405**, 112.
8. J.W. Jia, J. Crock, S. Lu, R. Croteau and X.Y. Chen, *Arch. Biochem. Biophys.*, 1999, **372**, 143.
9. M. Clastre, PhD Thesis, INP Toulouse, France, 1993.
10. M.M. Bradford, *Anal. Biochem.*, 1976, **72**, 248.
11. J.P. Ebang-Oke, G.M. de Billerbeck and C. Ambid, Proceedings of the 10th Weurman Flavour Research Symposium, Eds. J.L. Le Quéré and P.X. Etiévant, Lavoisier/Intercept, Paris, 2003, p. 321.

ISOLATION AND IDENTIFICATION OF PHENOLIC GLYCOSIDES FROM QUINOA SEEDS (*CHENOPODIUM QUINOA* WILLD)

Nanqun Zhu, Shengmin Sang, Robert T. Rosen and Chi-Tang Ho

Department of Food Science and Center for Advanced Food Technology, Rutgers University, 65 Dudley Road, New Brunswick, NJ 08901-8520

1 ABSTRACT

Quinoa has a long and distinguished history in South America, where quinoa has been cultivated since 3,000 BC. As reported, quinoa seeds are superior to traditional cereals in terms of their nutritional value, with protein contents ranging between 10 and 18% and with a fat content of 4.1 to 8.8%. Furthermore, this protein is of an exceptionally high quality, because of it is particularly rich in essential amino acids, such as histidine and lysine. Traditionally, the seeds are either mechanically abraded to remove the bran, where the saponins are predominantly located, or leached with water to debitter the seeds prior to use. In our continued investigation into the chemical components of debittered quinoa seeds, a mixture of two interchangeable isomers were isolated by semi-preparative HPLC column. Their structures were determined by detailed MS and NMR techniques, especially 2D NMR. This is the first report of these new phenolic glycosides in quinoa.

2 INTRODUCTION

Quinoa, *Chenopodium quinoa*, has been cultivated as a food crop for centuries in Latin America. Currently, descendants of the Inca Empire still use its seeds as an important component in their diet.[1] Its adaptation to cold, dry climates, the similarity to rice, and more importantly, the excellent nutritional qualities of its seeds make quinoa a crop of considerable importance around the world, especially in places that are currently limited as far as crop diversity and nutritional value are concerned.[2,3] As reported, quinoa seeds are superior to traditional cereals in terms of their nutritional value, with protein contents ranging between 10 and 18% and with a fat content of 4.1 to 8.8%. Furthermore, this protein is of an exceptionally high quality, because of it is particularly rich in essential amino acids, such as histidine and lysine, which are deficient in most other grain crops and necessary for proper amino acid nutrition in humans.[2,3] In addition to their almost perfect amino acid composition, quinoa seeds are rich in α-tocopherol, carotenes, calcium, phosphorus, iron and fiber.[2,3] The 400-year decline in quinoa production that began with the Spanish conquest has been reversed beginning this past decade as a result of their high nutritional value. The seeds can be used in the same way as rice or wheat, such as boiling

Analysis

or grinding into flour. They are now used to prepare pasta, puffed cereals, breads, cakes, beers, and animal feedstuffs.[2,3]

However, there is a big disadvantage related to the consumption of quinoa, namely bitterness, which is attributed to its saponins.[4] With the renewed interest in quinoa, over twenty triterpene saponins have been identified in quinoa seeds.[5-11] Traditionally, the seeds are either mechanically abraded to remove the bran, where the saponins are predominately located, or leached with water to debitter the seeds prior to use.[3]

Except saponins, very little is known about the phytochemical composition of quinoa. We have isolated five ecdysteroids from quinoa seeds.[12] Their structures were determined as ecdysterone, makisterone A, 24-*epi*-makisterone A, 24(28)-dehydromakisterone A, and 20,26-dihydroxyecdysone. Here we report the identification of two novel phenolic glycosides in quinoa seeds.

3 MATERIALS AND METHODS

3.1 Materials

Quinoa seeds were collected in Bolivia, but purchased from Quinoa Co. (Torrance, CA) in the United States. TLC was performed on Sigma-Aldrich (St. Louis, MO) silica gel TLC plates (250 μm thickness, 2-25 μm particle size), with compounds visualized by UV_{365nm} light and spraying with 10% (v/v) H_2SO_4/ethanol solution while being heated. Silica gel (130-270 mesh), Sephadex LH-20 and RP-18 (60 μm) (Sigma) were used for column chromatography. All solvents used were purchased from Fisher Scientific (Springfield, NJ). Positive APCIMS were obtained on a Micromass Platform II system (Micromass Co., Beverly, MA) equipped with a Digital DECPC XL560 computer for data analysis. ^{1}H NMR, ^{13}C NMR, ^{1}H-^{1}H COSY, HMQC, HMBC and NOESY spectra were recorded on a Varian U-500 Spectrometer (Varian Inc., Palo Alto, CA). Chemical shifts were expressed in parts per million (δ) using tetramethylsilane (TMS) as internal standard.

3.2 Extraction and Isolation of Quinoa Seeds

Dried seeds (4 kg) were extracted twice with 6 L 95% aqueous ethanol at room temperature. Each extraction lasted three days. The extracts were concentrated (120 g), suspended in water, and partitioned successively using hexane (20 g), ethyl acetate (10 g) and butanol (35 g).

The butanol fraction was subjected to column chromatography on Diaion HP-20 gel to give 4 (water, water-ethanol (3:7, v/v), water-ethanol (1:9, v/v) and acetone) fractions. Then the water-ethanol (1:9) fraction was separated into 4 fractions by Sephadex LH-20 column using 90% aqueous ethanol as the eluent. The first fraction was repeatedly separated by column chromatography (silica gel) and eluted with ethyl acetate-methanol-water-hexane (22.5:1:0.8:0.8, v/v/v/v) to give several fractions (B1-B11).

Fraction B2 was subjected to a silica gel column and eluted with ethyl acetate/methanol/water (15:1:0.8, v/v/v) to give five subfractions (I – V). A couple of chiral isomers (A and B) were isolated from subfractions II, III and IV of B2, by Microsob phenyl semi-preparative HPLC column. Fractions were obtained on a Waters 600E HPLC pump (Milford, MA) coupled to a Milton Roy Spectro monitor 3100 Variable wavelength detector (Riversa Beach, FL).

3.3 HPLC Conditions

Mobile phases used were A: 0.1 % of acetic acid in water; B: 0.1 % acetic acid in methanol; running as isocratic condition with 75 % of A and 25 % of B were used.

3.4 Quinoside A or Quinoside B

White Powder. ^{1}H NMR (in CD$_3$OD): 2.84 (1H, m, H-8a), 3.03 (1H, m, H-8b), 3.38 (2H, m, H-4' and H-5'), 3.55 (2H, m, H-5"), 3.58 (1H, m, H-2'), 3.65 (3H, s, H-11), 3.70 (1H, m, H-6'a), 3.74 (1H, m, H-7), 3.78 (1H, d, J = 9.5 Hz, H-4"), 3.88 (1H, dd, J = 12.0, 1.0 Hz, H-6'b), 3.95 (1H, d, J = 1.5 Hz, H-2"), 4.06 (1H, d, J = 9.5 Hz), 4.82 (1H, d, J = 7.5 Hz, H-1'), 5.45 (1H, brd, J = 1.5 Hz, H-1"), 6.96 (1H, dd, J = 8.6, 2.5 Hz, H-6) 7.05 (1H, d, J = 2.5 Hz, H-2) 6.80 (1H, d, J = 8.6 Hz, H-5).

4 RESULTS AND DISCUSSION

A mixture of two chiral isomers (A and B) was isolated using a semi-preparative HPLC column. During the separation, they were separated into two peaks. However, after collection, each peak changed back to the same two peaks as seen previously. This fact clearly suggests that these isomers are interchangeable, and only exist in the form of a mixture with a certain ratio (1:1 for A to B). Their structures were determined by detailed analysis by MS and NMR techniques, especially on 2D NMR, including ^{1}H-^{1}H COSY, HMQC, HMBC and NOESY spectra. Although these are considerable literature on quinoa seeds, this is the first report of these new phenolic glycosides in quinoa.

The mixture of A and B was isolated as a white powder. In APCI positive spectrum, it showed three important ions: m/z 516 [M-H$_2$O]$^+$, m/z 384 [M-H$_2$O-apiofuranosyl]$^+$, and m/z 516 [M-H$_2$O-apiofuranosyl-glucopyranosyl]$^+$. Combined with ^{13}C NMR and HMQC spectra, the molecular formula was deduced as C$_{22}$H$_{30}$O$_{15}$. Among a total of 22 carbon signals for each compound, 11 were assigned to the aglycone part and the other half signals were assignable to the oligosaccharide moiety.

Both the type of sugar units and the sequence of the oligosaccharide chain of quinoside A and B were established using a combination of literature data comparison and 2D-NMR experiments. The ^{1}H and ^{13}C NMR Spectra (shown in Table I) showed two sugar anomeric protons at δ 4.82 (1H, d, J = 7.5 Hz) and 5.45 (1H, brd, J = 1.5 Hz), and carbons at δ 102.0, 110.8, respectively. Starting from the anomeric protons of each sugar unit, all the ^{1}H and the ^{13}C resonances within each spin system were delineated. The first sugar was identified as β-D-glucose, with carbon signals at δ 102.0, 78.8, 78.6, 78.1, 71.4 and 62.5 ppm. Proceeding in the same way, the other sugar was identified as a β-D-apiofuranose, with carbon signals at δ 110.8, 80.7, 77.9, 75.4 and 65.5 ppm. Linkage of the sugar units was established from the following HMBC correlations: H-2' with C-1" and H-1" with C-2'. NOESY correlation of the sugar sequence yielded the same conclusion as above.

Similarly, the aglycone structure was determined by a detailed analysis of 1D and 2D NMR spectra. In the ^{1}H - ^{1}H COSY spectrum, the signals at δ 6.96 (1H, dd, J = 8.6, 2.5 Hz, H-6) coupled with the signals at 7.05 (1H, d, J = 2.5 Hz, H-2) and 6.80 ppm (1H, d, J = 8.6 Hz, H-5); the signal at δ 2.84 (1H, m, H-8a) coupled with the signal at δ 3.03 ppm (1H, m, H-8b) and 3.74 ppm (1H, m, H-7). In the HMBC spectrum, H-8 correlated with

C-3, C-9 and C-10; H-7 correlated with C-2, C-3, C-4, C-8, C-9 and C-10. The attachment of the disaccharide moiety to C-1 of the aglycone was deduced from HMBC spectrum, in which H-1' (δ 4.82) showed long-range correlation with C-1 (δ 155.1). The β linkage of the oligosaccharide was deduced from the anomeric proton signal at δ 4.82 ppm with a coupling constant of 7.5 Hz in the ^{1}H NMR. Therefore, the combined information above led to the total structure of A and B as shown in the Figure 1, which were named quinoside A (S configuration) and B (R configuration), respectively. It is interesting to see this kind of interchange between the chiral isomers.

Table 1 ^{13}C NMR spectra data of compound A and B

Position	Aglycone	Position	Sugar Moiety
1	155.1 s	1'	102.0 d
2	115.0 d	2'	78.8* d
3	131.5 s	3'	78.6* d
4	138.6 s	4'	71.4 d
5	111.1 d	5'	78.1* d
6	117.5 d	6'	62.5 t
7	44.1 d	1"	110.8 d
8	35.0 t	2"	77.9 d
9	173.1 s	3"	80.7 s
10	181.2 s	4"	75.4 t
11	52.5 q	5"	65.5 t

Both spectra were recorded in CD$_3$OD. * refers to the chemical shifts that are interchangeable.

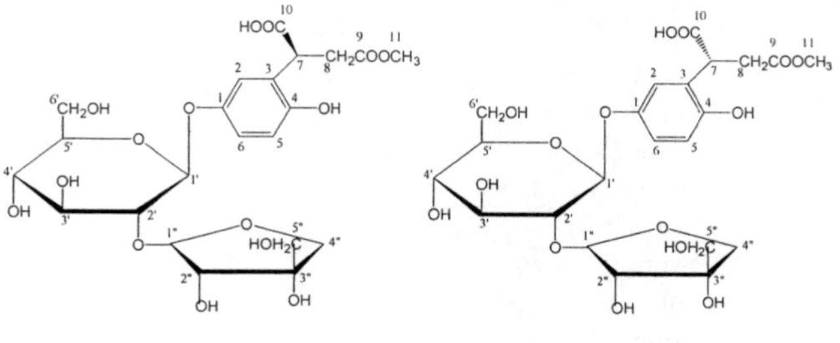

Figure 1 *Structures of quinoside A and quinoside B.*

5 CONCLUSIONS

Overall, our systematic study on debittered quinoa seeds has resulted in the isolation and identification of more than twenty compounds, including ecdysteroids, flavonol glycosides, triterpenes,[11-13] and two glycosides, and these provide a necessary basis for more applications of quinoa seeds in the future.

References

1. G. Schlick and D.L. Bubenheim. In *Progress in New Crops;* J. Janick, Ed., ASHS Press: Arlington, VA, 1996, pp. 632-640.
2. M. Koziok, *Food Comp. Anal.* 1992, **5**, 35-68.
3. C.L. Ridout, K.R. Price, M.S.Dupont, M.L. Parker and G.R. Fenwick, *J. Sci. Food Agric.* 1991, **54**, 165-176.
4. J.M. Gee, K.R. Price, C.L. Ridout, G.M. Wortley, R.F. Hurrel and I.T. Johnson, *J. Sci. Food Agric.* 1993, **63**, 201-209.
5. F. Mizui, R. Kasai, K. Ohtani and O. Tanaka, *Chem. Pharm. Bull.* 1988, **36**, 1415-1418.
6. F. Mizui, R. Kasai, K. Ohtani and O. Tanaka, *Chem. Pharm. Bull.* 1990, **38**, 375-377.
7. W.W. Ma, P.F. Heinstein and J.L. McLaughlin, *J. Nat. Prod.* 1989, **52**, 1132-1135.
8. B.R. Meyer, P.F. Heinstein, M. Burnouf-Radosevich, N.E. Delfel and J.L. McLaughlin, *J. Agric. Food Chem.* 1990, **38**, 205-208.
9. I. Dini, O. Schettino, T. Simioli and A. Dini, *J. Agric. Food Chem.* 2001, **49**, 741-746.
10. G.M. Woldemichael and M. Wink, *J. Agric. Food Chem.* 2001, **49**, 2327-2332.
11. N. Zhu, S. Sheng, S. Sang, J.W. Jhoo, N. Bai, M.V. Karwe, R.T. Rosen and C.-T. Ho, *J. Agric. Food Chem.* 2002, **50**, 865-867.
12. N. Zhu, H. Kikuzaki, B.C. Vastano, N. Nakatani, M.V. Karwe, R.T. Rosen and C.-T. Ho, *J. Agric. Food Chem* 2001, **49**, 2576–2578.
13. N. Zhu, S. Sheng, D. Li, E.J. LaVoie, M.V. Karwe, R.T. Rosen and C.-T. Ho, *J. Food Lipids* 2001, **8**, 37–44.

Analysis

POTENTIAL MIGRATION OF ORGANIC POLLUTANTS FROM RECYCLED PACKAGING MATERIALS INTO DRY FOOD

V.I. Triantafyllou[1], K. Akrida-Demertzi[2] and P.G. Demertzis[2]

[1]Department of Molecular Biology and Genetics, Democritus University of Thrace, GR-68100 Alexandroupoli, Greece
[2]Department of Chemistry, University of Ioannina GR-45110 Ioannina, Greece

1 ABSTRACT

Paperboard packages represent a large and constantly growing part of the food packaging industry. To protect the environment the use of recycled paper as food contact material has increased. The safety of recycled fiber-based materials for food contact applications is largely dictated by the ability of post-consumer contaminants to be absorbed into recycled materials and later released by the packaging material and trapped on the food. In the present work the migration of different organic surrogates from recycled paper into a dry food (high fat milk powder) has been investigated using a method based on solvent extraction and GC-FID quantification. Results showed that the extractive power of the food powder, under the experimental conditions used, is high and that migration occurs rapidly. The proportion of substances migrated to food was strongly dependent on the nature of the paper samples, the fat content of the food and the chemical nature and volatility of the migrant.

2 INTRODUCTION

In 1999, the CEPI (Confederation of European Paper Industries) member countries produced 85.2 million tons of paper and board. In volume terms, the EU's paper production is graded: 1) graphic paper, 50%; 2) packaging paper, 40%; 3) hygiene and specialty papers, 10%. This means that half of the European paper and board production is directly or indirectly concerned by food-contact standards.[1]

To protect the environment, new ways of recycling waste, including packaging materials, are appearing in the market. The environmental pressure and the general tendency of recycling place recycled paper and board (P&B) in the situation of being used as food contact material. Food-contact materials, including recycled fiber-based paper, have to comply with a basic set of criteria concerning safety. This means that recycled paper for food contact should not give rise to migration of components, which can endanger human health.

Recovered P&B may vary in origin and could include paper containing printing inks, adhesives, trace elements, waxes, fluorescent whitening agents and dyes, sizing agents, organochlorine substances, plasticizers, aromatic hydrocarbons, volatile organic compounds, curing and grease-proofing agents, amines, biocides and surfactants.[2-8] Therefore, chemicals from these sources may migrate into foodstuffs and potentially cause a risk to human health. In the last years, more and more publications about migration of contaminants from paperboard packaging materials into food and food simulants have been published to evaluate the suitability of recycled paperboard for direct food contact applications.[2, 9-13]

In the present work a considerable amount of data concerning the mobility of different organic pollutants from recycled paper packaging onto a dry food has been gathered. More specifically, the kinetics of migration of selected model contaminants (surrogates) from contaminated recycled paper packaging into a food powder was studied using a recently proposed method based on solvent extraction and gas chromatography-flame ionization detection (GC-FID) quantification.[2] Obtained results were discussed in terms of the possibility, from a safety point of view, to use recycled fibers for food contact applications.

3 MATERIALS AND METHODS

3.1 Reagents and Solutions

A set of 10 model substances or surrogates was selected which can be considered as representative with regard to chemical structure, polarity, molecular weight, volatility and functionality in the paper making process. The selected chemicals are the following: (1) o-xylene, (2) acetophenone, (3) n-dodecane, (4) naphthalene, (5) diphenyl ether, (6) 2,3,4-trichloroanisole, (7) benzophenone, (8) diisopropyl-naphthalenes (DIPN), isomeric mixture, (9) dibutyl phthalate (DBP), and (10) methyl stearate. All reagents were of analytical grade and were obtained from Sigma, Aldrich and Fluka. Solutions of these standards in ethanol (Merck) at appropriate concentrations were used for calibration. Quantification was achieved using an ethanol solution of BHT as internal standard. As dry food an infant milk powder with high fat content (27.7%) supplied by Nestlé, was used.

3.2 Paper Samples

For migration studies two paper samples were used, having different pulp, percentage of recycled matter, grammage and thickness. Their properties are listed in Table 1.

Table 1 Properties of the Paper Samples Used in the Present Study

Sample	Type	Pulp	Recycled (%)	Grammage $(g\ m^{-2})$	Density $(kg\ m^{-3})$	Thickness (μm)
P1	Fluting	NSSC[a] + 30% recycled	30	107,0	511	209
P2	Kitchen towel	recycled	100	46,7	248	188

[a] NSSC: neutral sulfite semi-chemical pulp.

Analysis

3.3 Analytical Procedures

3.3.1 Fortification of Paper Samples with the Surrogates. Strips of each of the paper samples with approximate dimensions of 6 cm×1 cm were placed in 22 ml septum glass vials. The vials were filled with 20 ml of contamination solution (solvent, absolute ethanol HPLC grade) having surrogate concentration of approximately 250 mg l^{-1}. The vials were sealed and left for approximately 2 hours in a horizontal position (paper strips were totally immersed into the contamination liquid) at room temperature. The contaminated paper strips were then removed and placed on a steel lattice in a ventilated hood at room temperature. After 15 min drying the samples were used for the migration test.

3.3.2 Determination of Contaminant Concentration in Paper Samples. The determination of the initial concentration of the sorbed surrogates was performed as follows: contaminated paper samples were cut in small pieces (1 cm×1 cm) and were placed in 5 ml vials with 4 ml of ethanol. Extraction was performed by gentle shaking of the vials for 1h at room temperature. The ethanol extracts were directly analyzed by GC.

3.3.3 GC Analysis. The GC unit was a Fisons 9000 series gas chromatograph equipped with an autoinjector and a FID detector. The separation column was a 30m×0.32mm internal diameter fused silica capillary DB-1 with a film thickness of 0.25 μm. The following GC parameters were kept constant: detector temperature, 290°C; injector temperature, 240°C; injection mode, split with split ratio ca.15 ml/min; injection volume, 1μl. Column temperature program, 60°C (3 min), from 60°C at a rate of 10°C/min to 270°C (3 min). Carrier gas, He, at a flow rate of 1.45 ml/min.

3.3.4 Kinetics of Migration in Contact with Food Powder. Kinetic migration studies were carried out with strips of both paper samples fortified with the mixture of surrogates as described previously. Each contaminated strip was placed into a 22 ml septum glass vial lying in horizontal position. Strips were covered with 0.75 g of milk powder. The exact experimental temperature/time exposure conditions were as follows:
- 70°C for 20, 40, 90, 120, 240 and 360 min
- 100°C for 10, 20, 30, 40, 60 and 120 min

Triplicate determinations were carried out. The determination of the amount of the desorbed surrogates in the dry food substrate under all temperature/time conditions was performed as follows: After removal of the respective paper strip the migrated surrogates were extracted from milk powder using 3 ml of ethanol as the extraction solvent. The extraction was carried out at room temperature for approximately 5 min under gentle agitation of the sample. The extraction was repeated twice. The combined extracts were concentrated under a stream of nitrogen and were then analysed by GC.

4 RESULTS AND DISCUSSION

4.1 Surrogate determination in the contaminated paper samples

Absolute ethanol was found to afford complete extraction of sorbed surrogates from both paper and food powder samples, under the applied experimental conditions. The concentrations of surrogates sorbed onto paper samples are presented in Fig. 1. The results show that, under the same sorption conditions (concentration of contamination solution and temperature), large differentiations existed in the sorbed amounts of contaminants for both paper types, as detailed elsewhere.[2]

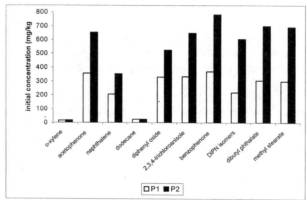

Figure 1 *Chart comparing the concentrations of surrogates determined in P1 (fluting) and P2 (kitchen towel) paper samples.*

4.2 Migration Studies

It has been demonstrated that migration from paper and board occurs even with dry foods. To evaluate if the recycled paper and board could be used as a food packaging material, migration studies of selected surrogates from contaminated paper and board into infant milk powder have been performed at two different temperatures in a wide range of storage time. The test conditions represent short-term exposures at high temperatures (rapid testing). It is important to know the transfer or migration capacity of these pollutants. A large number of experimental data have been obtained and used to evaluate the effects of high temperatures on the actual mass transfer during contact with foods.

The results of the time dependent migration for both paper samples at 70 and 100°C are given in Tables 2 and 3.

It was observed that short exposure times were enough to achieve equilibrium of migrants between paper and food since paper is porous and diffusion of contaminants proceeds mainly in air spaces within the paper matrix. Concerning the migration behavior of both paper samples, the concentrations of migrants into milk powder were generally increased as time and temperature increased. The high fat content of the food contributed to a more pronounced migration tendency.

Analysis

Table 2 *Time Dependent Migration of Surrogates from P1 Paper Sample into Infant Milk Powder at 70 and 100°C given as Relative Migration (% of Initial Concentration)*

Temperature: 70°C

Time (min)	20	40	90	120	240	360
o-Xylene	2.4	5.8	7.8	8.4	18.0	17.7
Acetophenone	8.2	12.7	18.4	35.5	40.3	40.7
Naphthalene	7.2	11.8	19.6	37.3	46.1	45.4
Dodecane	18.2	27.4	36.5	45.4	52.5	49.1
Diphenyl oxide	11.7	18.5	23.4	28.5	31.4	33.6
2,3,4-TCA	11.8	17.4	23.1	29.7	30.7	29.8
Benzophenone	8.8	15.7	21.4	28.7	30.4	32.4
DIPN isomers	44.9	57.3	61.7	65.8	68.7	70.2
Dibutyl phthalate	18.9	26.7	38.7	45.2	49.7	50.4
Methyl stearate	28.6	38.9	46.7	59.9	78.4	84.7

TCA: trichloroanisole
DIPN: diisopropyl-naphthalenes

Temperature: 100°C

Time (min)	10	20	30	40	60	120
o-Xylene	4.4	5.6	7.9	19.2	20.2	19.8
Acetophenone	7.6	15.7	24.9	33.7	48.5	54.8
Naphthalene	9.2	14.7	26.6	37.3	58.1	59.4
Dodecane	19.2	32.4	39.4	49.5	58.7	56.5
Diphenyl oxide	9.8	21.4	29.4	35.6	39.7	41.7
2,3,4-TCA	18.4	19.7	26.4	35.7	36.5	35.7
Benzophenone	6.9	14.7	25.7	36.4	40.4	39.1
DIPN isomers	38.5	48.7	56.4	62.1	72.8	77.4
Dibutyl phthalate	16.9	24.7	35.7	49.1	55.9	57.4
Methyl stearate	24.6	36.6	47.1	62.1	85.7	91.4

TCA: trichloroanisole
DIPN: diisopropyl-naphthalenes

Table 3 *Time Dependent Migration of Surrogates from P2 Paper Sample into Infant Milk Powder at 70 and 100°C given as Relative Migration (% of Initial Concentration)*

Temperature: 70°C

Time (min)	20	40	90	120	240	360
o-Xylene	4.8	5.9	8.2	19.6	20.4	20.2
Acetophenone	9.2	14.7	21.3	28.9	47.3	49.7
Naphthalene	10.2	19.8	28.6	39.3	48.1	47.6
Dodecane	27.2	39.4	49.5	61.4	61.5	60.7
Diphenyl oxide	11.8	23.4	31.4	39.4	44.8	46.4
2,3,4-TCA	21.4	23.4	29.2	38.9	39.5	37.9
Benzophenone	23.7	28.6	32.7	44.9	50.7	53.4
DIPN isomers	59.4	68.7	75.5	85.4	86.9	87.7
Dibutyl phthalate	35.7	46.8	59.4	65.2	73.7	76.4
Methyl stearate	38.6	58.7	77.1	82.4	90.7	92.4

TCA: trichloroanisole
DIPN: diisopropyl-naphthalenes

Temperature: 100°C

Time (min)	10	20	30	40	60	120
o-Xylene	4.2	6.3	8.4	15.8	21.9	22.3
Acetophenone	11.2	18.7	29.3	38.9	57.3	62.7
Naphthalene	10.2	21.8	29.6	41.3	65.1	60.6
Dodecane	29.2	42.4	56.5	63.4	61.5	63.7
Diphenyl oxide	9.3	15.7	27.4	36.4	51.8	54.4
2,3,4-TCA	29.4	32.7	38.2	42.9	45.4	48.9
Benzophenone	28.4	34.1	38.6	43.1	52.2	59.4
DIPN isomers	55.4	66.7	78.1	88.8	92.1	93.7
Dibutyl phthalate	38.7	49.4	60.4	74.1	83.9	87.2
Methyl stearate	47.6	59.7	75.6	88.4	94.7	93.4

TCA: trichloroanisole
DIPN: diisopropyl-naphthalenes

Comparing the results of the present study with those obtained in a previous study using Tenax as food simulant,[2] the following conclusions were discerned: As the selected organic surrogates have generally a higher solubility in fatty media the percentage migration values into milk powder with a fat content of 27.7% are higher than those into Tenax. Benzophenone for example is a highly fat-soluble compound and it is not surprising that milk powder in direct contact with paper samples pick up benzophenone at higher level compared to Tenax.

Migration values for DIPN, DBP and methyl stearate were also quite high, eventually higher than expected according to their molecular weights. Lipids encourage the absorption of these apolar, non-volatile compounds. Taking into account their chemical structure, probably the activation energy to break the steric hindrance has been surpassed at high temperatures, and consequently, the migration tendency is higher than expected for these compounds. The loss by volatilization from the heated food was also negligible for these contaminants due to their high boiling points. The most volatile compounds (e.g. acetophenone, o-xylene, and naphthalene) exhibit lower migration levels. In this case, it can be assumed that small amounts of surrogates were lost in the vapor phase. It is important to mention that not all the compounds released from the paper are trapped by the simulants or by the food, which in terms of real migration conditions is an advantage.

Concerning the migration behavior of the two paper samples, it was found that the migration from P2 samples was higher than that from P1 samples. This can be partially attributed to the fact that P2 paper samples exhibit a greater sorption capacity (higher initial surrogate concentrations) than to P1 samples. Moreover, P2 sample had a lower thickness, density and grammage than P1 and this could facilitate the release of sorbed compounds into food and/or food simulants. In other words, the migration of the surrogates to the food phase is higher in the case of thinner samples. Not only the grammage, density and thickness but also the paper composition and the chemical structure of the model compounds play a role in the extent of migration. Sample P1, for example, with only 30% of recycled pulp had a stronger chemical affinity with compounds, which have aromatic groups and some polarity. These compounds could have a stronger interaction with P1 paper sample (partially recycled pulp) and the migration levels from P1 into food contacting phase, even at high temperatures, were lower than those obtained from sample P2 (100% recycled pulp).

5 CONCLUSIONS

In this work, migration of selected model contaminants from paperboard into a fatty dry food was studied. Papers of different qualities were compared and the suitability, from a safety point of view, of recycled fibers for food-contact applications was evaluated. One major issue was to focus on the influence of temperature on the time dependent migration course. Useful conclusions were extrapolated about the migration behavior of several surrogates, which are representative of chemical structures and physical properties. Infant whole milk powder, a substrate with high fat content (27.7%), is an excellent adsorbent and due to its thermal stability can be used efficiently for migration studies at high temperatures. Such studies can be applied as rapid screening methods for anticipating the potential of certain contaminants from paper and board food contact materials to migrate into foods under regular conditions of packaging, distribution and storage.

Further work is required to check if the applied rapid test will give comparable results to migration tests using dry foods with different composition when tested under conditions representing shelf storage, e.g. longer term tests at ambient temperature over a period of 6-12 months. Corresponding investigations are in progress and results will be the objective of a forthcoming paper.

References

1. J.Y. Escabasse and D. Ottenio, *Food Addit. Contam.*, 2002, **19**, 79.
2. V.I. Triantafyllou, K. Akrida-Demertzi and P.G. Demertzis, *Anal. Chim. Acta*, 2002, **467**, 253.
3. H. Kim and S.G. Gilbert, *J. Food Sci.*, 1989, **54**, 770
4. P.A. Tice and C.P. Offen, *Tappi J.*, 1994, **77**, 149.
5. L. Castle, C.P. Offen, M.J. Baxter and J. Gilbert, *Food Addit. Contam.*, 1997, **14**, 35.
6. G. Ziegleder, *Packag. Technol. Sci.*, 2001, **14**, 131.
7. T. Sipiläinen-Malm, K. Latva-Kala, L. Tikkanen, M.L. Suihko and E. Skyttä, *Food Addit. Contam.*, 1997, **14**, 695.
8. M.L. Binderup, G.A. Pedersen, A.M. Vinggaard, E.S. Rasmussen, H. Rosenquist and T. Cederberg, *Food Addit. Contam.*, 2002, **19**, 13.
9. M. Sarria-Vidal, J. de la Montana-Miguélez and J. Simal-Gándara, *J. Agric. Food Chem.*, 1997, **45**, 2701.
10. B. Aurela, H. Kulmala and L. Söderhjelm, *Food Addit. Contam.*, 1999, **16**, 571.
11. J. Salafranca and R. Franz, *Deut. Lebensm.- Rundsch.*, 2000, **96**, 355.
12. W. Summerfield and I. Cooper, *Food Addit. Contam.*, 2001, **18**, 77
13. M. Boccacco-Mariani, E. Chiacchierini and C. Gesumundo, *Food Addit .Contam.*, 1999, **16**, 207.

Antioxidants and Health

ANTIOXIDANT CAPACITY OF PHENOLIC EXTRACTS FROM WILD BLUEBERRY LEAVES AND FRUITS

Marian Naczk[1], Ryszard Amarowicz[2], Ryszard Zadernowski[3], Ron Pegg[4] and Fereidoon Shahidi[4]

[1]Department of Human Nutrition, St. Francis Xavier University, Antigonish, NS, Canada, B2G 2W5; [2]Division of Food Science, Institute of Animal Reproduction and Food Research of the Polish Academy of Sciences, Olsztyn, Poland; [3]Faculty of Food Science, Warmia and Mazury University, Olsztyn Poland; [4]Department of Biochemistry, Memorial University of Newfoundland, St John's, NF, Canada A1B 3X9.

1 ABSTRACT

Leaves are by-products from mechanical harvesting of wild blueberries. To explore their use as a natural source of antioxidants, blueberry leaves were extracted with ethanol (95%, v/v) and acetone (70%, v/v). The dried crude extracts were then fractionated on a Sephadex LH-20 column. The extracts contained 181-213mg catechin equivalents/g of tannins and their antioxidant activity was similar to that of BHA is a β-carotene linoleate model system. Over 99% of DPPH was also scavenged by 100 µg of extract in each assay. The antioxidant activity of polyphenolic extracts was evaluated by the β-carotene-linoleate and DPPH radical assays.

2 INTRODUCTION

There is an increasing interest in replacing synthetic antioxidants such as butylated hydroxyanisole (BHA), butylated hydroxytoluene (BHT), propyl gallate (PG) and *tert*-bytylhydroxyquinone (TBHQ) with natural alternatives. Plant materials such as fruits, vegetables, spices, leaves, roots, and barks have been extensively evaluated as potential sources of natural antioxidants.[1-6]

The antioxidant activities of plant extracts are usually linked to the presence of anthocyanins, phenolic acids, flavonoids and tannins, among others.[7] Prior et al.[8] noticed an increase in antioxidant activity of blueberry extracts with maturity of berries. Later, Wang and Lin[9] evaluated the antioxidant activity of fruits and leaves of blackberry, raspberry and strawberry using oxygen radical absorbance capacity (ORAC). Extracts of phenolics from leaves exhibited significantly higher ORAC values than extract from corresponding berries. Recently, Ehlenfeldt and Prior[10] also reported that leaf tissue extracts from highbush blueberry displayed higher ORAC values than those of the corresponding fruit tissues.

The leaves of blueberry are by-products from mechanical harvesting of wild berries and these are not commercially utilized. In this contribution we discuss possible use of blueberry leaves as a potential source of natural food antioxidants. The antioxidant activity of blueberry leaf phenolics will also be compared to that displayed by fruit phenolics.

3 MATERIALS AND METHODS

3.1 Materials

Blueberry leaves, a by-product of mechanical harvesting of wild blueberry, and blueberry fruits were collected from a wild blueberry farm located in Antigonish County, Nova Scotia, Canada. The leaves were separated from other debris, dried at room temperature, and then stored in sealed polyethylene bags at -18°C. Blueberry fruits were packed in sealed polyethylene bags and stored at -18°C.

3.2 Extraction and Fractionation of Phenolics

Crude phenolics were extracted from both fruits and leaves with 95% (v/v) ethanol three times, at 50°C for 30 min at a solid-to-solvent ratio of 15:100 (w/v). The ethanolic extracts were combined, evaporated to near dryness under vacuum at 40°C, and lyophilized. Crude polyphenolics were also extracted from both the leaves and fruits with 70% (v/v) aqueous acetone, three times, at room temperature for 30 min at a solid-to-solvent ratio of 15:100 (w/v). The acetone extracts were combined, evaporated to near dryness under vacuum at 40°C, and lyophilized. The chlorophylls were removed from the leaf extracts as described by Amarowicz et al.[11] Selected crude phenolics extracts were fractionated into non-tannin and tannin fractions according to the procedure described by Strumeyer and Malin.[12]

3.3 Quantification of Phenolics

The total content of phenolic compounds in the extracts was estimated using the Folin-Denis reagent[13] and expressed as gallic acid equivalents per gram of extract.

3.4 Evaluation of Antioxidant Activity of Phenolic Extracts

The antioxidant activity of phenolic extracts of wild blueberry leaves was evaluated in a β–carotene-linoleate model system.[14] A methanolic solution (0.2 ml) containing 2 mg of extract was added to a series of tubes containing 5 mL of an emulsion of linoleate and β-carotene stabilized by Tween 40, prepared as described by Wanasundara et al.[15] A controlled experiment was carried out using 0.5 mg BHA. Immediately after the addition of the emulsion to tubes the zero-time absorbance at 470 nm was recorded. Samples were kept in a water bath at 50°C and their absorbances read over a 120 min period at 15 min intervals.

The scavenging effect of phenolic extracts from both blueberry leaves and fruits on α,α–diphenyl-β–picrylhydrazyl (DPPH) radical was monitored according to the method of Hatano et al.[16] An aliquot of (0.1 ml) methanolic solution containing 20-100 µg of phenolic extract of wild blueberry leaves was mixed with 2 ml of methanol and then added to a methanolic solution of DPPH (1 mM, 0.25 ml). The mixture was vortexed for 10 s, left to stand at room temperature for 30 min, and then its absorbance at 517 nm was recorded.

The scavenging effect of phenolic extracts from both blueberry leaves and fruits on 2,2'-azinobis-(3-ethylbenzothiazoline-6-sulfonate) (ABTS) radical ion was monitored using the method of van der Berg et al.[17] as modified by Kim et al.[18] The concentration of ABTS radical ion solution was adjusted to an absorbance of 0.50 at 734 nm. The ABTS radical ion scavenging activity of blueberry extracts was expressed in mM Trolox (6-hydroxy-2,5,7,8-tetramethylchroman-2-carboxylic acid) equivalent antioxidant capacity (TEAC) values per gram of extract. TEAC was expressed as a slope of a line reflecting the amount of Trolox equivalents per assay as a function of the amount of phenolic extract added to the reaction mixture. The ABTS radical ion solutions were freshly prepared each day.

3.5 Data Treatment

The results presented in Figures are mean values of at least three experiments (with three replicates per experiment). Statistical analysis of data (analysis of variance (ANOVA) test, t-test) was carried out using SigmaStat v. 2.03 (SSPS Science Inc., Chicago, IL).

4 RESULTS AND DISCUSSION

Phenolics were extracted from blueberry leaves and fruits with 70% (v/v) acetone and 95% (v/v) ethanol, respectively. These solvents systems are commonly used for extraction of phenolics from plant materials. The extraction yield of crude extracts was between 10.1 and 14.5%. The extraction of leaves yielded more crude extract than that of fruits (Table 1). The non-tannin phenolics were the predominant fraction of crude phenolic extract as tannin fraction comprised only from 2 (ethanol; fruits) to 13% (ethanol; leaves) (Table 1). Total phenolics content (TP) in the extracts from wild blueberry leaves and fruits was expressed as gallic acid equivalents/g of extracts. Gallic acid is commonly used as a standard for quantification of phenolic present in plant extracts. The TP of extracts obtained with ethanol was slightly higher than that of the acetone extract. The crude extracts from leaves contained ~10 fold more TP than corresponding fruits extracts (Table 1). This is in good agreement with data published by Ehlenfeldt and Prior.[10]

The effect of phenolic extracts of wild blueberry leaves on the coupled oxidation of β–carotene and linoleic acid was compared to that of BHA (Figure 1). The antioxidant activity of both tannin fractions and crude phenolic extracts was somewhat lower compared to that displayed by BHA. However, crude extract of blueberry leaves phenolics obtained using 70% (v/v) acetone was more effective ($P<0.05$) than the corresponding ethanolic extract. Stronger antioxidant activity of acetone extracts may be brought about by higher level of proanthocyanidins which are more soluble in acetone. On the other hand non-tannin fraction of phenolics exhibited a significantly weaker antioxidant activity than the corresponding tannin fractions and crude phenolic extracts.

Table 1 *Yield, total phenolics content and antioxidant activity, expressed as TEAC values, of various phenolic fractions obtained from blueberry leaves and fruits*

Sample	Solvent System	Phenolic fraction	Extract yield[a]	Total phenolic content[b]	TEAC[c]
Leaves	70% (v/v) acetone	Crude	14.2	218.0±3.2	2.96±0.08
		Non-tannin	12.8	189.0±2.6	2.28±0.11
		Tannin	1.1	523.0±9.0	6.42±0.28
Fruits	70% (v/v) acetone	Crude	10.7	31.3±0.4	0.37±0.01
		Non-tannin	10.2	18.8±0.2	0.26±0.01
		Tannin	0.2	348.0±1.6	5.03±0.15
Leaves	Ethanol	Crude	14.5	227.0±2.5	3.24±0.14
		Non-tannin	12.5	202.0±6.3	2.41±0.12
		Tannin	1.9	519.0±7.4	8.12±0.10
Fruits	Ethanol	Crude	10.1	23.6±0.9	0.28±0.01
		Non-tannin	9.8	18.3±0.1	0.25±0.01
		Tannin	0.2	371.0±10.6	4.79±0.08

[a] % of fresh sample weight; [b] expressed in mg gallic acid equivalents per gram of extract; [c] mM of Trolox equivalents per gram of extract.

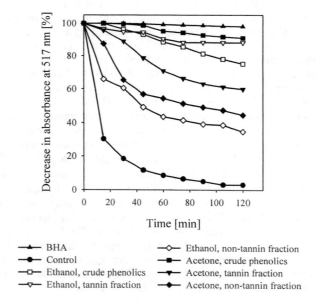

Figure 1 *Antioxidant activity of phenolic extracts from blueberry leaves in β-carotene-linoleate model system as measured by a decrease [%] in absorbance at 517 nm.*

The DPPH assay is commonly used to evaluate the ability of antioxidants to scavenge free radicals. The absorbance at 517 nm decreases as the reaction between the antioxidant molecules and the DPPH radicals progresses. The scavenging effect of the antioxidants on DPPH radical is influenced by their concentration, as well as molecular structure and kinetic behavior of phenolics involved.[19] Figure 2 displays the scavenging effects of crude phenolic extracts from blueberry leaves and fruits and their corresponding tannin and non-tannin fractions. No statistically significant difference (P>0.05) was noted between crude phenolic extracts in aqueous acetone and ethanol. The scavenging effect of crude extracts, at 100 µg/assay for leaf extracts and at >1 mg/assay fruit extracts, on DPPH radical was over 99%. Tannin fractions isolated from both leaf and fruit crude phenolic extracts exhibited stronger scavenging effects on DPPH radical than a corresponding non-tannin phenolics fractions and crude phenolic extracts. Only ≤40 µg tannin fraction/assay was needed to scavenge over 99% of DPPH radicals present in the reaction mixture. The strong radical scavenging effects displayed by various phenolic extracts from wild blueberry leaves and fruits suggest that these extracts may also possess a strong antimutagenic activity. Hochstein and Atallah[20] associated the ability of antioxidant to scavenge free radicals with its antimutagenic activity.

The scavenging effects of blueberry extracts were also evaluated using the ABTS assay described by van der Berg et al.[17] as modified by Kim et al.[18] Figure 3 shows the time course of scavenging of ABTS radical ion by the phenolics present in crude acetone extracts from both leaf and fruit as measured by the percentage of decolorization of the initial ABTS radical ion solution. These results indicated that reaction progressed rapidly during the initial one minute during which up to 40% of free radicals were scavenged. Following this, the rate of reaction was significantly decreased down. Similar kinetic responses were reported for extract of phenolics from Gala apple[19] as well as several pure phytochemicals.[18, 21, 22]

The determination of TEAC values for plant extracts is usually carried out at one concentration of extract and the decrease in the absorbance is used for the calculation of Trolox equivalents per gram of extract (TEAC values). In this study we calculated the TEAC values as a slope of lines depicting the relationship between the TEAC/assay and the amount of extract added to the reaction mixture (Figure 4). The slope value is a more meaningful measure of TEAC value of the extract than the measurement carried out at only one concentration of the extract. It is based on statistical analysis (linear regression) of experimental data involving the measurement of antioxidant activity of the extract at a minimum four different concentrations. The TEAC values for various phenolic extracts from blueberry leaves ranged from 0.28 (ethanolic extract from fruit) to 8.12 mM Trolox equivalents/g of extract (tannin fraction of ethanolic extract from fruit) (Table 1). The TEAC values calculated for crude phenolic extracts from leaf were 8-12 fold greater than those for crude extracts obtained from fruits. This result is in good agreement with that reported by Ehlenfeldt and Prior[10] who found ~25 fold difference in ORAC values between the phenolic extracts from highbush blueberry leaves and fruits. The difference in the TEAC values reported here may be due ~10 fold higher level of total phenolics in leaf extracts as compared to those in the corresponding fruit extracts. The antioxidant activity exhibited by tannin fractions isolated from both leaf and fruit was much stronger than that displayed by the corresponding non-tannin fractions and crude extracts. Bors et al.[22] and

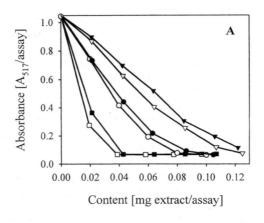

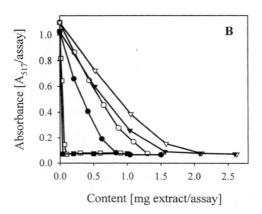

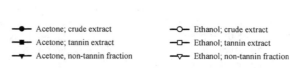

Figure 2 *Scavenging effects of phenolic extracts from blueberry leaves (A) and fruits (B) on α,α-diphenyl-β-picrylhydrazyl (DPPH) radical as measured by a change in absorbance values at 517 nm.*

Hagerman et al.[23] also observed that tannins displayed a stronger antioxidant activity than other phenolics. Moreover, the TEAC values of tannin fraction from crude phenolic extracts from fruits were similar to those of crude phenolic extracts from leaves. This suggests that the molecular structures of tannins present in leaves and fruits are probably very similar.

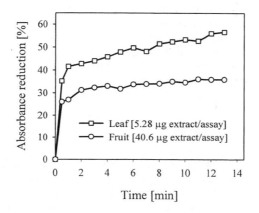

Figure 3 *The effect of reaction time course on the scavenging of ABTS radical ion by crude acetone extracts from blueberry leaves and fruits.*

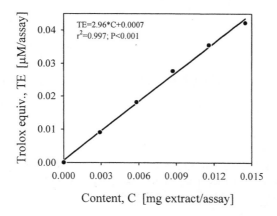

Figure 4 *Determination of TEAC value as a slope of the line depicting antioxidant activity of extract expressed as in Trolox equivalents per assay as a function of extract concentration in reaction mixture. The data for crude acetone phenolic extract from blueberry leaves are shown here.*

The antioxidant activities of various blueberry extracts as determined by both DPPH and ABTS assays correlated strongly with total content of phenolics present in them (Figure 5). The antioxidant potential of extract, measured by the DPPH assay, was expressed in μg of extract per assay required to lower the initial absorbance of DPPH solution by 50%. These correlations suggest that phenolics are mainly responsible for the antioxidant activities displayed by blueberry extracts. Thus, the total phenolic content may be used as an indicator antioxidative potential of blueberry extracts. Similar relationships have been reported for major phenolics in apple,[24] red wines,[25,26] herbal products,[27] sorghum and sorghum products[28] and berries.[29]

The antioxidant activities of blueberry extracts as measured by the ABTS assay correlated strongly with those obtained using the DPPH assay (Figure 6). Awika et al.[28] and Leong and Shui[30] also reported that DPPH and ABTS values correlated well for crude extracts from sorghum and its products and fruits, respectively. Thus, both assays are useful for determination of antioxidant potential of blueberry extracts.

5 CONCLUSIONS

The antioxidant activity of phenolics extracted from wild blueberry leaves is similar or even slightly better than that exhibited by phenolics extracted from blueberry fruits. The tannin fractions displayed much stronger free radical scavenging effects than corresponding non-tannin phenolic fractions. This is in a good agreement with the literature data.[22,23] The crude phenolic extracts from leaves are much richer sources of tannins than crude phenolic extracts from fruits. Thus, our findings indicate that blueberry leaves may be considered as a good source of potent natural antioxidants. More research is still needed to determine both composition and molecular structures of phenolics present in wild blueberry leaves.

Acknowledgment

We thank the Natural Sciences and Engineering Research Council (NSERC) for research fundings.

References

1. C.-C. Chyau, S.Y. Tsai, P.-T. Ko, and J.-L. Mau, *Food Chem.* 2002, **78**, 483.
2. M. Ellnain-Wojtaszek, Z. Kruczynski, and J. Kasprzak, *Food Chem.* 2002, **79**, 79.
3. P. Picerno, T. Mencherini, M.R. Lauro, F. Barbato, and R. Aquino, *J. Agric. Food Chem.* 2003, **51**, 6423.
4. F. Shahidi and M. Naczk, 2003. *Phenolics in Food and Nutraceuticals,* CRC Press, Boca Raton, FL, Chapter , p. .
5. P. Sidduhuraju, and K. Becker, *J. Agric. Food Chem.* 2003, **51**, 2144.
6. E.R. Sherwin, 'Antioxidants' in *Food Additives,* eds., L.R. Branen, Marcel Dekker, New York, NY, 1990, Chapter 18, pp. 139-193.

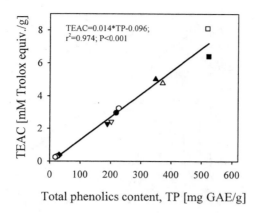

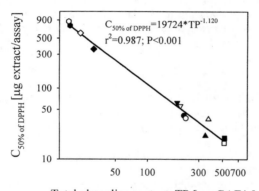

- ● Leaves, crude phenolics
- ■ Leaves, tannin fraction
- ▼ Leaves, non-tannin fraction
- ◆ Fruits, crude phenolics
- ▲ Fruits, tannin fraction
- ● Fruits, non-tannin fraction

Figure 5 *Correlation between the TEAC (A) and DPPH (B) values and total phenolics content (expressed as gallic acid equiv.(GAE)/g extract) in extracts from blueberry leaves and fruits. Filled symbols represent acetone extracts while those unfilled methanol extracts.*

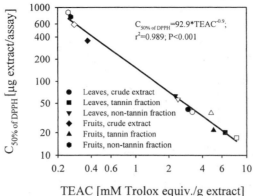

Figure 6 *Correlation between the antioxidant activities of phenolic extracts from blueberry leaves and fruits as determined by the ABTS and DPPH assays. Filled symbols represent acetone extracts while those unfilled methanol extracts.*

7. G. Cao, E. Sofic, and R.L. Prior, R.L. *J. Agric. Food Chem.* 1997, **44**, 3426.
8. R.L.Prior, G. Cao, A. Martin, E. Sofic, J. McEwen, C. O'Brien, N. Lischner, M. Ehlenfeldt, W. Kalt, G. Krewer, and C.M. Mainland, *J. Agric. Food Chem.* 1998, **46**, 2686.
9. S.Y. Wang and H.-S. Lin, *J. Agric. Food Chem.* 2000, **48**, 140.
10. M.K. Ehlenfeldt and R.L. Prior, *J. Agric. Food Chem.* 2001, **49**, 2222.
11. R. Amarowicz, R.B. Pegg, P. Rahimi-Moghaddam, B. Barl, B., and J.A.Weil, *Food Chem.* 2004, **84**, 551.
12. D.H. Strumeyer and M.J. Malin, *J. Agric. Food Chem.* 1975, **23**, 909.
13. T. Swain and W.E. Hillis, *J. Sci. Food Agric.* 1959, **10**, 63.
14. H.E. Miller, *J. Am. Oil Chem. Soc.* 1971, **48**, 91.
15. U. Wansundara, R. Amarowicz, and F. Shahidi, *J. Agric. Food Chem.* 1994, **42**, 1285.
16. T. Hatano, H. Kagawa, T. Yasuhara, and T. Okuda, *Chem. Pharm. Bull.* 1998, **36**, 2090.
17. R. van den Berg, G.R.M.M. Haenen, H. van den Berg, and A. Bast, *Food Chem.* 1999, **66**, 511.
18. D.-O. Kim, K.W. Lee, H.J., Lee, and C.Y. Lee, *J. Agric. Food Chem.* 2002, **50**, 3713.
19. C.A. Rice-Evans, N.J. Miller, and G. Papanga, *Free Rad. Biol. Med.* 1996, **20**, 933.
20. P. Hochstein and A.S. Atallah, *Mutat. Res.-Fund. Mol. M.* 1988, **202**, 363.
21. R.Re, Pellegrini, N., Proteggente, A., Pannala, M. Yang, and C. Rice-Evans, *Free Radic. Biol. Med.* 1999, **26**, 1231.

22. W. Bors, C. Michel, K.Stettmaier, *Arch. Biochem. Biophys.* 2000, **374**, 347.
23. A.E. Hagerman, K.M. Riedl, G.A. Jones, K.N. Sovik, N.T. Ritchard, P.W. Hartzfield, and T.K. Riechel, *J. Agric. Food Chem.* 1998, **46**, 1887.
24. K.W. Lee, Y.J. Kim, H.J. Lee, and C.Y. Lee, *J. Agric. Food Chem.* 2003, **51**, 6516.
25. E.N. Frankel, A.L.Waterhouse, and P.L. Teissedre, *J. Agric. Food Chem.* 1995, **43**, 890.
26. P. Simonetti, P. Pietta, and G. Testolin, *J. Agric. Food Chem.* 1997, **45**, 1152.
27. O.A. Zaporozets, O.E. Krushynska, N.A. Lipkovska, and V.N. Barvinchenko, *J. Agric. Food Chem.* 2004, **52**, 21.
28. J.M. Awika, L.W. Rooney, X. Wu, R.L. Prior, and L. Cisneros-Zevallos, *J. Agric. Food Chem.* 2003, **51**, 6657.
29. S. Sellappan, C.C. Akoh, and G. Krewer, *J. Agric. Food Chem.* 2002, **50**, 2432.
30. L.P. Leong and G. Shui, *Food Chem.* 2001, **76**, 69.

ANTIOXIDATIVE ACTIVITY OF CRUCIFEROUS VEGETABLES AND THE EFFECT OF BROCCOLI ON EDIBLE OIL OXIDATION

Ryoyasu Saijo[1,3], Rong Wang[2], Keiko Saito[1], Reiko Nakata[1], Satoko Ofuji[1], Tomoko Inoue[1], Yuko Mori[1], Miwa Motoki[1] and Yoko Tabata[1]

[1]Faculty of Education, Kagawa University, Saiwai cho 1-1, Takamatsu City, Kagawa Pref., 760-8522, Japan, [2]Department of Food and Nutrition Science, Jinan University, No.13 Shungeng Road, Jinan, Shandong, 250002, China, [3] Present address: Otowa cho 15-15-307, Shizuoka City, Shizuoka Pref., 420-0834, Japan

1 ABSTRACT

Antioxidative activities of Cruciferous vegetables were measured by beta-carotene discoloration method, i.e., preventive activity of β-carotene degradation coupled with linoleic acid peroxide, and by peroxide value (PV) assessment, i.e., preventive activity of formation of linoleic acid peroxide. In order to investigate the activities 80% ethanol-soluble fraction was prepared from seven kinds of the vegetables. The stronger antioxidative activity of Cruciferous vegetables was found in flower-cluster of rape and broccoli by beta-carotene discoloration method. The activities of cauliflower, Japanese white long radish (daikon), cabbage and Chinese cabbage were weaker than the above two vegetables. By use of PV method broccoli showed stronger activity. Effects of broccoli extracts on weight changes in soybean oil and whale oil at 80 °C were measured; it was found that the weight increases were retarded by addition of broccoli extracts. PV of olive oil and sesame oil increased linearly at 100 °C, but the increases of PV were reduced by broccoli.

2 INTRODUCTION

Oxidation sometimes imparts an undesirable effect on food taste, flavor and color, and may also influence the nutritional value and safety of foods. Vegetables are important sources of vitamins A, C and E, minerals such as Ca, P, Fe, K, and dietary fiber. In addition to their nutritional value recent studies have revealed that vegetable intake also helps human body functions by rendering antioxidative activity, cancer prevention and ageing prevention etc.[1,2] Therefore, it is important to protect foods and the human body from oxidation.

In Japan, the most consumed vegetables are Japanese white long radish (daikon), followed by cabbage, Chinese cabbage, onions, cucumber, tomatoes, carrots and eggplant. The first 3 kinds of vegetables belong to the Cruciferous family of vegetables.

Although antioxidative activity of some vegetables has been studied by different authors [3,4,5], no detailed studies have been reported on Cruciferous vegetables. The objectives of this study were to examine antioxidative activity of Cruciferous vegetables and to test their effect on autoxidation of edible oils.

3   MATERIALS AND METHODS

3.1   Vegetables

Vegetables used were broccoli, cabbage, cauliflower, Chinese cabbage, flower-clusters of rape and Japanese white long radish (daikon), all purchased from a local supermarket in Takamatsu City, Kagawa Pref., over 3 year period (1997 to 2000).

3.2 Preparation of 80% ethanol-soluble fraction from vegetables

Antioxidative activity was measured in 80% ethanol-soluble fraction prepared from the selected vegetables. Fresh vegetables, 10 g, were macerated by homogenization with 80% ethanol solution (ethanol-water; 8:2, v/v) and then filtered through cotton gauze. The residue was extracted for several times with the ethanol solution and total 100 ml of ethanol-soluble fraction was prepared as vegetable extract.

3.3 Antioxidative activity

3.3.1 β-carotene discoloration method. β-carotene discoloration coupled with oxidation of linoleic acid was measured as reported by others[6,7,8] and our previous study.[9] A mixture of approximately 1.4 ml of β-carotene solution (100 mg of β-carotene/100 ml of chloroform), 0.2 ml of linoleic acid solution (20 g of linoleic acid/100 ml of chloroform) and 1.0 ml of Tween-40 solution (10 g of Tween-40/100 ml of chloroform) were prepared in a 200 ml Erlenmeyer flask. Chloroform was evaporated under a flow of nitrogen and then 100 ml of water were added to the flask and the residue was dissolved. This linoleic acid-β-carotene solution, 90 ml, was mixed with 8 ml of 0.2 M phosphate buffer (pH 6.8) to prepare substrate solution, 98 ml. It is desirable to adjust the absorbance at 470 nm of the substrate solution to around 2.0 in order to obtain reproducible results. The ethanol-soluble fraction of vegetables, 0.1 ml, which corresponded to 10 mg of vegetables, was added to 4.9 ml of the substrate solution and incubated at 55^0C in each test tube. Absorbance at 470 nm of the resulting reaction mixture was measured every 5 min. Water was used as a control. The antioxidative activity of 3(2)-*tert*-butyl-4-hydroxyanisole (BHA) (10 ppm) was also tested. The typical time courses of change at 470 nm absorbance of water (A), BHA (B) and sample (X) are illustrated in Figure 1. The antioxidative activity of sample was expressed by the following equation.

Antioxidative activity (ppm BHA equivalent) = A-X/A-B×10 ppm BHA

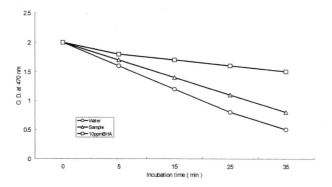

Figure 1 *Model schemes of absorbance change in water, sample and BHA in beta-carotene discoloration method*

3.3.2 Peroxide value (PV) assessment (ferric thiocyanate method).[10,11,12] Methyl esters of palmitic acid, stearic acid, oleic acid, linoleic acid and linolenic acid, 100 mg each, were oxidized in a test tube at 50 ^{0}C and PV was determined every 24 h for 96 h.[10] Ethanol, 60 ml, was added to the test tubes to dissolve the fatty acid methyl esters and 0.2 ml of the ethanol solution was used. A mixture of 9.4 ml of 80% ethanol, 0.2 ml of 30% ammonium thiocyanate solution and 0.2 ml of 0.02 M ferrous chloride in 3.5% hydrochloric acid solution was added to 0.2 ml of the ethanol solution dissolving the fatty acid methyl esters. After 3 min the absorbance at 500 nm, the maximum absorbance of ferric thiocyanate, was recorded.[11,12] The time courses of the 5 fatty acid methyl esters are shown in Figure 2. Figure 2 indicates that PV in methyl palmitate, methyl stearate and methyl oleate were all quite small during the experimental period, however, PV in methyl linoleate and methyl linolenate, having two and three diene conjugation respectively, increased and reached maximum values at 72 h and 48 h, respectively Since methyl linoleate was linearly oxidized until 72 h, the effect of vegetable extracts on the time course of methyl linoleate was used to investigate antioxidative activity of the vegetable extracts.

Antioxidants and Health

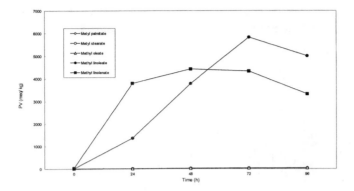

Figure 2 *Changes in PV of fatty acid methyl esters during oxidation at 50 °C.*

3.4 Chemicals

Methyl palmitate, methyl stearate, methyl oleate, methyl linoleate and methyl linolenate, and soybean oil, olive oil, sesame oil, whale oil and beef fat were reagent grade.

4 RESULTS

4.1 Antioxidative activity of Cruciferous vegetables

4.1.1 β-carotene discoloration method. An ethanol-soluble fraction was prepared from selected Cruciferous vegetables and their antioxidative activity measured. The results are shown in Figure 3. Flower-cluster of rape (in April only) and broccoli (both in April and June) showed the strongest activities. The activities of cauliflower, Japanese white long radish (daikon), cabbage and Chinese cabbage were at the same level and weaker than the above two vegetables.

4.1.2 Peroxide value (PV) assessment. Ethanol-soluble fraction of vegetables, 50 μl, which corresponded to 5 mg of fresh vegetables, was added to 100 mg of methyl linoleate in a test tube and kept at 50°C. Water as a control and 100 ppm BHA were also used in all experiments. PV was determined every 6 h for 48 h and the effect of vegetable extracts on oxidation of methyl linoleate examined (Figure 4).

Figure 4 indicates that addition of all vegetable extracts to methyl linoleate reduced the formation of peroxide during 48 h. The most effective vegetable was broccoli, followed by cabbage, cauliflower, Chinese cabbage, and Japanese white long radish (daikon). For the experiments carried out in October flower-cluster of rape was not available. From the results obtained by the methods used in the work broccoli showed stronger antioxidative activity among Cruciferous vegetables tested.

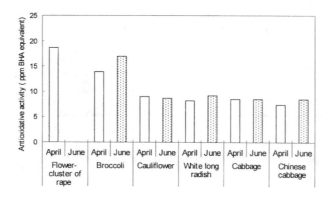

Figure 3 *Antioxidative activity of Cruciferous vegetables at two different times (April or June) (β-carotene oxidative method).*

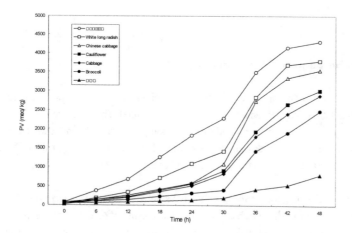

Figure 4 *Effect of various kinds of Cruciferous vegetables on oxidation of methyl linoleate at $50\,^{0}C$ (October harvest).*

4.2 Effect of broccoli extracts on weight increase of edible oil

Several kinds of edible oil such as soybean oil, olive oil, sesame oil, whale oil and beef oil, ca. 1 g each, were kept in a 50 ml beaker at 80°C and those weights was examined during autoxidation of 196 h. Weight of soybean oil and whale oil was increased 3-4% in 48-72 h, respectively. On the other hand the weight of olive oil, sesame oil, and beef oil was not changed during 96 h (data not shown). Ethanol-soluble fraction of broccoli with 10 times greater concentration of PV assessment was prepared; 2 ml or 4 ml of the fraction (corresponding to 2 g or 4 g of the fresh vegetables respectively) was added to soybean oil and whale oil to observe weight change in the oils. As seen in Figure 5, the weight of soybean oil was increased at 64 h and reached a maximum at 88 h before decreasing. The weight increase of the oil was further delayed and reduced by addition of 2 ml of the broccoli extracts and again further by addition of 4 ml of the extracts during the experimental period (196 h). Addition of broccoli extracts to whale oil also prevented weight increase of the oil during the experimental period.

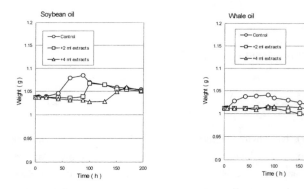

Figure 5 *Effect of broccoli extracts on weight of soybean oil and whale oil at 80 °C.*

4.3 Effect of broccoli on PV of oil during autoxidation

Three kinds of edible oil, i.e., olive oil, sesame oil and beef oil, were kept in a 200 ml beaker at 100°C on boiling water and changes in PV of the oil were investigated every 3 h. Three experiments were carried out as follows: 1) 20 g of the oil and 160 ml of water, 2) 20 g of the oil, 20 g of broccoli and 140 ml of water, 3) 20 g of the oil and 160 ml of 20 ppm BHA. The oil, 0.1 g, was sampled every 3 h and dissolved in 10 ml of 99.5% ethanol and 0.2 ml of ethanol was used for PV assessment (same method as for methyl linoleate). As shown in Figure 6, PV of olive oil (left Figure) and PV of sesame oil (middle Figure) increased linearly but the increases of PV was reduced by addition of broccoli or 20 ppm BHA. In beef oil (right Figure) PV increased and no effect of broccoli or 20 ppm BHA was observed.

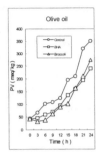 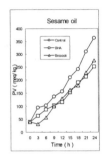 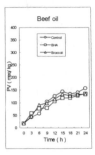

Figure 6 *Changes in PV of olive oil, sesame oil and beef oil at 100 ^{0}C.*

5 CONCLUSION

The stronger antioxidative activity of Cruciferous vegetables was found in flower-cluster of rape and broccoli by beta-carotene discoloration method. The activities of cauliflower, Japanese white long radish (daikon), cabbage and Chinese cabbage were weaker than the above two vegetables. By use of PV method, broccoli showed stronger activity. Effects of broccoli extracts on weight changes in soybean oil and whale oil at 80^{0}C were measured; it was found that the weight increase was retarded by addition of broccoli extracts. PV of olive oil and sesame oil increased linearly at 100 ^{0}C, but the increases of PV were reduced by broccoli.

Antioxidants contained in vegetables are reported to be flavonols,[13,14] isoflavonols, catechins, phenylpropanoids,[15,16] ascorbic acids, isothiocyanates,[17] thiol compounds, citric acid and malic acid as water-soluble substances and tocopherols and carotenoids[18] as lipid-soluble substances. The antioxidants in Cruciferous vegetables shown in this experiments seem to be components of 80% ethanol-soluble (Figures 3, 4 and 5) and water-soluble (Figure 6) substances. Studies to determine which specific components contribute to antioxidative activity in Cruciferous vegetables remain to be determined.

References

1. T. Osawa, A. Yoshida, S. Kawakishi, K. Yamashita and H. Ochi, in *Oxidative Stress and Aging*, eds., R. G. Cutler, L. Packer, J. Bertram & A. Mori, Birkhauser Verlag Basel/Switzerland, 1995, 367-377.
2. H. Ochi, R-Z Cheng, S. Sri Kantha, M. Takeuchi and N. Ramarathnam, in *BioFacters,* eds., L. Flohe, E. Niki and W. J. Whelan, ISO Press, Amsterdum/The Netherlands, 2000, **13,** 195-203.
3. T. Tsushida, M. Suzuki, and M.Kurogi, *Nippon Shokuhin Kogyo Gakkaishi* (in Japanese), 1994, **41,** 611.
4. S. Nishibori and K. Namiki, *Nippon Shokuhin Kogyo Gakkaishi* (in Japanese), 1998, **45,** 144.
5. K. Azuma, K. Ippoushi, H. Higashio and J. Terao, *J. Sci. Food Agric.,* 1999, **79,** 2010.
6. G. J. Marco, *J. Am. Oil Chem. Soc.*, 1968, **45,** 594.
7. H. E. Miller, *J. Am. Oil Chem. Soc.,* 1971, **48,** 91.
8. D. E. Pratt and P. M.Birac, *J. Food Science*, 1979, **44,** 1720.
9. M. Kawase, R. Wang, T. Shiomi, R. Saijo and K. Yagi, *Biosci. Biotechnol. Bochem.,* 2000, **64,** 2218.
10. □□Nose and N.Fujino, *Nippon Shokuhin Kogyo Gakkaishi* (in Japanese), 1982, **29,** 507.
11. H. Mitsuda, K. Yasumoto and K. Iwami, *Eiyo to Shokuryo* (in Japanese), 1966, **19,** 210.
12. □. Osawa and M. Namiki, *Agric. Biol. Chem.,* 1981, **45,** 735.
13. W. Bors, W. Heller, C. Michel and M. Saran, *Method in Enzymology*, 1990, **186,** 343.
14. C. Kandaswami and E. Middleton, Jr. in *Adv. Exp. Med. Biol.*, eds., D. Armstrong, Plenum Press, New York, 1994, **366,** 351-376.
15. J. K. Nielsen, O. Olsen, L. H. Pedersen and H. Sørensen□*Phytochmistry,* 1984, **23,** 1741.
16. Y. Uda, Y. Ozawa, M. Takayama, K. Suzuki and Y. Maeda, *Nippon Shokuhin Kogyo Gakkaishi,* 1988, **35,** 360.
17. N. J. Miller, J. Miller, L. P. Candeias, P. M. Blamley, C. A. Rice-Evans, *FEBS Letters,* 1996, 384, 240.
18. C. Manesh and G. Kuttan, *J. Exp. Clin. Cancer Res.,* 2003, **22,** 193.

BIOLOGICAL STUDIES AND ANTIOXIDATIVE ACTIVITY FOR WHITE TRUFFLE FUNGUS (*TUBER BORCHII*)

Emad S. Shaker

Agricultural Chemistry Department, College of Agriculture, Minia University, Minia, Egypt

1. ABSTRACT

The white desert truffle fungus *Tuber borchii* has been evaluated by chemical proximate analysis. *Tuber borchii* was shown to have 18.9% protein, 6.21% lipids, 9.33% fiber and 52.81% carbohydrates in the dry weight. The phenolic content was 0.93% as tannic acid in the dry weight. At pH 8.5, the total extractable truffle protein (ETP) was 17.1%, while at pH 3.7, the truffle protein isolate (TPI) was 65.7% in the freeze dried matter. The amino acids (except isoleucine) were not limiting. The rat nutritional bioassay showed that the protein efficiency ratios for the extracts were close to that for the control. Either 5% ETP or 5% TPI decreased liver weight, decreased AST activity, but did not decrease ALT activity. Total lipids and total cholesterol were also decreased in rats.

The antioxidative activity (AOA) for the methanol and hexane white truffle extracts showed significant activity up to 12 days. The AOA of the methanol and the hexane extracts for 7 days storage were 73.17, and 64.33%, respectively compared to that for vitamin E (91.5%). The AOA for the 10 day storage of methanol extract (94.2%) was higher than that for vitamin E (88.27%). The hexane extract showed prooxidative activity past 16 days, while, the AOA for the methanol extract was (5.02%) past 24 days of storage. The high antioxidative activity for the methanol extract may due to the phenolic content 0.53% and the reducing power of the extract. The methanolic extract proved to inhibit some pathogenic fungi. On the other hand, the hexane extract has low phenolic content (0.02%), no reducing power, and less protection against the pathogenic fungi.

2. INTRODUCTION

Desert truffles are a complex growth of mycorrhizal hypogeous fungi including several species of the genera *Balsamia, Terfezia, Tirmania,* and *Tuber*. They are distributed in semi-arid and arid conditions.[1] Wild dessert truffles are extensively grown in Matrouh, North Sinai and Red Sea governorates, and the areas averaged from 400,000 to 600,000 fed. The income averaged from LE 10-20 million to LE 60-120 million depends on many factors.[2] The most important desert truffles species are those in the genera *Terfezia* and

Tuber that are highly appreciated for their commercial value. Such truffles are particularly enjoyed in Mediterranean and Arabic peninsula countries.[3] Morphological studies have been carried out with desert truffles,[4] while no biochemical investigations for their metabolism have been performed.

Tuber magnatum, Tuber borchii (white truffles) and other species belong to the genus Tuber are subterranean fungus that are highly appreciated for their unique and characteristic aroma. The white truffle edible fungi (*Tuber borchii*) grows in the northern western Egyptian desert. Their culinary and commercial value is mainly due to their organoleptic properties which increases the economic value of such edible fungi.[5] The genes were expressed and identified to understand the *Tuber borchii* morphogenesis.[6] A molecular method was described for identification the ectomycorrhizae which belong to the white truffles (*T. magnatum, T. dryophilum, T. maculatum* and *T. borchii*).[7]

The chemical composition and nutritive values of truffles *Terfezia claveryi* and *Tirmania nivea* were studied.[3] All essential amino acids are present in ideal levels as recommended by FAO/WHO,[8] except leucine and lysine which are limiting amino acids in *T. claveryi*, and valine in *T. nivea*. However, the nutritive value depends on the quality of protein and different food components such as high moisture, high fat and high protein content can result in false determinations of protein quality.[9] Samples of *Terfezia claveryi* Chatin ascocarps (a wild edible fungi) that grows freely in desert regions were collected (from Saudi Arabia[10]) and analyzed. *T. claveryi* ascocarps contained 16% total protein in the dry weight, 28% total carbohydrate, 4% total crude fibre and 2% total crude fat. *T. claveryi* was rich in minerals and carbohydrates. Nine saturated and 4 unsaturated fatty acids and 19 amino acids were also detected.

Few literature reports mention the potent odor compounds for truffles. The predominant sulfur compounds in white truffle aroma are dimethyl sulfide and bis (methylthio) methane.[11] Truffle melanins (allomelanins) have been detected in the peridium and the gelba tissues.[12] A noncovalent association between polyphenolics and polyquinone biopolymer was found. The pathway of melanin biosynthesis utilized poly ketide-derived quinones through a tyrosinase and laccase catalysed oxidative step.[13]

The aim of this study is to evaluate the nutritional quality of *T. borchii* protein and the content essential amino acids. Feeding experiments using albino rats was carried out to study the effect on the liver function and cholesterol. The antioxidative activity for the methanol and hexane truffle extracts, their phenolic content and reducing power were investigated. The antifungal activity of the same extracts against some pathogenic fungi was also tested.

3. MATERIALS AND METHODS

Desert white truffles (*Tuber borchii*) were collected from Sidi-Barani (Matrouh governorate) in winter 2003. The fruit bodies were frozen immediately until tested. The samples were freeze dried and then defatted by ice-cold acetone in a blender and filtered under vacuum. The residue was kept at 4 °C for protein analysis.

3.1 Proximate Analysis

Another portion of freeze fruit bodies of *T. borchii* were analyzed according to the AOAC method.[14] All determinations were conducted in triplicates. Crude protein was calculated as:

The total nitrogen X 6.25

3.2 Determination of Phenolic Substances

Phenolic compounds were extracted from 5.0 g freeze dried sample by refluxing with 50 ml of methanol containing 1% HCl for 4 hr. Polyphenols were determined as tannic acid equivalent.[15] The phenolic content (%) of the methanol and hexane extracts (0.05 ml) was also determined as tannic or gallic acid equivalents.

3.3 Extractability and Solubility of Truffle Protein

Defatted samples of 5.0 g freeze dried sample were suspended in 50 ml of distilled water and stirred for 5 min. The pH was adjusted to the desired value by adding 1 N NaOH and/or 1 N HCl, and stirred for 30 min. The pH was adjusted from 2 to 10 units, and the supernatant for each unit was measured at 280 nm to identify the protein solubility curve. Extraction of the truffle protein (ETP) at maximum solubility of pH was carried out. The precipitation of the soluble protein at isoelectric point (PI) for the minimum solubility was done to obtain the truffle protein isolate (TPI).

3.4 Amino Acid Analysis

Amino acid composition of both ETP and TPI were determined by hydrolyzing the samples with 6 N HCl at 110°C for 24 hr.[16] Determinations were carried out on the hydrolyzed sample by an automatic amino acid analyzer model ALFA plus 4151 LKB Biochrom. at the Central Laboratory, Faculty of Agriculture, Cairo University. Tryptophane was estimated as well.[17]

3.5 Feeding Experiment

Nutritional bioassay was performed using adult male albino rats (8 weeks of age). The rats were divided into 4 groups, and each 5 rats were fed as following: control group rats fed on diet consist of 10% casein, 70% starch, 10% corn oil, 4.8% cellulose, 4% salt mixture, 1% vitamin mixture and 0.2% cholin chloride. Groups 2, 3 and 4 were fed tested protein instead of casein. Group 2 was fed on 5% ETP + 5% casein. Group 3 was fed on 5% TPI + 5% casein and Group 4 was fed 5% ETP + 5% TPI without casein.

Food and water allowed *ad libitum* and body weight gain changes and food consumed were recorded after 4 weeks. Then rats were sacrificed and the liver was dissected and weighed. Serum samples were collected to estimate serum aspartic transaminase (AST) and alanine transaminase (ALT) activities.[18] Protein efficiency ratio (PER) was calculated as follows:

PER=Body weigh gain (g) / Protein intake (g)

Lipid profile was determined for total lipid (g/dl) and total cholesterol (mmol/l), respectively.[19,20]

3.6 Antioxidative Activity

About 10 g of ground freeze dried sample (*T. borchii*) were extracted with 100 ml methanol and stirred for 2 h. The sample was filtered and evaporated under vacuum to about 5 ml. The residue after filtration was extracted in the same fashion using 100 ml hexane for 2 hr, then filtered and evaporated. 0.5 ml from each of the methanol and hexane extracts were tested using the antioxidant linoleic acid method.[21] α–Tocopherol (0.5 ml, 500 ppm) was used in one sample instead of the truffle extract. A control without any additive was also used in the analysis. The absorbance at 500 nm was recorded at time intervals using ferrous chloride and ammonium thiocyanate method. Antioxidative activity (AA) was expressed as % inhibition relative to the control using:[22]

$$AA = [(\text{degradation rate of control} - \text{degradation rate of sample}) / \text{degradation rate of control}]\ 100$$

3.7 Reducing Power

The reducing power of truffle extracts was determined.[23] Truffle extracts (0.05 ml) in distilled water were mixed with phosphate buffer (2.5 ml, pH 6.6) and potassium ferricyanide [$K_3Fe(CN)_6$] (2.5 ml, 1%). The mixture was incubated at 50 °C for 20 min. Trichloroacetic acid (2.5 ml, 10%) was added to the mixture, which was then centrifuged at 300 rpm for 10 min. The upper layer of the solution (2.5 ml) was mixed with distilled water and $FeCl_3$ (0.5 ml, 0.1 %), and the absorbance was measured at 700 nm. Increased absorbance of the reaction mixture indicated increased reducing power.

3.8 Antifungal Activity

The antifungal activity of the truffle extracts was determined against pathogenic fungi;[35] *Fusarium oxysporum*, *Mucor haemalis*, *Macrophomixa phaseolina* and *Thielaviopis basirala*. These pathogenic fungi infect *Sesamum indicum* L. causing root rot and wilt diseases. Mycelium disks (5 mm diameter) were used from the growing edge of 3 day-old cultures of each treated fungus. These fungi were inoculated onto the plates to test the effect of the extracts on the linear growth of the tested fungi. The inhibition % (I%) for the truffle extracts was calculated using:

$$I\% = [(\text{growth fungus of control} - \text{linear growth fungus of sample}) / \text{growth fungus of control}]\ 100$$

3.9 Statistical Analysis

Significant differences between means of 5 replicates were tested by T. test one way (ANOVA) sps package at $p < 0.05$

4. RESULTS AND DISCUSSION

Chemical composition of the desert white truffle (*Tuber borchii*) is shown in Table 1. The moisture content was 75.32%, while the crude protein in the dried matter was 18.9%. The crude protein for the Egyptian white truffle (*Tuber magnatum*)[24] and that for *Terfezia claveryi* (Saudi truffle)[3] were higher than tested truffle. The dessert truffle ash and the crude fiber were relatively higher than these for Egyptian and Saudian ones. The tested truffle contains 0.93% phenolic compounds as tannic acid, which is less than that for the Egyptian one.

Table 1 *Proximate Analysis of Tuber borchii*

Content %	Desert truffle	Egyptian truffle[24]	Saudi truffle[3]
Moisture (%)	75.32	76.5	75.44
Crude protein (g/100 g dry weight)	18.9	20.12	24.96
Crude lipids (g/100g dry w)	6.21	6.13	4.2
Crude fiber (g/100 g dry w)	9.33	8.18	7.02
Total ash (g/100g dry w)	11.82	11.24	6.39
Phenolics (g/100 g dry w) as tannic acid	0.93	1.05	Not determined
Crude carbohydrates (calculated by the differences)	52.81	53.28	Not determined
C/N ratio	2.79	2.65	Not determined

4.1 Extractability and Solubility of Truffle Protein

Figure 1 shows that protein extraction from defatted matter of truffle as a function of pH has a maximum solubility at pH 8.5. The total extractable truffle protein (ETP) contained 17.1% protein. The minimum solubility at pH 3.7 (isoelectric point) showed 65.7% protein in the truffle protein isolate (TPI). That is in agreement with the white truffle protein for *T. magnatum*.[24]

4.2 Amino Acid Content

The essential amino acid content for both of ETP and TPI extracted protein are shown in Table 2. Except for isoleucine, no amino acids were limiting and their levels were higher than that determined in the reference protein.[8] Therefore, *T. borchii* protein could be combined with a legume protein to yield a high nutritive food containing essential amino acids. These extracted proteins are similar to that recorded for animal protein and higher than that recorded for soybean protein.[25,26]

On the other hand, the pecan truffle protein is deficient in methionine and contains a higher percentage of cystine compared with the fruiting bodies of several edible fleshy fungi.[27]

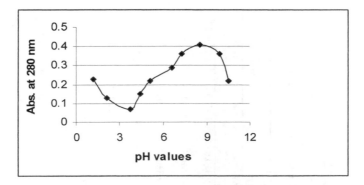

Figure 1 Solubility Curve for T. borchii Protein at Different pH Values

Table 2 *Essential Amino Acids of The Tested Proteins as g/100 g Crude Protein*

Amino acids	ETP	TPI	Soybean protein[26]	FAO/WHO (1973)
Lysine	5.6	5.5	6.1	5.5
Methionine	2.9	2.9	2.5	3.5
Cysteine	1.0	1.1		
Leucine	7.1	7.2	7.7	7.0
Isoleucine	3.2	3.1	4.7	4.0
Phenylalanine	3.5	3.7	8.7	6.0
Tyrosine	2.5	3.0		
Valine	5.2	5.1	4.8	5.0
Threonine	4.1	4.2	3.7	4.0
Tryptophane	1.1	1.2	1.2	1.0
Total essential AA	36.2	37.0	39.4	36.0
Limiting AA	Isoleucine	Isoleucine	Sulfur AA, Thr., Val.	---

4.3 Nutritional Bioassay

The results which are shown in Table 3 indicate that rats fed on 5% ETP + 5% casein; and TPI + 5% casein had significant increase in body weight gain (BWG) comparing to rats fed on control diet, or even on soybean protein diet.[28] The highest BWG amount was for ETP/casein. On the other hand, no significant differences have been found for protein efficiency ratio (PER) for the tested proteins and the control diet. However, the mixture of ETP and TPI showed a significant decrease in PER compared to control.

No or only slight significant difference has been detected in liver weight/100 g body weight and liver functions (AST, ALT activities) for diets containing 5% ETP or TPI compared to control. In contrast, the mixture of ETP and TPI showed a significant increase value for AST and ALT activities compared to control. Thus, the diet of 5% ETP or TPI + 5% casein was considered an ideal formula containing high nutritive value ingredients without any disadvantages or harmful effect on liver function in the experimental animals.

Table 3 Expermintal Feeding Values for Extracted Truffle Protein (ETP) and Truffle Protein Isolate (TPI)

Diets	Gain weight (g)	Food consume (g)	Protein intake (g)	PER	Liver weight (g)	Liver weight /100 g weight	Liver Function		Lipid profile	
							AST	ALT	T. lipid g/dl	T. cholest mmol/l
Control	76.2±0.08	381.3±2.1	38.1±0.25	2.0±0.0	7.86±0.2	4.43±0.11	21.6±0.0	19.3±0.05	4.95±0.2	1.81±0.17
5% ETP	81.5±0.09	409.8±1.8	41.0±0.14	1.99±0.12	7.80±0.17	4.33±0.09	21.4±0.15	19.6±0.1	4.89±0.33	1.70±0.02
5% TPI	78.2±1.1	397.3±1.7	39.7±0.29	1.97±0.08	7.78±0.15	4.41±0.0	21.3±0.07	19.3±0.12	4.9±0.25	1.65±0.0
10% mixture	80.6±1.1	450.2±1.1	45.0±0.02	1.79±0.03	8.03±0.0	4.44±0.0	22.7±0.05	20.2±0.22	5.32±0.51	2.03±0.15
Soyprotein[28]	75.9±0.05	383.2±0.7	38.3±0.07	1.98±0.03	8.30±0.0	4.8±0.25	17.2±1.3	17.9±1.3	4.92±0.25	1.26±0.1

The Values are The Mean of 5 Rats (8 weeks age) ± SD Analyzing by ANOVA ($p<0.05$).

Lipid profile data showed a slightly significant decrease in total lipids and/or total cholesterol for the rat group fed on 5% ETP or TPI compared to control. Therefore, 5% ETP or TPI are considered as a cholesterol lowering factor in serum rats. On the other hand, 10% mixture of ETP + TPI showed slightly significant increase for total lipids and cholesterol.

For these reasons, truffle protein is suggested as a food additive source up to 5% concentration of the total protein in a diet, and as medicinal food source. Soybean protein showed a sharp decrease in serum cholesterol compared to control or truffle protein.[28] Our results were in agreement with those which showed that 5% truffle protein (*T. melanosporum*) is considered an optimal and healthy food additive.[29] Generally, truffles thought to be a healthy food for their low calories, low fat and rich in proteins.[27]

4.4 The Antioxidative Activity

Few researchers have investigated the antioxidative activity for fungus. The antioxidant and glutathione dependent enzymatic endowment of Tuber truffles have been studied.[30] Superoxide dismutase, catalase, glutathione peroxidase Se dependent, glutathione reductase, glyoxalase 1 and glyoxalase 2 are expressed and correlated with the microaerobic metabolism, growth rate and mycorrhizal symbiosis of truffles. They also found a very low or undetectable glutathione S-transferase activity. In another study, two major antioxidative compounds were isolated from wild mushroom,[31] *Suillus bovines*, and their activity to BHA and tocopherol were compared. The identified antioxidants were variegatic acid and diboviquinone.

In our study, both the methanol and hexane truffle extracts showed a satisfactory activity up to 12 days (Figure 2). The AOA of the methanol and the hexane extracts were 73.17, and 64.33%, respectively. The activity for vitamin E was 91.5% at the same period of storage. For the first 10 days of storage, the antioxidative activity of the methanol extract (94.2%) was higher than that for vitamin E (88.27%). The AOA was dramatically decreased after 14 days of storage. The hexane extract showed prooxidative activity after 16 days of storage at 37°C. The methanol extract showed a very weak activity (5.02%) up to 24 days of storage, while the activity of vitamin E was (21.51%) at the end of storage time. The antioxidative activity for the methanol extract may be due to the total phenolic content or to the flavonoids.

Our results were in agreement with those which showed higher antioxidative activity for raw *Picoa juniperi* and *Terfezia claveryi* than for frozen truffles.[32] The activity has been attributed to the content of polyphenols and flavonoids, while the reduction in activity to the early formation of Maillard reaction products with oxidant properties.[32]

4.5 Phenolic Content of Truffle Extracts

The phenolic content for the methanol extract was much higher than the hexane extract as shown in Figure 3, measured as tannic acid and gallic acid. The phenol compounds are the most important factors for the antioxidative activity. The phenolic content in dry truffle (Table 1) was 0.93%, where it was 0.53 and 0.02% for the methanol and hexane fractions, respectively calculated as tannic acid. The phenomenon of suppression of inhibitory activity after the fractionation has been observed for other antioxidants.[33]

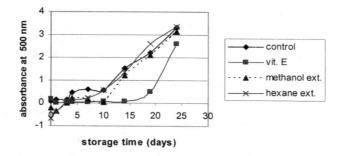

Figure 2   *The Antioxidative Activity of Methanol and Hexane White Truffle Extracts.*

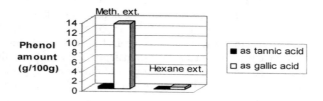

Figure 3   *The Phenolic Content (g/100g dry weight) of Methanol and Hexane White Truffle Extracts.*

4.6   Reducing Power of Truffle Extracts

As shown in Figure 4, the reducing power (reported as absorbance at 700 nm) of hexane and methanol extracts truffle were 0.0 and 0.052, respectively. The methanol extract has a moderate reducing power, and that correlated with antioxidative results. The reducing power gives indication for the electron donor activity for the methanol extract. The antioxidative activity is due to the ability to react with free radicals and convert them to more stable products and to terminate the radical chain reaction.[34]

4.7   Antifungal Activity of Truffle Extracts

The methanolic extract was more effective (37.5, 50%) than the hexane extract (12.5, 25%) especially for *F. oxysporum* and *T. basirola*, respectively as shown in Figure 5. The methanolic and hexane extracts had no effect on *M. phaseolina* and *M. haemalis*. The antifungal activity for the white truffle methanolic extract was correlated with that of the high antioxidative activity, phenolic content and reducing power for the methanolic extract. These facts may give a suggestion about the importance of truffle extracts in protection of *Sesamum indicum* L. from root rot and wilt.[35]

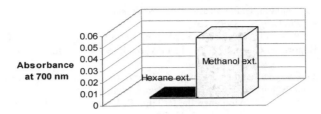

Figure 4   *The Reducing Power of Methanol and Hexane White Truffle Extracts.*

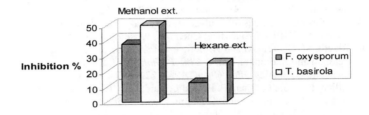

Figure 5   *The Antifugal Activity of Methanol and Hexane White Truffle Extracts.*

More research should be done in the near future to investigate the compounds responsible for the antioxidative activity, agricultural and the health aspects of the white truffle fungus *Tuber borchii*. As a new nutritional source, research should also be concentrated on the preservative role of the truffle and incorporation of truffles into food products for flavors and extenders of storage time.

Acknowledgment

The author would like to thank Prof. A. Galal, professor of Plant Pathology, Faculty of Agriculture, Minia Univ., Egypt for his assistance in the antifungal activity method.

References

1. M. Honrubia, A. Cano and C. Molina-Ninirola, *Persoonia*, 1992, **14**, 647.
2. H. El-Khouli, *Egyptian J. Appl. Sci.*, 2000, **15**, 107.
3. W. Sawaya, A. Al Shalhat, A. Al Sogair and M. Al Mohamed, *J. Food Sci.*, 1985, **50**, 450.
4. G. Moreno, R. Galan and A. Ortega, *Criptogamic Mycol.*, 1986, **7**, 201.
5. R. Staples, *Trends Plant Sci.*, 2001, **6**, 140.
6. I. Lacourt, S. Duplessis, S. Abba, P. Bonfante and F. Martin, *Appl. Environm. Microbiol.*, 2002, **68**, 4574.

7   L. Bertini, D. Agostini, L. Potenza, I. Rossi, S. Zeppa, A. Zambonelli and V. Stocchi, *New Phytologist.*, 1998, **139**, 565.
8   FAO/WHO, 1973, In FAO Nutritional. Meeting Report Series No. 52, WHO Technical Report Series No. 522, Food and Agriculture Organization, Rome. Bluath
9   H. Chang, G. Gatignani and H. Swaisgood, *J. Food Biochem.*, 1992, **16**, 133.
10 H. Bokhary and S. Parvez, *J. Food Comp. Anal.*, 1993, **6**, 285.
11 F. Pelusio, T. Nilsson, L. Montanarella, R. Tilio, B. Larsen, S. Facchetti and J. Madsen, *J. Agric. Food Chem.*, 1995, **43**, 2138.
12 F. De Angelis, A. Arcadi, F. Marinelli, M. Paci, D. Botti, G. Pacioni and M. Miranda, *Phytochem.*, 1996, **43**, 1103.
13 M. Miranda, A. Bonfigli, O. Zaviri, A. Ragnelli, G. Pacioni and D. Botti, *Plant Sci.*, 1992, **81**, 175.
14 AOAC, Official Methods of Analysis of the Analytical Chemists 15th ed. Association of Official Analytical Chemists, Washington, C.D. 1990.
15 T. Swain and W. Hills, *J. Sci. Food Agric.*, 1959, **10**, 63.
16 AOAC, Official Methods of Analysis of the Analytical Chemists 12th ed. Association of Official Analytical Chemists, Washington, C.D. 1975.
17 O. Blauth, M. Choreinski and H. Berhee, *Anal. Biochem.*, 1963, **8**, 69.
18 S. Reitman and S. Frankel, *Am. J. Clin. Pathol.*, 1957, **28**, 56.
19 J. Folch and G. Stanley, *Biol. Chem.*, 1957, **222**, 297.
20 C. Allian, L. Poon, C. Chan, P. Fu and C. Paul, *Clin. Chem.*, 1974, **20**, 470.
21 H. Mitsuda, K. Yasumoto and K. Iwami, *Eiyoto Shokuryo*, 1966, **19**, 210.
22 M. Ogata, M. Hoshi, K. Shimotohno, S. Urano and T. Endo, *J. Am. Oil Chem. Soc.*, 1997, **74**, 557.
23 M. Oyaizu, *Jpn. J. Nutr.*, 1986, **44**, 307.
24 M. Mahmoud, *Minia J. Agric. Res. Develop.*, 2003, **23**, 475.
25 E. Danell and D. Eaker, *J. Sci. Food Agric.*, 1992, **60**, 333.
26 T. Morita and S. Kirryama, *J. Food Sci.*, 1993, **58**, 1393.
27 L. Beuchat, T. Brenneman and C. Dove, *Food Chem.*, 1993, **46**, 189.
28 S. Latif, *J. Agric. Sci. Mansoura Univ.*, 2003, **28**, 6221.
29 M. El-Sayed, M. Ghazy, G. Abd El-Naem and M. Mahmoud, Conference of Role of Biochemistry in Environment and Agriculture, Cairo Univ., Agric. Chem., 24-26 Feb. 2004.
30 F. Amicarelli, A. Bonfigli, S. Colafarina, A. Cimini, B. Pruiti, P. Cesare, M. Ceru, C. Ilio, G. Pacioni, M. Miranda and C. di-Ilio, *Mycolog. Res.*, 1999, **103**, 1643.
31 A. Kasuga, Y. Aoyagi and T. Sugahara, *J. Food Sci.*, 1995, **60**, 1113.
32 M. Murcia, M. Martinez-Tome, A. Jimenez, A. Vera, M. Honrubia and P. Parras, *J. Food Prot.*, 2002, **65**, 1614.
33 E. Shaker, M. Ghazy and T. Shibamoto, *J. Agric. Food Chem.*, 1995, **43**, 1017.
34 G. Yen and H. Chen, *J. Agric. Food Chem.*, 1995, **43**, 27.
35 A. Galal and E. Abdo, *Egyptian Phytopathol.*, 1996, **24**, 1.

STRUCTURED LIPIDS CONTAINING HIGHLY UNSATURATED OMEGA-3 FATTY ACIDS

S.P.J. Namal Senanayake[1] and Fereidoon Shahidi[2]

[1]Martek Biosciences Corporation, 555 Rolling Hills Lane, Winchester, KY 40391
[2]Department of Biochemistry, Memorial University of Newfoundland, St. John's, Newfoundland, A1B 3X9 Canada

1 ABSTRACT

Structured lipids (SL) are triacylglycerols (TAG) with mixtures of short, medium and/or long chain fatty acids esterified to the same glycerol molecule. SL may provide an effective means of delivering fatty acids with particular characteristics desirable either in nutrition or in specific diseases. In this contribution, enzyme-assisted synthesis of structured lipids, sources of fatty acids for synthesis of structured lipids and their health effects are discussed.

2 INTRODUCTION

Structured lipids (SL) are defined as triacylglycerols (TAG) restructured or modified to change the fatty acid composition and/or their positional distribution in glycerol molecules by chemical or enzymatic processes.[1] There are several possible methods to synthesize them;[2] hydrolysis and esterification, interesterification, lipase interesterification, traditional chemical methods and genetic manipulation. However, only enzymatic interesterification and genetic manipulation can create targeted structured lipids, resulting in specific placement of fatty acids on the glycerol backbone. The other methods produce randomized structured lipids. Enzymatic interesterification has become the preferred method for structured lipid synthesis because reaction conditions tend to be milder and less side reactions may occur. Furthermore, some lipases are highly specific, allowing targeting of the fatty acids to a particular position on the glycerol molecule. Structured lipids provide an effective means for delivering fatty acids with particular properties desirable either in overall nutrition or in specific diseases of humans.[3]

3 ENZYMATIC SYNTHESIS OF STRUCTURED LIPIDS

Several enzyme-catalyzed reactions, namely esterification, interesterification, alcoholysis and acidolysis may be used to produce structured lipids. However, the methods of choice depends on the types of substrates available.[1]

(a) Esterification:
$$RCOOH + R'OH \rightarrow RCOOR' + H_2O$$

(b) Interesterification:
$$RCOOR' + R''COOR^* \rightarrow RCOOR^* + R''COOR'$$

(c) Alcoholysis:
$$RCOOR' + R''OH \rightarrow RCOOR'' + R'OH$$

(d) Acidolysis:
$$RCOOR' + R''COOH \rightarrow R''COOR' + RCOOH$$

Structured lipids may be produced using lipases either in the presence or absence of an organic solvent. Lipases used for this purpose include immobilized lipases such as *Candida antarctica* and *Mucor miehei* as well as nonimmobilized lipases such as *Pseudomonas sp.* The positional specificity of *Candida antarctica* depends on the type of reactants. In some reactions, *Candida antarctica* shows sn-1,3 positional specificity, whereas in other reactions the lipase functions as a nonspecific enzyme.[4] Lipase from *Mucor miehei* attacks at the primary hydroxyl positions (1 and 3) of triacylglycerol molecule preferentially and are said to be sn-1,3 specific.[5] On the other hand, lipase from Pseudomonas sp. functions as a nonspecific enzyme. Enzyme-catalyzed reactions have many advantages such as milder processing conditions and the posiibility of regio-, stereo- and fatty acid specificity.[6]

3.1 Enzymatic incorporation of ω3 fatty acids into plant oils to produce structured lipids

Incorporation of ω3 PUFA into plant oils has been successfully achieved using enzyme-catalyzed reactions.[7-13] Huang and Akoh[9] studied the ability of immobilized lipases IM60 from *Mucor miehei* and SP435 from *Candida antarctica* to modify the fatty acid composition of soybean oil by incorporating ω3 PUFA into the soybean oil. The interesterification was carried out with free fatty acid and ethyl esters of EPA and DHA as acyl donors. With free EPA as acyl donor, IM60 gave higher incorporation of EPA than SP435. However, when ethyl esters of EPA and DHA were the acyl donors, SP435 gave higher incorporation of EPA and DHA than IM60.

The ability of lipase PS30 from *Pseudomonas sp.* to modify the fatty acid profile of melon seed oil by incorporation of oleic acid (18:1ω9) was investigated.[14] Oleic acid content increased from 13.5 to 53%, while linoleic acid (18:2ω6) content decreased from 65 to 33%. In another study, Huang et al.[9] reported on the incorporation of EPA into crude melon seed oil by two immobilized lipases, IM60 from *Mucor miehei* and SP435 from *Candida antarctica* as biocatalysts. Higher EPA incorporation was obtained using EPA ethyl ester than using EPA free fatty acid for both enzyme-catalyzed reactions. Increasing the molar ratio of acid or ester to TAG, significantly increased EPA incorporation, especially when EPA ethyl ester was used.

Akoh et al.[15] used two immobilized lipases, nonspecific SP435 from *Candida antarctica* and sn-1,3 specific IM 60 from *Mucor miehei*, as biocatalysts for the restructuring of trilinolein to incorporate EPA and DHA with ethyl esters (EEPA and EDHA, respectively) as acyl donors. With EEPA as acyl donor, the total EPA product yields with IM60 and

SP435 were 79.6 and 81.4%, respectively. However, with EDHA as acyl donor and IM60 and SP435 as biocatalysts, the total DHA product yields were 70.5 and 79.7%, respectively.

Recently, EPA and capric acid (10:0) have been incorporated into borage oil using two immobilized lipases, SP435 from *Candida antarctica* and IM60 from *Rhizomucor miehei* as biocatalysts.[13] Higher incorporation of EPA (10.2%) and 10:0 (26.3%) was obtained with IM60 lipase, compared to 8.8 and 15.5%, respectively, with SP435 lipase.

It has been shown that sn-1,3 specific lipase from *Mucor miehei* is capable of incorporating EPA and DHA into borage and evening primrose oils. Table 1 shows the fatty acid composition of borage and evening primrose oils, before and after acidolysis with EPA and DHA by *M. miehei* enzyme. In this study, the substrate mole ratio of 1:0.5:0.5 (oil/EPA/DHA) was used because the incorporation of EPA and DHA was satisfactory at this mole ratio. Predominant fatty acids found in borage oil before lipase-catalyzed interesterification were 18:2ω6 (38.4%) and 18:3ω6 (24.4%). The concentration of these acids were comparable to those reported by Akoh and Sista[10] and Akoh and Moussata.[13] After interesterification reaction, 18:2ω6 and 18:3ω6 decreased by 9.6 and 2.1%, respectively. The amount of EPA and DHA incorporated into borage oil was 11.8 and 11.0%, respectively. The ratio of ω3 PUFA to ω6 PUFA increased from 0.002 to 0.45. On the other hand, the main fatty acid found in evening primrose oil before interesterification reaction was 18:2ω6 (73.6%). The content of 18:3ω6 found in this oil was 9.1%. After interesterification reaction, the content of 18:2ω6 was decreased by 18.6%. However, the content of 18:3ω6 decreased only marginally (0.6% decrease). The amount of EPA and DHA incorporated into evening primrose oil was 12.3 and 10.5%, respectively. The ω3/ω6 PUFA increased from 0 to 0.36. Ju et al.[12] incorporated ω3 PUFA into the acylglycerols of borage oil. They have selectively hydrolyzed borage oil using immobilized *Candida rugosa* lipase and then used this product with ω3 PUFA for the interesterification reaction. The total content of ω3 and ω6 PUFA in acylglycerols was 72.8% following interesterification. The contents of GLA, EPA and DHA were 26.5, 19.8 and 18.1%, respectively. The corresponding ω3/ω6 ratio changes from 0 to 1.09 after modification.

Previously, Akoh and Sista[10] modified the fatty acid composition of borage oil using EPA ethyl ester, with an immobilized nonspecific SP435 lipase from *Candida antarctica* as a biocatalyst. The highest incorporation (31%) was obtained with 20% SP435 lipase. At a substrate mole ratio of 1:3, the corresponding ratio of ω3 to ω6 PUFA was 0.64. Under similar conditions, Akoh et al.[11] were able to increase the ω3 PUFA (up to 43%) of evening primrose oil with a corresponding increase in the ω3/ω6 ratio from 0.01 to 0.6. Sridhar & Lakshminarayana[7] were able to effectively modify the fatty acid composition of groundnut oil by incorporating EPA and DHA using a sn-1,3 specific lipase from *Mucor miehei* as the biocatalyst. The content of EPA and DHA incorporated into groundnut oil were 9.5 and 8.0%, respectively.

3.2 Sources of fatty acids for structured lipid synthesis

Different classes of unsaturated lipids of the ω3, ω6 and ω9 fatty acids are included in structured lipids to promote health and nutrition. Table 2 illustrates the levels of the various ω fatty acids that are considered optimum by nutritional experts. The clinical advantages of structured lipids are derived from the short-, medium- and long-chain fatty acids and the uniqueness of the structured lipid molecule itself. Many of these effects are due to the differences in metabolic fate of the various fatty acids. It is important to consider the

Table 1. Fatty acid composition of borage (BO) and evening primrose oils (EPO) before and after modification by Mucor miehei lipase

Major fatty acids[a]	Unmodified BO	Modified BO[b]	Unmodified EPO	Modified EPO[b]
16:0	9.81 ± 0.12	6.97 ± 0.25	6.16 ± 0.09	4.50 ± 0.02
18:0	3.12 ± 0.26	2.60 ± 0.10	1.72 ± 0.12	1.27 ± 0.05
18:1	15.2 ± 0.74	11.6 ± 0.31	8.65 ± 0.56	6.43 ± 0.42
18:2ω6	38.4 ± 0.89	28.8 ± 0.56	73.6 ± 0.91	55.0 ± 0.85
18:3ω6	24.4 ± 0.89	22.3 ± 0.73	9.12 ± 0.38	8.55 ± 0.20
20:1	4.09 ± 0.25	3.05 ± 0.22	ND[c]	ND[c]
20:5ω3	ND[c]	11.8 ± 0.20	ND[c]	12.3 ± 0.07
22:1	2.49 ± 0.10	1.86 ± 0.11	ND[c]	ND[c]
22:6ω3	ND[c]	11.0 ± 0.57	ND[c]	10.5 ± 0.38
ω3/ω6 ratio	0.002	0.45	0.0	0.36

[a]As area percentage
[b]Reaction conditions: The reaction mixture contained 300 mg oil, 53.5 mg EPA, 58.2 mg DHA, 350 units of enzyme and 3 ml hexane. The reaction mixture was incubated at 35°C for 30 h in an orbital shaking water bath at 250 rpm. Results are averages of triplicate determinations from different experiments.
[c]Not detected

metabolism of fatty acids to understand their physiological effects. The parent compounds of the ω3 and ω6 series are α-linolenic (18:3ω3) and linoleic (18:2ω6) acids. They are metabolized by a series of alternating desaturation and elongation. Linoleic acid is desaturated and elongated to arachidonic acid. Arachidonic acid is a precursor of eicosanoids,[16] especially, prostaglandin G_2 (PGI_2) and leukotrienes$_4$. In general, the metabolism of ω3 PUFA that facilitates prevention and treatment of the diseases has been addressed by considering changes in the eicosanoids in the circulatory system. Since eicosanoids are ultimately derived from PUFA provided by the diet, it is clear that quantitative and qualitative changes in the supply of dietary PUFA will have a profound effect on the production of eicosanoids. There is an emerging consensus that ω3 fatty acids are essential nutrients.

Two families of fatty acids with well studied health benefits are medium chain and ω3 fatty acids.

Table 2. *Optimum levels of omega fatty acids for structured lipids*

Omega fatty acid	Suggested levels for clinical use
ω3	2-5% enhances immune function, reduces blood clotting, lowers serum TAG, reduce risk of coronary heart disease
ω6	3-4% for essential fatty acid requirements
ω9	Monounsaturates for balance of long-chain fatty acids

(Source: ref. *18*)

3.3 Medium-chain fatty acids

Triacylglycerols containing medium-chain fatty acids are often the basis for structured lipids.[2] Unlike traditional fatty acids, medium-chain triacylglycerols go directly into the portal vein and are rapidly converted into energy. The bulk of our dietary fat intake consists of triacylglycerols with long-chain fatty acids (>C_{12}); only 3% coming from triacylglycerols with short (C_2-C_4)- and medium (C_6-C_{12})-chain fatty acids.[17] Short-chain fatty acids are found mainly in bovine milk. Medium-chain fatty acids are present in vegetable oils such as palm kernel and coconut oils.[3] However, because medium-chain triacylglycerols alone do not contain essential fatty acids, they need to be used with long-chain triacylglycerols containing essential fatty acids to provide balanced nutrition in enteral and parenteral nutrition products.[2] Clinical nutritionists have taken advantage of medium chain triacylglycerol's simpler digestion to nourish individuals who can not utilize long-chain triacylglycerols. Any abnormality in the numerous enzymes or processes involved in the digestion of long-chain triacylglycerols can cause symptoms of fat malabsorption. Thus, patients with diseases like Crohn's disease, cystic fibrosis and colitis have shown improvement when medium-chain triacylglycerols is included in their diets.[18]

3.4 Omega-3 fatty acids

The omega-3 fatty acids are also the subject of considerable nutritional research. These fatty acids are defined as having their first C-C double bond, three carbons from the methyl end. ω3 fatty acids are also known as essential fatty acids and must be provided in the food because they cannot be easily manufactured within the body. α-Linolenic acid, EPA and DHA belong to this group. α-Linolenic acid is found in linseed, canola and soybean oils while EPA and DHA are found in marine oils such as fish and seal oils. EPA and DHA are made by algae and hence enter the food chain through the animals that feed on algae such as fish.[19]

Many of the physiological effects attributed to ω3 fatty acids intake relate to their role in eicosanoid production. Eicosanoids are locally acting hormone-like substances and affect blood pressure, blood clotting, immune and inflammatory responses. They include the prostaglandins, prostacyclins, leukotrienes and thromboxanes.[20] When cells use different starting materials (either ω3 or ω6 FA) different eicosanoids are produced. The ω6 fatty acids are converted to arachidonic acid which generally is a precursor for eicosanoids that promote platelet aggregation, while eicosanoids from ω3 fatty acids decrease platelet aggregation.[21] In the cell there is a competition between two precursors and high amounts of ω6 (i.e. linoleic acid) inhibits the conversion of α-linoleic acid to EPA. Because of this, the key factor in regards to the consumption of these classes of FAs is the ratio in which they are consumed. Typical western diets contain 10-30 g/day of linoleic acid and 0.5-1.0 g/day of α-linolenic acid (the primary ω3 FA in Western diets). With current dietary patterns, the observation that α-linolenic acid supplementation fails to increase blood EPA concentration is easily explained.[21]

Docosahexaenoic acid has also garnered considerable attention by itself since a number of studies have shown it to be very important in early neurological development and visual development. DHA is one of the major components of the gray matter of the brain,[22] the phospholipids of the retina,[3] the testes and sperm,[23] and fish oil.[21] DHA deficiency is a problem in preterm infants that have inadequate fat stores. Even if breast fed the mother's milk will not contain enough DHA for normal visual development, thus some form of supplementation is required.[24,25]

Retro-conversion of DHA to DPA and EPA in humans has been documented.[26] After ingestion of ethyl esters of DHA, the DHA and EPA in plasma phospholipids was increased but DPA showed little change.[27] However, ingestion of DHA increased the levels of DHA, DPA and EPA in the phosphatidylcholine (PC) and phosphatidylethanolamine (PE) fractions of platelets. Blood platelet aggregation was significantly decreased by ingestion of DHA, supporting the view that dietary ω3 PUFAs may alleviate certain forms of cardiovascular dysfunction.

The dietary ω6/ω3 PUFA ratio has been supported to be in the range of 4:1 to 10:1.[28] The western diet currently contains an average of 1.7 g ω3 PUFA per day of which 90% is ALA.[29] It has been recommended that the daily intake of ω3 PUFA should be increased to 3.0 g/day of which 1g should include EPA and DHA. The recommended daily intake of ω3 PUFA in Canada is a minimum of 0.55 g/1000 kcal or 0.5% of energy.[30]

4 STRUCTURED LIPIDS IN HUMAN NUTRITION AND HEALTH PROMOTION

Structured lipids containing medium-chain and long-chain essential fatty acids meet the nutritional needs of patients and those with special dietary requirements. When medium-chain fatty acids such as capric and caproic acids are consumed, they are not incorporated into chylomicrons and are therefore not likely to be stored, but will be used for energy. They are readily oxidized in the liver and constitute a highly concentrated source of energy for premature babies and patients with fat malabsorption disease. Jandazek et al.[31] showed that a structured lipid containing octanoic acid at the sn-1 and sn-3 positions and a long-chain fatty acids at the sn-2 position is more rapidly hydrolyzed and effectively absorbed than a typical long-chain triacylglycerol. Mendez et al.[32] compared the effects of a structured lipid (produced by interesterifying fish oil with medium chain fatty acid) with a physical mixture of fish oil and medium chain triacylglycerols and found that the structured lipid improved the nitrogen balance in mice, possibly because of the modified absorption rates of structured lipids. In another study, a positive nitrogen balance, higher serum albumin concentration and increase body weight were observed in injured rats fed with structured lipid made from safflower oil and medium chain fatty acids.[33] McKenna et al.[34] observed the enhance absorption of linoleic acid in cystic fibrosis patients fed structured lipid containing long-chain fatty acids and medium chain fatty acids.

In a series of studies, Hamam and Shahidi[35-37] reported on the inclusion of capric acid into algal oils containing highly unsaturated fatty acids. Capric acid was inserted primarily in the sn-1 and sn-3 positions of the triacylglycerol molecules (Table 3). Despite lowering the degree of unsaturation in the resultant products, their oxidative stability was less than those of the starting lipids. This was true for both primary products of oxidation, represented by conjugated diene values (Figure 1), and secondary oxidation products, represented by the 2-thiobarbituric acid reactive substances (TBARS) values (Figure 2) using a single cell oil reach in docosahexaenoic acid (DHASCO) as the starting material. Removal or alteration of endogenous antioxidants, such as tocopherols, was considered as the underlying explanation for this observation. Subjecting of the reactants to the same experimental steps in the absence of any enzyme showed lower stability of the resultant material with unaltered fatty acid composition (Figures 1 and 2).

Table 3. *Fatty acid composition of DHASCO triacylglycerols and their positional distribution before and after modification with capric acid*

Fatty acid (weight %)	Before modification		After modification	
	Total	Sn-2 (% distribution)	Total	Sn-2 (% distribution)
C10:0	0.46	41.3	10.2	6.0
C12:0	3.46	36.7	3.03	26.7
C14:0	12.9	17.5	10.0	25.3
C16:0	10.5	28.4	7.9	45.2
C18:1	26.6	36.1	26.3	41.3
C18:2	1.43	35.2	1.21	32.6
C22:6	37.1	23.3	37.1	41.4

Source: ref. 35.

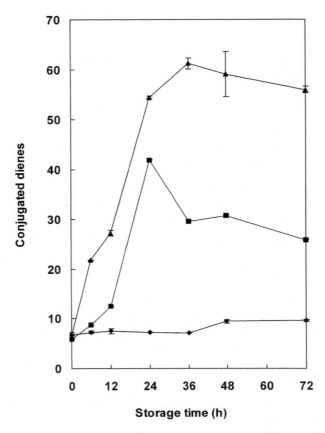

Figure 1. *Conjugated diene values of (■) modified DHASCO in the presence of lipase and (▲) without lipase as well as (♦) the control unmodified stored under Schaal oven conditions at 60°C. (Source: ref. 35)*

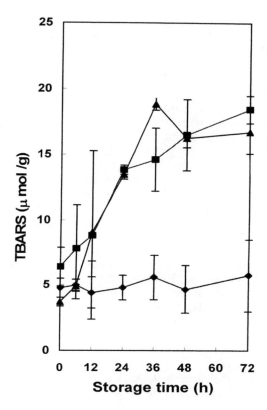

Figure 2. Thiobarbituric acid reactive substances (TBARS) values of (■) modified DHASCO in the presence of lipase and (▲) without lipase as well as (♦) the control unmodified stored under Schaal oven conditions at 60°C. (Source: ref. 35)

IMPACT (produced by Novartis Nutrition, Menneapolis, MN) is a medical product containing structured lipids. It is produced by interesterifying high-lauric oil with high-linoleic acid oil. This product is used for patients who have suffered from trauma or surgery, sepsis or cancer.[2]

SALATRIM is a structured lipid made from TAG containing short chain fatty acids (e.g., tributyrin) and long chain saturated TAG (e.g., hydrogenated rapeseed oil). This product has the potential to reduce substantially the energy value of foods into which it is incorporated while still retaining desirable textural properties. In a human clinical trial in which 17 individuals received a diet containing 22% of energy from SALATRIM for 7 days, a metabolizable energy value of this product of 4.9 kcal/g and absorption of stearic acid ranging from 63 to 70% were reported.[38]

Omega-3 fatty acids may be incorporated into structured lipids to promote health and nutrition. These fatty acids have several health benefits to combat arthritis, thrombosis, cardiovascular disease, diabetes and cancer.

4.1 Arthritis

Beneficial effects derived from consumption of a diet high in ω3 and ω6 fatty acids have been reported in arthritic patients. However, EPA increased the incidence of collagen-induced arthritis in mice.[39] In another study, arthritic patients had significant improvement in morning stiffness and number of tender joints, when consuming EPA supplements compared to placebo in a double blinded, crossover study.[40]

4.2 Thrombosis

Blood clotting involves the clumping together of platelets into large aggregates and is triggered when the endothelial cells lining the artery walls are damaged. This formation of blood clots is known as thrombosis. If the platelet membranes are rich in the long chain ω3 PUFA, this promotes the formation of certain eicosanoids such as prostacyclin I_3 and thromboxane A_3. These do not trigger platelet aggregations as much as the corresponding eicosanoids, prostacyclin I_2 and thromboxane A_2, formed from the ω6 PUFA. Therefore long chain ω3 PUFA can help to reduce the tendency for blood to clot.

4.3 Cardiovascular disease

Interest in fish oils was originally stemmed from the observations made in the 1960's of near complete absence of heart disease among Greenland Eskimos despite their use of a diet very high in fat and cholesterol.[24] It was later found that the blood of these Eskimos contained a high concentration of EPA and DHA from a diet rich in fish and seal meat. Recent work suggests that ω3 intake from fish consumption in conjunction with a low fat diet is most beneficial in terms of reducing cardiovascular disease.[41]

Cardiovascular disease is the leading cause of death in industrialized countries.[42] Recent research indicates that the long chain ω3 PUFA, especially EPA and DHA, may be effective in reducing the clinical risk of cardiovascular disease by favourably altering lipid and haemostatic factors (such as bleeding time and platelet aggregation).[43] Dietary supplements of EPA, DHA and other ω3 PUFA are also recommended to lower the risks of cardiovascular disease and to improve the general overall health of humans.

4.4 Diabetes

Dietary supplementation diabetic subjects with long-chain PUFA from if fish oil has been studied more extensively than that of other fatty acids. The two predominant PUFA in fish oil are EPA and DHA. Possible effects of ω3 fatty acids on persons with diabetes are being studied. Recently, the development of insulin resistance in normal rats fed a high fat (safflower oil) diet was found to be prevented by the partial replacement of the linoleic with EPA and DHA from fish oil.[44] From human studies, it is clear that in diabetic subjects, ω3 fatty acids appear to have some beneficial effects on lipid metabolism and may decrease the severity of cardiac disorder and lower the incidence of coronary artery disease.[45]

4.5 Cancer

Ling[46] reported that a structured lipid made from fish oil and medium-chain triacylglycerol was able to decrease tumor growth in mice. In another study, the tumor growth rate was slowed in the rats fed with structured lipids containing medium chain fatty acids and fish oil.[35] In contrast to the tumor-promoting effects of diets high in fat, diets high in fish oil failed to promote the development of tumors in rats.[19] Reddy and Maruyama[47] also pointed out that diets containing high levels of fish oil have been shown to inhibit or suppress tumor growth in animal models. With special concentrations of the fish oils have been effective in destroying some cancer cells, it is not currently known whether such results would be reproducible with humans, or what side effects there might be. It is known that ω3 fatty acids play an important role in the growth of certain cells in the human body, but the mechanisms involved in the cancer treatment are incompletely understood.

References

1. K. Lee, C.C. Akoh, *Food Rev. Int.* 1998, **14**, 17-34.
2. B.F. Haumann, *INFORM*, 1997a, **8**, 1004-1011.
3. C.C. Akoh, *INFORM*, 1995, **6**, 1055-1061.
4. Novo Nordisk Biochem North America, Inc., Product Sheet. B 606c-GB, Franklinton, NC, 1994.
5. N.N. Gandhi, *J. Am. Oil. Chem. Soc.* 1997, **74**, 621-634.
6. A.G. Marangoni, D. Rousseau, *Trends Food Sci. Technol.* 1995, **6**, 329-335.
7. R. Sridhar, G. Lakshminarayana, *J. Am. Oil Chem. Soc.* 1992, **69**, 1041-1042.
8. K. Huang, C.C. Akoh, M.C. Erickson, *J. Agric. Food Chem.* 1994, **42**, 2646-2648.
9. K. Huang, C.C. Akoh, *J. Am. Oil Chem. Soc.* 1994, **71**, 1277-1280.
10. C.C. Akoh, R.V. Sista, *J. Food Lipids* 1995, **2**, 231-238.
11. C.C. Akoh, B.H. Jennings, D.A. Lillard, *J. Am. Oil Chem. Soc.* 1996, **73**, 1059-1062.
12. Y. Ju, F. Huang, C. Fang, *J. Am. Oil Chem. Soc.* 1998, **75**, 961-965.
13. C.C. Akoh, C.O. Moussata, *J. Am. Oil Chem. Soc.* 1998, **75**, 697-701.
14. C.O. Moussata, C.C. Akoh, *J. Am. Oil Chem. Soc.* 1997, **74**, 177-179.
15. C.C. Akoh, B.H. Jennings, D.A. Lillard, *J. Am. Oil Chem. Soc.* 1998, **72**, 317-1321.
16. L.M. Braden, K.K. Carroll, *Lipids* 1986, **21**, 285-288.
17. A. Hashim, V.K. Babayan, *Am. J. Clin. Nutr.*, 1978, **31**, 5273-5276.
18. J.P. Kennedy, *Food Technol.* 1991, **11**, 76-83.

19. H. Groom, *Nutr. Food Sci.*, 1993, **6**, 4-8.
20. G.M. Wardlaw, Perspectives in Human Nutrition, 3rd edn., Mosby-Year Book Inc., St. Louis, MO, 1996, p.112.
21. R. Rice, *Lipid Technology*, 1991, **23**, 12-116.
22. O. Ward, *INFORM* 1995, **6**, 683-688.
23. P. Langholz, P. Andersen, T. Forskov, W. Schmidtsdorff, *J. Am. Oil Chem. Soc.* 1989, **66**, 1120-1123.
24. M.A. Crawford, *Am. J. Clin. Nutr.* 1997, **66** (suppl), 1032S-1041S.
25. B.F. Haumann, *INFORM*, 1997b, **8**, 428-447.
26. R.G. Ackman, W.M.N. Ratnayake. 1989. Fish oils, seal oils, esters and acids-are all forms of omega-3 intake equal? In Health effects of Fish and Fish oils. ed. R.K. Chandra, pp. 373-393, ARTS Biomedical Publishers & Distributors, St. John's, NL.
27. J.E. Kinsella, 1990. Sources of omega-3 fatty acids in human diets, In Omega-3 fatty acids in health and disease. ed. R.S. Lees and M. Karel, pp. 157-200. Marcel Dekker, Inc., New York.
28. L. Caston, S. Leeson, *Poultry Sci.* 1990, **69**, 1617-1620.
29. M.E. Van Elswyk, *Nutr. Today* 1993, **28**, 21-28.
30. A.O. Ajuyah, K.H. Lee, R.T. Hardin, J.S. Sim, *Poultry Sci.* 1991, **70**, 2304-2314.
31. R.J. Jandacek, J.A. Whiteside, B.N. Holcombe, *Am. J. Clin. Nutr.* 1987, **45**, 940-945.
32. B. Mendez, P.R. Ling, N.W. Istfan, V.K. Babayan, B.R. Bistrain, *J. Parent. Enteral Nutr.* 1992, **16**, 545-551.
33. K.T. Mok, A. Maiz, K. Yamakazi, J. Sobrado, V.K. Babayan, L. Moldawen, B.R. Bistrian, G.L. Blackburn, *Metabolism* 1984, **33**, 910
34. M.C. McKenna, V.S. Hubbard, J.G. Pieri, *J. Pediatr. Gastroenterol. Nutr.* 1985, **4**, 45.
35. F. Hamam, F. Shahidi, *J. Agric. Food Chem.* 2004, **52**, 2900-2906.
36. F. Hamam, F. Shahidi, *J. Am. Oil Chem. Soc.* 2004, **81**, 887-892.
37. F. Hamam, F. Shahidi, *J. Food Lipids* 2004, **11**, 147-163.
38. J.W. Finley, L.P. Klemann, G.A. Leveille, M.S. Otterburn, C.G. Walchak, *J. Agric. Food Chem.* 1994, **42**, 495-499.
39. J.D. Prickett, D.E. Trentham, D.R. Robinson, *J. Immunol.* 1984, **132**, 725-729.
40. J.M. Kremer, W. Jubiz, A. Michalek, R.I. Rynes, L.E. Bartholomew, J. Bigaouette, M. Timchalk, D. Beeler, L. Lininger, *Ann. Int. Med.* 1987, **106**, 497-502.
41. T.A. Mori, L.J. Beilin, V. Burke, J. Morris, J. Ritchie, *Arterioscler. Thromb. Vasc. Biol.*, 1997, **17**, 279-286.
42. I.S. Newton, *INFORM* 1996, **7**, 169-177.
43. G. Hornstra, 1989. Effects of dietary lipids on some aspects of the cardiovascular risk profile. In *Lipids and Health*. Ed. Ziant, G., Elsevier, New York, NY, p. 39.
44. L.H. Storlien, E.W. Kraegen, D.J. Chisholm, G.L. Ford, D.G. Bruce, W.S. Pascoe, *Science* 1987, **237**, 885-888.
45. S.J. Bhathena, In *Fatty Acids in Foods and Their Health Implications*. Ed. Chow, C.K., Marcel Dekker, Inc., New York, NY, pp. 823-855.
46. P.R. Ling, N.W. Istfan, S.M. Lopes, V.K. Babayan, *Am. J. Clin. Nutr.* 1991, **53**, 1177-1184.
47. B.S. Reddy, H. Maruyama, *Cancer Res.* 1986, **46**, 3367-3370.

HEPATIC ACUTE-PHASE RESPONSE TO 3-ALKYL-2-PHENYL-2-HYDROXY-MORPHOLINIUM CATIONS

F. M. Fouad[1], O. A. Mamer[1], F. Sauriol[2], A. Lesimple[1], F. Shahidi[3] and *G. Ruhenstroth-Bauer[4]

[1]The Biomedical Mass Spectrometry Unit, 1130 Pine Avenue West, McGill University, Montreal, QC, Canada H3A 1A3
[2] Department of Chemistry, Queen's University, Kingston, ON, Canada K7L 3N6
[3]Department of Biochemistry, Memorial University of Newfoundland, St. John's, NL, Canada A1B 3X9
[4]The Max-Planck-Institüt für Biochemie, München, Germany, Deceased 2004

1 ABSTRACT

Morpholine, a local anesthetic and antiseptic is known for its toxicity and mutagenicity to humans. A facile synthesis of 3-alkyl-2-phenyl-2-hydroxymorpholinium cations was employed and products were used in a study of the acute-phase response (APR) to their long-term ingestion by rats. The chain length of the 3-alkyl groups ranged between 0 and 16 carbons as confirmed by FAB mass spectrometry and ^{13}C-NMR as well as protons, NMR. These morpholinium cations, especially those with long alkyl chains, are expected to show inhibition of mitochondrial carnitine palmitoyltransferase. The morpholinium cations were subcutaneously administered to rats (1 mmole/kg) to follow the kinetics of hepatic APR to a single application. The APR kinetics resembled those of CCl_4 which suggests that the toxicity of morpholinium ions is most likely mediated by leukocyte endogenous mediator and/or kinins and that their possible elimination by binding to APR proteins, is followed by regression of the spectrum of APR to aseptic control. Furthermore, consumption of drinking water supplemented with 1 mM morpholinium cations over a period of one month resulted in a hepatic APR characteristic of the cirrhosis developed upon consuming thioacetamide or a combination of acetaminophen and ethanol. Cirrhotic livers of rats failed to regenerate to their original weight when 75% partially hepatectomized due to irreversible damage to hepatic infrastructure.

2 INTRODUCTION

Carnitine, a quantitatively important compound in muscle tissue is reversibly acylated by acyl-coenzyme A (Ac-CoA) and stimulates fatty acid oxidation in liver homogenate. Thus, carnitine facilitates the transport of activated fatty acids across the mitochondrial membrane which is a fundamental function within the cell. Carnitine and fatty acid esters of carnitine are excellent oxidizable substrates for isolated mitochondria.[1] Furthermore, the isolated mitochondria showed to embrace an external and internal carnitine palmitoyltransferase which reversibly catalyze the acylation of carnitine. The observation that oxidation of short- and medium-chain fatty acids is carnitine independent while long-chain fatty acids require carnitine assisted transportation into michondrial matrix prior to oxidation suggests a cellular filtration and/ or recognition barrier to distinguish between fatty acids of various chain lengths. Therefore, carnitine facilitated transport of long-chain fatty acids such as C22:1 for β-oxidation at the mitochondrial level which are poorly oxidized,[2] and may require partial peroxisomal oxidation, that is chain length shortening. In view of the varying ability of mitochondria from various animal tissue to oxidize acylcarnitine of different chain lengths, it became inevitable to suggest that more than one carnitine acyltransferase existed.[3] According to chain length specificity, three carnitine acyltransferase, namely acetyltransferase, octanoyltransferase and palmitoyltransferase for C2-3, C6-10 and C14-16, respectively, were identified.

Considering the key role of carnitine in fatty acid oxidation, various degrees of pathological conditions are expected to develop in response to a reduced cellular concentration of carnitine. The common feature for reduced level of carnitine is accumulation of triacylglycerols within the cell where fatty acids accumulating due to inhibited oxidation would alternatively undergo esterification, i.e., lipidosis. In general, heart failure in diphtheria is associated with lipidosis and reduced cellular carnitine.[4] In-vivo administration of carnitine to diphtheritic animals revealed a protective effect and prolonged survival against *Neisseria diphtheria* toxin, which induces lipidosis and decreases the ability to oxidize fatty acids by heart homogenate.[5] Toxin-induced loss of carnitine is mainly due to inhibition of protein synthesis of a carnitine carrier in the cell membrane[6] as well as loss of carnitine produced by constriction of the aorta[7] and sever ischemia of the heart.[8] In severe liver cirrhosis, the carnitine content in plasma and other tissue is 25-33% that of healthy individuals.[9] Accordingly, the effect of a range of synthetic organic chemicals on cellular function was studied in terms of intracellular and plasma carnitine levels to yield conflicting results, e.g., failure to modify the hypolipidemic effects by mildronate treatment.[10] In a fasting rat model, fatty liver and hypertriacylglyceridemia are easily induced by the administration of an inhibitor of fatty acid oxidation, 3-amino-4-trimethylaminobutyric acid.[11] In addition, etomoxir, an irreversible carnitine palmitoyltransferase I inhibitor, has been shown to be a potent hypoglycemic chemical in laboratory animals and humans.[12] Recently, 2-propylpentanoic acid, a hepatotoxic chemical used for treatment of epilepsy, showed inhibition of carnitine biosynthesis.[13]

Morpholine, a local anesthetic and antiseptic agent, is known for its toxicity and mutagenicity to humans. Thus, it was deemed essential to prepare morpholinium cations with long chain alkyl groups and inquire into the course of its hepatic toxicity. For this purpose, a facile one-step preparation of 3-alkyl-2-phenyl-2-hydroxymorpholinium cations was adopted to study the acute-phase response (APR) to their long-term ingestion by rats. The chain length of the 3-alkyl group ranged from 0 to 16 carbons as cnfirmed by FAB mass spectrometry and ^{13}C-NMR as well as proton NMR. These morpholinium cations, especially those with long alkyl chains are expected to show inhibition of mitochondrial carnitine- palmitoyltransferase.

3 EXPERIMENTAL

3.1 Chemical and Reagents

Chemicals and reagents were purchased from Sigma-Aldrich Canada Ltd. (Oakville, ON) and used as received. Compounds 2-bromohexanoic, 2-bromodecanoic, 2-bromododecanoic, 2-bromooctanoic and 2-bromostearic acids (0.01 mole) were allowed to react with thionyl chloride (0.01 mole) for 2 h at 95°C. The acid chloride was diluted with anhydrous benzene (200 ml) and anhydrous aluminum chloride (0.02 mole) were added with stirring at 0°C for 1 h. The reaction mixture was refluxed for 1h and processed to isolate the corresponding α-alkylphenacyl bromide. Commercially available 4-substituted phenacyl bromides and prepared α-alkyl derivatives (0.005 mole) were mixed with stirring for 30 min at 0°C with dimethylethanolamine (0.005 mole) solution in anhydrous tetrahydrofuran (10 ml) and refluxed for 1-9 hours. Structures of final products were confirmed by FAB mass spectrometry, ^{13}C-NMR and ^{1}H-NMR (Table 1). Depending on the phenacyl bromide used, morpholinium yields ranged between 45 and 60%.

3.2 Morpholinium Cations-Induced APR in Healthy and Diabetic Rat Model

Animals were handled according to animal care guidelines. In the first set of experiments, female Sprague-Dawley aseptic rats (100-120 g) or 35days post alloxan-induced diabetes, were subcutaneously (s.c.) injected with morpholinium cation (1 mmole/L/kg body weight) and allowed free access to rat chow and tap water. At predetermined time intervals, blood of three control and anesthesized experimental animals was collected and serum was separated and kept frozen at -20°C until assayed by two-dimensional immunoelectrophoresis (2D-IEP) as described earlier.[14]

In the second set of experiments, female aseptic Sprague-Dawley rats (100-120 g) were fed standard rat chow, allowed water ad libitum. Rats were allowed drinking water supplemented with morpholimiun cation (1 mM/L) for a period of one month. Ten aseptic animals were kept untreated during the duration of the experiment to serve as control. At predetermined time intervals, blood of three control and experimental animals under ether anesthesia, was collected and serum was separated and kept frozen at -20°C until assayed by 2D-IEP.

3.3 Partial Hepatectomy of Morpholinium Cations-Induced Cirrhotic Liver

Livers of aseptic female rats turned cirrhotic due free access to morpholinium cations derivatives in drinking water (1 month) were partially hepatectomized, i.e., 75% surgical removal of liver mass. At predetermined intervals of time, blood was withdrawn to separate serum for 2D-IEP and remnant hepatic tissue was removed and weighed.

4 RESULTS AND DISCUSSION

The preparation of 3-alkyl-2-phenyl-2-hydroxymorpholinium cations was a facile procedure, allowing further inquiry into the hepatic integrity of model animals with possible extrapolation to similar effects on humans. ^{1}H-NMR (Table 1) as well as FAB mass spectrometry and ^{13}C-NMR were used to confirm the structures of the final products.

In this context, similar to carbon tetrachloride, intraperitoneal administration of 3-alkyl-2-phenyl-2-hydroxymorpholinium cations at 1 mmole/kg into rats led to instantaneous neurotic convulsions followed by immediate death of the animal. Therefore, saline solution of 3-alkyl-2-phenyl-2-hydroxymorpholinium cations was administered subcutaneously (s.c.) into the rear hind muscle of the rat. In the first set of experiments, a single s.c. administration of morpholinium cations to aseptic female rats evoked an APR (Table 2) characteristic to an inflammatory stimulation by compounds such as carbon tetrachloride[15] with regression to levels of healthy controls within 9 days. The peak APR was manifested on the second day post administration and regression began on the 7th day. On the 9th day APR plasma proteins attained levels within 15% of healthy abundance. Thus, either hepatocytes ceased to export APR plasma proteins into circulation upon expiry of the half life time of the inflammatory signal or alternatively APR proteins were consumed to conjugate with most of the morpholinium cations or their metabolites prior to elimination.

In the second set of experiments, long-term free access to consumption of morpholinium cations in drinking water precipitated liver damage obvious to the naked eye whereas all rats developed fatty livers whose APR were similar to those of thioacetamide[16] or acetaminophen/alcohol treated rats, thermally oxidized lipids- or CCl_4-azathioprine-induced cirrhosis models.[17,18] Long term consumption of morpholinium cations in drinking water most likely depleted hepatocytes of basophilic precursors required to initiate hepatic response to physical or chemical injury as cellular resources were prioritized in favor of DNA mitotic activity as a prelude to regeneration, if possible at all, until the damaged liver could recover its original weight. Notwithstanding functional prioritization and increased functional capacity of the morpholinium cations-injured hepatocytes permitted biosynthesis and release of an atypical-bland APR for the 30 day period of experiment (Table 3). As observed with earlier liver insufficiency models,[16-18] morpholinium cations-injured hepatocytes failed to regenerate due either to depletion of hepatocytes of basophilic resources required to initiate hepatic mitosis or to permanent damage to sites releasing hepatopietin, or both of them.

In conclusion, it is advantageous if risks and damaging effects of long term consumption or exposure to certain synthetic chemicals and suspected toxic pollutants are weighed in terms of spectral specifications of hepatic APR plasma proteins and the capacity of pathological liver to regenerate. Thus, the APR was typically exaggerated upon s.c. administration of 3-alkyl-2-phenyl-2-hydroxymorpholinium cations into alloxan-induced diabetic rats (Table 4). These elevated APR proteins could potentially cause CHD[20] which is seen to amplify in diabetic individuals yielding a wide range of CHD. Accordingly, introduction of new chemicals intended for particular pharmaceutical applications should be treated with great circumspection until proven not to adversely affect hepatic function.

Table 1 ^{1}H-NMR of de novo prepared 3-alkyl-2-phenyl-2-hydroxymorpholinium cations

						3-Alkyl	
$(CH_2)n$	5-CH_2	6-CH_2	3-Phenyl	3-CH	CH_3	$(CH_2)n$	$N^+(CH_3)_2$
0	4.12	4.36	7.1 - 7.9	4.77	0	0	3.08
3	3.46	4.36	7.1 - 7.9	4.22	0.82	1.41 - 2.78	3.07
5	3.52	4.34	7.1 - 7.9	4.03	0.86	1.27 - 2.66	3.08
9	3.44	4.26	7.1 - 7.9	4.42	0.87	1.26 - 2.63	3.09
15	3.44	4.28	7.1 - 7.9	4.42	0.88	0.89 - 2.68	3.08

Table 2 Percent change in the APR proteins immunoprecipitates relative to healthy control rats upon subcutaneous administration of 3-alkyl-2-phenyl-2-hydroxymorpholinium cations into rats

Protein	Time (days)					
	1	2	3	5	7	9
Alb	-9.3	-12.4	-8.6	-5.4	-3.0	-2.9
Pre-Alb	n.a	-39.3	-27.9	-15.3	n.a	-5.1
Ag	200.7	137.2	92.1	36.0	12.9	7.4
C3	3.6	18.4	27.6	8.5	2.9	8.0
Tf	0.7	1.1	-3.6	31.2	25.2	9.9
Hg	91.8	138.6	66.5	76.5	23.2	10.8
α-Lp	1.8	32.3	15.3	10.2	9.6	3.6
At	92.7	122.1	53.0	38.3	17.9	6.3

Table 3 Percent change in APR proteins immunoprecipitates relative to healthy control rats induced by enriching drinking water with 3-alkyl-2-phenyl-2-hydroxymorpholinium cations

Protein	Time (days)					
	4	8	12	18	24	30
Alb	-1.0	-5.4	-6.3	-5.1	-4.8	-5.7
Pre-Alb	n.a	-11.6	-7.8	-13.5	n.a	-8.1
Ag	21.6	37.2	52.2	33.0	32.4	17.1
C3	9.6	17.1	19.8	18.3	22.8	18.0
Tf	0.3	3.6	6.6	3.2	5.4	7.5
Hg	11.8	138.6	66.5	76.5	23.2	10.8
α-Lp	1.2	2.7	5.4	9.9	9.6	10.2
At	18.6	21.3	23.1	21.9	23.9	20.1

Table 4 Percent change in APR proteins immunoprecipitates relative to healthy controls upon subcutaneous administration of 3-alkyl-2-phenyl-2-hydroxymorpholinium cations into diabetic rats.

Protein	Time (days)					
	0diabetic	1	3	5	7	9
Alb	-31.3	8.1	-9.3	-19.2	-26.1	-21.9
Pre-Alb	15.6	-29.1	-7.9	-21.3	-10.2	-6.3
Ag	-20.7	117.6	199.2	136.0	98.9	147.4
C3	-53.6	218.1	145.3	98.5	92.9	48.0
Tf	-21.6	51.1	33.9	39.9	15.9	29.9
Hg	-51.9	218.7	364.9	116.1	132.3	40.5
α-Lp	-32.4	36.3	44.1	20.1	19.2	7.5
At	22.5	52.2	21.3	40.5	27.9	3.3

Values in tables above are averages of three experiments and uncertainties in values reported are: α-1-Acid glycoproteins (Ag; ±25%), Haptoglobin (Hg; ±12%), α-1-Antitrypsin (At, 13) and Transferrin (Tf, 11%), Albumin (Alb; ±7%), Pre-albumin (pre-Alb; ±7%).

5 ACKNOWLEDGEMENT

One of us (O.A.M.) is grateful to the Canadian Institutes of Health Research for financial support.

References

1. J.D.M. McGarry and D.W. Foster, *Annu. Rev. Biochem.*, 1980, **49**, 395-420.
2. J. Bremer and K.R. Norum, *J. Lipid Res.*, 1982, **23**, 243-56.
3. J. Bremer. *J. Biol. Chem.*, 1962, **237**, 2228-2231.
4. B. Wittels and R. Bressler, *J. Clin. Invest.*, 1964, **43**, 630-637.
5. D.R. Challoner and H.G. Prols, *J. Clin. Invest.*, 1972, **51**, 2071-2076.
6. P. Molstad and T. Bohmer, *Biochim. Biophys. Acta.*, 1981, **641**, 71-78.
7. B. Wittels and J.F. Spann, Jr. *J. Clin. Invest.*, 1968, **47**, 1787-1794.
8. A. Shug, J.H. Thomsen, J.D. Flots, N. Bittar, M.I. Klein, J.R. Koke and P.J. Huth, *Arch. Biochem. Biophys.*, 1978, **187**, 25-33.
9. D. Rudman, C.W. Sewell and J.D. Ansley, *J Clin Invest.* 1977, **60**, 716-723.
10. M. Tsoko, F. Beauseigneur, J. Gresti, J. Demarquoy and P. Clouet, *Biochimie.*, 1998, **80**, 943-948.
11. H. Maed, M. Fujiwara, K. Miyamoto, H. Hamamoto and N. Fukuda, *J. Nutr. Sci. Vitaminol.*, 1966, **42**, 469.
12. S. Steiner, D. Wahl, B.L.K. Mangold, R. Robison, J. Raymackers, L. Meheus, N.L. Anderson and A. Cordier, *Biochem. Biophys. Res. Commun.*, 1996, **218**, 777-782.
13. V. Farkas, I. Bock, J. Cseko and A. Sandor, *Biochem. Pharmacol.*, 1996, **52**, 1429-1433.
14. F.M. Fouad, R. Scherer, M. Abd-El-Fattah and G. Ruhenstroth-Bauer, *Eur. J. Cell Biol.*, 1980, **21**, 175-179.
15. M. Leonhardt and W. Langhans, *Physiol. Behav.*, 2004 Dec. 30, **83**(4), 645-51).
16. F.M. Fouad, M. Goldberg and G. Ruhenstroth-Bauer, *J. Clin. Chem. Clin. Biochem.*, 1983, **21**, 203-208.
17. F.M. Fouad, F. Shahidi and O.A. Mamer, *J. Toxicol. Environ. Health,* 1995, **46**, 217-232.
18. F.M. Fouad, O.A. Mamer and F. Shahidi, *J. Toxicol. Environ. Health,* 1996, **47**, 601-615.
19. J. George, *Semin Liver Dis.*, 2002, **22**(2), 169-83.
20. A.R. Folsom, F.J. Nieto, P. Sorlie, L.E. Chambless, D.Y. Graham, *Circulation*, 1998, **98**, 845-850.

STIMULATION OF HEPATIC ACUTE-PHASE RESPONSE BY STRESS, SUCROSE POLYESTER AND ZOCOR IN ANIMAL MODEL

F. M. Fouad[1], O.A. Mamer[1], F. Sauriol[2], A. Lesimple[1], F. Shahidi[3], M. Khayyal[4], M. Hasseeb[5] and G. Ruhenstroth-Bauer[6]

[1]The Biomedical Mass Spectrometry Unit, 1130 Pine Avenue West, McGill University, Montreal, QC, Canada H3A 1A3; [2]Department of Chemistry, Queen's University, Kingston, ON, Canada K7L 3N6; [3]Department of Biochemistry, Memorial University of Newfoundland, St. John's, NL, Canada A1B 3X9; [4]Department of Pharmacology and Parasitology, Al Azhar University, Cairo, Egypt; [5]Faculty of Medicine, Alexandria University, Alexandria, Egypt; [6]The Max-Planck-Institüt fur Biochemie, München, Germany, Deceased 2004

1 ABSTRACT

Evidence associated low-grade inflammation yielding elevated levels of acute-phase response (APR) proteins such as serum amyloid A, fibrinogen, CRP and haptoglobin in response to *H. pylori* and *C. pneumoniae* infection to initiation of gastritis and ischemic heart diseases. Sucrose polyesters (SPE) as substitutes for natural vegetable oils are expected to decrease essential fatty acid intake. Although adipose tissue may be a link as it is a possible site for the biosynthesis and release of inflammatory mediators, such as IL-6, it remains unrealistic to consume excessive proportions of SPE enriched diets to barely manipulate serum's triacylglycerol (TAGs) and cholesterol. In this context, SPE, the lipophilic solute contained within the dietary lipid solvent micelle, is excreted undigested together with bulky TAGs. In addition, the role of stress in health status and use of zocor for lowering cholesterol are discussed in this context.

2 INTRODUCTION

Circulating level of serum cholesterol together with high blood pressure are considered the main causative factors for high mortality associated with coronary heart disease (CHD). Accordingly, food and pharmaceutical industry strive to introduce substitutes in fat-rich diets that could reduce intake of cholesterol or other medications to reduce high levels of circulating cholesterol such as sucrose polyester (SPE) and zocor (simavastatin), respectively. Nonetheless, recent reports accentuated that human stomach is a major host for *Helicobacter pylori* (*H. pylori*) that may inflict mucosal injury, gastritis, and development of gastric carcinoma, among others, leading to elevation of acute-phase response (APR) proteins and hence CHD. Similarly, societal stress is envisaged as a contributory factor to CHD. As *H. pylori* colonizes the stomach, spread of the infection must be either through fecal-oral or oral-oral means.[1] In addition, *H. pylori* infection can adversely affect the nutritional status of children, thus retarding their growth. Early reports[2,3] suggested chronic infection with either *C. pneumoniae* or *H. pylori* could cause elevations in APR proteins and fibrinogen to precipitate CHD. Moreover, infectobesity is entertained as the obesity of infectious origin[4]. As expected, the many years of associating cholesterol with CHD dictated understanding mechanism(s) of the initial release of toxic lipophiles from triacylglycerols to provide keys to manipulate their absorption into mammalian tissue and circulation. Thus, micelles were proposed as vehicles for importing lipophiles from the fat matrix into the aqueous phase for export to animals' circulation.[5-7] This prompted the drive to develop SPE and chemically manipulate butter.[8] The safety of SPE was supported by short term experimentations on animal models and human volunteers.[9]

SPE was proposed as a zero-calorie fat substitute which would retard absorption of lipids and other lipophilies remaining in an oil phase in the lumen and secreted almost unchanged with the feces,[10,11] thus expected to lower the concentration of plasma cholesterol. However, recently published data pertaining to clinical short-term trials negated the above deductions and led us to interpret the results in terms of hepatic injury.[9,12,13]

This report documents the hepatic APR to (a) cholesterol lowering products such as lipolysis-resistant synthetic lipids such as SPE or zocor and (b) scheduled and unscheduled environmental stress in an animal model. In this context, (a) the failure of lipolysis-resistant lipids to retard rates of lipolysis of naturally occurring triacylglycerols and (b) i.p. administration of zocor evoked sharp APR with substantial hepatic necrosis, call for the relevance of both for human consumption. The pertinent parallel discussed is the association of societal stress and bacterial-induced APR proteins to cardiac dysfunction. The integrity of gastric mucosa is likely to be compromised by societal stress and excessive intake of aspirin and anti-rheumatoid drugs or when diets are contaminated with $H.$ $Pylori$ which initiate gastritis[14,15] in a chain reaction yielding CHD.

3 EXPERIMENTAL

3.1 One-Step Preparation of Lypolysis-Resistant Lipids

Polyols reacted with different proportions of a single or a mixture of fatty acid chlorides at 95°C to yield a mixture of lipid esters whose structures were confirmed by detailed mass spectral analyses. These pure synthetic lipids resisted lipolysis and, when added 5-10% by weight to butter oil, did not influence the rate of lypolysis of the resultant mixed lipid. Hepatic APR to these lipolysis-resistant lipids was established upon subcutaneous administration into healthy rabbits at 250 mg kg^{-1}; blood was collected at predetermined time interval, and separated serum was kept at -20°C for two-dimensional immunoelectrophoresis (2D-IEP).

3.2 Psychological-Induced APR in Animal Model

A total of 48 male aseptic rats, 100±10 g each, were acclimatized to laboratory conditions for 2 day with free access to drinking water and chew, prior to experimentation. Rats were randomly selected, 12 in each group; one group being the control. Animals were subjected to 12 h day light and 12 h darkness cycles.

3.3 Unscheduled Psychological Stress

In a group of 12 animals, rats were offered chew and water twice per day for 10 min while playing at different intervals of time a recorded sound of cats in fight for 10-30 min, 6-9 times during the day either prior to, post or in the presence of chew and water. In a second set of experiments, 12 rats were exposed to a similar treatment but during the darkness of the night. While under ether anaesthesia, at predetermined time intervals, blood was drawn from the vena cava of 3 individuals and serum was separated and frozen at -20°C for 2D-IEP (Table 1).

3.4 Scheduled Psychological Stress

In the third group of 12 animals, rats were freely offered chew and water while playing at 9am, 1 and 6pm to a recorded sound of cats in fight for 20 min. Under ether anaesthesia, at same predetermined time intervals blood was drawn for the vena cava and serum was separated and frozen at -20°C for 2D-IEP (Table 2). At the same predetermined time intervals, in both unscheduled and scheduled psychological stress experimentations, rats' stomach was isolated, opened washed with 4°C saline and visually examined for signs of ulcerations such as redness and consistency of mucosal barrier.

3.5 Zocor-Induced Hepatic APR

On a daily basis for a period of 30 days at 9 am, zocor suspended in saline was intraperitoneally (i.p.) administered into rabbits, 50 mg kg^{-1} body weight. At predetermined intervals of time blood was withdrawn to separate plasma by centrifugation for 2D-IEP. For histological assay of tissue damage, kidney, heart and hepatic tissues were concomitantly collected and kept in a formaldehyde solution.

4 RESULTS AND DISCUSSION

4.1 Hepatic-Induced Acute-Phase Response to Societal Stress

Humans are considered a vulnerable medium to *Helicobacter pylori* and societal stress. Atherosclerosis is generally accepted as an inflammatory disorder in the arterial wall.[16] Early stages of atherosclerosis are characterized by subendothelial lipid accumulation, endothelial activation and leukocyte adhesion to the endothelium resulting in infiltration of macrophages and T-lymphocytes into arterial intima. Levels of inflammatory mediators IL-6 and TNF-α, mainly biosynthesized by macrophages endothelial cells and adipocytes, respectively, were significantly elevated in CHD patients.[17,18] Nonetheless, association of CHD to APR was diluted because the assumption "*Chlamydia* infection contributes to a proinflammatory state in CHD patients but not in healthy individuals" contradicts the basic conceptualization of the genesis of APR[19] and suggests other sites for the biosynthesis of tumor necrosis factor-α (TNF-α). Thus selecting a specific serum protein as a marker to the development of a given pathological condition while multi-inflammatory stimuli are in progress is a complex task. In a rat model, synthesizing a second stimulus while the APR of the earlier insult is in regression, did not manifest itself additively but evoked a general depression of the primary APR to allow erroneous interpretation as healing.[20] Thus, concomitant administration of inflammatory stimuli yields a depressed APR spectrum indicative of the limited capacity of hepatic rough endoplasmic reticulum-bound ribosomes to contain the damaging effect of the two toxins.[20] Most interesting is the observation that *C. pneumoniae* infection of rabbit model subsisting on medium lipidemic diet enriched with 0.25% cholesterol, significantly developed increased atherosclerosis lesions.

On the other hand circulating reactive proteins (CRP), serum amyloid A (SAA), fibrinogen (Fb) and heptoglobin (Hg) increased in diabetic individuals versus healthy subjects which agrees with earlier deduction pertaining to the consumption of excess serum glucose to biosynthesize serum glycoproteins in atypical APR spectrum suggesting post ribosomal addition of poorly controlled glucose residues onto the nascent or recycled polypeptide chains as an alternative pathway for its metabolism in carageenan, partially hepatectomized and cholera intoxicated diabetic rat models.[20] It is our contention, these observed elevation in APR proteins could be the classical APR to injuries emanating from long term contracting *H. pylori* or CHDs, i.e., an after event phenomenon and not necessarily the biological factor triggering CHD. Therefore, it is hard to envisage a pathogenetic role of *H. pylori* in the aetiology of atherosclerosis if we consider men living in rural China with higher prevalence of *H. pylori* infection than in American counterparts are 16.7 times less likely than US men to die of coronary heart disease.[21-23] Baschetti[24] rightly concluded that the proposed association of *H. pylori* infection with atherosclerosis is welcomed by those who are reluctant to accept that CHD is due to the western "nutritional extravagance" which leads many to adopt unnaturally high fat diets.

The second parallel is societal stress which is expected to yield psychological aberrations and neurobiological activation of a physiological alarm chain of events as a rate limiting step to decoding psychological stress into physiological injuries inducing hepatic ARP, thus instigating CHD. Therefore, to conclude this present study, a rat model was used. In an unscheduled stress rat model, aseptic animals were irregularly offered chew and water twice per day for 10 min while playing at random time intervals, 6-9 times per day for 10-30 min each, a recorded sound of fighting cats either prior to, post or during availability of chew and water. Exposure of rats to unscheduled psychological stress evoked APR in terms of increased relative concentrations of Ag (α-1-acid glycoproteins; +63%), Hg (+21%) and Fb (+90%) and decreased relative concentrations of albumin (Alb; -9%) and pre-Alb (-8%) which chronically lasted for the duration of the experiment, 15 days (Table 1). The decreased APR as of the 7th day does not necessarily imply healing but rather regressed biosynthesis and/or degradation of APR proteins in response to introducing pathological injuries to psychological load. Most interesting is the observed steady increase in the concentration of serum glucose by up to 20% on day 15 of the experiment. This may be indicative of the appreciable catabolism of hepatic intracellular and skeletal proteins and glycoproteins to provide hepatic intracellular pool with free amino acids and glucose to biosynthesize defensive APR proteins. Repeated episodes of psychologically-induced low-grade inflammation are assumed to activate the innate immune system to lay type-2 diabetes and atherosclerosis, among others. By comparison, synthesis of psychological stress during the dark hours of the night expressed relatively higher APR (Table 1) which could be explained in terms of the ability of the animal's biomedical host defence to adapt to and internalise unscheduled repeated neuro-inflammations in the absence of visual manifestation of the psychological impetus. The phenomenon of internalisation of repeated psychological rage was further confirmed as APR to scheduled psychological stress regressed as of day 7 of the experiment, which suggests minimization of associated injuries allowing full expression of APR to neuro-provocation based on simple inflammation (Table 2).

Although animals used were raised and kept according to aseptic conditions, they reacted positively under various experimental conditions of psychological stimulation displaying a different spectrum of APR intensities to neuro-inflammation (Tables 1 & 2), implying concurrent association with varied degrees of injuries proportional to conditions and frequency of applied psychological stimulus.

Table 1 *Percent changes in the Area of Immunoprecipitate APR protein in Healthy Control Rats Induced by Unscheduled Psychological Stress in Daylight and Night-Time.**

Protein	Daylight				Night			
	1d	3d	7d	15d	1d	3d	7d	15d
Ag	63	234	189	153	99	290	201	145
Hg	21	36	18	12	27	45	27	15
Fb	90	150	111	81	111	189	120	72
At	51	72	72	42	63	108	102	33
Alb	-9	-15	-18	-15	-8	-15	-14	-14
Pre-Alb	-8	-12	-15	-9	-8	-9	-11	-15
Glucose	1.04	1.07	1.16	1.20	1.06	1.12	1.16	1.22

*Uncertainties are: ±-1-acid glycoproteins (Ag; 0 ± 21%); heptoglobin (Hg, ± 9%), fibrinogen (Fb, ± 25%), ±-1-antitrypsin (At, ± 12%); albumin (Alb, ± 5%); and pre-albumin (Pre-Alb, ± 6%).

Table 2 *Percent Changes in the Area of Immunoprecipitate APR protein to Healthy Control Rats Induced by Scheduled Psychological Stress in Daylight**

Protein	1d	3d	7d	15d
Ag	72	183	129	63
Hg	24	33	15	15
Fb	92	120	91	63
At	54	63	51	42
Alb	-10	-10	-15	-12
Pre-Alb	-7	-15	-11	-12
Glucose	1.04	1.08	1.19	1.15

* Footnotes to Table 1 apply.

4.2 Acute-Phase Response to Subcutaneous (s.c.) Administration of Lipolysis-Resistant Lipids:

Since SPE is not readily available to independent investigators a facile synthesis of lipolysis resistant lipid(s) was attempted. Various polyols were reacted with different proportions of a single or a mixture of fatty acids chlorides at 95°C to yield a mixture of esters. Regardless of the degree of esterification, all prepared synthetic lipids failed to undergo lypolysis with the triacylglycerol enzymatic assay medium. FAB mass spectrometry showed that all synthetic lipids underwent the same pattern of fragmentation. Mannitol di-, tri-, tetra-, and pentalaurate esters, upon reaction of 0.06 mole lauroyl chloride and 0.01 mole mannitol were produced in the proportions 2.6, 2, 1.5 and 1, respectively. A 2-fold reduced concentration of lauroyl chloride (0.03 mole) with 0.01 mole mannitol produced respective proportions of 1.35, 0.93, 1.2 and 1. When a mixture of 0.03 moles of lauroyl (L) and palmitoyl (P) chlorides were allowed to react with mannitol, the following wide proportions of ester products were obtained: (residues:

relative yield) L_2, 1.6; LP, 3.2; P_2, 1; L_3, 1.7; L_2P, 3.7; LP_2, 3.9; P_3, 1.7; L_4, 1.1; L_2P_2, 3.6; LP_3, 1.7; L_3P_2, 3.2; L_2P3, 2.3; and P_4L, 1. Based upon the relative sizes of the two acid chlorides, most likely this product distribution was influenced by a steric hindrance effect which affected the biological esterification of glycerol. This may justify the necessity of fractionation of butter oils to different fractions suitable for application in the food industry.[25-27] However, one of these butter oil fractions[25] contained a substantial proportion of long- chain saturated fatty acids associated with cardiac disease.[28]

Under the same conditions, when arabitol (0.01 mole) was reacted with equal concentrations of lauroyl and palmitoyl chlorides in two separate reactions (0.025 and 0.015 moles each), the following products were obtained: (residues: relative yields in 0.025 and 0.015 runs) L, 14/8.7; P, 15.2/8.7; L_2, 11.2/9.3; LP, 18/15.7; P_2, 8/5.7; L_3, 1.6/1; PL_2, 4/3.7; P_2L, 2.4/1.7; P_3, 1; and PL_3, in trace amount in accordance with an intuitive expectation of the product distribution.

^{13}C-NMR of these lipid mixtures supported the FAB-MS data. NMR spectrum revealed, regardless of the degree of esterification in the case of mannitol palmitate(s), the terminal methyl groups at palmitate residue had the lowest peak at δ 0.8-0.9 whereas the methylene hydrogens of C_α to the carbonyl group shifted to δ 3.7-3.9, the most acidic at δ 4.7-5.1 was the C2 hydrogen of mannitol carrying a palmitoyl residue. Methylene hydrogens of palmitoyl residue ranged between δ 1.3 and 1.5.

In view of the endless list of newly developed drugs and fat replacers such as SPE that affect blood serum lipids and lipoproteins and avoid bias, independent studies are needed.[30] Thus, the present reporting is deemed an essential prelude to future investigation of viability of hepatocytes conditioned to long-term ingestion of SPE in healthy and pathological animal models according to previously reported schemes.[19] Our preliminary results on ad libitum long-term consumption of chew supplemented with SPE, rats secreted pasty stools with concomitant depletion of gastric mucous and development of gastric inflammation.[28] Although, SPE was reported to be a safe food material which would not precipitate measurable health side effects,[31-35] individuals consuming 40 g SPE/day experienced increased excretion of fecal fat to levels observed in patients with steatorrhea caused by malabsorption syndrome.[36] As consumption of SPE can cause false positive results and erroneous diagnosis, it was recommend to health specialists to inquire into a patient's intake of fat substitute enriched diets and ensure that these diets are not being consumed excessively or substituting nutrient-dense foods that are naturally low in fat.[37,38] This prompted the basic study model of hepatic induced-APR in animal model to s.c. administration of SPE.[38] Subcutaneous administration of any of the lipolysis-resistant lipids, into rabbit model, evoked a typical APR reaction to inflammation, reduction in Alb 7±2, elevation in Ag 207±15, At 138±15, Fb 192±18, C3 proteins 60±9, and Hg 154±15% maximizing at 32 h and remaining at almost a sate of inflammation for a period of 7-days with a mild declining slope of -5 to -10% as the inflammatory stimulant was not metabolized. These results lend support to the published data pertaining to the possibility of olestra inducing-liver damage in pigs. AST decreased on the 12th week of feeding on diets enriched with 0.25-5.5% pure SPE or mixed with vitamins to increase on the 26th week.[36] Under the same conditions, ALT decreased to almost half and barely increased to 0.8-fold its initial values on the first 12th and 26th week experimentation, respectively. This is indicative of hepatic irregular pathology leading to necrosis as AST and ALT are elevated in circulating serum upon disintegration of hepatocytes. Furthermore, 8 weeks of experimentation in humans is too short to allow irrefutable conclusions. A parallel observation has been noted for liver damage induced by chronic ingestion of alcohol.[39] Re-examining same available data revealed that SPE had no significant effects on the serum concentration of triacylglycerols, total lipids and cholesterol in humans (Tables 3, 4). According to present results, within the experimental error of 10-15%, addition of 5 or 10% of mannitol palmitate polyesters did not influence the rate of *in-vitro* lipolysis or the net concentration of butter triacylglycerol. Accordingly, a parallel plausible rational can be drawn considering an average diet of a North American individual to weigh roughly 1 kg per day of which approximately 200 g is fat, it is more acceptable to assume the daily ingested 50 g SPE as the liophilic solute in the matrix of dietary lipids. SPE would be then released undigested from dietary lipid micelles without drastically changing the proportion of the absorbed dietary lipids or that of the cholesterol.

4.3 Zocor-Induced Acute-Phase Response

Similarly, i.p. administration (50 mg/kg) of saline emulsion of zocor into rabbits was found to evoke a typical APR inflammation with progressive necrosis of hepatocyte where rabbits developed focal hepatic necrosis with mononuclear and Kupffer cell aggregations, peripherolobular severe hydropic degenerated and necrotic hepatic cells and numerous hepatic fatty vacuolations with nuclear pyknosis and congestion on the 2nd, 3rd and 4th week, respectively, on ip administration saline emulsion of zocor's. There was corresponding elevation of Fb, Ag, Hg and at 300 ±25, 660$\pm$27, 200$\pm$17 and 163$\pm$23%, respectively, while Alb was reduced by 5$\pm$10%. By extrapolating the hypothesis of associating APR proteins to CHD, Zocor which is supposed to reduce levels of circulatory cholesterol, may evoke APR protein-associated CHD. Intermuscular edema in heart muscle on the 3rd week of ip administration of zocor into rabbits may suggest weakening of the heart muscle as the causative factor for CHD, not necessarily elevated blood pressure and concentration of circulating cholesterol. In brief, the more effective approach to combating obesity would be a general decrease in dietary intake of animal fat, and its replacement with vegetable and marine oils.

Table 3 *Serum Cholesterol and Triacylglycerol (mmole/L) of Individuals Fed 18 g/d SPE or placebo**

Week	Cholesterol			Triacylglycerol		
	SPE + TA	SPE	Placebo	SPE + TA	SPE	Placebo
0	5.14 $\pm$ 0.05	5.14 $\pm$ 0.14	5.20 $\pm$ 0.11	1.21 $\pm$ 0.07	1.21 $\pm$ 0.06	1.09 $\pm$ 0.07
2	4.99 $\pm$ 0.12	5.02 $\pm$ 0.12	5.39 $\pm$ 0.12	1.21 $\pm$ 0.08	1.20 $\pm$ 0.07	1.23 $\pm$ 0.07
4	5.07 $\pm$ 0.11	5.06 $\pm$ 0.13	5.29 $\pm$ 0.12	1.22 $\pm$ 0.09	1.24 $\pm$ 0.08	1.15 $\pm$ 0.06
6	5.04 $\pm$ 0.11	5.14 $\pm$ 0.13	5.34 $\pm$ 0.12	1.20 $\pm$ 0.09	1.27 $\pm$ 0.07	1.04 $\pm$ 0.06
8	5.11 $\pm$ 0.10	5.09 $\pm$ 0.12	5.35 $\pm$ 0.12	1.31 $\pm$ 0.09	1.36 $\pm$ 0.10	1.06 $\pm$ 0.06
10	5.03 $\pm$ 0.11	4.91 $\pm$ 0.11	5.17 $\pm$ 0.12	1.29 $\pm$ 0.11	1.24 $\pm$ 0.08	1.09 $\pm$ 0.07
12	4.97 $\pm$ 0.11	4.87 $\pm$ 0.11	5.29 $\pm$ 0.13	1.20 $\pm$ 0.07	1.32 $\pm$ 0.09	1.11 $\pm$ 0.06
14	4.93 $\pm$ 0.11	5.01 $\pm$ 0.12	5.27 $\pm$ 0.11	1.36 $\pm$ 0.13	1.22 $\pm$ 0.08	1.08 $\pm$ 0.06
16	4.83 $\pm$ 0.10	4.85 $\pm$ 0.11	5.14 $\pm$ 0.11	1.26 $\pm$ 0.12	1.21 $\pm$ 0.08	1.18 $\pm$ 0.09

*SPE + TA = Sucrose polyester + tocopheryl acetate. Adapted from ref. 9.

Table 4 Changes in Serum Cholesterol and Triacylglycerol (mmole/L) of Individuals Consuming Increased Concentration of SPE for 8 Wk.

SPE g/d	Wk 0	Wk 4	Wk 8
	Cholesterol		
0	4.53 ± 0.18	4.37 ± 0.23	4.40 ± 0.18
8	4.63 ± 0.18	4.24 ± 0.23	4.32 ± 0.18
20	4.63 ± 0.16	4.19 ± 0.23	4.16 ± 0.16
32	4.55 ± 0.18	4.63 ± 0.23	4.23 ± 0.18
	Triacylglycerol		
0	1.02 ± 0.10	1.04 ± 0.11	0.96 ± 0.11
8	0.98 ± 0.10	1.12 ± 0.11	1.02 ± 0.11
28	0.96 ± 0.09	1.02 ± 0.11	1.03 ± 0.11
32	1.22 ± 0.09	1.19 ± 0.11	1.25 ± 0.11

* Adapted from ref. 9.

References

1. M. Woodward, C. Morrison and K. McColl, *J. Clin. Epidemiol.*, 2000, **53**, 175-181, Leong R. W.-L., Sung J. J. Y. Review article: Helicobacter species and hepatobiliary diseases. *Aliment Pharmacol Ther* 2002, **16**, 1037-1045.
2. P. Patel, D. Carrington, D.P. Strachan, E. Leatham, P. Goggin, T.C. Northfield, M.A. Mendall, *The Lancet* 1994, **343**, 1634-1635.
3. C. Martin-de-Argila, D. Boixeda, R. Canton, J.P. Gisber, A. Fuertes, *The Lancet* 1995, **346**, 310.
4. N.V. Dhurandhar, *J. Nutr.* 2001, **131**, 2794S-2797S.
5. F.H. Mattson, R.A. Volpenhein, *J. Biol. Chem.* 1964, **239**, 2772-2777.
6. V. Surpuriya, W.I. Higuchi, *Biochem. Biophys. Acta.* 1972, **290**, 375-383.
7. R.J. Jandacek, *Drug Metab. Rev.* 1982, **13**, 695-714.
8. G.P. Rizzi, H.M. Taylor, *J. Am. Oil Chem. Soc.* 1978, **55**, 398-401.
9. Consult range of articles in *J. Nutr. 1997*; 127, 1539S-1728S specifically Schlagheck, T.G., Riccardi K.A., Zorich N.L., Torri S.A., Dugan L.D., Peters J.C. *J. Nutr.* 1997; 127: 1646S-1665S; Daher G.C., Cooper D.A., Zorich N.L., King D., Riccardi K.A., Peters J.C. *J. Nutr.* 1997; 127: 1694S-1698S and Cooper D.A., Berry D.A., Spendel V.A., Kiorpes A.L., Peters J.C. *J. Nutr.* 1997; 127:1555S-1565S.
10. C.J. Glueck, R.J. Jandacek, M.T. Ravi Subbiah, L. Gallon, R. Yunker, C. Allen, E. Hogg, P.M. Laskarzewski, *Am J Clin Nutri* 1980, **33**, 2177-2181.
11. C.J. Glueck, F.H. Mattson, R.J. Jandacek, *Am. J. Chem. Nutr.* 1979, **32**, 1636-1644.
12. R.J. Roberts, R.D. Leff, *Clin. Pharmacol. Theor.* 1989, **43**, 299-304.
13. E.G. Sletten, D. Hollander, V. Dadufalza, *Acta Vitaminol Enzymol* 1985, **7**, 49-53.
14. F.M. Fouad, W.D. Marshall, P.G. Farrell, P. Prehm, *J. Toxicol. Environ. Health* 1993, **39**, 355-74.
15. F.M. Fouad, D. Waldron-Edward, *Hoppe-Seyler's Z Physiol. Chem.* 1980, **361**, 703-713.
16. R. Ross, *N. Engl. J. Med.* 1999, **340**, 115-126.

17   P. Barath, M.C. Fishbein, J. Cao, J. Berenson, R.H. Helfant, J.S. Forrester, *Am. J. Cardiol.* 1999, **65**, 297-303.
18   A. Schumacher, I. Seljeflot, A.B. Lerkerod, L. Sommervoil, J.E. Otterstad, H. Arnesen, *Atherosclerosis* 2002, **164**, 153-160.
19   F.M. Fouad, O.A. Mamer, F. Shahidi, *Medical Hypotheses* 1996, **47**, 157-177.
20   F.M. Fouad, M. Goldberg, G. Ruhenstroth-Bauer, *J. Clin. Chem. Clin. Biochem.* 1983, **21**, 203-208; Fouad F.M., Farrell P.G., Marshall W.D., Scherer R., Ruhenstroth-Bauer G. *J. Toxicol. Environ. Health* 1992, **36**, 43-57 and Fouad F.M., Marshall W.D., Farrell P.G., FitzGerlad S. *J. Toxicol. Environ. Health* 1993, **38**, 1-18.
21   J.L. Ma, W.C. You, M.H. Gail et al. *Intl. J. Epidemiol.* 1998, **27**, 570-573.
22   A.R. Folsom, F.J. Nieto, P. Sorlie, L.E. Chambless, D.Y. Graham, *Circulation* 1998, **98**, 845-850.
23   T.C. Campbell, B. Parpia, J. Chen, *Am. J. Cardiol.* 1998, **82**, 18T-21T.
24   R. Baschetti. *Atherosclerosis* 1999, **147**, 190.
25   V.A. Amer, D.B. Kupranyez, B.E. Baker. *J. Am. Oil Chem. Soc.* 1985, **62**, 1551-1557.
26   F.M. Fouad, F.R. van de Voort, W.D. Marshall, P.G. Farrell, *J. Am. Oil Chem. Soc.* 1990, **67**, 981-988; F.M. Fouad, F.R. van de Voort, W.D. Marshall, P.G. Farrell, *J. Food Lipids* 1993, **1**, 119-141.
27   F.M. Fouad, F.R. van de Voort, W.D. Marshall, P.G. Farrell, *J. Food Lipids* 1994, **1**, 195-219.
28   F.M. Fouad, O.A. Mamer, F. Sauriol, F. Shahidi, *J. Toxicol. Environ. Health* 1998, **1**, 149-179.
29   T. Sharmanov, V. Maksimenko, G. Servetnik-Chalaia, S. Abdraimova, *Vopr Pitan.* 1988, **4**, 45-48.
30   K.C. Klontz, Measurement of Selected Fecal Parameters in Subjects Consuming Increasing Levels of Olestra (FAP 3997, vol 289) Office of Premarket Approval (HFS-216), December 26, 1995, **12**, 321-326.
31   G.S. Allgood, D.J. Kuter, K.T. Roll, S.L. Taylor, N.L. Zorich, *Regul. Toxicol. Pharmacol.* 2001, **33**, 224-233.
32   B. Cook, D. Cooper, D. Fitzpatrick, S. Smith, D. Tierney, S. Mehy, *Clin. Positron Imaging* 2000, **4**, 150.
33   R.E. Patterson, A.R. Kristal, J.C. Peters, M.L. Neuhouser, C.L. Rock, L.J. Cheskin, D. Neumark-Sztainer, M.D. Thornquist, *Arch Intern. Med.* 2000, **160**, 2600-2604.
34   J. McRorie, J. Kesler, L. Bishop, T. Filloon, G. Allgood, M. Sutton, T. Hunt, A. Laurent, C. Rudolph, *Am. J. Gastroenterol.* 2000, **95**, 1244-1252.
35   J. McRorie, S. Brown, R. Cooper, S. Givaruangsawat, D. Scruggs, G. Boring, *Ailment Pharmacol. Ther.* 2000, **14**, 471-477.
36   D.A. Cooper, J. Curran-Celentano, T.A. Ciulla, B.R. Hammond, R.B. Danis, L.M. Pratt, K.A. Riccardi, T.G. Filloon, *J. Nutr.* 2000, **130**, 642-647.
37   D. Neumark-Sztanier, A.R. Kristal, M.D. Thornquist, R.E. Patterson, M.L. Neuhouser, M.J. Barnett, C.L. Rock, L.J. Cheskin, P. Schreiner, D.L. Miller, *J. Am. Diet Assoc.* 2000, **100**, 198-204.
38   G. Blackburn, *Health News* 2000, **6**, 4.
39   E. Rubin, H. Rottenberg, *Fed. Proc.* 1982, **41**, 2465-2471.

THE FUNCTIONALITY OF BACKWHEAT SOUR JUICE

Kozo Nakamura,[1] Chiho Nakamura,[1] Shigeyoshi Maejima,[2] Masataka Maejima,[3] Michiaki Watanabe,[3] Mayumi Shiro[1] and Hiroshi Kayahara[1]

[1]Graduate School of Agriculture, Shinshu University, 8304 Minamiminowa, Nagano 399-4598, JAPAN
[2]Kaiyoubokujyou Co., Ltd., 614 Naka, Kitsuki, Oita 873-0012, JAPAN
[3]Fujikohgei Co., Ltd., 333 Akoji, Fujinomiya, Shizuoka 418-0078, JAPAN

1 ABSTRACT

A major flavonol in buckwheat grains, rutin is reported to have a function in blood pressure regulation. In the germination of buckwheat for 20 h, the content of rutin was increased around two times. We can use this germinated buckwheat with 2-3 mm bud in place of normal buckwheat. After 6-10 days germination, the buckwheat sprout grows 7-10 cm with yellowish green cotyledon and rubious hypocotyls. The sprout in 6 days contains over 10 times more rutin without negative taste than in the nongerminated one. Four flavonols are produced during the germination. Recently, a rubious juice was produced from buckwheat sprouts by lactic fermentation. This sour juice possesses several medical properties.

2 INTRODUCTION

A major flavonol in buckwheat grains, rutin is reported to have a function in blood pressure regulation. In the germination of buckwheat for 20 h, the content of rutin was increased around two times. We can use this germinated buckwheat with 2-3 mm bud in place of normal buckwheat. After 6-10 days germination, the buckwheat sprout grows 7-10 cm with yellowish green cotyledon and rubious hypocotyls. The sprout in 6 days contains over 10 times more rutin without negative taste than in nongerminated one. Four flavonols are produced during the germination.[1,2] Recently, rubious sour juice was produced by lactic fermentation of buckwheat sprouts grown to 7-10 cm within 6-10 days of germination. This "buckwheat sprout sour juice" is marketed as a food product in Japan and narratives from consumers suggest that "buckwheat sprout sour juice" possesses medical properties, such as improvement of blood flow, antiallergenic effects, promotion of hair growth, and blood pressure regulation. We analyzed flavonols containing the "buckwheat sprout sour juice" and demonstrated the blood pressure decrease by *in vitro* and *in vivo* experiments.

3 METHOD AND RESULTS

3.1 Preparation of Buckwheat Sprout Sour Juice

Buckwheat seeds were soaked in water and planted on a hydroponic bed for budding. The buckwheat sprouts were harvested at about 15 centimeters in height. After removal of the roots, the collected sprouts were washed well and subsequently crushed in a mixer. The crushed material was squeezed into a juice, which was filtered through a properly meshed filter to remove uncrushed debris. About 0.8 to 1.0 litere juice was obtained from 10 kg of buckwheat sprouts. The juice was stored in a refrigerator for 3 to 4 days at about 5 to 10°C and was allowed to undergo fermentation using lactobacillus and/or yeast. Once the fermentation process had been initiated, the juice was frozen at about −20 to −30°C. When completely frozen, the juice was thawed by placing it in a vial at room temperature. This spontaneous thawing allowed the lactobacillus and/or yeast fermentation to resume as the juice thawed. The fermentation was completed within about 10 to 15 days. The buckwheat sprout sour juice prepared by this method has rubious color, characteristic odor and sour taste (Figure 1).

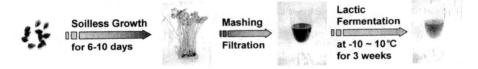

Figure 1 *Prepration method for buckwheat sprout sour*

3.2 Analysis of Flavonoids in Buckwheat Sprout Sour Juice

Buckwheat sprout contains 7-10 times more rutin and four flavonols; isoorientin, orientin, vitexin, isovitexin. In the "buckwheat sprout sour juice", although the four flavonols remained, the rutin was almost completely eliminated. Quercetin, an aglycon of rutin, and another unknown flavonol were found instead of rutin. HPLC analytical conditions as follows: column; TSK-GEL, ODS120A (150×f 4.6 mm), mobile phase; B 0-100% in 30 min (A; CH_3CN with 0.1% TFA and B; H_2O with 0.1% TFA), flow rate; 1.0 ml/min, column temperature; 40°C, detection; 380 nm (Figure 2).

3.3 Test for the Presence of Allergens

The Designated Material Detection Kit (MORINAGA) was used in conjunction with ELISA assay using FASTKIT ELISA kit (NIPPON HAM) according to the method described in the notification of Japanese Ministry of Health, Labor and Welfare, entitled "Detection method for the designated allergic food materials (egg, milk, wheat, buckwheat, and peanuts)" to test the buckwheat sprout sour juice composition for the presence of allergens. No allergic substances were detected either by the Designated Material Detection Kit (MORINAGA) or by FASTKIT ELISA assay kit (NIPPON HAM). These results indicate that the buckwheat sprout sour juice according to the present invention is free of allergens that can cause buckwheat allergy and is thus safe for consumption.

352   Food Flavor and Chemistry: Explorations into the 21st Century

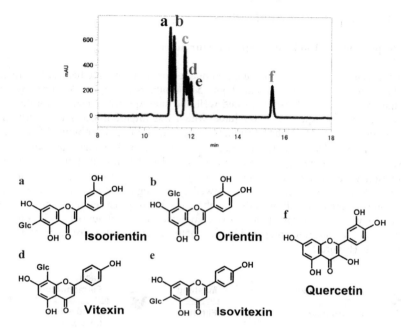

Figure 2 HPLC analysis and structural formula of flavonoids in buckwheat sprout sour juice.

2.4 Angtiotensin Converting Enzyme (ACE) Inhibition Activity of the Buckwheat Sprout Sour Juice

Increase of blood pressure in our body is regulated by ACE as shown in Figure 3. Inhibition of ACE reduce the production of angiotensin II which is the major compound to increase blood pressure and does not reduce the bradykinin which is one of compound to decrease blood pressure.

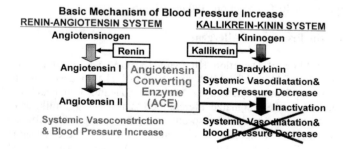

Figure 3 Basic mechanism of blood pressure

The ACE inhibitory activities of samples were measured as follows. The buckwheat sprout sour juice was extracted with distilled water at room temperature. The extract was diluted 1.5, 2, 2.5, 5, and 7 fold. The ACE inhibitory activity of each dilution was assayed by the method of Cushman.[3] The 30 μl sample solution was mixed with 250 μl of 7.6 mM Hip-His-Leu dissolved in 0.1 M sodium phosphate buffer (pH 8.3) containing 0.608 M NaCl. The mixture was incubated for 5 min at 37°C. The reaction was initiated by addition of 100 μl of 6 mU ACE, and the mixture was incubated for 30 min at 37°C. The reaction was stopped by addition of 250 μl of 1N HCl. The hippuric acid liberated by ACE was extracted with ethyl acetate. The mixture of was concentrated for HPLC analysis. The concentration of component that inhibits 50% of ACE activity (IC_{50}) was expressed by plotting the content of hippuric acid and the percent of ACE inhibitory (Figure 4).

Undiluted buckwheat sprout sour juice aged 2 month and 10 month inhibited around 90% of ACE activity. And 50% ACE inhibitory activity of the 2 month aged and 10 month aged samples were 2 fold and 4 fold diluted solutions, respectively. Overall, the buckwheat sprout sour juice has a strong ACE inhibitory activity and long term aging of the buckwheat sprout sour juice decrease the activity.

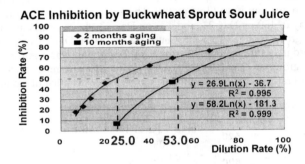

Figure 4 *ACE inhibition test results for buckwheat sprout sour*

3.5 The Effects of Buckwheat Sprout Sour Juice on Blood Pressure of SHR Rat

It has been reported that long-term administration of fructose to rats brings about an increase in their blood pressure and body fat.[4] In view of this, we examined the effects of the buckwheat sprout sour juice on disease conditions caused by excessive sugar intake and high blood pressure. The animal experiment was carried out by breeding SHR rats for 43 days.

3.5.1. Test Animal and Test Groups. SHR/Izm rats, aged 7 weeks, were purchased from a breeder (Japan SLC, Nakaizu branch and Inasa branch) and were acclimatized for 10 days. Subsequently, the 8-week-old rats were divided into the following groups for testing: standard group; SHR rats (6 rats/group) were given tap water and were allowed to feed freely on CE-2 powder feed (Japan Clea), control group; SHR rats (6 rats/group) were given a 25wt% fructose solution prepared by dissolving fructose in tap water and were allowed to feed freely on the powder feed, test group; SHR rats (6 rats/group) were given a 25wt% fructose solution prepared by dissolving fructose (Wako Pure Chemical Industries) in tap water and were allowed to feed freely on a mixture of the feed with 1wt% of the freeze dried buckwheat sprout sour juice.

3.5.2. Experimental technique/information. The freeze dried buckwheat sprout sour juice was mixed with the feed and was orally administered. The animals in each group were allowed to feed freely on the respective feed every day for 6 weeks and drink as much water (tap water or fructose solution) as they want. The feed and water were supplied on Day 0, Day 5, Day 8, Day 12, Day 15, Day 19, Day 22, Day 26, Day 29, Day 33, Day 36, Day 40, and Day 43. Heart rate and blood pressure (systolic, diastolic, and mean pressures) of each animal were measured by a noninvasive technique (tail-cuff method) on Day 0, Day 21, and Day 42 using a noninvasive auto-sphygmomanometer (BP-98A, Softron Co., Ltd.). Rats were immobilized in a restrainer, and temperature was maintained at 37°C. Measurements were taken over a 5- to 15-minute period during which the temperature was kept constant.

3.5.3. Experimental Results. The variation in heart rate is shown in Figure 5. As can be seen, a significant variation was observed in the heart rate of the group administered with the buckwheat sprout sour juice (Test group). The average heart rate in this group was 439 (per min) on Day 0, 460 on Day 21, and 473 on Day 42. In comparison, the average heart rate in Control group was 453 (per min) but rose to 494 on Day 21 and further to 501 on Day 42. Accordingly, the average heart rate of Test group was significantly lower than that of Control group on Day 21 (**P<0.05). While the difference was insignificant on Day 42, the heart rate of Test group still tends to be lower than that of Control group.

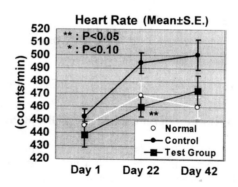

Figure 5 *Results of animal experiment of Heart Rate.*

Variations in systolic blood pressure, diastolic blood pressure, and mean blood pressure are shown in Figure 6, respectively. As can be seen from the results shown in the figures, no significant difference was observed between Standard group and Test group in any of systolic blood pressure, diastolic blood pressure, and mean blood pressure. In comparison, the systolic blood pressure was significantly lower in Test group than in Control group on Day 21 (*p<0.05). Also, the diastolic blood pressure and the mean blood pressure were significantly lower in Test group than in Control group on Day 21 and Day

42 (**p<0.05, *p<0.10). These results indicate that the "buckwheat sprout sour juice" has beneficial effect on heart rate and hypertension of SHR rat in the disease conditions caused by excessive sugar intake.

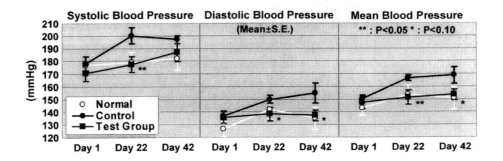

Figure 6 *Results of animal experiment of Blood Pressure*

4 CONCLUSION

"Buckwheat sprout sour juice" with rubious color, characteristic odor and sour taste acts as an efficient ACE inhibitor and a blood pressure regulation effect was demonstrated in animal experiments. In addition, a heart rate decrease effect of "buckwheat sprout sour juice" was also demonstrated by animal experiments. These results suggest that the buckwheat extract composition of the present invention has an ability to reduce some of the risk factors for life style-related diseases, including high blood pressure and apoplectic stroke.

References

1. M. Watanabe, Y. Ohshita and T. Tsushida, *J. Agric. Food Chem.*, 1997, **45**, 1039.
2. M. Watanabe, *J. Agric. Food Chem.*, 1998, **46**, 839.
3. D.W. Cushman, H.S. Cheung, E.F. Sabo and M.A. Ondetti, *Biochem.*, 1977, **16**, 5484.
4. Y. Morimoto, M. Sakata, A. Ohno, T. Maegawa and S. Tajima, *Folia Pharmacol. Jpn.*, 2001, **117**, 77.

FUNCTIONALITY ENHANCEMENT IN GERMINATED BROWN RICE

Kozo Nakamura,[1] Su Tian,[2] and Hiroshi Kayahara[1]

[1] Graduate School of Agriculture, Shinshu University, 8304 Minamiminowa, Nagano 399-4598, JAPAN
[2] United Graduate School of Agriculture Sciences, Gifu University, Gifu 501-1193, JAPAN.

1 ABSTRACT

"Germinated whole rice" is brown rice soaked in water just until it begins to bud. During the germination, enzymatic hydrolysis of starch and protein produces sweet sugars and flavorful amino acids respectively. And decomposition of the outer layer of bran makes its texture softer. So we enjoy germinated whole rice as our staple in much the same cooking method as white rice. The germination makes up several defects of brown rice as a daily food, and also adds effective functionality to it. We studied phenolic compounds in germinated whole rice. Phenolic compounds including two novel compounds in rice were identified and their contents were determined. Increase of phenolic compounds in germinated whole rice was directly connected to the enhancement of functionality.

2 INTRODUCTION

"Germinated brown rice (GBR)" is brown rice soaked in water just until it begins to bud. During the germination, enzymatic hydrolysis of starch and protein produces sweet sugars and flavorful amino acids respectively. And decomposition of the outer layer of bran makes its texture softer. So we enjoy GBR as our staple in much the same cooking method as white rice. The germination makes up several defects of brown rice as a daily food, and also adds effective functionality to it. We studied phenolic compounds in GBR. Phenolic compounds are known to reduce the risk of coronary heart disease and cancer.[1,2] Eating grain including rice is an excellent way to take phenolic compounds.[3] Phenolic compounds are found in rice bran layer and are lost in the process of threshing and polishing. GBR with whole bran layer is expected to be a good source of phenolic compounds. Eleven phenolic compounds, including two new compounds in rice, were identified and their structures were determined.[4] We carried out quantitative analysis during germination and considered the relationship between the increase of phenolic compounds in GBR and the enhancement of functionality of rice, such as superoxide scavenging activity and tyrosinase inhibitory activity.

3 METHODS AND RESULTS

3.1 Preparation of GBR and Rice Flour

Rice is one of the major staple cereals all over the world, especially in Asia. There are two types of rice, *Japonica* and *Indica*. *Japonica* is round short grain and becomes sticky after being cooked. *Japonica* is suitable for sushi and other healthy Japanese foods.

Indica is long grain and is less sticky than *Japonica*. We used No1 Japonica rice Koshihikari for this research. Brown rice (*Oryza* L., koshihikari) was grown in Nagano, Japan, and white rice was obtained by polishing the brown rice using a household rice-cleaning mill. Germination was carried out in a germination appliance according to the following procedure: 400 g of brown rice was cleaned and soaked in 2000 ml water at 32°C for 21 h, until a 0.5 to 1 mm bud developed (Figure 1). The lyophilized and ground GBR was sieved to 42 mesh. Brown rice and white rice were also cleaned, lyophilized and powdered.

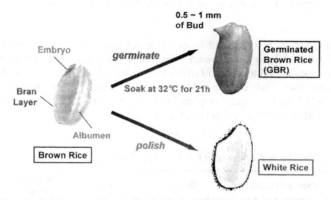

Figure 1 *Preparation of germinated brown rice (GBR)*

3.2 Comparison of Nutrients in GBR and in Brown Rice

White rice is our chief diet and is usually boiled to eat using a commercial rice cooker. Cooking of white rice is easy and it is finished in around 40 minutes. Cooked white rice is smooth, soft and has a plain taste, so it is goes well in any dish. However, the nutritional value and food functionality of white rice are very low because the nutritious bran layer and embryo are taken off. Brown rice has a whole bran layer and embryo. The nutritional value is much higher than that of white rice. Cooked brown rice has a chewy texture and often its characteristic smell spoils our appetite. We can soften brown rice using a pressure cooker, but this appliance is expensive and not common. Hence the majority give up cooking and eating brown rice. GBR overcame the problem. It can be cooked in an ordinary rice cooker in much the same way as white rice and is soft enough to chew even for children. Besides, this GBR is much more nutritious than brown rice (Figure 2).

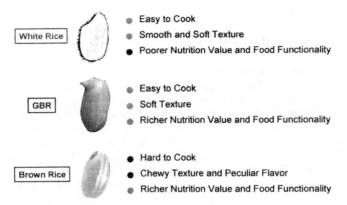

Figure 2 *Characteristics of white rice, brown rice and GBR*

Nutritional values in GBR and brown rice are listed in Table 1. GBR is lower in calories and carbohydrate than brown rice. Nutrients with food functionality, lipids, dietary fiber, vitamins, minerals and GABA are increased in GBR. The GABA content is 10 times more than that of brown rice. GABA affects blood pressure regulation, improvement of blood flow and acts as an inhibitory neurotransmitter in our brain. This is one of the benefits of GBR as a functional food.

Table. 1 *Comparison of Nutrients in GBR and in Brown Rice* [5]

Nutrient (120g)		Brown Rice	GBR	Nutrient Index
Energy	(cal)	409	330	0.8
Protein	(g)	6.5	6.5	1
Lipids	(g)	1.1	2.6	2.4
Carbohydrate	(g)	93.2	70.1	0.8
Dietary fiber	(g)	0.7	2.6	3.7
Vitamin E	(mg)	0.4	1.6	4.0
Vitamin A	(mg)	0.10	0.30	3.0
Calcium	(mg)	6	101	1.7
Magnesium	(mg)	30	98	3.3
γ-Aminobutyric (GABA)	(mg)	1.2	12.0	10.0

3.2 Analysis of Phenolic Compounds in GBR

In our HPLC analysis of an ethanol extract of rice, two larger unknown peaks were observed. We isolated and determined these unknown compounds **1** and **2**.

3.2.1 Identification of Compound 1 and Compound 2.[1] From 820 g of rice bran, 16.3 mg of compound **1** and 5.4 mg of compound **2** were isolated. Acid hydrolysis of compound **1** or **2** gave mixtures in 1: 1: 1 ratio of ferulic acid: D-glucose: D-fructose or in 1: 1: 1 ratio of sinapinic acid: D-glucose: D-fructose, respectively. The cationic ESI-MS spectrum showed a $[M + Na]^+$ peak at m/z 541 and fragment ions at m/z 177 $[M - 2\ \text{hexose} + H]^+$ for compound **1** or three major peaks at m/z 571 $[M + Na]^+$, m/z 369 [M-hexose $+H]^+$, and m/z 207 [M-2 hexose $+H]^+$ for compound **2**. Based on the spectroscopic data from ^{1}H-, ^{13}C- and 2D-NMR, compound **1** and **2** were confirmed to be 6'-*O*-(*E*)-feruloylsucrose and 6'-*O*-(*E*)-sinapoylsucrose (Figure 3). These chemical structures are consistent with the literature,[6] although this is the first identification of compounds **1** and **2** from rice.

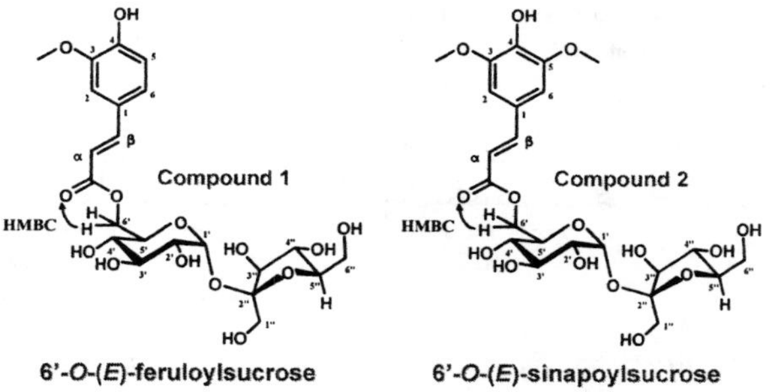

Figure 3 *Structures of isolated Compounds 1 and 2*

3.2.2 Extraction of Soluble Phenolic Compounds. Soluble phenolic compounds in white rice, brown rice, and GBR flour (5 g each) were extracted with 70% ethanol (4 × 50 ml, 10 min each) according to the method of Ayumi et al.[7] Each extract was pooled and evaporated to about 10 ml at 30°C under reduced pressure and was then lyophilized to dryness. Dry samples were dissolved in 2 ml of 15% methanol and subjected to HPLC analysis.

2.2.3 Extraction of Insoluble Phenolic Compounds. Two germs of white rice, brown rice and GBR flour were extracted with hexane (4 × 50 ml) and 70% ethanol (4 × 50 ml) in order to remove fat and soluble phenolic acids, respectively. The residue was hydrolyzed with 1M sodium hydroxide (2 × 100 ml, 2 h each) at room temperature with stirring under nitrogen.[8] The clear supernatants were pooled and acidified with 4 N HCl to pH 1, and were then extracted four times with ethyl acetate (200 ml each). The ethyl acetate fractions were evaporated to dryness, and the phenolic acids were dissolved in 5 ml of 15% methanol and analyzed by HPLC.

2.2.4 HPLC Analytical Condition. Phenolic compounds in white rice, brown rice, and GBR were estimated by reversed-phase HPLC. A 20-μl aliquot of sample solution was separated using a Shimadzu HPLC system equipped with a diode array detector on a C_{18}-ODS analytical column (Waters). The mobile phase consisted of acetonitrile (B) and purified water with 0.1% trifluoroacetic acid (TFA) at a flow rate of 0.8 ml/min. Gradient elution was performed as follows: from 0 to 5 min, linear gradient from 5% to 9% solvent

B; from 5 to 15 min, 9% solvent B; from 15 to 22 min, linear gradient from 9% to 11% solvent B; and from 22 to 35 min, linear gradient from 11% to 18% solvent B. Column temperature was set at 40 °C. Hydroxybenzoates were detected at a wavelength of 280 nm and hydroxycinnamates at 325 nm. Nine phenolic compounds were identified by comparing their relative retention times and UV spectra with authentic compounds.

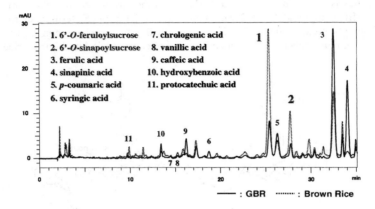

Figure 4 *HPLC chromatogram of phenolic compounds in brown rice and GBR*

2.2.5 Soluble Phenolic Compounds Analysis. Typical HPLC chromatogram traces of soluble phenolic compounds in brown rice, and GBR are shown in Figure 4. The content of soluble phenolic compounds in brown rice and GBR is summarized in Figure 5. Our results showed that soluble phenolic compounds contained free phenolic acids and hydroxycinnamate sucrose esters. Feruloylsucrose and sinapoylsucrose were found to be the major soluble phenolic compounds in brown rice, along with ferulic acid. During germination, an approximately 70% decrease was observed in feruloylsucrose and sinapoylsucrose content, while free ferulic acid content increased, becoming the most abundant phenolic compound. Moreover, the content of sinapinic acid in GBR increased to nearly 10 times as much as that in brown rice. Levels of other soluble phenolic compounds identified were generally low. With regard to total content of soluble phenolic compounds, brown rice and GBR were found to have significantly higher levels than white rice. Antioxidant effects of hydroxycinnamates depend on their available absorption in the gut.[9] Zhao et al.[10] reported that the form of hydroxycinnamates in the diet has an effect on its absorption. Free ferulic acid was absorbed almost completely, while one of the conjugated phenolic compounds with a relatively simple structure present in cereal grains, feruloylarabinose, showed lower bioavailability in rat than free ferulic acid. The increase in free phenolic compound content in GBR suggests that it has a higher potential bioavailability and thus a higher antioxidant potential.

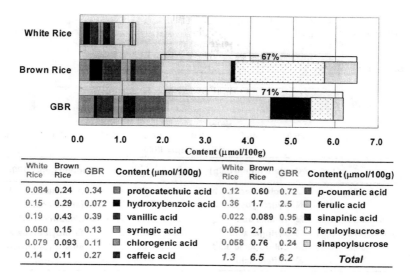

Figure 5 Contents of soluble phenolic compounds in white rice, brown rice and GBR

2.2.6 Insoluble Phenolic Compounds Analysis. The content of insoluble phenolic compounds which are bound to polysaccharides in the cell wall, of white rice, brown rice, and GBR is summarized in Figure 6. The content of insoluble phenolic compounds in rice was found to be significantly higher than that of soluble phenolics, and that in GBR and brown rice was substantially higher than that in white rice. This could be because phenolic compounds are mostly located in the bran layer of rice grains. The most abundant insoluble phenolic compound in white rice, brown rice and GBR was ferulic acid, followed by *p*-coumaric acid, and the content of other bound phenolic compounds was lower. This result is in agreement with the literature.[7,11] In GBR, a 1-2-fold increase in all insoluble phenolic compounds was observed when compared with brown rice. In the cell wall, phenolic compounds, particularly hydroxycinnamates, are ester linked to insoluble fiber, polysaccharides and lignin components.[12] The increase in phenolic compounds in GBR could be explained as an increase in the free forms with hydrolysis, due to dismantling of the cell wall during germination. The increase in insoluble phenolic compounds may increase the availability of hydrolyzable insoluble phenolic compounds during germination of brown rice.

3.3 Taste of Rice Phenolic Compounds

We investigated the taste of the commercially 9 available phenolic compounds. Purity of all chemicals was over 98%. Results of our taste examinations showed that astringency or bitter were the main tastes in 32 μM solutions. Their threshold values are very low, and the taste intensity increases gradually with concentration (Table. 2). We also tasted weak sweetness in the hydroxycinnamates solutions. The astringency of phenolic compounds and their low threshold values were reported in Ref. 13.

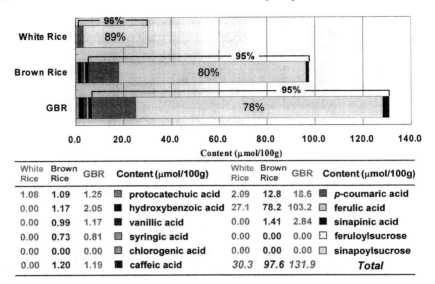

Figure 6 *Contents of insoluble phenolic compounds in white rice, brown rice and GBR*

Then we prepared sample solutions containing 9 phenolic compounds in the same ratio as white rice, brown rice and GBR. The phenolic contents are shown in Figure 7. Phenolics concentration of white rice, brown rice and GBR samples were 1.2, 3.6 and 5.4 µM respectively. The concentration of the white rice phenolics solution was below the threshold of each phenolic compound. But the mixed solution did have a taste. The concentrations of brown rice and GBR phenolic solutions were near the phenolics' threshold, so these had a weak astringent flavor. Synergism of phenolics flavor was reported in Ref. 14. In our study, the thresholds of these mixed solutions were lower than that of each phenolic solution. The taste intensity was not so strong, and so the effect of the synergism of rice phenolics is estimated to be weak. Basedon these taste examinations, we concluded that the effect of negative tastes from phenolic compounds in GBR were limited.

Table 2 *Tastes of Rice Phenolic Compounds*

	Taste (32μM)	Threshold (μM)	(ppm)
▨ protocatechuic acid	ASTRINGENCY/BITTER	3.9	20 [13]
■ hydroxybenzoic acid	ASTRINGENCY	3.9	40 [13]
■ vanillic acid	ASTRINGENCY	2.6	30 [13]
▨ syringic acid	ASTRINGENCY	2.6	240 [13]
▨ chlorogenic acid	ASTRINGENCY/BITTER	2.6	-
■ caffeic acid	ASTRINGENCY/SWEET	2.6	90 [13]
▨ *p*-coumaric acid	ASTRINGENCY/SWEET	2.6	40 [13]
☐ ferulic acid	ASTRINGENCY/SWEET	2.6	90 [13]
■ sinapinic acid	ASTRINGENCY/SWEET	2.6	Insolv. [13]

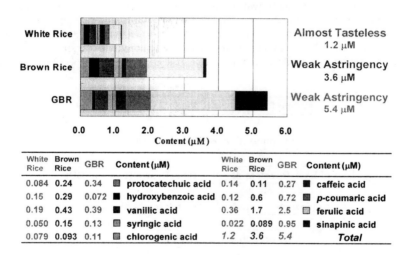

White Rice	Brown Rice	GBR	Content (μM)	White Rice	Brown Rice	GBR	Content (μM)
0.084	0.24	0.34	▨ protocatechuic acid	0.14	0.11	0.27	■ caffeic acid
0.15	0.29	0.072	■ hydroxybenzoic acid	0.12	0.6	0.72	▨ *p*-coumaric acid
0.19	0.43	0.39	■ vanillic acid	0.36	1.7	2.5	☐ ferulic acid
0.050	0.15	0.13	▨ syringic acid	0.022	0.089	0.95	■ sinapinic acid
0.079	0.093	0.11	▨ chlorogenic acid	*1.2*	*3.6*	*5.4*	**Total**

Figure 7 *Tastes of rice phenolic compounds solutions of white rice, brown rice and GBR*

3.4 Superoxide Scavenging Activity of GBR

Active oxygen species are always generated in our body. Around 2% of indrawn oxygen is converted to active oxygen species. Active oxygen species have strong toxicity and may severely injure our cell structure (Figure 8). We have enzymes to eliminate these active oxygen species but they are not perfect. Antioxidant compounds are effective and essential for our health. Phenolic compounds have potent antioxidant properties and free radical scavenging capabilities.[15,16]

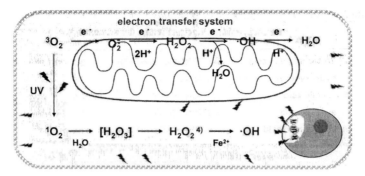

Figure 8 *Production of active oxygen species in cell*

3.4.1 Preparation of Rice Methanol Extract and Superoxide Scavenging Activity Assay. 10 g of germinated whole rice, brown rice and white rice flours were extracted three times with 200 ml of methanol for 24 hours at room temperature with stirring. The extracted mixture was then filtered and the filtrate was evaporated to dryness at 30°C under vacuum. The residue was used for the examination of scavenging activity of superoxide (Figure 9). The average extracts of white rice, brown rice and GBR were 109.6, 326.7 and 269.6, respectively.

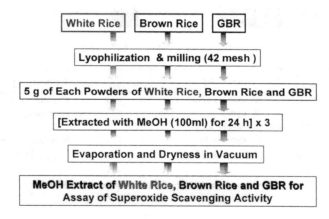

Figure 9 *Preparation of samples for assay of superoxide scavenging activity*

In order to determine if changes of phenolic acids in germinated whole rice have an effect on scavenging activity of superoxide anion radical, the scavenging activity IC_{50} of methanol extracts was measured with the method of cytochrome c. Increased absorbance at 550 nm with the reduced cytochrome c was measured and the scavenging activity was found in the formula in Table 3. The results are shown in Figure 10. Every extract had superoxide scavenging capability, and the IC_{50} values of GBR, brown rice and white rice were 4.2, 5.0, 5.1 mg/ml, respectively. Among the three rice extracts, GBR extract had the most efficient superoxide scavenging capability. Interestingly, superoxide scavenging capabilities of brown rice and white rice extracts were the same. The capability of GBR extract was 1.3 times stronger than that of brown rice and white rice.

Table 3 *Measurement Procedure of Superoxide Scavenging Activity*

Regents	Blank (ml)	Sample (ml)	Sample Control (ml)
0.6mM EDTA-2Na Phosphate Buffer (pH7.8)	0.5	0.5	0.5
0.3mM H_2O	0.5	0.5	0.5
	1.0	1.0	1.3
Sample in DMSO	—	0.3	0.3
DMSO	0.3	—	—
Incubation at 25°C for 20 min			
Cytochrome c	0.5	0.5	0.5
Xanthine oxidase	0.3	0.3	—
Measurement of the increase of absorbance at 550.0nm at 25°C for 1 min (ΔA)			

Scavenging Activity(%) = $[1 - (\Delta As - \Delta As')/ \Delta Ab]$

ΔAb: ΔA of Blank
ΔAs: ΔA of Sample
$\Delta As'$: ΔA of Sample Control

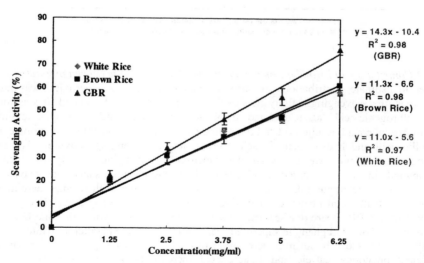

Fig. 10 *Superoxide scavenging activity of MeOH extracts of rice samples*

When we estimate superoxide scavenging capability of rice samples, the amount of the extract must considered, in addition to IC_{50}. The rice superoxide scavenging indexes in Table 4 shows estimated superoxide scavenging capability of each rice sample. Even though IC_{50} of white rice extract and brown rice extract are the same level, the rice superoxide scavenging index of brown rice was 3 times stronger than that of white rice. The extract of GBR was more efficient than brown rice extract, and the IC_{50} of GBR extract was smaller than that of brown rice extract, so their rice superoxide scavenging indexes are the same. In conclusion, we could receive the benefit of superoxide scavenging activity by eating a bowl of brown rice or a bowl of GBR instead of 3 bowls of white rice.

Table 4 *Comparison of Superoxide Scavenging Activity in Rice Samples*

	Extract* (mg/10g)	Extract Index**	IC_{50} (mg/ml)	Rice Superoxide Scavenging Index***
White Rice	109.6 ± 3.4	1.0	5.1	1.0
Brown Rice	326.7 ± 14.4	3.0	5.0	3.0
GBR	269.6 ± 22.8	2.5	4.2	3.0

* n = 4   ** Means the ratio of extract (standard; white rice = 1.0)
*** Means the ratio of Extract Index/IC_{50} (standard; white rice = 1.0)

3.4.2 Contribution of Phenolic Compounds to the Rice Superoxide Scavenging Activity. In order to estimate contributions of phenolic compounds to the superoxide scavenging activities, the scavenging activity IC_{50} of 9 rice phenolics were measured. It was found that chlorogenic acid, caffeic acid and sinapic acid showed stronger scavenging activities. Protocatechuic acid, ferulic acid and *p*-coumaric acid showed moderate, and syringic acid, vanillic acid and hydroxybenzoic acid showed lower scavenging activities (Table 5). Using these results, we estimated the superoxide scavenging activity of phenolic compounds in rice. The X axis in Figure 11 shows the phenolic superoxide scavenging index using concentration/IC_{50}. A combined area of phenolics with stronger activities is very high in the total phenolic superoxide scavenging index. Sinapic acid is especially important in GBR superoxide scavenging activity. In the case of white rice and GBR, the proportion of rice and phenolic superoxide scavenging index are parallel. In white rice and GBR, phenolics can be an important group of superoxide scavengers. We can say that efficient phenolics, caffeic and sinapic acid increase is one of the reason for the enhancement of superoxide scavenging activity in GBR.

Table 5 *Superoxide Scavenging Activities of Rice Free Phenolic Compounds*

IC_{50} (mM)*		IC_{50} (mM)	
1.7	chlorogenic acid	30.5	ferulic acid
2.0	caffeic acid	60.8	syringic acid
2.4	sinapinic acid	84.8	vanillic acid
15.4	protocatechuic acid	110	hydroxybenzoic acid
29.2	p-coumaric acid		

* Measurement results by cytochrome c method

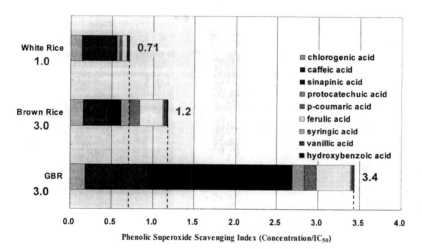

Fig. 11 *Phenolic superoxide scavenging index in white rice, brown rice and GBR*

3.5 Tyrosinase Inhibitory Activity of GBR

We investigated tyrosinase inhibitory activities of phenolics in white rice, brown rice and GBR. Many studies have reported that phenolic compounds inhibit tyrosinase. Tyrosinase is an enzyme to synthesize black melanine as shown in Figure 12. We expect that GBR has a whitening effect by inhibiting tyrosinase activity.

Methanol extracts of white rice, brown rice and GBR prepared in the manner described in Figure 9 were used for the tyrosinase inhibitory test. The increase of absorbance at 475 nm with produced quinones was measured, and the tyrosinase inhibitory activity was found in the formula in Table 6.

Figure 13 shows the results of Tyrosinase inhibitory test for the rice extracts. Every rice extract had tyrosinase inhibitory activity. IC_{50} values of GBR, white rice and brown rice were 5.7, 9.2, 16.8 mg/ml, respectively. GBR extract had the most efficient tyrosinase inhibitory capability. Tyrosinase inhibitory capability of white rice extract came next and brown rice extract last. The capability of GBR extract was 1.6 times stronger than that of white rice and 3 times stronger than that of brown rice.

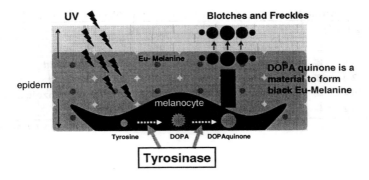

Figure 12 *The main role of tyrosinase in the synthesis melanine*

Table 6 *Measurement Procedure of Tyrosinase Inhibitory Activity*

Regents	Blank	Blank Control	Sample	Sample Control
McIlvaine Buffer (Na$_2$PO$_4$-Citrate, pH6.8)	0.9	0.9	0.9	1.9
0.3mM Tyr *	1.0	1.0	1.0	—
Sample in 5% DMSO	—	—	1.0	1.0
DMSO	1.0	1.0	—	—
Incubation at 30°C for 10 min				
Tyrosinase*	0.1	—	0.1	0.1
1 M NaN$_3$ aq.	—	1.0	—	—
Incubation at 30°C for 20 min				
Tyrosinase*	—	1.0	—	—
1 M NaN$_3$ aq.	0.1	—	0.1	0.1
Measurement of absorbance at 475nm at 25°C				

Inhibitory Activity(%) = $[1 - (S - Sc)/(B - Bc)] \times 100$

B: Blank
Bc: Blank Control
S: Sample
Sc: Sample Control

* They were dissolved in McIlvaine Buffer (pH6.8)

Based on the results of the tyosinase inhibitory test for the rice extracts, we estimate the rice tyrosinase inhibitory index in a similar manner as the superoxide scavenging index (Table 7). Even though IC$_{50}$ of brown rice extract is the most inefficient among them, the tyrosinase inhibitory index was bigger than that of white rice, because there was three times more extract. Tyrosinase inhibitory index of GBR and brown rice were 4.0 and 1.6. Judging from this point, we could say GBR is the best. In conclusion, we can expect the same level of tyrosinase inhibition by using 1 bowl of GBR, 2.5 bowls of brown rice and 4 bowls of white rice.

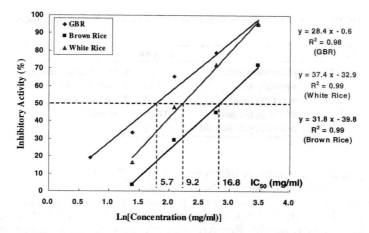

Figure 13 *Tyrosinase inhibitory activities of methanol extract of white rice, brown rice and GBR*

Table 7 *Tyrosinase Inhibitory Activities of Rice Free Phenolic Compounds*

	Extract* (mg/10g)	Extract Index**	IC$_{50}$ (mg/ml)	Rice Tyrosinase Inhibitory Index***
White Rice	109.6±3.	1.0	9.2	1.0
Brown Rice	326.7±14.	3.0	16.8	1.6
GBR	269.6±22.	2.5	5.7	4.0

* n = 4   ** Means the ratio of extract (standard; white rice = 1.0)
*** Means the ratio of Extract Index/IC$_{50}$ (standard; white rice = 1.0)

The nine rice phenolics had tyrosinase inhibitory capability and 50% inhibitory concentration listed in Table 8. Their activities were categorized as strong, moderate and low. Sinapic acid and chlorogenic acid have not only stronger superoxide scavenging activities, but also stronger tyrosinase inhibitory activities. GBR contains more efficient sinaptic acid than both white rice and brown rice. Ferulic acid is larger, but it had a relatively lower tyrosinase inhibitory capability.

Table 8 *Tyrosinase Inhibitory Activities of Rice Phenolic Compounds*

IC$_{50}$ (mM)		IC$_{50}$ (mM)	
0.25	p-coumaric acid	1.86	hydroxybenzoic acid
0.37	syringic acid	1.42	caffeic acid
0.49	sinapinic acid	6.14	vanillic acid
0.51	chlorogenic acid	12.6	ferulic acid
1.02	protocatechuic acid		

We estimated the tyrosinase inhibitory activity of phenolic compounds in rice. The X axis in the graph on Figure 14 shows phenolic tyrosinase inhibitory index using concentration/IC$_{50}$. As you can see, area of phnolics with stronger activities, p-coumaric and sinapic acid, are very high in total tyrosinase inhibitory index and both are important for tyrosinase inhibitory activity. Rice tyrosinase inhibitory indexes are shown on the left side in the Figure. The proportion of rice and phenolic tyrosinase inhibitory index are parallel. Phenolics can be an important group of tyrosinase inhibitors. We can say that rice phenolics play an important role to inhibit tyrosinase in rice, and efficient tyrosinase inhibiting phenolic, sinapic acid increase is one of the reasons for the enhancement of tyrosinase inhibitory activity in GBR.

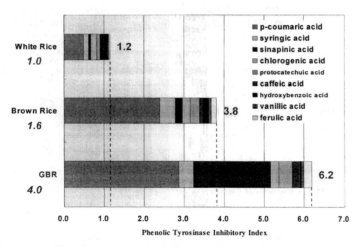

Figure 14 *Phenolic tyrosinase inhibitory index in white rice, brown rice and GBR*

4 CONCLUSION

Soluble and insoluble phenolic compounds including newly isolated 6'-O-(E)-feruloylsucrose and 6'-O-(E)-sinapoylsucrose from white rice, brown rice and GBR were quantitatively analyzed by HPLC. The results showed that the content of insoluble phenolic compounds was significantly higher than soluble phenolics in rice, while almost all compounds identified in GBR and brown rice were more abundant than those in white rice. The two sucrose esters were found to be the major soluble phenolic compounds in brown rice. During germination, their decrease was observed, while the two hydroxycinnamate contents increased significantly. The content of sinapinic acid increased to nearly 10 times as much as that in brown rice. The total content of insoluble phenolic compounds increased in GBR. As the result of an increase in free phenolic compounds, superoxide scavenging activity and tyrosinase inhibitory activity were increased in GBR. Thus germination of brown rice can be a method to improve its health-related benefits and make it more nutritious, functional and flavorful.

References

1. I. Martinez-Valverde, M.J. Periago and G Ros, *Arch. Latinoam. Nutr.* 2000, **50**, 5-18.
2. H.L. Newmark, *Adv. Exp. Med. Biol.* 1996, **401**, 25-34.
3. A. Scalbert and G. Williamson,. *J. Nutr.* 2000, *130*, 2073-2085.
4. S. Tian, K. Nakamura and H. Kayahara, *J. Agric. Food Chem.*, in press.
5. H. Kayahra, K. Tsukahara and T. Tatai, In *Food Flavors and Chemistry: Advances of the New Millennium*, (Eds. by A. H. Spanier, F. Shahidi, T. H. Parliament, C. Mussian, C-H. Ho and E. T. Contis), pp. 546-551, Athenaueun Press, 2001.
6. T. Miyase, H. Noguchi and X.M. Chen, *J. Nat. Prod.* 1999, **62**, 993-996.
7. H. Ayumi, M. Masatsune and H. Seiichi, *Food Sci. Technol. Res.* 1999, **5**, 74-79.
8. E. Nordkvist, A.C. Salomonsson and P. Aman, *J. Sci. Food Agric.* 1984, **35**, 657-661.
9. P.A. Kroon and G. Williamson, *J. Sci. Food Agric.* 1999, **79**, 355-361.
10. Z.H. Zhao, Y. Egashira and H. Sanada, *J. Agric. Food Chem.* 2003, **51**, 5534-5539.
11. M. Bunzel, E. Allerdings, V. Sinwell, J. Ralph and H Steinhart, *Eur. Food Res. Technol.* 2002, **214**, 482-488.
12. C.B. Faulds and G. Williamson, *J. Sci. Food Agric.* 1999, **79**, 393-395.
13. J. A. Maga and K. Lorenz,. *Cereal Science Today*, 1973, **18 (10)**, 326-328, 350.
14. M. Naczk, R. Amarowicz, and F. Shahidi, In *Developments in Food Scinece*, 597-613.
15. A.D. Wentworth, L.H. Jones, P. Wentworth, Jr., K.D. Janda and R. Lerner, *PNAS*. 2000, **97**, 10930–10935.
16. F. Shahidi and P.K. Wanasundara, *Crit. Rev. Food Sci. Nutr.* 1992, **32**, 67-103.

Quality

COMPARISON OF THE INFLUENCE OF HYDRODYNAMIC PRESSURE (HDP) – TREATMENT AND AGING ON BEEF STRIPLOIN PROTEINS.[§]

A.M. Spanier[1,3], T.M. Fahrenholz[2], E.W. Paoczay[2] and R. Schmukler[3]

[1] Spanier Science Consulting LLC, 14805 Rocking Spring Drive, Rockville, MD 20853, USA
[2] USDA, ARS, BA, Animal & Natural resources Institute (ANRI), Food Technology and Safety Laboratory (FTSL), Bldg 201, BARC-East, Beltsville, MD 20705, U.S.A.
[3] Pore[2] Bioengineering Inc., 13905 Vista Drive, Rockville, MD 20853, U.S.A.

1 ABSTRACT

Hydrodynamic pressure (HDP) treatment and aging effects were examined. Striploins (48 h postslaughter), were divided into control, C, and matching HDP sections. HDP-treatment occurred on day 0 only. Warner-Bratzler shear (WBS) declined during aging (0→5→8 days): 5.65→4.40→4.14 kg_f for C; 4.82→3.87→3.81 kg_f for HDP). WBS differences between C and HDP were significant ($P<0.05$) only on day 0. Thirteen peaks were identified by reverse phase high performance liquid chromatography (RP-HPLC). Peak 5 increased with aging ($P<0.05$) while peak 7 decreased with aging and differs between C and HDP ($P<0.05$). Thus, aging and HDP effects on meat can be monitored with these two soluble protein peaks.

2 INTRODUCTION

The meat industry has identified inconsistency in tenderness as a high priority problem that it needs to address if it is to respond to the consumer's expectations of consistency in the products they purchase.[1,2,3] In just over a decade, the meat industry has accelerated its adoption of several new technologies to meet the consumers' expectations for consistency in the tenderness of the meat products they purchase. Among the new technologies under examination is hydrodynamic pressure (HDP)-treatment. HDP-treatment passes a shock wave at supersonic speed through water and any object in the water that is an acoustical match.[4] Meat has a close acoustical match (mechanical impedance with water) since it is composed of approximately 75% water.[5] HDP-treatment has been shown to be effective in instantaneously tenderizing meat and meat products [6,7,8,9,10] and has been reported to lead to a decrease in levels of several microbes associated with meat. [11,12,13]

Tenderization of meat is a result of the conversion of muscle to edible meat and involves the degradation of muscle proteins beginning shortly after the animal is

---

[§] Mention of brand or firm names are necessary to report factually on available data and does not constitute an endorsement by any of the authors or their employers over other products or manufacturers of products of a similar nature not mentioned.

slaughtered.[14] However, the tenderization process is highly variable and is thought to be due to, but not limited to, several ante- and postmortem factors including, animal age, breed, sex, muscle cut, feeding regimen, method of slaughter, etc [2,14,15] as well as to complex biochemical changes and interactions that occur during the conversion of muscle to edible meat.[16,17,18,19,20] If scientists and the meat industry are to be able to increase the consistency of meat tenderness, then they will be required to have a detailed understanding of the processes that affect meat tenderness. This study was designed to examine similarities and differences in tenderness resulting from aging and from hydrodynamic pressure (HDP)-treatment as reflected by meat proteins.

3 MATERIALS AND METHODS

3.1 Meat

A total of 15 fresh, never frozen beef striploins were obtained from a local supplier and used within 42 to 48 h (experimental day 0) post-mortem. Samples were USDA Select grade, Angus, U.S. Yield Grade 2 and 3. Each striploin was sectioned into two 14 cm lengths from the rib end; the anterior (rib) section was used as the control sample, (C), while the posterior section was used as the paired HDP-treated sample. Beef striploin samples were labelled and then vacuum packaged (Cryovac®/Sealed Air Corporation; Saddle Brook, NJ). After packaging, samples were shrink-wrapped in a water bath set to 86.5 °C. Packaged samples were held at refrigerated temperature (4° C).

Hydrodynamic pressure (HDP)-treatment followed the procedure described by Solomon.[6] HDP-treatment used a 98.4 L plastic explosive container (PEC, a household garbage container; Rubbermaid Inc., Wooster, OH) which was suspended and filled with water that was cooled by addition of chipped ice. The supersonic pressure shock wave was induced using 100 g of binary explosive in the shape of a hot-dog suspended 30.5 cm above the top of the meat. Packaged striploins with their fat-side facing upward were secured to a 1.25 cm thick, 40.6 mc diameter steel plate that fit into the bottom of the PEC. The steel plate was seated on the bottom of the PEC. The charge was detonated by a licensed handler of explosives. The samples were HDP-treated on day 0; HDP-treatment was performed on 5 separate occasions using 3 individual striploins at each treatment.

After HDP-treatment, each HDP and C sample had a 4 – 4.6 cm thick section cut from its anterior end; the remaining portions of the C and the HDP samples were repackaged and aged at 2 to 4 °C for 5 and 8 days prior to analysis. Using a fabricated Plexiglass cutting block, a 2.5 cm thick steak was cut from the anterior end of each HDP and C group. The first slice of each section was used for texture analysis by Warner-Bratzler shear force (WBS) determination. The next slice (approximately 1 cm) was used for subsequent protein analysis.

3.2 Shear Force Determination (Warner-Bratzler Shear, WBS)

C and HDP-treated striploin steaks (*longissumus* muscle) were cut at 0, 5, and 8 days for shear force analysis. All steaks were grilled to an internal end-point temperature of 71 °C monitored using a NIST certified thermometer (Model HH21, Omega Engineering,

Stanford, CT) using an indoor/outdoor electric grill (George Forman model GGR50B, 1600 watt, Salton, Inc., Mt. Prospect, IL). Steaks were turned midway between the initial and final temperature of 71 °C.

WBS determination (kg_f) was made from a minimum of 12 cores using a 1.27 cm diameter coring device. All cores were removed parallel to the muscle fiber direction. Each core was sheared once perpendicular to the fiber orientation using a WBS test cell attachment (1.8 mm thick blade with inverted V-notch) mounted on a texture measurement system (TMS-90; Food Texture Corp., Chantilly, VA) with a cross head speed of 250 mm/min).

3.3 Homogenization and Subcellular Fractionation

All samples were maintained on ice at all times before homogenization. Samples were trimmed of all visible fat and connective tissue. Each steak was longitudinally divided in half (from lateral end to medial end of the striploin steak (see other article in this volume): the lower half was minced into chunks that were further mixed by shearing for 5 sec with a handheld blender (Braun MR 370; Braun Inc., Woburn, MA) while the upper section of the steak was frozen for possible future use. The minced sample served as a homogeneous representation of the entire steak.

A sample (5 g) was removed from the homogeneously dispersed steak and was mixed with 45 mL of homogenization media [HM consisting of 0.05 M Tris (Trizma acid) with 1.5 mM dithiothreitol (DTT) and 1.5 mM tetrasodium ethylenediaminetetraacetic acid (EDTA) with its pH adjusted to 7.0] and placed into a stainless steel homogenization vessel as described previously.[9] Tissue disruption was in a 100 ml capacity holding cup using a Waring Blendor. Homogenization speed was controlled by rheostat (Powerstat™, Superior Electric Co., Bristol, CN) at a setting of 65 for 10 sec. The resulting homogenate was filtered through two layers of cheese cloth to remove any residual fat and connective tissue. The homogenate was labelled as total homogenate (TH) and dispersed into two aliquots: 30 mL for fractionation and ~ 10 ml either saved or discarded.

The TH was subjected to differential centrifugation (Sorvall RC-2B, Sorval Instruments, Inc., Newtown, CN) at 4,000 rpm (10,800 x g) at 4°C. The pellet was discarded and the final supernatant solution ("*sol*") was saved for analysis by reverse phase (RP) – high performance liquid chromatography (HPLC) and for analysis of protein concentration.

3.4 Estimation of Protein Content in Subcellular Fractions

Protein content in the soluble fractions ("*sol*") of the "C" and the "HDP"-treated striploins was determined on diluted samples using a 96-multiwell, UV plate (Costar #3635, Corning Inc., Corning, NY). Protein levels were assessed using bovine serum albumin (BSA) as standard by recording the A_{230} (absorbency at 230 nm) with a SPECTRAFLOURPlus™ multiwell plate reader (Tecan US, Inc., Research Triangle Park, NC). "*Sol*" fractions were adjusted to a final concentration of 8.7 mg/ml using HM.

3.5 Reverse Phase (RP) High Performance Liquid Chromatography (HPLC)

3.5.1. Sample preparations: Samples for RP-HPLC were obtained from the soluble fractions ("*sol*") of C and HDP groups. Samples were examined after 0, 5, and 8 days of storage. A 300K Omega Nanosep® centrifugal filter (Pall Life Science, Ann Arbor, MI) waste prewashed with 500 µl of water and centrifuged at high speed for 12 m (Dual Speed Microcentrifuge, Thomas Scientific, Woodbridge, NJ) before use with "*sol*" samples. The waste water was disposed of after centrifugation and 500 µl of thawed diluted "*sol*" sample [100 µl of sample combined with 400 µL of filtered water (final 1.75 mg/ml)] were added to the washed filter and centrifuged (12 m high speed, Dual Speed microcentrifuge). A 200 µl portion of the filtrate was injected into the HPLC for analysis (see below).

3.5.2. RP-HPLC conditions: Two hundred microliters (200 µl) of sample were injected into a 200 µl loop. RP-HPLC used a Varian dual pump HPLC system (Varian Instrument Co., Walnut Creek, CA) with a Supelco Discovery Bio widepore C8 guard column (2 cm x 4 mm x 5 µm, Supelco, Bellefonte, PA) attached to a Supelco Discovery Bio widepore C8 column (25 cm x 4.6 mm x 5 µm). The Star Chromatography Workstation™ (Version 5.52; Varian, Walnut Creek, CA) as used to run the HPLC and to collect the data. Sample concentrations were detected using a Varian ProStar® photodiode array detector (PDA, Varian, Walnut Creek, CA) set to record several wave lengths including 280, 254, and 225 nm. The two HPLC solutions were labelled as A and B and were composed of 0.1% triflorotetraacetic acid in water (TFA; Aldrich Chemical Co. Inc., Milwaukee, WI) and 0.09% TFA in acetonitrile (EM Scientific, Gibbstown, NJ), respectively. Total run time for each RP-HPLC sample was 42 m including a built-in wash step. Flow rate for all runs was 0.47 ml/m. To insure that the column was completely cleared of sample, each injection of "sol" sample was followed by a separate injection of 200 µL water followed by a 42 m HPLC run. The HPLC run included a step gradient as listed in Table 1.

Table 1 *Gradient and method used for HPLC*

Time (minutes)	Gradient [A : B] A = 0.1% TFA in water B = 0.09%TFA in acetonitrile
00.00 → 12.00	60% : 40% → 50% : 50%
12.00 → 16,00	50% : 50% → 20% : 80%
16.00 → 26.00	20% : 80% → 20% : 80%
26.00 → 30.00	20% : 80% → 60% : 40%
30.00 → 42.00	20% : 80% → 60% : 40%

3.6 Data Preparation and Statistical Analysis

Chromatograpic data was converted into ASCII format using Varian's Star Chromatography Workstation™ (Version 5.52; Varian, Walnut Creek, CA). ASCII files were transferred to an Excel® spreadsheet (Microsoft Corp.; Redmond, WA) on a computer dedicated to data analysis. Composite chromatograms for each treatment (HDP,

C) and each day (0, 5, 8) were prepared at each of 3 wavelengths (225, 254, 280 nm). Chromatograms were visually examined and several peaks chosen for statistical analysis.

Separate univariate 2-way repeated measures ANOVA analysis (SAS Institute, Inc., Cary, NC) was conducted for each response variable. The alpha level for means was set to 0.05. Repeated values are all data values measured on the same experimental unit.

4. RESULTS AND DISCUSSION

4.1 Shear Force (WBS)

Tenderness of HDP-treated striploins (WBS = 4.82 kg_f) was significantly different (P<0.05 than that of the matching control, C (5.65 kg_f), at day 0 (Figure 1). This is not unexpected since HDP-treatment has been shown to instantaneously tenderize beef products.[6, 7, 8, 9, 10]

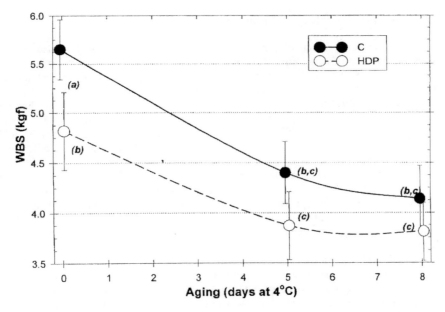

Figure 1 *Effect of aging (refrigerated storage at 4°C) on the tenderness [WBS (kg_f)] of C and matching HDP-treated beef striploins (N=15). A 3rd order-regression of the data from the 15 striploins indicates that the slope of the control (C) is -0.3505 kg_f/d while the slope of the matched pair HDP-treated sample is -0.2980 kg_f/d. The y-intercept for the C sample is 5.6359 while the y-intercept for the HDP-group is 4.8319. The r2 for both C and HDP is 1.0000.*

Since aging has been demonstrated to cause an increase in meat tenderness, [16, 17, 19, 20] both untreated C-steaks and HDP-treated steaks had their WBS values determined after storage at 2-4°C for 0, 5, and 8 days (Figure 1). The rate of change in WBS of HDP and C were significantly different (P<0.05 from each other over the 8 days aging period (slope of HDP: -0.2980 kg_f/d, slope of C: -0.3505 kg_f/d, slope). One possible explanation for the different rates of change in tenderness is that the HDP pressure shockwave had already altered some of the proteins in the HDP-treated loins so that these proteins were no longer available for fragmentation and degradation by endogenous tissue proteinases during the aging process.

One can observe that the change in meat texture between 0 and 5 days was much faster than the change in texture between 5 and 8 days in the C and HDP samples (Fig. 1) suggesting that the aging related changes in tenderness ... presumably by proteolysis ... had caught up to each other in the two groups. The last suggestion gains support by noting that the C and the HDP sample were different from each other at day 0 (P<0.05), but were not different at day 5 or day 8 (P>0.05) (Figure 1).

4.2 Proteins of the Soluble ("sol") Fraction

Thirteen peaks were found originally by reverse phase high performance liquid chromatography (RP-HPLC) using a C8 column. Initial observation of the chromatograms at 254 nm (Figure 2) indicated two rather high absorbing peaks with the first falling between 6.6 – 7.6 m and the second band between 8.3 – 8.5 m. The presence of several peaks in the chromatogram (9-21 m) that absorbed at lower values (see insert in Figure 2 for expansion of absorbency values) along with the observation of differences in the composite chromatogram of HDP-C absorbency, suggests that other wavelengths should be examined in addition to 254 nm. The other two wavelengths examined were 225 and 280 (Figures 3 & 4). Examination of the distribution of peaks that appeared at different wavelengths, in different C and HDP samples, and a different aging times, led to selection of 13 peaks labelled 1-10 with 3 additional peaks between peak 5 and 6 labelled as peak 5A, 5B and 5C (Table 2). Separation of the chromatogram into this number of peaks was a critical decision, since many of the peaks only appeared in certain "Trt" (HDP and C) groups, while other peaks seemed to split into multiple peaks at different storage (or aging) times.

For statistical purposes each peak had its absorbency at 5 retention times averaged, this is the peak RT alone plus two absorbency values on either side of the selected retention time. Each peak was described as a "weighted peak mean" by obtaining 5 individual absorbency values (called points A to E - that were multiplied by a factor (1, 2, or 3) and then added as shown: 1A + 2B + 3C + 2D + 1E. The resulting sum was divided by 9 to give a mean peak absorbency value that was used for statistical analysis of each sample. Means for each peak are shown in Figure 5.

Meat is a dynamic chemical system that is continuously changing with aging. Thus, as a meat ages, it is logical to predict that the level and concentrations of several meat proteins will change. Since the peaks separated by RP-HPLC represent the soluble portion of the meat proteins, several peak ratios as well as individual peaks were examined for changing patterns during aging. Some of this data is shown in Tables 3 and 4. Tables 3 and 4 show those peaks and those peak ratios showing a significant treatment (HDP versus C) and/or aging (0 to 8 days) main effect and those that show as "HDP x aging" effect.

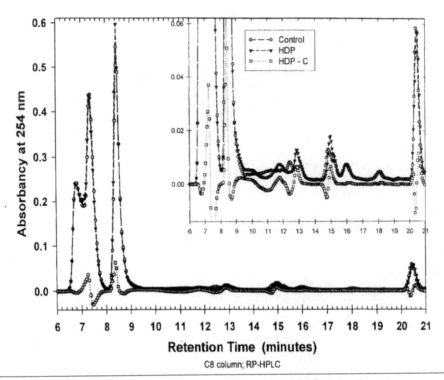

Figure 2 *Composite chromatograms of 15 individual 200 µL samples of beef striploin soluble fractions ("sol") from HDP-treated, matched-pair C, and HDP-C beef striploins. Reverse phase high performance liquid chromatography (RP-HPLC) was performed using a widepore, C8 column [25 cm x 4.6 mm x 5 µm]. Chromatograms were prepared of the C, HDP-treated, and the HDP-C with different Y-axis values to highlight the differences between control and HDP-treated groups.*

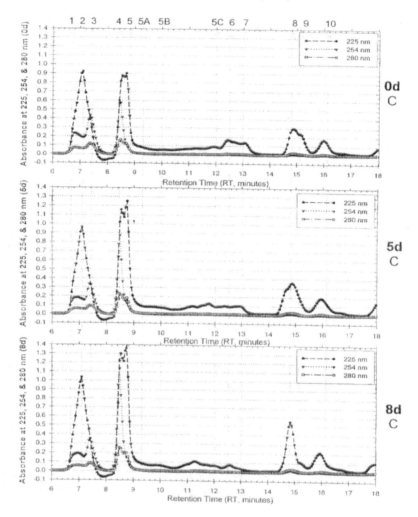

Figure 3 Record of composite chromatograms of C8, RP-HPLC taken at 3 different wavelengths (225, 254, and 280 nm). Chromatographic samples were obtained from the soluble fraction of 15 individual control (C) beef striploins stored for 0 days (top), 5 days (middle), and 8 days (bottom).

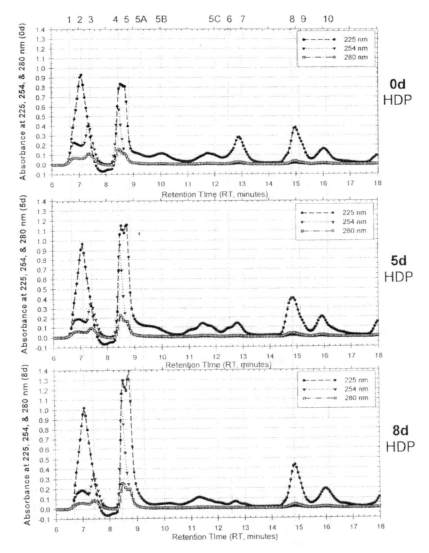

Figure 4 *Record of composite chromatograms of C8, RP-HPLC taken at three different wavelengths (225, 254, and 280 nm). Chromatographic samples were obtained from the soluble fraction of 15 individual HDP-treated (HDP) beef striploins stored for 0 days (top), 5 days (middle), and 8 days (bottom).*

Table 2 Range and mean retention time (minutes) for C8-RP-HPLC peaks in Figures 3 and 4

Peak number (C8 RP-HPLC)	Range Retention time (minutes)	Mecn Peak retention time (minutes)
1	6.56 – 6.77	6.7
2	7.15 -7.31	7.2
3	7.36 – 7.57	7.5
4	8.27 - 8.48	8.4
5	8.64 – 8.85	8.8
5A	9.17 – 9.31	9.3
5B	9.92 – 10.13	10.0
5C	11.94 – 12.05	12.0
6	12.43 – 12.65	12.5
7	12.85 – 13.07	13.0
8	14.72 – 14.93	14.8
9	14.99 – 15.15	15.1
10	15.95 – 16.16	16.1

The ratio p5/p1+p2+p3 shows a significant aging effect (increasing ratio) at 254 and 280 nm, but not at 225 nm (Table 4). Other peak ratios show a decrease ($P<0.05$) as the samples are aged longer (Table 3). These latter ratios include peaks p10/p5, p7+p9+p10/p5, and p7+p9/p5 that show an aging effect ($P<0.05$) at 254 nm and p2+p4+p7+p8+p9/p3+p5+p10 and p3+p6/p1+p5+p8+p10 that show a significant ($P<0.05$) aging effect at 280 nm (Table 3).

Certain peaks (possibly representing individual proteins) such as peak #5 and peak #7 show significant differences with aging for 0 to 8 days. For example, at 225 and 280 nm, peak #5 shows a significant ($P<0.05$) increase with aging (Figures 3-5); however, peak #5 does not show s significant difference at 254 nm (Table 4). This suggests that the composition of this peak may be quite different from a peak that shows a change at 254 nm. Peak #7 shows a significant ($P<0.0001$) aging effect (it decreases with aging), a significant treatment or HDP effect ($P= -0.0397$) and a significant "aging x HDP" effect ($P = -0.0249$) at all wavelengths examined (Table 3). These observations may prove to be particularly important because it suggests that peak #7 may be capable of being used to distinguish between C and HDP while peak #5 may be able to be used as an indication of how many days a sample has been aged.

Examination of individual peaks has been initiated using other analytical techniques (not shown here) since the possibility exists that each peak separated by RP-HPLC may represent more than one protein. Preliminary comment may be made regarding the larger peak seen at 254 nm between peaks 4 and 5 (Figures 2-4); capillary electrophoresis (CE, not shown) of this peak from pooled RP-HPLC samples has indicated that two proteins are present, one at ~110 kD molecular weight and the other at >300 kD. The CE observation closely follows a similar observation made in both myofibrillar and soluble fractions in HDP and soluble samples; the SDS-PAGE shows a significant increase in the ~110 kD protein in the HDP-treated samples (see other chapter this volume). As meat becomes more tender during aging there are changes in protein composition such as an increase in a 30 kD protein[21,22] and an increase in a 36 kD protein[23]; similar evidence has been observed in this laboratory using control striploins aged for 17 days (data not shown).

The protein bands seen on SDS-PAGE (see other chapter this volume) have shown clear differences between C and HDP-treatment; these proteins have an estimated molecular weight of 160 kd, 153 ks, ~110 kd and around 11.9 kD; however, at this time, the absolute identity of these bands has not been determined. The 160 kD and 153 kD bands have an electrophoretic mobility equivalent to that of various myosin heavy chain isoforms[21,24,25] and perhaps 'm-protein' which has a mass of 165 kd.[26]

Locker[27] and Locker and colleagues[28,29] have found a breakdown product referred to as 'band-B' which migrated on SDS-PAGE just below titin. Band-B is thought to be derived from unidentified material unable to penetrate the gel.[27] No such protein bands were seen in our earlier work using either homogeneous or gradient SDS gel electrophoresis.[9] However, at this time, we can not rule out the possibility of the existence of band-B in our RP-HPLC.

During post-mortem aging, fragments of higher molecular mass proteins form, e.g. the myosin head component, S1 (light meromyosin), has been recovered in extracts of conditioned beef.[30] Light meromyosin is a lower Mr than α-actinin (~97 kD) and does migrate near α-actinin on homogeneous gels where it is difficult to clearly distinguish these two bands particularly on overloaded gels. At this time we only have preliminary data that indicates that the molecular weights of the material in the RP-HPLC peaks 4 and 5 are ~110 and >300 KD, suggesting that we are not looking at light meromyosin.

A protein of approximately ~110 kD molecular mass was found in the HDP-treated striploins (Figures 3 & 4). This protein peak increased with aging from 0 to 8 days (or as the meat became more tender as the WBS suggested; see peak 5 in Figure 4). Recently investigators[21] have shown that on aging of bovine muscle a 110 kD fragment appeared in myofibrillar fractions. Additionally, Casserly and associates[31] have recently shown by sequence analysis that the 110 kD myofibrillar protein fragment showed strong homology with C-protein (~140 kD). Thus, there is reason to believe that the ~110 kD protein observed in HDP-treated striploins in this current study could be the same 110 kD protein found by Casserly and others. This suggestion gains support if one considers the function of C-protein which forms cylindrical structures around the packed tails of myosin molecules acting as a clamp to hold and stabilize the backbone of the thick filament. Furthermore, C-protein binds myosin heavy chain, F-actin, and native thin filaments.[32]

O'Halloran and others[21] indicated that the 110 kD fragment appears in much smaller quantities than the 30 kD fragment. These observations are similar to those of others who have shown a relationship between tenderness and protein composition.[14, 25, 26, 33, 34] Since HDP-treatment improves meat tenderness (WBS) value) it is reasonable to suggest that observed change in the quantity of the ~110 kD protein (Figures 3 & 4) is related to the tenderness of the striploin. His suggestion gains added support from the work of Zuckerman and Solomon[10] who have shown that HDP-treatment affects the physical structure of treated striploins by disrupting the Z-line and the A-I band junctures. The data in Tables 3 and 4 show that there is clearly a relationship between aging and at least one resolvable protein peak (peak #5) and with the ratio of meat protein peaks and aging.

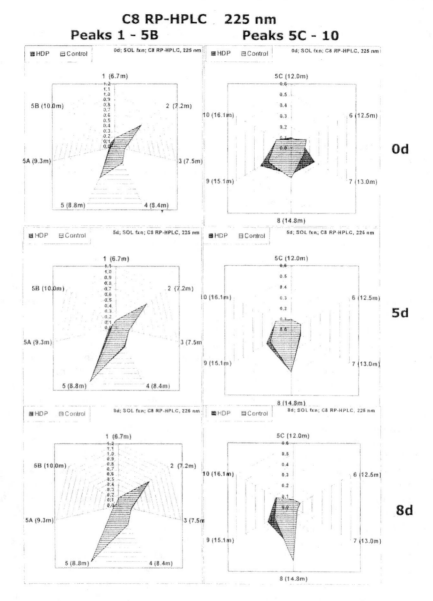

Figure 5 *Spider plots of C8, RP-HPLC of peaks 1 → 5B (left column) and peaks 5C → 10 (right column). Top, middle and bottom rows refer to storage time at 2–4°C for 0d, 5d, and 8d, respectively.*

Table 3 Two-way repeated-measures Anova analysis of response variables in control, C, and HDP-treated beef striploins. Unless otherwise listed alpha is 0.05

Variable or Observations	Wavelength (nm)	Effect (or Mean) days[a] treatment ("trt")[b], days x treatment (dxt)[c]	Days	Significance
peak 7	225 254 280 280/254	Days "Trt" d x t	0, 5, 8	<0.0001
WBS		Days "Trt" d x t	0 5 8	A B B
p7	225	Days	0 5 8	A B C
p7	254 & 280	Days	0 5 8	A B B
p7	225 280 / 254	Trt	HDP C	A B
Peak ratios				
p10 / p5	254	0d mean: 0.0963 5d mean: 0.0757 8d mean: 0.0523	0 5 8	A B C
p7+p9+p10 / p5	254	0d mean: 0.3946 5d mean: 0.2519 8d mean: 0.1460	0 5 8	A B C
p7+p9 / p5	254	0d mean: 0.2983 5d mean: 0.1762 8d mean: 0.0937	0 5 8	A B C
p2+p4+p7+p8+p9 / p3+p5+p10	280	0d mean: 1.2707 5d mean: 1.0235 8d mean: 0.8280	0 5 8	A B C
p3+p6 / p1+p5+p8+p10	280	0d mean: 0.9541 5d mean: 0.7165 8d mean: 0.4422	0 5 8	A B C

[a] "days" refer to the aging for 0, 5, and 6 days.
[b] "Trt" refers to the HDP-treatment versus the control (C) striploins.
[c] "days x treatment" refer to aging x HDP effect
p = peak

Table 6 Two-way repeated-measures Anova analysis of response variables in control, C, and HDP-treated beef striploins. Unless otherwise listed alpha is 0.05. Reverse "day" relationship

Variable or Observations	Wavelength (nm)	Effect (or Mean) days^a treatment ("trt")^b, days x treatment (dxt)^c	Days	Significance
Peak				
p5	225	8d mean: 1.1973	8	A
		5d mean: 1.0274	5	B
		0d mean: 0.6324	0	C
	280	8d mean: 0.1758	8	A
		5d mean: 0.1490	5	B
		0d mean: 0.0873	0	C
Peak ratios				
p5 / p1+p2+p3	254	8d mean: 1.3570	8	A
		5d mean: 1.0970	5	B
		0d mean: 0.6687	0	C
	280	8d mean: 1.8318	8	A
		5d mean: 1.3379	5	B
		0d mean: 0.9763	0	C

^{a} "days" refer to the aging for 0, 5, and 6 days.
^{b} "Trt" refers to the HDP-treatment versus the control (C) striploins.
^{c} "days x treatment" refer to aging x HDP effect

The data presented in this report indicate clearly that the distribution and relative quantities of several proteins in the soluble fraction of homogenates of HDP-treated beef striploins are related to the degree of tenderness of striploin sample; furthermore, this relationship may be closely tied to the ~110 kD protein. At this time, the origin or source of this protein is unknown, but we suggest that it may be a result of fragmentation of larger meat proteins[14,25,26,33-35] such as, but not limited to, titin, nebulin, myosin, connectin, or C-protein. Alternatively, the peaks that appear after HDP-treatment may result from bound-disruptions in other high molecular mass protein complexes such as myosin and acting with known functional bonding interaction and banding patterns.[25]

Since HDP-treatment of meat has been shown to lead to visual structural-alterations[10] and to chemical alterations of proteins in homogenized meat,[9] these proteins or their altered parent proteins in more tender cuts may be more susceptible to the physical shearing caused by homogenization. This HDP-induced change in meat's susceptibility to physical perturbations could also potentially include an enhanced susceptibility to chemical perturbations such as proteolytic digestion by endogenous tissue proteases such as cathepsins B, D, H, and L, calpains, and exoproteases or altered reactivity to oxidation which could enhance or detract from the initial tenderization effect of HDP-treatment. This potential response to chemical perturbations must be explored in future experiments because of its potential impact on meat quality and to further examine the relationship, if any, between HDP-treatment and post-mortem aging.

5 CONCLUSION

These data of C and HDP-treated striploins show that there is one RP-HPLC peak from striploin soluble fractions that is related to aging for 0 to 8 days (Peak #5) and another peak related to HDP-treatment (Peak #7). Continued research on the effect of hydrodynamic pressure on meat will provide information essential to developing HDP or other pressure-related technologies into viable, commercially applicable processes that will attract a high level of consumer confidence. It is also hoped that it will lead to developing under-exploited markets for new products that produce value-added, high quality meat with desirable palatability, extended shelf-life and improved safety.

References

1   R.L. Alsmeyer, J.W. Thornton and R.L. Hiner. *J. Anim. Sci.*, 1965, **24**, 526-530.
2   J.B. Morgan, J.W. Savell, D.S. Hale, R.K. Miller, D.B. Griffin, H.R. Cross and S.D. Shackelford, *J. Anim. Sci.*, 1991, **69**, 3274-4283.
3   S.D. Shackelford, T.L. Wheeler and M. Koohmaraie, *J. Anim. Sci.*, 1997, **75**, 175-176.
4   H. Kolsky, In,, *Stress Waves in Solids*, 1980, Dover Publishing Company, NY
5   J.F. Price and B.S. Schweigert, In, *The Science of Meat and Meat Products*. 1978, Food and Nutrition Press Inc., Westport, CT.
6   M.B. Solomon, 51st *Annu Recip. Meat Conf.*, 1998, **51**, 171-176.
7   M.B. Solomon, J.S. Eastridge, H. Zuckerman and W. Johnson, In, *Proc 43 Intl. Cong. Meat Sci. Technol.* 1997, 121-124
8   A.M. Spanier, B.W. Berry and M.B. Solomon, *J. Muscle Foods*, 2000, **11**, 183-196
9   A.M. Spanier and, R.D. Romanowski, *Meat Sci.*, 2000, **56**, 193-202
10 H. Zuckerman and M.B. Solomon, *J. Muscle Foods*, 1998, **9**, 419-426
11 H.R. Gamble, M.B. Solomon and J.B. Long, *J. Food Protect.*, 1998, **61**, 637-639.
12 S. Moeller, D. Wulf, D. Meeker, M. Ndife, N. Sundararajan and M.B. Solomon, *J. Anim. Sci.* 1999, **77**, 2119-2123.
13 A.M. Williams-Campbell and M.B. Solomon, *J. Food Prot (research note)*, 2002, **65**, 571-574.
14 D.E. Goll, M.L. Boehm, G.H. Geesink and V.F. Thompson, 50th *Annu. Recip. Meat Conf.*, 1997, **50**, 60-67.
15 E. Dransfield, *Meat Sci.*, 1994, **36**:105-121.
16 A. Asghar and A.M. Pearson, In *Advances in Food Research*, 1980, **26**, 53-213.
17 C. Faustman, In: *Muscle Foods: Meat, Poultry and Seafood Technology*, 1994 Chapman and Hall, New York pp 63-78,

18 E.J. Huff-Lunegran and S.M. Lonegran, In, *Quality Attributes of Muscle Foods*, 1999, Kluwer Academic/Plenum Press, New York pp 229-252.
19 C. Valin and A. Ouali, In, *New Technology for Meat and Meat Products*, 1992, P Audet Tijdschriften B.V., The Netherlands, pp 163-179.
20 M. Koohmaraie, M.P. Kent, S.D. Shackelford, E. Veiseth and T.L. Wheeler. *Meat Sci.* 2002, **62**, 345-352.
21 G.R. O'Halloran, D.J. Troy and D.J. Buckley. *Meat Sci.*, 1997, **45**, 239-251.
22 A. Ouali. *J. Muscle Food,* 1990, **1**, 126-165.
23 M.A. MacBride and F.C. Parish, Jr., *J Food Sci.*, 1977, **42**, 1627-1629.
24 D.L. Hopkins and J.M. Thompson, *Meat Sci.*, 2000, **57**, 19-22.
25 D.L. Hopkins, P.J. Littlefield and J.M. Thompson, *Meat Sci*, 2000, **56**, 19-22.
26 R.M. Robson, E. Huff-Lonergan, F.C. Parrish, Jr., C.-Y. Ho, M.H. Stomer, T.W. Huiatt, R.M. Bellin and S.W. Sernett, *50th Annu, Recip. Meat Conf.*, 1997 **50**, 43-52.
27 R.H. Locker, *J. Food Microstruct*, 1984, **3**, 17-24.
28 R.H. Locker, D.J.C. Wild, *Meat Sci.*, 1984 **10**, 207-233.
29 R.H. Locker, D.J.C. Wild, *Meat Sci.*, 1984 **11**, 89-108.
30 L.D. Yatess, T.R. Dutson, J. Caldwell and Z.L. Carpenter, *Meat Sci*, 1983, **9**, 157-179.
31 U. Casserly, S. Stoeva, W. Voelter, A. Healy and D.J. Troy, In, *44th IcoMST (Vol B115)*, 1998, Barcelona, Spain pp 726-727.
32 D.O. Furst, U. Vinkomeier and K. Weber. *J Cell Sci*, 1992, **102**, 769-778.
33 J.D. Fritz and M.L. Greaser, *J Food Sci*, 1991, **56**, 607-610.
34 R.W. Purchas, X. Yan and D.G. Hartley, *Meat Sci*, 1999, **51**, 135-141.
35 P.F.M. van der Ven, G. Schaart, H.J.E. Croes, P.H.K. Jap, L.A. Ginsel and F.C.S. Ramaekers, *J Cell Sci.*, 1993, **106**, 749-759.

HYDRODYNAMIC PRESSURE (HDP)-TREATMENT: INFLUENCE ON BEEF STRIPLOIN PROTEINS.[§]

A.M. Spanier[1,2] and T.M. Fahrenholz[3]

[1] Spanier Science Consulting LLC, 14805 Rocking Spring Drive, Rockville, MD 20853, USA;
[2] Pore^2 Bioengineering Inc., 13905 Vista Drive, Rockville, MD 20853, U.S.A.
[3] USDA, ARS, BA, Animal & Natural resources Institute (ANRI), Food Technology and Safety Laboratory (FTSL), Bldg 201, BARC-East, Beltsville, MD 20705, U.S.A.

1 ABSTRACT

This study examined the influence of hydrodynamic pressure (HDP; an effective tenderizing technology) on beef proteins. Beef striploins were obtained 48 hours postmortem (experimental day 0). Sample tenderness (Warner-Bratzler shear, WBS) in HDP samples (7.12 kg$_f$) improved 12.5% ($P<0.05$) over their controls (C, 8.14 kg$_f$). Sodium dodecyl sulphate-polyacrylamide gel electrophoresis (SDS-PAGE) of C and HDP striploins showed significantly different changes ($P<0.0001$) in myofibrillar proteins such as myosin, actin, 110 kD protein, and a group called >actin. Data suggested that HDP-treatment most likely disrupted protein bonds (perhaps C-protein) associated with the structural integrity of myosin and actin, thereby, having a positive effect on meat tenderness.

2 INTRODUCTION

Morgan and others[1] and Love[2] indicated that among the organoleptic properties of meat, tenderness is considered the most important quality characteristic for consumers. Texture is typically measured instrumentally using the Warner-Bratzler shearing device, but compression, tensile strength, penetrometry and sensorial bite tests have also been used.[3] Meat texture depends on the state of the interaction of various muscle components at any moment in time after slaughter. This is particularly true for myofibrillar and connective tissue, thereby making it generally difficult to evaluate meat texture.[4]

After an animal is slaughtered and during the conversion of muscle to meat, several mechanisms begin to affect muscle structures and, thereby, tenderness. The conversion of muscle to meat and the tenderization process typically are a result of the activity of several pH-dependent enzyme systems[5] that degrade the myofibrillar structure.[6] Connective

---

[§] Mention of brand or firm names are necessary to report factually on available data and does not constitute an endorsement by the authors over other products or manufacturers of products of a similar nature not mentioned.

tissue, considered by many researchers as a main limiting element in meat tenderness, is also partially degraded during postmortem aging.[7] Unfortunately, all of the factors affecting meat texture are not completely understood.

It has been suggested that cooking meat prior to evaluation of texture hinders the true assessment of tenderness.[8,9] Rheological studies are typically conducted on cooked meat and these findings then related to sensory characteristics. From the perspective of understanding the physico-chemical changes induced by a particular treatment (in this case HDP) it is extremely important to assess the chemical characteristics of meat components prior to and without the influence of heating. Since variation in tenderness has been reported along the surface of individual steaks[10] and since this variation is greatly minimized by HDP-treatment,[11,12] the objective of this current investigation was to examine the effect of HDP-treatment on the variation in protein structure, and, if possible, to determine if HDP had a different effect at different locations across the surface of the beef striploin (*longissimus* muscle) steak [lateral (*Lt*), central (*Cn*), medial (*Md*)]. Meat protein profiles in HDP and C treatment groups were examined by densitometric analysis of stained gels resolved by SDS-PAGE.

3 MATERIALS AND METHODS

3.1 Striploin Samples [for myofibrillar fractions]

Experimental samples were obtained from a local supplier and used within 42 - 48 h postmortem. Samples were USDA Select grade, U.S. Yield Grade 2 and 3. Striploins were sliced to ~15 cm lengths from the anterior (rib end); the anterior section was used for HDP-treatment, while the posterior section was used as a paired control, C. Steaks (2.5 cm thick) were cut from the adjacent (facing ends) of each HDP and C group and used for protein analysis. The next immediate steak (from C and HDP-treated groups) were sectioned and used for Warner-Bratzler shear force (WBS) determination. Striploin samples for HDP-treatment were vacuum packaged in yellow 'bone-guard' bags (Cryovac® / Sealed Air Corporation; Saddle Brook, NJ) and shrink-wrapped in a water bath set to 86.5EC.

3.2 Striploin Samples [for soluble fractions]

A separate series of experiments was performed to examine the proteins in the soluble ("*sol*") fractions of C and HDP groups. In these experiments, it was desired to determine whether the "*sol*" fraction would prove useful in assessing differences between HDP-treated and C steaks since earlier experiments by Spanier and Romanowski[13] indicated that HDP treatment seem to shift proteins from the myofibrillar fraction to the soluble ("*sol*") fraction. It was also hoped that these experiments might show if there was a possible relationship between initial meat texture and meat's response to HDP treatment.

Experiments using the "*sol*" fraction necessarily had to be performed slightly differently than the steaks used for the myofibrillar fractions described above. The "*sol*" fraction experiments included 4 individual striploins chosen from a group of 36 striploins that had been screened for their WBS shear value on the date of purchase and then frozen until needed. The 4 striploins were selected because of their initial WBS screening values of 4, 6, 8, and 10 kg$_f$. The selected samples reflected an even change in WBS shear value and the samples were thawed overnight in a refrigerator prior to homogenization and fractionation. As with the 3-core experiments, the C and HDP samples were obtained from the same steak and were treated similarly. Soluble fractions were prepared as described below. The process of frozen-storage (-20 °C, about 2 months) followed by overnight thawing (4 °C) prior to treatment and analysis led to a change in the WBS value of 'C' samples from 4, 6, 8, and 10 kg$_f$ at screening to 3.6, 5.5, 7.5, and 8.2 kg$_f$, respectively, after freeze/thawing.

3.3 Hydrodynamic Pressure (HDP)

HDP-treatment essentially followed the procedure described previously by Solomon.[14] HDP-treatment used a 98.4 L plastic explosive containers (PEC, a household garbage container; Rubbermaid Inc., Wooster, OH) filled with water and maintained cool by addition of chipped ice. The PEC was suspended above the floor. Striploins were placed fat side up on top of a 1.25 cm thick, 40.6 cm diameter steel plate and then lowered into the bottom of the PEC. The explosive charge was 100 g of binary explosive in the shape of a hot-dog (shape in earlier work was more cuboidal) suspended 30.5 cm above the top of the meat.

Once prepared as described above, the charge was detonated by a licensed handler of explosives. HDP-treatment was performed on 3 separate occasions. The first of the loins had been exposed twice to HDP-treatment instead of once due to a suspected misfire; however, since the purpose of these experiments was to examine the effect of HDP (and not necessarily the magnitude of the HDP effect) all 3 samples were examined.

C and HDP-treated samples ('trt' group) were sliced 2.5 cm thick. Mirror images of core-location (lateral, *Lt*, central, *Cn*, medial, *Md*) were prepared from facing steaks. Steaks within each treatment (HDP and C) group were analyzed for WBS value (facing steaks from within the same striploin) and protein composition (adjacent steak from within the same treatment group). After removal of the 3-cores, the remainder of each steak was used to obtain a WBS value from the mean of 15 cores.

3.4 Shear Force Determination (Warner-Bratzler Shear, WBS)

Control and HDP-treated striploin steaks (*longissimus* muscle) were grilled to an internal end-point temperature of 71EC monitored using Type J, iron-constantan thermocouples attached to a multipoint recording potentiometer (Speedomax, Model 1650, Leeds and Northrup, North Wales, PA). Steaks were turned midway between the initial and final temperature of 71EC. Grilling occurred on an indoor/outdoor electric grill (George Forman, model GR-20, 820 watt, Salton, Inc., Mt. Prospect IL). Cores for WBS determination (Figure 1; a total 15/steak including the *Lt*, *Cn*, and *Md* cores) were removed from each steak using a 1.27 cm diameter coring device. All cores were made parallel to the muscle fiber orientation. Each core was sheared once at right angles to the fiber orientation using a WBS test cell with 3.18 mm thick blade mounted on a texture measurement system (TMS-90, Food Texture Corp., Chantilly, VA) using a flat bottom blade with an inverted V-shape to surround the meat core during shearing. The cross-head speed was set to 25 cm/min.

3.5 Homogenization and Subcellular Fractionation.

All samples were kept cool (temperature maintained on ice) at all times before portioning for homogenization. All samples were trimmed of visible fat and connective tissue. Cores used for homogenization were 1.7 cm diameter and were obtained from each steak using a freshly sharpened #12, copper cork borer. Central (*Cn*), lateral (*Lt*) and medial (*Md*) cores were trimmed as needed with a scalpel to yield a 5.0 gm sample from steaks of the C and HDP group (Figure 1). The first core sample, the *Cn* core, was obtained at the crossover point midway between the lateral and medial end obtained from midway between the bone and subcutaneous sides of the steak. *Lt* cores were obtained by sampling at the crossover point midway between the lateral and center of the central core and the midway between the bone and subcutaneous side of the steak; *Md* core samples were obtained similarly, but from the medial end of the steak.

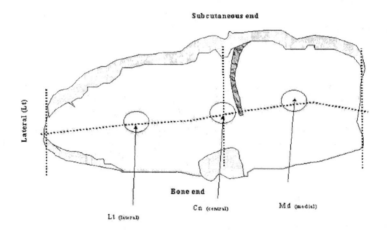

Figure 1 *Schematic diagram of the location form which 3-cores were removed from a 2.5 cm thick striploin steak. Cores were analyzed for their tenderness level by Warner-Bratzler shear (WBS) and protein composition determined using sodium dodecyl sulfuate – polyacrylamide gel electrophoresis (SDS-PAGE).*

The 5 gm core samples were minced and placed into stainless steel homogenization vessels containing 45 ml of cold (4 EC) homogenization medium [HM: 0.05 M Tris (Trizma acid), 1.5 mM dithiothreitol (DTT), 1.5 mM tetrasodium ethylenediamine-tetraacetic acid (EDTA) with 0.001 M sodium azide; pH = 7.0]. Samples were homogenized as reported earlier[13] using a homogenization protocol that resulted in maximum tissue disruption without producing undo fragmentation of the HDP-treated meat. Tissue disruption occurred in a 100 ml capacity holding cup using a Waring

Blender. Homogenization speed was controlled by rheostat (Powerstat™, Superior Electric Co., Bristol, CN) at a setting of 65 for 10 seconds. The resulting homogenate was filtered through two layers of cheese cloth to remove any residual fat and connective tissue. The homogenate was labelled as total homogenate (TH) and dispersed into two aliquots: 30 ml for use in fractionation studies and the remaining filtrate (~10 ml) either saved or discarded.

The TH was subjected to differential centrifugation in a refrigerated centrifuge (Serval RC-2B, Sorval Instruments, Inc., Newtown, CN) at 4,000 rpm (10,800 x g) at 4 EC for 10 min. The supernatant was decanted into a centrifuge tube and labelled. The pellet containing the myofibrillar material was re-suspended by vortex to one-half its original volume using cold "wash" solution [0.1 M sodium chloride, 0.001 M sodium azide, 0.5% Triton X-100) and again centrifuged once more at 4,000 rpm (10,800 x g) for 10 min at 4 EC. The resulting pellet (labelled as *MF*) was resuspended 3 times to one-half its original volume with myofibril isolation buffer [MFIB: 0.01 M phosphate buffer, 0.001 ethyleneglyco-*bis*tetraacetic acid (EGTA), 0.002 M magnesium chloride, 0.1 M potassium chloride, adjusted to pH 7.0] and centrifuged to remove residual detergent and azide. The final *MF* pellet was resuspended with 10-15 ml of a 1:1 mixture of MFIB:glycerol (to serve as a cryoprotectant) and saved. The original supernatant solution was centrifuged at 20,000 rpm (48,200 x g) for 30 minutes at 4 EC. The pellet was discarded and the final supernatant solution ("*sol*") saved for analysis.

3.6 pH Measurement.

Replicate pH assessments of the *Lt, Cn*, and *Md* core sections from each striploin was obtained for each C and HDP sample and recorded. The pH was measured using three different approaches. The first pH measurements were made on the total homogenate (TH), the second was made by placing a pH probe (6 mm polymer, Cole-Parmer Instruments, Vernon Hills, IL) attached to a laboratory pH meter (Omega; Cole-Parmer Instruments, Vernon Hills, IL) directly into a small slit made into the steak, and the third in the same manner as the 2nd except using a portable pH probe (Cole Parmer Instruments, Vernon Hills, IL). The pH measured directly in the meat was approximately 0.1 pH-unit greater than that measured in the total homogenate.

3.7 Analysis of Protein.

Protein content in the "*MF*" and "sol" fractions was determined on appropriately diluted samples (samples diluted so that the absorbency reading would be in linear range) samples using bovine serum albumin (BSA) as standard. Protein levels were assessed by reading the absorbency at 230 nm (A_{230}) using a 96-well UV multiwell plates (Costar #3635, Corning Inc, Corning, NY) in a multiwell plate reader (SECTRAFLOURPlus, Tecan, US, Inc, Research Triangle Park, NC).

3.8 Sodium Dodecyl Sulfate (SDS) - Polyacrylamide Gel Electrophoresis (PAGE)

Proteins were separated electrophoretically on polyacrylamide gels in the presence of SDS using the discontinuous gel procedure of Laemmli and Favre[15]. The SDS-PAGE system was used to compare control and HDP-treated samples. The system consisted of a commercially available Tris-HCP 4-20% acrylamide gradient for myofibrillar (MF) samples and 10-20% acrylamide gradient for "sol" samples (Criterion System®, Bio-Rad Labs, Hercules, CA); gels contained built-in sample-combs. These gels were found to give more consistent results than gradient or homogeneous gels prepared in the laboratory.

All SDS-PAGE samples were rendered soluble by mixing a known concentration of sample 1:1 (v/v) in a 2.0 ml capacity, capped, plastic conical centrifuge tube with rendering medium (GOOP+: 0.5 M Tris/HCl, pH 6.8, 10% SDS, 20% glycerol, 5% 2-mercaptoethanol, 0.1% bromphenol blue). Samples in GOOP were boiled for 6-10 min in capped tubes (to prevent sample loss) to inactivate proteolytic enzymes and render proteins soluble. Samples were loaded into each well at the concentration noted in the figure. Gradient acrylamide gels were subjected to a constant voltage (200 V) for 45 min to 1 hour powered by a Power Pac 3000 (Bio-Rad Labs, Hercules, CA). The electrophoresis unit was placed in an ice bath to minimize temperature-induced production of artifacts in the gels during electrophoresis. Gels were removed from their sandwiched position between the glass holding plates and stained with 0.05% Coomassie blue R-250 in 7% acetic acid:40% methanol overnight at room temperature. Gels were stained and destained in a mixture of 7% acetic acid and 15% methanol until the gel background was clear. A mixture of protein standards having a progressive increase of 10 kiloDaltons (kD) per protein was used (10kD protein ladder, Gibco BRL, Grand Island, NY). Densitometric scanning of wet, stained, gels was performed using a densitometer (Bio-Rad Model 710, Hercules, CA); semiquantitative assessment of protein peak area was performed using Quantity 1™ software (Bio-Rad Model 710, Hercules, CA).

3.9 Statistical Analysis.

Statistical examination of the data from *MF* fractions was performed using two-way repeated measures ANOVA for univariate analysis (SAS Institute, Inc., Cary, NC); the alpha level for means was set at 0.05. Experiments using soluble fractions (sol) were assessed using "repeated measures ANCOVA" (SAS Institute, Inc., Cary, NC).

4.0   RESULTS AND DISCUSSION

A 12.5% improvement in tenderness is seen in HDP-treated steaks (WBS of C = 8.14 kgf; WBS of HDP = 7.12kgf; p<0.05). Statistical analysis of the data suggests that there may be a variation in tenderness across the surface of both control, (C) and HDP-treated steaks; for example, the medial section appears somewhat tougher than the lateral or central section. Tenderness change from lateral (*Lt*) | central (*Cn*) | medial (*Md*) section of the steak averaged 7.1 kgf to 8.4 kgf, for the C and averaged 6.3 kgf to 7.4 kgf, for the HDP-treated steaks. SDS-PAGE analysis of the protein of the C and HDP samples show that myosin, actin and a band termed <actin did not show a significant location (*Lt, Cn, Md*) effect yet the 110 kD protein band that showed a significant "Trt x section" (HDP and C by core location) interaction effect (p=0.0348). The 110 kD protein band also showed a "Trt'" (HDP against C; p<0.0001) and "Section" (*Lt, Cn, Md*; p=0.0015) main effect.

Because the proteolytic enzymes of meat show a wide range of pH dependent activity, the pH of both C and HDP groups were assessed. A large difference in pH is observed between either locations (*Lt, Cn, Md*) or treatments (HDP and C) one might suggest that any observed differences might be a result of proteolytic activity, whereas, if the pH is similar in locations and treatment groups it is likely that the observed differences are due to the HDP treatment itself. The pH value in both treatment groups averaged 5.4 with a significant difference ($p<0.05$) between the two treatments (HDP = 5.46; C = 5.43). The difference in pH between treatments was possibly a result of the difference in pH observed between the medial (*Md*), central (*Cn*) and lateral (*Lt*) sections of the steak (pH of Md = 5.42, Lt and Cn each = 5.46; $P<0.05$). However, based on the optimum pH of most endogenous meat proteases, it is unlikely that the observed differences in WBS were a result of differences in proteolytic activity based on a 0.03 difference in pH.

SDS-PAGE was performed on myofibrillar fractions (*MF*) obtained from each repetition; samples were obtained from different sections (*Lt, Cn, Md*) of each treatment (HDP and C). Four of the protein bands (Figure 2) were selected for analysis via densitometry and subsequent statistical analysis. Results of the scan are shown in Figure 3 for four proteins tentatively identified as myosin, a 110 kD protein, actin, and several bands with a molecular weight below that of actin (<actin). Two-way repeated measures ANOVA was conducted for each response variable with "Trt" (HDP or C), "Section" (*Cn, Lt, or Md*), and a "Trt x Section" interaction as factors. Analysis of the variables showed that all 4 proteins exhibited a treatment, "Trt," effect of $p<0.0001$. Three of the proteins

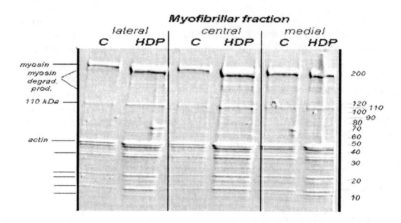

Figure 2 *Gradient (4-20%) electropherogram of myofibrillar (MF) enriched fractions (10,800 x g pellet) obtained from three individual cores of control, C, and hydrodynamic pressure (HDP)-treated beef striploins (duplicate gels run for 3 indicidual C and HDP striploins). Each lane contained 5 µg of protein (A_{230}). Gels were stained with Coomassie brilliant blue R-250 and scanned using a BioRad 710 densitometer. The electropherogram (SDS-PAGE gel) is presented in an "embossed" format to facilitate visualizaion.*

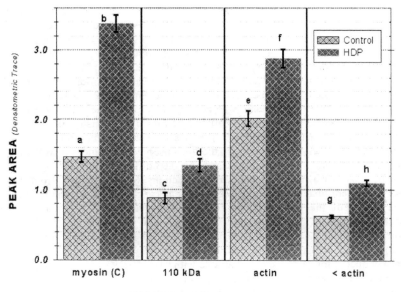

Figure 3 Mean ± standard deviation of densitometric scens (BioRad 710 densitometer using Quantity One software) of proteins on SDS-PAGE gels of control, C, and HDP-treated striploins. Protein was obtained from myofibrillar (MF) fractions obtained from three locations (lateral "Lt", central, "Cn", Medial, "Md", see Fig 1) on three steaks from three individual striploins. Protein bands are labelled as myosin, a protein of approximately 110 kD mass, actin, and bands of proteins of lower MW than actin (<actin); some ot the <actin bands may represent tropomyosin, troponin-T, degradation products of actin or other muscle proteins, etc.

TABLE 1: Means comparison at alpha = 0.05 for the 110 kD protein in beef striploin myofibrillar (MF) fractions.

Trement	Location in striploin	Means estimate	Significance group @ $p<0.05$
C	Md	1.046	a
	Cn	0.846	b
	Lt	0.746	b
HDP	Md	1.480	c
	Cn	1.380	c, d
	Lt	1.180	a. d

Abbreviation key: "C" = control treatment group; "HDP" = hydrodynamic pressure treated group'; Location in steak (as per Fig. 1): "Md" = medial; "Cn" = central; "Lt" = latera; Significance is at the level of $p<0.05$, if the lower-case letters are different.

(identified as myosin, actin and <actin) showed only a positive "Trt" effect, while the 110 kD protein showed both a significant "Section" effect (p=0.0015) and a "Trt x Section" effect (p=0.0348; Table 1). All HDP-groups (all sample reps and all sections, Lt, Cn, and Md) exhibited a broader appearing bands for both myosin and actin suggesting that HDP-treatment may have torn and disrupted several bonds associated with these two proteins. This latter suggestion is supported by the observed presence of protein bands often associated with fragmentation products of actin (<actin) and myosin[16] in the HDP-samples.

Close examination of Figure 2 reveals bands in the region of 160 kD and about 153 kD, these bands have an electrophoretic mobility similar to that of various myosin heavy chain isoforms;[17] these bands are also of similar molecular mass to 'm-protein' which has a mass of ~165 kD.[18] On the other hand, earlier investigations of HDP-treated meat[13] revealed that hydrodynamic shock-wave treatment of beef striploins led to a change in the localization and to a reorganization of meat proteins from one subcellular compartment to another, for example, from the myofibrillar (MF) compartment to the soluble (sol) compartment. A separate series of experiments was performed to examine the proteins in C and HDP soluble ("sol") fractions. In these experiments, it was desired to determine if the "sol" fraction would also prove useful in assessing differences between HDP-treated and C steaks. Furthermore, if possible, it was hoped that these experiments might show if there was a relationship between initial meat tenderness and the HDP-response.

A question to be asked is if the change in peptide patterns is a function of pre-HDP (control) tenderness (WBS). The answer may be seen in the SDS-PAGE of the "sol" fraction seen in Figure 4. Statistical analysis of the data determined that there was a significant difference ($P<0.05$) between HDP and C steaks; however, that analysis of four striploins was insufficient to adequately test the hypothesis concerning a possible relationship between initial meat tenderness and HDP-response. An investigation with adequate numbers of C samples having a broad range in WBS values merits further attention. On the other hand, important information was obtained from the SDS-PAGE gels and is presented in Figure 5.

Visual examination of Figure 5 suggests clearly that as the samples' WBS decreases (tenderness increases) from 8.2→7.6→5.5→3.8 kg$_f$, the sum of the peak-difference volumes (HDP$_{Vol}$ - C$_{Vol}$) becomes greater (28.1→39.4→48.9→109.8, respectively) with a rate of change in HDP-C/WBS kgf of 360.2 (2nd order regression analysis of slope) having an r^2 of 0.9704. It should be noted that densitometry of stained gels is a measure of the amount of stain bound to a specific protein. Since the amino acid composition of most proteins is different and since Coomassie blue R-250 only binds to certain amino acids, exact quantization of differences between two different proteins can not be performed (Tal and others[19]) ... unless you have a standard curve of each individual protein. However, staining can be used to analyze differences in the same protein even if it is from a different treatment group on the same gel. Since Figure 5 shows the "sum" of the changes, it suggests that more than one protein is being affected by the HDP-shock treatment and that, perhaps, the proteins in more tender cuts may be more susceptible to the treatment than the same proteins in tougher cuts.

Statistical-examination of individual proteins shows that they have varying responses to HDP-treatment. In some cases, there is no correlation between the HDP-C protein concentration and the initial WBS (slope). For example, the 11.9 kD protein has P=0.1233 for the slope of the change in HDP-C with WBS, while another protein (the 155 kD band) has a slope that is significantly different at P=0.0037 for HDP-C vs WBS. It is not unreasonable to expect a variation in the response of individual meat proteins to HDP-induced shock wave treatment since, by virtue of the their differing amino acid compositions, each cellular protein has a different cellular function ranging from structural

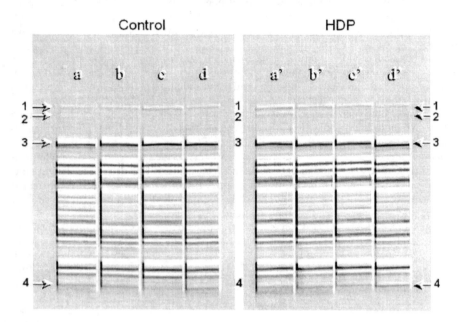

Figure 4 *SDS-PAGE gradient (10-20%) electropherogram of proteins from the soluble ("sol") fraction (post 48,200 x g supernatant solution) of control, C, (TOP LEFT) and HDP-treated beef (TOP RIGHT) striploin steaks. Gels are presented as match pair treatments (C and HDP) from four separate striploins selected based on their Warner-Bratzler shear (WBS) values at screening (10.0, 8.0, 6.1, and 4.0 kg_f). As a result of aging and freeze/thawing, the WBS values of the samples had declined between the time of purchase and the initial screening and the date of the experiment (WBS on the day of treatment was 8.2, 7.6, 5.5 and 3.8 kg_f, respectively). The lanes on the left electropherogram are labelled "a", "b", "c", and "d" representing the matching HDP-treated samples (WBS post HDP-treatment was 6.7, 6.7, 5.3, and 3.6 kg_f, respectively). Each lane contained 40 μg of protein (A_{230}). Gels were stained with Coomassie brilliant blue R-250 and scanned using a BioRad 710 densitometer. Arrows numbered 1, 2, 3, and 4 point to proteins with molecular mass of approximately 160 kD, 153 kD, 110 kD, and ~11.0 kD, respectively. The electropherogram is presented in an "embossed" format to facilitate visualization.*

Quality

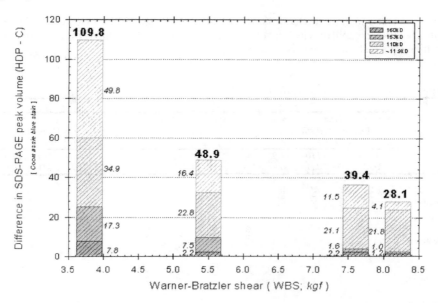

Figure 5 *Each bar shows both the individual (# to side of bar) and the summed (# on top of bar) difference in peak volumes (HDP – C) determined from densitometric scans of replicate SDS-PAGE gels. Gels were prepared from the soluble fractions of four separate striploins. The striploins had a range in WBS values on the day of the experiment from ~4 kg_f (tender) to ~8 kg_f (tough). Each bar represents the sum of the mean differences (HDP-C) for 4 proteins in the soluble fractions, i.e., 160 kD, 153 kD, 110 kD, and ~11.9 kD.*

to functional. Following this logic, each protein should respond somewhat differently to HDP-induced shock waves. Statistical evaluation of the data shows clearly that the response of four different meat proteins, that is, a 160 kD, a 153 kD, a 110 kD, and a ~11.9 kD protein, show variable response to HDP treatment such as no significant response (P=0.3141), significant response (P=0.0016 and P=0.0374), and a marginally significant response (P=0.0636), respectively (Table 2). Thus, three of the four proteins examined in the *sol* fraction showed a positive response to HDP-treatment.

Soluble fractions of meat contain a broad range of material around 11.9 kD (Figure 4). The current working hypothesis is that HDP-treatment causes fragmentation of higher molecular weight proteins; unfortunately SDS-PAGE is not the best method available to assess protein composition in lower molecular weight ranges. Therefore, no further attempt will be made to discuss the soluble fraction, low molecular weight proteins in this communication. Subsequent communications will include studies using procedures such as reverse phase high performance liquid chromatography (RP-HPLC) [see other chapter in this book] and capillary electrophoresis (CE) to examine the changes in meat proteins resulting from HDP-treatment as well as aging.

Table 2 Linear analyses of covariance

Treatment	Probability											
	Type 3 Tests of fixed effects [Pr > F]				Least Squares Means (LSM) [Pr > |t-|]				Differences of Least Squares Means (LSM) [Pr > |t-|]			
	160 kD	153 kD	110 kD	11.9 kD	160 kD	153 kD	110 kD	11.9 kD	160 kD	153 kD	110 kD	11.9 kD
No treatment	n.s. .3141	.0016	.0374	.0636								
C treatment					.0028	.0004	<.0001	.0066				
HDP treatment					.0011	<.0001	<.0001	.0010				
HDP – C treatment									n.s. .2786	.0030	.0066	.0669

n.s. = not significant   kD = kilodaltons (a measure of molecular weight)

HDP-treatment of beef striploins has been shown to improve the WBS in beef striploin samples (an average of 12.5% in these experiments with $P<0.05$) and also alters the level of several proteins in treated samples. Thus, it is reasonable to presume that there is a relationship between the change in protein composition and the tenderness of HDP-treated samples. Data suggest that this relationship may be closely tied to the levels of a ~110 kD protein since this protein appears in greater concentrations in HDP-treated striploins (Figures 2-5) of both the *MF* and the *sol* fractions. The ~110 kD protein or protein fragment seems to increase as the meat became more tender or WBS shear force decreased (Figures 4 & 5).

Recently, O'Halloran and colleagues[20] showed that during the aging of bovine muscle and as meat became more tender, a 110 kD fragment appeared in the myofibrillar, MF, fraction. Additionally, Casserly's laboratory recently showed[21] that a 110 kD myofibrillar protein fragment showed strong homology with C-protein (~140 kD) as shown by sequence analysis. There is a reasonable reason to suspect that the ~110 kD protein observed in the HDP-treated striploins could be the same 110 kD protein discussed by Casserly and colleagues.[21] The suggestion that the ~110 kD protein found in both the aging and the HDP studies are related, gains support if one considers the function of C-protein. C-protein forms cylindrical-like structures around the packed tails of myosin molecules acting as a clamp to hold and stabilize the backbone of the thick filament. C-protein also binds myosin heavy chain, F-actin, and native thin filaments.[22]

At this time, the current data does not show the origin or source of the 110 kD protein. However, the data suggest clearly that the 110 kD protein may be a result of fragmentation of larger meat proteins following HDP-treatment since less of the material is seen in the paired and untreated "C" samples. This is similar to the fragmentation of high molecular weight proteins during aging as suggested by others.[18,23-27] The high molecular weight proteins that fragment have been suggested to be, but are not limited to, titin, nebulin, myosin, and connectin. Alternatively, the 110 kD protein may result from bond-disruptions in other high molecular weight protein complexes such as myosin and actin with known functional bonding interaction and banding patterns.[17]

5.0 CONCLUSIONS

It is clear from these studies of beef striploins that HDP-treatment causes the meat to become more tender and that the tenderizing effect is most likely mediated through alteration of several meat proteins. Continued research on the effect of hydrodynamic pressure (HDP) on meat proteins will provide information essential for developing HDP or other related technologies into viable, commercially applicable, processes that will attract a high level of consumer confidence. It is also hoped that such research will lead to developing under-exploited markets for new products that produce value-added, high quality meat with desirable palatability, extended shelf-life and improved safety.

Acknowledgments

The authors graciously acknowledge the technical assistance of several individuals during various portions of this project: Dr. Brad Berry, Mrs. Janice Callahan, Ms. Janet Eastridge, and Mr. Ernie Paroczay. We gratefully acknowledge the assistance of Dr. Bryan Vinyard, Driector, BARC Biometrics Consulting Center. Sincere thanks are also extended to Dr. Morse Solomon who is licensed to handled and detonate the explosives used in this investigation. Sincere gratitude is extended to Dr. Robert (Bob) Schmukler of Pore^2 for opening his laboratory for this work. This manuscript is dedicated to the memory of Dr. Robert (Bob) Romanowski who passed away in February 2000.

References

1. D.G. Morgan, J.W. Savell, D.S. Hale, R.K. Miller, D.B. Griffin, H.R. Cross and S.D. Shackelford, *J. Anim. Sci.*, 1991, **69**, 3274-3283.
2. J. Love, In, *Quality Attributes and Their Measurement in Meat, Poultry and Fish Products*. 1994, Blackie Academic & Professional, Glasgow, pp. 337-358.
3. J. Lepetit and J. Culioli, *Meat Sci.,* 1994, **36**, 203-237
4. J. Lepetit, P. Salé and A. Ouali, *Meat Sci.*, 1986, **16**, 161-174.
5. E.J. Huff-Lonegran and S.M. Lonegran. In, *Quality Attributes of Muscle Foods.,* 1999, Kluwer Academic/Plenum Press New York. pp. 229-252
6. P. Roncalés, G.H. Geesink, R.L.J.M.Van Laack, I. Jaime, J.A. Beltrán and V.M.H. Barnier, *Expression of Tissue Proteinases and Regulation of Protein Degradation Related to Meat Quality*. 1995 Utrecht: ECCEAMST, pp. 311-332

7   T. Nishimura, A. Liu, A. Hattori and K. Takahashi, *J. Anim. Sci.*, 1998, **76**, 528-532.
8   M.M. Campo, P. Santolaria, C. Sañudo, J. Lepetit, J.L. Olleta, B. Panca and P. Albertí, *Meat. Sci.*, 2000, **55**, 371-378.
9   H.O. Hultin, *J. Chem. Edu.*, 1984, **61**, 289-298.
10 H.. Zuckerman and M.B. Solomon, *J. Muscle Foods*, 1998, **9**, 419-426.
11 A.M. Spanier, B.W. Berry and M.B. Solomon, *J. Muscle Foods*, 2000, **11**, 183-196.
12 H. Zuckerman, B.W. Berry, J.S. Eastridge and M.B. Solomon, *J. Muscle Foods*, 2002, **13**, 1-12.
13 A.M. Spanier and R.D. Romanowski, *Meat Sci.*, 2000, **56**, 193-202.
14 M.B. Solomon, 51st *Annu. Recip. Meat Conf.*, 1998, **51**, 171-176.
15 U.K. Laemmli and M. Favre, *J. Molec. Biol.*, 1973, **80**, 575-599.
16 A.M. Spanier, *Studies of the Lysosomal Apparatus in Progressive Muscle Degeneration Induced by Vitamin E - Deficiency.* Ph.D. Thesis, 1977 Rutgers, The State University of New Jersey.
17 D.L. Hopkins and J.M. Thompson, *Meat Sci.*, 2000, **57**, 1-12.
18 R.M. Robson, E. Huff-lonergan, F.C. Parrish Jr,, C-Y Ho, M.H. Stromer, T.W. Huiatt, R.M. Bellin and S.W. Sernett, 50th *Annu. Recip. Meat Conf.*, 1997, **50**, 43-52.
19 M. Tal, A. Silberstein and E. Nusser, *J. Biol. Chem.*, 1984, **260**, 9976-9980.
20 G.R. O'Halloran, D.J. Troy and D.J. Buckley, *Meat Sci.*, 1997, **45**, 239-251.
21 U. Casserly, S. Stoeva, W. Voelter, A. Healy and D.J. Troy, 44th *IcoMST (Vol. B115)*, 1998, pp. 726-727.
22 D.O. Furst, U. Vinkomeier and K. Weber, *J. Cell Sci.*, 1992, **102**, 769-778.
23 J.D. Fritz and M.L. Greaser, *J. Food Sci.*, 1991, **56**, 607-610.
24 D.E. Goll, M.L. Boehm, G.H. Geesink and V.F. Thompson, 50th *Annu. Recip. Meat Conf.*, 1997, **50**, 60-66.
25 D.L. Hopkins, P.J. Littlefield and J.M. Thompson, *Meat Sci.*, 2000, **56**, 19-22.
26 R.W. Purchas, X. Yan and D.G. Hartley, *Meat Sci.* 1999, **51**, 135-141.
27 P.F.M. van der Ven, G. Schaart, H.J.E. Croes, P.H.K. Jap, L.A. Ginsel and FCS Ramaekers, *J. Cell Sci.*, 1993, **106**, 749-759.

USE OF CAPILLARY ELECTROPHORESIS (CE) TO ASSESS THE INFLUENCE OF HYDRODYNAMIC PRESSURE (HDP)-TREATMENT AND AGING OF BEEF STRIPLOIN PROTIENS. A METHOD FOR ASSESSMENT OF MEAT TENDERNESS.[§]

A.M. Spanier[1,2] and T.M. Fahrenholz[3]

[1] Spanier Science Consulting LLC, 14805 Rocking Spring Drive, Rockville, MD 20853
[2] Pore² Bioengineering Inc., 13905 Vista Drive, Rockville, MD 20853
[3] USDA, ARS, BA, Animal & Natural Resources Institute (ANRI), Food Technology and Safety Laboratory (FTSL), Bldg 201, BARC-East, Beltsville, MD 20705

1 ABSTRACT

This study used capillary electrophoresis (CE) to compare the effects of hydrodynamic pressure (HDP)-treatment and aging on meat proteins since tenderness (measured by Warner-Bratzler Shear, WBS) developed by HDP and aging are thought to be due to changes in meat protein composition. Striploins were divided into control, C, and matching HDP sections. WBS (kg_f) for C at 0, 5, and 8 days was 5.65, 4.40, 4.14; HDP-samples were 4.82, 3.87, and 3.81 kg_f on day 0 only. WBS change was slower in HDP-treated samples (-0.2980 kg_f/day) than in C (-0.3505 kg_f/day). Difference in tenderness was significant ($P<.05$ at day 0, but only marginally different ($P<0.06$) over the aging period. CE identified 10 peaks: 1-5, 7, 9, and 10 showed significant aging change ($P<0.01$), while peaks 6 and 7 showed significant ($P<0.05$) HDP effect. Two peaks correlate with either aging (peak 5) or with HDP-treatment (peak 7). Data should lead to development of rapid method of predicting beef quality (tenderness) and possibly to development of treatments to ensure uniform and consistent meat tenderness.

2 INTRODUCTION

The U.S. meat industry has identified solving the problem of inconsistent meat tenderness as a top priority,[1,2,3,4] if they are to meet consumer meat quality expectations. It is known that tenderization of meat is a result of events that occur during the conversion of muscle to meat; the tenderization involves the degradation of muscle proteins beginning almost immediately after the animal is slaughtered.[5,6,7] However, the tenderization process is highly variable as a result of several ante- and post-mortem factors including, but not limited to, animal age, breed, sex, muscle cut, feeding regimen, method of slaughter,[1,3,5,8,9,10] as well as to complex biochemical changes and interactions that occur during the

---

[§] Mention of brand or firm names are necessary to report factually on available data and does not constitute an endorsement by the authors over other products or manufacturers of products of a similar nature not mentioned.

conversion of muscle to edible meat.[1,6,10,11,12,13,14,15,16]

Among the technologies currently[6] available to tenderize meat is hydrodynamic pressure (HDP)-treatment. HDP-treatment passes a pressure shockwave through water at supersonic speed; the shock wave also passes through any object in the water that is an acoustical match.[17] Meat, which is approximately 75% water[18] is thus a close acoustical match to water. Experiments using HDP-treatment have shown the method not only to led to a decrease in levels of several microbes associated with meat,[2,19,21] but also to be effective in instantaneously tenderizing meat and meat products[20,22,23,24,25,26] and making the tenderness across a steak[26,27] or across an entire striploin[24] more consistent.

Fragmentation of several larger meat proteins via proteolysis or other physical means has been shown to occur;[6,28-45] this change in protein compartmentation from a myofibrillar fraction to the sarcoplasmic fraction has been demonstrated to occur in several laboratories[6,46,47] and frequently in this laboratory.[24,25,27,28] Correlation between proteolytic breakdown and beef myofibrillar (MF) proteins during aging,[24,25,27,29] and changing tenderness[47] have been made and have shown clear redistribution of proteins from the MF to the "soluble" or "*sol*" fraction. This led us to hypothesize that we should be able to determine beef tenderness by using an antigen/antibody immunoassay involving the direct swabbing or sampling of the surface or directly from meat drippings (representative of a soluble fraction). Unfortunately, the senior author's forced disability retirement from his laboratory research position did not permit him to complete the latter to his satisfaction. However, initial research has shown excellent correlation between meat tenderness and specific meat proteins

If the meat industry is to develop and utilize methods that respond to the consumers' desire for product consistency, then we must obtain a thorough and detailed understanding of the processes that affect meat tenderness. Thus, the objective of this study was to use to use capillary electrophoresis (CE) to study the changes in meat proteins resulting from aging and from hydrodynamic pressure (HDP)-treatment and to examine the correlation of these changes with Warner-Bratzler shear (WBS).

3 MATERIALS AND METHODS

3.1 Meat

A total of 15 fresh, never frozen beef striploins were obtained from a local supplier and used within 42-48 h (experimental day 0) postmortem. Samples were USDA Select grade, Angus, U.S. Yield Grade 2 and 3. Each striploin was sections into two 14 cm lengths from the rib end; the anterior (rib) section was used as the control sample, C, while the posterior section was used as the paired HDP-treated sample. Beef striploin samples were labelled and then vacuum packaged (Crytovac®/Sealed Air Corporation; Saddle Brook, NJ). After packaging, samples were shrink-wrapped in a water bath set to 86.5°C. Packaged samples were held at refrigerated temperature (4°C).

3.1.1 *Hydrodynamic pressure (HDP)-treatment* followed the procedure described by Solomon.[22] HDP-treatment used a 98.4 L plastic explosive container (PEC, a household garbage container; Rubbermaid Inc., Wooster, OH) which was suspended and filled with water that was cooled by addition of chipped ice. The supersonic pressure shock wave was induced using 100 g of binary explosive in the shape of a hot-dog suspended 30.5 cm above the top of the meat. (Figure 1).

Figure 1 *Side view (top) and top-to-bottom view (bottom) of the 98.4 L plastic explosive container (PEC), Rubbermaid Inc., Wooster, OH). Water and cooling ice chips are omitted from the photo to permit a more clear presentation. Note: the "dummy" charge in the photo is for visualization only and is not the "hot-dog" shape described in the text; the dummy charge represents earlier experiments designed to determine the best format of explosive charge to utilize.*

Packaged striploins with their fat-side facing upward were secured to a 1.25 cm thick, 40.6 cm diameter steel plate. The steel plate was seated on the bottom of the PEC. The charge was detonated by a licensed handler of explosives. The samples were HDP-treated on day 0; HDP-treatment was performed on 5 separate occasions using 3 individual striploins at each treatment (Figure 2).

After HDP-treatment, each HDP and C sample had a 4.0 to 4.6 cm thick section cut from its anterior end; the remaining portions of the C and DP samples were repackaged and aged at 2 to 4 °C for 5 and 8 days prior to analysis. Using a fabricated cutting block (Plexiglas) a 2.5 cm thick steak was cut from the anterior end of each HDP and C group. The first slice of each section was used for texture analysis by Warner-Bratzler shear force (WBS) determination (below). The next slice (approximately 1 cm) was used for subsequent protein analysis.

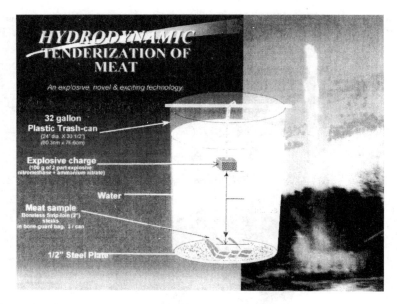

Figure 2 *Diagramatic representation of P.E.C. with meat, explosive charge, water, ½" thick steel plate, and 32 gallon P.E.C. (Note: charge is presented as a "Block" but was actually in the shape of a "hot-dog" for the experiments. To the RIGHT of the diagram is a photo of how the water looks as it was propelled out of the PEC after charge detonation (much like a bullet leaving the barrel of a rifle).*

3.2 Shear Force Determination (Warner-Bratzler Shear, WBS)

Control, C, and HDP-treated striploin steaks (*longissimus* muscle) were cut at 0, 5 and 8 day for shear force analysis/determination. All steaks were grilled to an internal end-point temperature of 71°C monitored using a NIST certified thermometer (Model HH21, Omega Engineering, Stanford, CT) using an indoor/outdoor electric grill (George Forman model GGR50B, 1600 watt, Salton, Inc., Mt. Prospect, IL). Steaks were turned midway between the initial and final temperature of 71°C.

WBS determination (kg_f) was made from a minimum of 12 cores using a 1.27 cm diameter coring device. All cores were removed parallel to the muscle fiber direction. Each core was sheared once perpendicular to the fiber orientation using a WBS test cell attachment (1.8 mm thick blade with inverted V-notch) mounted on a texture measurement system (TMS-90; Food Texture Corp., Chantilly) with a cross-head speed of 250 mm/m.

3.3 Homogenization and Subcellular Fractionation.

3.3.1 All samples were kept cool (temperature maintained on ice) at all times before portioning for homogenization. Each steak was divided longitudinally in half (from lateral end to medial end of the striploin[28]: the lower half was minced into chunks that were further mixed by shearing for 5 seconds with a handheld blender (Braun MR-370; Braun, Inc., Woburn, MA) while the upper section of the steak was frozen for possible future usage. The minced sample served as a homogeneous representation of the entire steak.

3.3.2 An individual control, C, or HDP-treated sample (5 g core) was removed from the homogeneously dispersed steak and was mixed with 45 ml of cold (4 EC) homogenization medium [HM: 0.05 M Tris (Trizma acid), 1.5 mM dithiothreitol (DTT), 1.5 mM tetrasodium ethylenediamine-tetraacetic acid (EDTA) with 0.001 M sodium azide; pH 7.0] and placed into a stainless steel homogenization vessel as described previously.[25] Tissue disruption was in a 100 ml capacity holding cup using a Waring Blendor. Homogenization speed was controlled by rheostat (Powerstat™, Superior Electric Co., Bristol, CN) at a setting of 65 for 10 seconds. The resulting homogenate was filtered through two layers of cheese cloth to remove any residual fat and connective tissue. The homogenate was labelled as total homogenate (TH) and dispersed into two aliquots: 30 ml for fractionation and ~ 10 ml either saved or discarded.

The TH was subjected to differential centrifugation (Sorvall RC-2B, Sorval Instruments, Inc., Newton, CT) at 4000 rpm (10,800 x *g)* at 4°C for 15 min). The supernatant solution was decanted into a centrifuge tube. The soluble fraction was centrifuged at 20,000 rpm (48,200 x *g*) for 30 min at 4°C. The pellet was discarded and the final supernatant solution (*"sol"*) was saved for analysis by reverse phase (RP) – high performance liquid chromatography (HPLC) and for analysis of protein concentration.

3.4 Estimation of Protein Content in Subcellular Fractions.

Protein content in the soluble fractions (*"sol"*) of the C and HDP-treated stiploins was determined on diluted samples using a 96-multiwell, UV late (Costar #3635, Corning Inc., Corning, NY). Protein levels were assessed using bovine serum albumin (BSA) as standard by recording the A_{230} (absorbency at 230 nm) with a SPECTRAFLOURPlus™ multiwell plate reader (Tecan, US, Inc., Research Triangle Park, NC). "Sol" fractions were adjusted to a final concentration of 8.7 mg/ml using HM.

3.5 Capillary Electrophoresis.

3.5.1 *Sample preparation*: Samples for capillary electrophoresis (CE) were obtained from the soluble fraction (*"sol"*) of C and HDP groups after 0, 5, or 8 days of storage. The proteins in all *"sol"* fraction samples were rendered soluble by mixing the sample 1:1 (v/v) in a 2.0 ml capacity, capped, plastic conical centrifuge tube with rendering medium (GOOP+: 0.5M Tris/HCl, pH 6.8, 10% SDS, 20% glycerol, 5% 2-mercaptoethanol, 0.1% bromphenol blue). Samples in GOOP wee boiled for 8-10 min to inactivate proteolytic enzymes and render proteins soluble. All samples were diluted to a final concentration of 1.03 mg / ml containing 1 μl/100 l of Orange G reference marker (Beckman Instruments, Inc., Fullerton, CA) was added to the rendered material. Fifty microliters (50 μl), of diluted orange-G containing sample solution was placed in a 400 μl sample vial. The sample, sample vial and sample loader were capped with a septum and placed into an autosampler for capillary electrophoresis (P/ACE™ System 5000 Series, Beckman Instruments, Inc. Fullerton, CA).

3.5.1.1 *CE protocol*: A SDS-coated capillary column (100 μm I.D. x 27 cm long; Beckman Instruments, Inc. Fullerton, CA) was used to separate and resolve individual proteins in the "*sol*" fractions. Absorbency of electrophoretically separated material was measured at 214 nm. Each CE run was preceded with a 1 minute rinse with 1N HCl followed by a 3 minute rinse with gel buffer (Beckman Instruments, Inc, Fullerton, CA). All samples were pressure (not electro-kinetically) injected for 30 seconds at approximately 0.5 psi. Electrophoretic separation was at 8.1 kV, 20°C (constant) for 19 minutes. All samples were prepared and run in triplicate so that 270 individual CE runs were made of 2 treatments (HDP & C) x 3-storage times (0, 5, and 8 days) of 15 striploins (2x3x15x3 = 270). The autosampleer carousel holding the samples also contained positions for gel buffer, 1N HCl, and empty vial for deposit of rinse.

3.6 Data Preparation and Statistical Analysis

CE data was converted into ASCII format using a Beckman P/ACE Station Chromatography Data System (Beckman Instruments, Inc., Fullerton, CA). ASCII files were transferred to an Excel® spreadsheet (Microsoft Corp.; Redmond, WA) on a computer dedicated to data analysis. Composite chromatograms for each treatment (HDP, C) and each day (0, 5, 8) were prepared. Chromatograms were visually examined and several peaks chosen for statistical analysis.

Separate univariate 2-way repeated measures ANOVA analysis (SAS Institute, Inc., Cary, NC)[48] was conducted for each response variable. The alpha level for means was set at 0.05. Repeated values are all data values measured on the same experimental unit.

4 RESULTS AND DISCUSSION

Earlier studies[28,29] (other chapters this proceedings) designed to examine the change in beef protein composition resulting from HDP-treatment and from aging have principally utilized sodium dodecyl sulphate (SDS) polyacrylamide gel electrophoresis (PAGE) and reverse phase high performance liquid chromatography (RP-HPLC); each of these methods has its own advantage and inherent limitations. Because there are limits to the analysis, we also examined fresh, 5 day and 8 day aged beef from untreated (C) and HDP-treated beef striploins.

As with any analytical technique, a set of samples of known molecular weights and composition should be examined and a standard curve developed. A standard curve using myosin (205 kD), β-galactosidase (116.02 kD), phospholipase b (97.4 kD), bovine serum albumin (66 kD), ovalbumin (45 kD), carbonic anhydrase (29 kD), and lactalbumin (14.2 kD). Figure 3 shows the standard curve plotted as a 1st order linear regression (top) and a 2nd order linear regression (bottom). Both of these graphs were used to estimate the amount of protein in the 10 peaks observed in our capillary electrophoresis analysis of C and HDP soluble fractions (Table 1, aged for 0, 5, and 8 days).

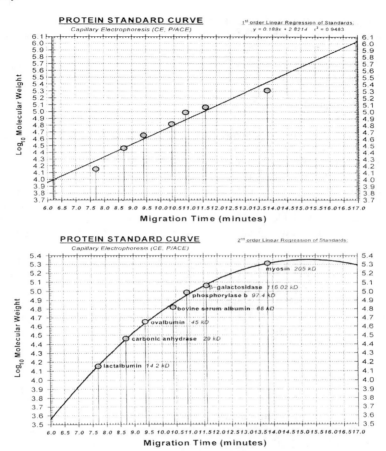

Figure 3 *Standard curve generated from seven known proteins on a P/ACE™ system 5000 with an SDS-coated capillary column (100 μm I.D. x 27 cm long; Beckman Instruments, Inc., Fullerton, CA, UA). Standards include the following protein: lactalbumin (14.2 kD), carbonic anhydrase (29 kD), ovalbumin (45 kD), bovine serum albumin (66 kD), phosphorylase b (97.4 kD), β-galactosidase (115.2 kD) and myosin 205 kD) with absorbency measured at 214 nm. TOP: 1st order regression of standard with $r^2 = 0.9483$. BOTTOM: 2nd order regression with $r^2 = 1.0000$.*

Electropherograms presented in Figure 4 show the resolution of protein material obtained from control (upper) and HDP (lower) soluble fractions from striploins aged 0, 5, or 8 days. Data on peak area (triplicate CE runs of each treatment at each storage time) were examined using SAS for statistical analysis.[48] Ten peaks were resolved (Figure 4) having an average peak molecular weight shown in Table 1.

Two-way repeat-analysis ANOVA of the CE response variable in control and HDP-treated soluble fractions (N=15C; N=15HDP) show that all but peaks CE-6 and CE-8 show significant difference with aging from 0 to 8 days aging (Table 2, column 2). Only peaks CE-6 and CE-7 (Table 2, column 3) show significant difference as a result of HDP0-treatments. None of the peaks show a statistically significant relationship in the "days x treatment" group (Table 2, column 4). Of note is that CE-5 appears to be related only to aging while CE-6 appears to be related to HDP-treatment only. Peak CE-7 is significant for both aging (Table 2, column 2) and HDP-treatment (Table 2, column 3).

The CE peak profile were examined statistically for "means" comparison at an alpha of 0.05. The response variable assessed by CE were from C and HDP-treated beef striploin soluble ("*sol*") fractions. Analysis by tw0-way repeated ANOVA is seen in Table 3.

CE-7 shows a significant difference (alpha 0.05) for both effects, i.e. days and treatment (Table 3, column 2, row 1) and treatment (HDP/C; Table 3, column 3, row 1). CE-4 shows a significant reduction in size from 0 to 8 days aging (Table 3, columns 5 & 6, row 4), while CE-3 and CE-9 show a small yet significant increase from 0 – 8 days (Table 3, columns 5 & 6 data, rows 5 & 6, respectively). Combined peaks CE-1, CE-2, CE-5 and CE-7, show small but significant increases for 0 to 5 days only (Table 3, columns 5 & 6, row 6).

5 CONCLUSIONS

It is clear from these studies (this manuscript and [28, 29]) of beef striploins that HDP-treatment causes the meat to become more tender and that the tenderizing effect is most likely mediated through alteration of several meat proteins. Continued research on the effect of hydrodynamic pressure (HDP) on meat proteins will provide information essential for developing HDP or other related technologies into viable, commercially applicable, processes that will attract a high level of consumer confidence. It is also hoped that such research will lead to developing under-exploited markets for new products that produce value-added, high quality meat with desirable palatability, extended shelf-life and improved safety. Last, these data have identified specific protein peaks that can be utilized as indicators of a meat's level of tenderness and a meat's aging period. Specifics will necessarily have to be determined on various muscle types and from different breeds (future manuscript) to determine the specific standard curve for the specific muscle and breed.

Table 1 Molecular weight of ten peaks resolved by capillary electrophoresis[a] (CE) of soluble fractions from C and HDP striploin[a] at 0, 5, and 8 days of aging.

Peak Number	Migration time (min)	Molecular Weight (kD)[b]	
		1st order regression	2nd order regression
1	8.03	17.7	18.2
2	8.67	25.7	26.3
3	8.86	30.9	30.9
4	9.13	35.5	36.3
5	9.53	42.7	46.8
6	9.84	49.0	55.0
7	10.23	57.5	69.2
8	11.02	79.4	93.3
9	12.39	151.4	154.9
10	16.87	1023.3	199.5

[a] capillary electrophoresis used a P/ACE™ system 5000 with an SDS-coated capillary column (100 µm I.D. x 27 cm long; Beckman Instruments, Inc. Fullerton, CA, USA)used to separate and resolve individual proteins in "sol" fractions. Absorbency of electrophoretically separated material was measured at 214 nm.

[b] molecular weight in kD (kilodaltons) were based on the use of the following protein standards: lactalbumin (14.2 kD), carbonic anhydrase (29 kD), ovalbumin (45 kD), bovine serum albumin (66 kD), phosphoryalse b (97.4 kD), β-galactosidase (115.2Kd and myosin (2105 kD). The "range" of MW (kD) is based on using values from both the 1st and 2nd order regression (see Figure 3)

Table 2 Two-way repeated-measures ANOVA analysis of response variable measured with capillary electrophoresis (CE) in control, C, and HDP-treated beef striploin soluble fractions.

Response (CE peak #)	Significance (better than 0.10)		
	Days (0, 5, and 8 days)	Treatment ("Trt") (HDP and C)	days x treatment
CE-1	0.235	n.s.[a]	n.s.
CE-2	<0.0001	n.s.	n.s.
CE-3	<0.0001	n.s.	n.s.
CE-4	<0.0001	n.s.	n.s.
CE-5	0.0132	n.s.	n.s.
CE-6	n.s.	0.0547[b]	n.s.
CE-7	0.0031	0.0229	n.s.
CE-8	n.s.	n.s.	n.s.
CE-9	<0.0001	n.s.	n.s.
CE-10	<0.0001	n.s.	n.s.

[a] n.s. = not significant below 0.10
[b] 0.05 – 0.06 are considered 'moderately' significant

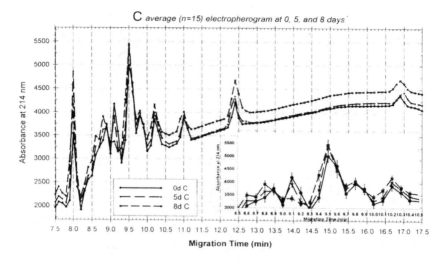
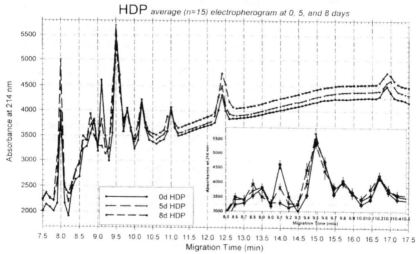

Figure 4 *Capillary electropherograms of control, C, (top) and HDP-treated (BOTTOM) beef striploin "sol" fractions at 0, 5, and 8 days storage (0 day = 2 days post-mortem). Ten peaks were determined to be important after statistical analysis and represent proteins with the migration time shown in Table 1*

Table 3 Two-way repeated-measures ANOVA analysis. Alpha level for means comparisons is 0.05. Response variable measured with capillary electrophoresis (CE) in control, C, and HDP-treated beef striploin soluble fractions.

Response	Effect (days, Trt)	Trt	Days	Estimate of mean ± standard error	Mean significance (MS) group
CE-7	Trt	HDP		7.17 ± 0.15	A
		C		6.67 ± 0.15	B
WBS	days		0	5.23 ± 0.24	A
			5	4.23 ± 0.24	B
			8	3.97 ± 0.24	B
CE-4	days		0	14.82 ± 0.28	A
			5	8.12 ± 0.54	B
			8	4.36 ± 0.28	C
CE-3	days		8	11.87 ± 0.18	A
			5	11.35 ± 0.18	B
			0	10.39 ± 0.18	C
CE-9	days		8	7.66 ± 0.27	A
			5	6.24 ± 0.27	B
			0	3.96 ± 0.27	C
CE-1 CE-2 CE-5 CE-7	days	CE-1	8	19.38 ± 0.54	A
		CE-2		6.95 ± 0.08	
		CE-5		28.53 ± 0.63	
		CE-7		7.30 ± 0.19	
		CE-1	5	18.23 ± 0.54	A
		CE-2		6.84 ± 0.20	
		CE-5		27.85 ± 0.63	
		CE-7		7.05 ± 0.19	
		CE-1	0	17.25 ± 0.54	B
		CE-2		6.21 ± 0.08	
		CE-5		25.93 ± 0.63	
		CE-7		6.40 ± 0.19	

Acknowledgments

The authors graciously acknowledge the technical assistance of several individuals during various portions of this project: Dr. Brad Berry, Mrs. Janice Callahan, Ms. Janet Eastridge, and Mr. Ernie Paroczay. We gratefully acknowledge the assistance of Dr. Bryan Vinyard, Driector, BARC Biometrics Consulting Center. Sincere thanks are also extended to Dr. Morse Solomon who is licensed to handled and detonate the explosives used in this investigation. Sincere gratitude is extended to Dr. Robert (Bob) Schmukler of Pore[2] Bioengineering Inc. for opening his laboratory for this work. This manuscript is dedicated in part to the memory of Dr. Robert (Bob) Romanowski who passed away in February 2000.

References

1. M. Koohmaraie, M.P. Kent, S.D. Shackelford, E. Veiseth and T.L. Wheeler, *Meat Sci.*, 2002, **62**, 345-352.
2. R.L. Alsmeyer, J.W. Thorton and R.L. Hiner, *J. Anim. Sci.*, 1965, **24**, 526-530.
3. J.B. Morgan, J.W. Savell, D.S. Hale, R.K. Miller, D.B. Griffin, H.R. Cross and S.D. Shackelford *J. Anim. Sci.*, 1991, **69**, 3274-3283.
4. S.D. Shackelford, T.L. Wheeler and M. Koohmaraie, *J. Anim. Sci.*, 1997, **75**, 175-176.
5. D.E. Goll, M.L. Boehm, G.H. Geesink and V.F. Thompson, 50th *Annu. Recip. Meat Conf.*, 1997, **50**, 60-67.
6. M. Koohmaraie, *Proc. Recip. Meat Conf.*, 1994, **36**, 93-104.
7. T.L. Wheeler and M. Koohmaraie, *J. Anim. Sci.*, 1994, **72**, 1232-1238.
8. E. Dransfield, *Meat Sci.*, 1994, **36**, 105-121.
9. M.A. Ilian, J.D. Morton, M.P. Kent, C.E. LeCounteur, J. Hickford, R. Crowley and R. Bickerstaffe *J. Anim. Sci.* 2001, **79**, 122-132.
10. M. Koohmaraie *Proc. Recip. Meat Conf.*, 1995, **48**, 69-75.
11. A. Asghar and A.M. Pearson, In: *Advances in Food Research*, Vol. 26, 1980, Academic Press, NY, pp. 53-213.
12. C. Faustman, In: *Muscle Foods: Meat, Poultry and Seafood Tech.*, 1994, Chapman and Hall, NY pp. 63-78
13. E.J. Huff-Lonegran and S.M. Lonegran, In: *Quality Attributes of Muscle Foods*, 1999, Kluwer Academic/ Plenum Press, NY, pp. 229-252.
14. C. Valin and A. Ouali, In: *New Technol for Meat and Meat Prod.*, 1992, P AUdet TIjdschriften B.V., The Netherlands, pp. 163-179.
15. R.M. Robson, E. Huff-Lonergan, F.C. Parrish, Jr., D-Y. Ho, M.H. Stromer, T.W. Huiatt, R.M. Bellin and S.W. Sernett, *Proc. Recip Meat Conf.*, 1997, **50**, 43-52.
16. Ouali, *J. Muscle Foods*, 1990, **1**, 126-165.
17. H. Kolsky, In: Stress Waves in Solids, 1980, Dover Pub. Co., NY.
18. J.F. Price and B.B. Schweigert, In: *The Science of Meat and Meat Products*, 1978, Food and Nutrition Press, Inc., Westport, CT
19. H.R. Gamble, M.B. Solomon and J.B. Long, *J. Food Protect.*, 1998, **61**, 637-639.
20. S. Moeller, D. Wulf, D. Meeker, M. Ndife, N. Sundararajan and M.B. Solomon, *J. Anim. Sci.*, 1999, **77**, 2119-2123.
21. A.M. Williams-Campbell and M.B. Solomon, *J. Food Protect.*, 2002, **65**, 571-574.
22. M.B. Solomon, 51st *Annu. Recip. Meat Conf.*, 1998, **51**, 171-176.
23. M.B. Solomon, J.S. Eastridge, H. Zuckerman and W. Johnson, *Proc. 43 Intl. Cong. Meat Sci. Technol.*, 1997, pp 121-124.
24. A.M. Spanier, B.W. Berry and M.B. Solomon, *J. Muscle Foods*, 2000, **11**, 183-196.
25. A.M. Spanier and, R.D. Romanowski, *Meat Sci.*, 2000, **56**, 193-202.
26. H. Zuckerman and M.B. Solomon, *J. Muscle Foods*, 1998, **9**, 419-426.
27. H. Zuckerman, B.W. Berry, J.S. Eastridge and M.B. Solomon, *J. Muscle Foods*, 2002, **13**, 1-12.
28. A.M. Spanier and T.M. Fahrenholz, In, *Recent Advances in Food and Flavor Chemistry*, 2005, this book.

29 A.M. Spanier, T.M. Fahrenholz, E.W. Paroczay, R. Schmukler and T.M. Fahrenholz, In, *Recent Advances in Food and Flavor Chemistry*, 2005, this book.
30 Ouali, *J. Muscle Foods*,1990, **1**, 126-165.
31 G.R. O'Halloran, D.J. Troy and D.J. Buckley, *Meat Sci.*, 1997, **45**, 239-251.
32 R.H. Locker, *J Food Microstruct*, 1984, **3**, 17-24.
33 R.H. Locker and D.J.C. Wild, *Meat Sci.*, 1984, **10**, 207-233.
34 R.H. Locker and D.J.C. Wild, *Meat Sci.*, 1984, **11**, 89-108.
35 M.S. MacBride and F.C. Parish, Jr., *J. Food Sci.*, 1977, **42**, 1627-1629.
36 R.W. Purchas, X. Yan and D.G. Hartley, *Meat Sci.*, 1999, **51**, 135-141.
37 D.O. Furst, U. Vinkomeier and K. Weber, *J. Cell Sci.*, 1992, **102**, 769-778.
38 J.D. Fritz and M.L. Greaser, *J Food Sci.*, 1991, **56**, 607-610.
39 U. Casserly, S. Stoeva, W. Voelter, A. Healy and D.J. Troy, In: *44th ICoMST* (Vol. **B115**), 1998, Barcelona, Spain, pp 726-727.
40 D.L. Hopkins and J.M. Thompson, *Meat Sci*, 2000, **57**, 1-12.
41 D.L. Hopkins, P.J. Littlefield and J.M. Thompson, *Meat Sci.*, 2000, **56**, 19-22.
42 P.F.M. van der Ven, G. Schaart, H.J.E. Croes, P.H.K. Jap, L.A. Ginsel and F.C.S. Ramaekers *J. Cell Sci.,* 1993, **106**, 749-759.
43 L.D. Yates, T.R. Dutson, J. Caldwell and Z.L. Carpenter, *Meat Sci.*, 1983, **9**, 157-179.
44 K.R. Rowe, S.M. Maddock, E. Lonegan and E. Huff-Lonergan, *J. Anim Sci.*, 2001, **79** (Suppl), 20-23.
45 S.M. Lonergan, E. Huff-Lonergan, B.R. Wiegand and L.A. Anderson, *J. Musc. Foods*, 2001, **12**, 121-136.
46 O.K. Taitatlonis, S. Stoeva, H. Echner, A. Bainfes, L. Margomenou, H.L. Katsouias, D.J. Troy, W. Voeiter, M. Papamichail and P. Lymbert, *J. Immuno. Mtds*, 2002, **206**, 141-149.
47 J.C. Sawdy, S.A. Kaiser, N.R. St-Pierre and M.P. Wick, *Meat Sci.*, 2004, **67**, 421-426.
48 SAS Institute Inc., 1996, *SAS Users Guide to the Statistical Analysis System*. Cary, NC, SAS Inst., Inc.

CHANGES IN PROTEIN DISTRIBUTION IN BEEF *SEMITENDINOSUS* MUSCLE (ST) IN SAMPLES SHOWING VARYING RESPONSE TO HYDRODYNAMIC PRESSURE (HDP)-TREATMENT.[§]

A.M. Spanier[1,2] and T.M. Fahrenholz[3]

[1] Spanier Science Consulting LLC, 14805 Rocking Spring Drive, Rockville, MD 20853, U.S.A.
[2] Pore² Bioengineering Inc., 13905 Vista Drive, Rockville, MD 20853, U.S.A.
[3] USDA, ARS, BA, Animal & Natural resources Institute (ANRI), Food Technology and Safety Laboratory (FTSL), Bldg 201, BARC-East, Beltsville, MD 20705, U.S.A.

1 ABSTRACT

Hydrodynamic pressure (HDP)-treatment is a technology that has shown promise in addressing problems in meat tenderness. On occasion, some samples do not respond nor even show a negative response to the shock wave treatment. In this study we examined semitendinosus (ST) muscle (inside top round) from Canadian, A-grade beef carcasses. Control, C, and HDP-treated SI homogenates were prepared and exposed to centrifugal fractionation to isolate myofibrillar (MF) and soluble ("*sol*") fraqctions. ST protein profiles were examined by sodium dodecyl sulfate (SDS) polyacrylamide gel electrophoresis (PAGE); level of tenderness was assessed by determination of Warner-Bratzler shear (WBS) values. Differences in protein distribution between C and HDP-treated samples were found in both the MF and "*sol*" fraction. However, the main proteins related to HDP-treatment were found in the MF fraction, i.e., a 200 kD and a 110 kD sample ($P<0.01$ and $P<0.05$, respectively). The data strongly implicate the shock-wave treatment of meat with probable disruption of c-protein.

2 INTRODUCTION

Morgan and others[1] suggest that meat tenderness has a profound affect on a consumer's perception and acceptance of the product. Thus, a major challenge for the meat industry is to market tender products without the necessity of utilizing lengthy, conventional aging. Several procedures have appeared over the years in an attempt to address tenderness issues including, but not limited to, electrical stimulation, proteolytic enzyme treatments, infusion with calcium, pH control, skeletal alteration, blade tenderization, hydrostatic pressure (HPP) and, more recently, hydrodynamic pressure (HDP).

[§] Mention of brand or firm names are necessary to report factually on available data and does not constitute an endorsement by the authors over other products or manufacturers of products of a similar nature not mentioned.

A fundamental approach to studying a very practical problem in meat science, such as meat tenderness, would be to learn as much as possible about the mechanisms by which a particular technology works. One such method, the HDP-process, is based on the theory that a pressure-induced shock wave passes through water and any object in the water.[2,3] Unlike traditional aging which removes Z-lines, as the HDP shockwave traverses the meat, the sarcomeres are torn at the Z-line and A-I band juncture[3] aiding to make the meat more tender.

The objective of this study was to obtain chemical data that might provide information useful to understanding how HDP-treated beef becomes more tender than its untreated control, C. Sodium dodecyl sulfate (SDS) polyacrylamide gel electrophoresis (PAGE) was utilized to examine the protein profiles of myofibrils isolated from control and HDP-treated ST muscle.

3 MATERIALS AND METHODS

3.1 Meat

3.1.1 Semitendinosus (ST, inside top round) meat was selected from Canadian A grade carcasses. Carcass weights ranged from 865 – 1098 lbs, fat thickness was from 3 to 8 mm and *longissimus* muscle area from 13.4 to 21.2 sq. in. for the carcasses providing the ST muscle. At four days post-mortem, ST muscles were removed from carcasses, vacuum packaged and shipped for arrive at 5 days post-mortem.

Control, C, and HDP-treated inside top round steaks (*semitendinosus* muscle, ST) were cut 3.2 cm thick; cuts were from each of the two facing steaks. The next immediate 2.5 cm section of the control and the HDP-treated group were sectioned after treatment and were used for determination of Warner-Bratzler shear force values (WBS). Beef samples to be HDP-treated were placed in yellow 'bone-guard' bags (Cryovac® / Sealed Air Corp., Saddle Brook, NJ) and shrink-wrapped in a water bath set approximately to 86.5 °C.

3.1.2 *Hydrodynamic pressure (HDP)-treatment:* HDP- shockwave treatment followed the procedure described earlier (this book) using a 98.4 L plastic explosive container (PEC, a household garbage container; Rubbermaid Inc., Wooster, OH). The explosive charge was 100 g of binary explosive in the shape of a hot dog suspended 30.5 cm above the top of the meat. Water and meat temperature were maintained at approximately 4 to 6 °C by addition of chipped ice. The PEC was suspended approximately 30 cm above a concrete floor. Samples were placed on top of a steel plate (1.25 cm thick, 40.6 cm diameter). Samples were placed "fat-side-up" on the steel. The PEC was filled with water to a height of 15.3 cm above the explosive charge.

3.2 Shear Force Determination (Warner-Bratzler Shear, WBS)

All steaks were grilled to an internal end-point temperature of 71°C monitored using a NIST certified thermometer (Model HH21, Omega Engineering, Stanford, CT, USA) using an indoor/outdoor electric grill (George Forman model GGR50B, 1600 watt, Salton,

Inc., Mt. Prospect, IL). Steaks were turned midway between the initial and final temperature of 71°C.

Steaks in this group contained control rounds and were cut to a thickness of 3.2 cm. C and HDP-steaks were grilled to an internal end-point cooking temperature of 71 °C using an indoor electric grill (George Forman model GR-20, 820 watt, Salton, Inc., Mt. Prospect, IL). The sample's internal temperature was monitored using Type J, iron-constantan thermocouples attached to a multipoint recording potentiometer (Speedomax, Model 1650, Leeds and Northrup, North Wales, PA). Steaks were turned midway between the initial and final temperature of 71°C as described above.

Cores (a minimum of 6 per steak) were removed from each steak using a 1.27 cm diameter coring device. Cores were made parallel to the muscle fiber orientation. Each core was sheared once at right angles to the fiber orientation using a WBS test cell with 3.18 mm thick blade mounted on a texture measurement system (Food Texture Corp., model TMS-90, Chantilly, VA) using a 3.18 mm thick flat bottom blade with an inverted V-shape to surround the meat core during shearing. The cross-head speed was set to 25 cm/min.

3.3 Homogenization and Subcellular Fractionation.

3.3.1 Beef samples were kept cool (temperature maintained with ice) at all times before portioning for homogenization. Cores 1.7 cm thick (~5.0 g) were made using newly sharpened #12 cork borer. All steaks were trimmed of visible fat and connective tissue.

The 5 g core was removed from the homogeneously dispersed steak and was mixed with 45 ml of cold (4 EC) homogenization medium [HM: 0.05 M Tris (Trizma acid), 1.5 mM dithiothreitol (DTT), 1.5 mM tetrasodium ethylenediamine-tetraacetic acid (EDTA) with 0.001M sodium azide; pH 7.0] and placed into a stainless steel homogenization vessel as described previously.[5] Tissue disruption was in a 100 ml capacity holding cup using a Waring Blendor. Homogenization speed was controlled by rheostat (Powerstat™, Superior Electric Co., Bristol, CN) at a setting of 65 for 10 seconds. The resulting homogenate was filtered through two layers of cheese cloth to remove any residual fat and connective tissue. The homogenate was labelled as total homogenate (TH) and dispersed into two aliquots: 30 ml for fractionation and ~10 ml either saved or discarded.

The TH was subjected to differential centrifugation (Sorvall RC-2B, Sorval Instruments, Inc., Newton, CT) at 4000 rpm (10,800 x g) at 4 °C for 10 min.). The supernatant solution was decanted into a centrifuge tube and labelled S_1. The pellet containing the myofibrillar material was resuspended by vortex to one-half the original volume with cold wash solution [0.1 M sodium chloride, 0.001 M sodium azide, 0.5% Triton X-100] and again centrifuged at 4,000 rpm (10,800 x g) for 10 min at 4°C. The resulting pellet (P) was resuspended 3-times to one-half its original volume with myofibril isolation buffer [MFIB: 0.01 M phosphate buffer, 0.001 M ethyleneglyco-*bis*tetraacetic acid (EGTA), 0.002 M magnesium chloride, 0.1 M potassium chloride, adjusted to pH 7.0 as needed] and centrifuged to remove residual detergent and azide. The final MF pellet was resuspended with 10 to 15 ml of a 1:1 mixture of MFIB:glycerol (to serve as cryoprotectant) and saved. The supernatant solution (S_1) was centrifuged at 20,000 rpm (48,200 x g) for 30 min at 4°C. The pellet was discarded and the final supernatant solution (S) saved.

3.4 Estimation of Protein Content in Subcellular Fractions.

Protein concentration in homogenized fractions was determined on appropriately diluted samples using bovine serum albumin (BSA) as standard.[6] Protein levels were assessed by reading the absorbance at 230 nm (A_{230}). Analysis was performed using 96-well UV multiwell plates (Costar #3635, Corning Inc., Corning, NY) in a multiwell plate reader (SPECTRAFLOURPlus, Tecan, US Inc., Research Triangle Park, NC).

3.5 Electrophoresis.

Proteins were separated electrophoretically on polyacrylamide gels in the presence of SDS using the discontinuous gel procedure first described by Laemmli and Favre.[6] The system was used for comparison of the control and HDP-treated sample and consisted of a commercially available 4-20% acrylamide gradient of Tris-HCL (Criterion system®, Bio-Rad Labs., Hercules, CA) containing built-in sample combs.

All SDS-PAGE samples were rendered soluble by mixing the sample 1:1 (v/v) in a 2.0 ml capacity, capped, plastic conical centrifuge tube with rendering medium (GOOP+: 0.5 M Tris/HCl, pH 6.8, 10% SDS, 20% glycerol, 5% 2-mercaptoethanol, 0.1% bromphenol blue). Samples in GOOP were boiled for 4 to 10 min to inactivate proteolytic enzymes and render proteins soluble. Samples were loaded in each well at the total concentrations of proteins noted in each figure. Acrylamide gels were subjected to a constant voltage (200V) for 45 min to 1 hour powered by a Power Pac 3000 (Bio-Rad Labs, Hercules, CA). The electrophoresis unit was placed in an ice bath to minimize temperature-induced production of artefacts in the gels during electrophoresis. Gels wee removed from their sandwiched position between the glass holding plates and stained with 0.05 Coomassie blue R-250 in 7% acetic acid:40% methanol overnight at room temperature. Gels were stained and destained in a mixture of 7% acetic acid and 15% methanol until the gel background was clear. A mixture of protein standards was used (10 kD protein ladder, Gibco BRL, Grand Island, NY). Densitometric scanning of the wet stained SDS-PAGE gel was performed using a densitometer (Bio-Rad, Model 710, Hercules, CA) with peak area assessed with Quantity 1™ software (Bio-Rad, Model 710, Hercules, CA).

3.6 Data Preparation and Statistical Analysis

Data was examined statistically using the Student's t-test.

4 RESULTS AND DISCUSSION

HDP-treatment frequently gives variable responses to the shockwave treatment. For example, sometimes we will see a significant improvement, sometime absolutely no improvement, and even more striking, sometimes get a product that is even less tender than its untreated control; we have named these three responses to HDP-treatment as responder, non-responder, and negative responder, respectively. Such variation in the improvement in tenderness are shown in (Table 1) with improvements of 24.5% (responder), 1.0% (non-responder), and -3.5% (negative responder) for steaks.

Table 1 *Warner-Bratzler shear (WBS) value (kg_f) of control, C, and hydrodynamic pressure (HDP)-treated beef <u>semitendinois</u> muscle.*

Sample Identification[a]	WBS (mean $kgf \pm s.d.$[b])		response category
	C	HDP	
UG-13	6.62 ± 0.89	5.00 ± 1.52	Responder
UG-12	6.94 ± 0.74	6.87 ± 0.71	non-responder
UG-02	6.90 ± 0.72	7.16 ± 0.55	negative responder

[a] UG = University of Guelph
[b] scanned with BioRad 710 densitometer using Quantity 1 software.

Statistical evaluation indicted that these response variables were not statistically significant (P>0.05) within the same steak (Figures 1 and 2) with the same treatments, i.e. a control, C, steak cored in several locations showed the same response at all locations. Lack of difference by locations is supported by the lack of variation in SDS-PAGE banding profiles of the sections (lateral, *Lt*; central, *Cn*; and medial, *Md*); in this case no significant difference (P>0.05) in protein profile within a treatment within a steak, e.g. in each rep (*semitendinosus* steaks #1, 2, and 3); in this case no significant difference (P>0.05) in protein profile within a treatment within a steak, e.g. in each rep (*semitendinosus* steaks #'s 1, 2, and 3) there were no differences seen between control-Lt and control-Cn, between control-Lt and control-Md, and between control-Cn and control-Md (P>0.05) and similarly for the HDP-treated samples. However, when one examined all of the cores (15/steak) for each of the reps in responsive steaks and then compared tenderness between treatments (HDP and C) one observes a significant improvement (P<0.05) in tenderness (12%, not shown) after HDP-treatment. This observation is consistent with earlier observations of variability in tenderness within a steak.[4,7]

SDS-PAGE was performed on myofibrillar (MF) fractions (Figure 1) and MF and soluble, "*sol*", fractions (Figure 2) obtained from each 'rep'; samples were obtained from a different section (lateral, *Lt*, central, *Cn*, and medial, *Md*) of each treatment (C and HDP) in responsive steaks. Several major bands were chosen for densitometric analysis: myosin, a 95-105 kD protein, actin, and lower MW bands (< actin). In all steaks (rep) and in all sections (*Lt*, *Cn*, *Md*) of each 'rep', HDP-treatment has visually thickened the band width (density) of myosin and actin, suggesting that the pressure-induced shockwave of the HDP-treatment tore and disrupted several bonds of these contractile proteins. This suggestion gains support from the observation of bands typically associated with fragmentation products of actin and myosin.[8] Densitometric scans of the electropherograms revealed some differences and some similarities in each steak and between similar sections of each steak that had each lane loaded with the same amount of protein. A clearly visible difference in the banding intensity of each of the three samples was evident (Figure 1), suggesting that each steak demonstrated a different resistance to the mechanical shearing of homogenization and to HDP-treatment. Statistical analysis indicated that there was a significant difference between C and HDP proteins (P<0.05).

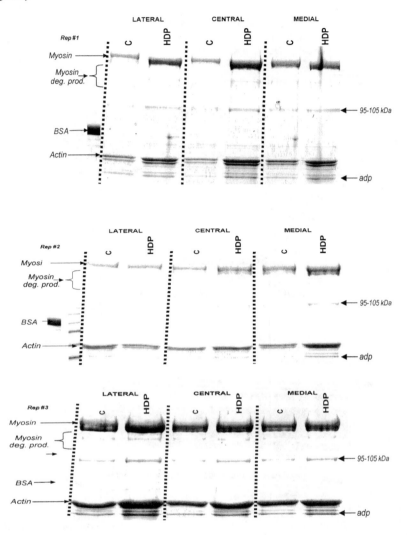

Figure 1 *Representative SDS-PAGE of a section (lateral, Lt, central, Cn, medial, Md) from 3 different sections in 3 different steaks that are responsive to HDP-treatments. The basic difference seen is not in the banding pattern, but rather in the total quantity of band; all lanes contain the same amount of loaded protein material. Within each group, the HDP always shows a more broad staining pattern than its matching control, C.*

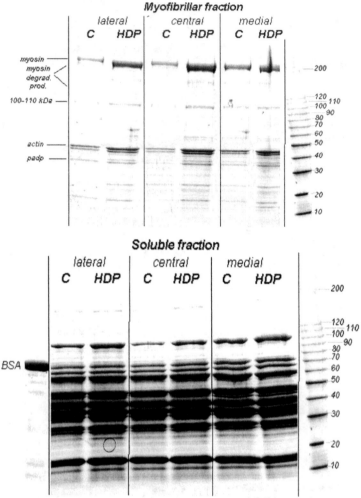

Figure 2: *Electropherogram (SDS-PAGE) of representative control (C) and hydrodynamic pressure (HDP) - treated "responsive" beef. TOP graph represents the electropherogram of the myofibrillar fraction (MF) while the BOTTOM electropherogram represents the soluble ("sol") fraction. C and HDP pairs are obtained from the lateral, central and medial section of the steak as described in text.*

SDS-PAGE analysis seen in Figure 2 (top) showed bands appearing in the HDP-responsive meat in the region of 160 kD and 153 kD; these bands have an electrophoretic mobility similar to that of various isoforms of myosin heavy chain[9] or perhaps 'M-protein' that has a mass of 165 kD.[10] Unfortunately, staining of these bands on the SDS-PAGE gels is variable (Figures 1 and 2) depending upon the location from which the core sample was taken in a particular steak. The soluble fractions of responsive meat samples (Figure 2, bottom) contained a broad range of material around 10 kD; this material is thought to represent various fragments of several higher molecular weight proteins that have been affected by the HDP-treatment. Figure 3 Plot of the mean and ± standard deviation of densitometric scans of four selected proteins from the MF gel of responsive meat samples (Figure 2).

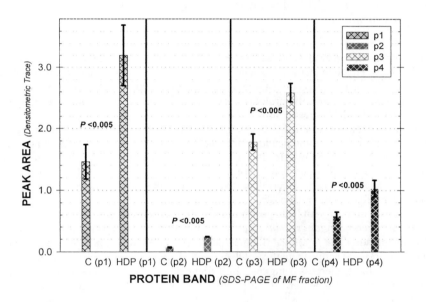

Figure 3: *Plot of the mean and ± standard deviation of densitometric scans of four selected proteins from the MF gel of responsive meat samples (Figure 2). Myosin is labelled as 'p1', a second protein of approximately 100 - 110 kD mass is labelled 'p2', a third protein band that is most likely actin is labelled 'p3' and the group of proteins that may represent a presumptive actin degradation products (padp) is labelled as 'p4'. (see Figure 2). Level of significance is shown graph at $P<0.05$ or better.*

Table 2 *Density of two protein peaks from SDS-PAGE analysis of semitendinosus muscle.*

Peak[c] ID	Mean peak area ± s.d.[a]								
	Negative responder			Non-responder			Responder		
	C	HDP	p[b]	C	HDP	p	C	HDP	p
MF 200 kD	12.8 ± 0.8	10.7 ± 0.7	n.s.	11.0 ± 0.6	12.3 ± 0.6	n.s.	9.2 ± 0.4	14.6 ± 0.6	0.01
sol 102 kD	1.1 ± 0.2	1.4 ± 0.3	n.s.	1.4 ± 0.2	1.1 ± 0.3	n.s.	1.5 ± 0.4	3.1 ± 0.3	0.05

[a] *scanned with Bio-Rad 710 densitometer using Quantity – 1 software.*
[b] p based on student-t test.
[c] protein origin from MF or sol fraction and apparent molecular weight
n.s. = not significant

These data and other data presented in this volume shows that there is a direct relationship between the tenderness associated with post-mortem aging and that associated with HDP-treatment (see preceding three articles in this volume). However, as seen in Figure 4, the response of specific proteins varies in groups that respond, don't respond, or negatively respond to HDP-treatment.

Some discussion is warranted to try to explain both responses, ie., post-mortem aging (earlier chapters in this communication) and shockwave treatment (this communication and the earlier three communications). Discussion will focus on observations from the data published in the literature on post-mortem aging since there is little such data available for HDP-treatment other than that of the laboratories these authors have associated themselves with in the past.

Olson and others[11] studied post-mortem changes in myofibrillar proteins of conventionally aged beef, *longissimus* muscle and found that there was a close association of Z-disk degradation and myofibril fragmentation with WBS values and sensory derived tenderness scores. Shortly thereafter, Olson and Parrish[12] developed a method to objectively measure tenderness using myofibril fragmentation; this method was named the myofibril fragmentation index or MFI. Subsequent work by others;[13-16] yielded additional support for this strong relationship. These investigators observed microscopically that during post-mortem aging as tenderness improved there was an increased myofibrillar fragmentation in addition to a visible loss of Z-disk thought to be a result of degradation by a calcium activated factor (CAF).

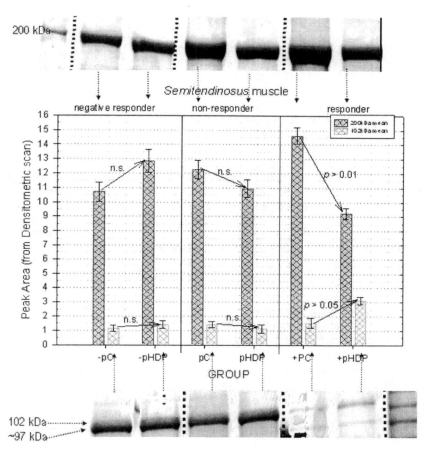

Figure 4 SDS-PAGE gel profiles of a section <u>semitendinosus</u> (top round) muscle from animals that that show a negative response (left), no response (middle), and full response (right) to HDP-treatment. SDS-PAGE gels were 4-20% gradient. Each lane had the identical amount of protein added acrylamide. The UPPER electropherogram shows the distribution of a 200 kD protein while the LOWER electropherogram shows the distribution of a 102 kD or ~97 kD protein peak(s). Data is as seen in Table 2.

Davey and Gilbert[17], Olson and others[11], and finally Olson and Parrish[12] (1977) used existing knowledge of muscle structure and developed a method now known as MFI or myofibrillar fragmentation index. Since its development, the MFI method has been used by many investigators to explain shear force (tenderness) in beef. Recently, Spanier and Romanowski[5] expanded upon and modified this method to include HDP beef samples. They noted that the more tender samples (lower shear force value) contained less protein in their myofibrillar fraction and more in their soluble fraction than did the tougher strip loin steaks. These changes were thought to be a result of alterations in the meat's physical structure such that the tissue became more susceptible to the shear stresses of homogenization. Others have shown that the tenderness associated with postharvest aging also leads to a change in meat structure,[11, 15, 17, 18] and protein composition,[10, 19, 20, 21] Based on the information above, it was reasonable to presume that similar structural alterations might be common in meat tenderized by HDP and meat subjected to postmortem aging.

Ouali[23] studying the effect of postmortem aging on meat tenderness observed that a 30 kD protein peak (SDS-PAGE) would increase as meat aged and became more tender. It was suggested by Ouali[11] that the 30 kD protein was a result of the degradation of troponin-T which has been shown to decline when exposed to calcium activated neutral proteases[11, 23, 24]. Proteolytic fragments were observed within a few hours post-slaughter and were seen to increase as storage time increased; such fragments were evident on SDS-PAGE as bands between tropomysin (about 36 kD) and myosin light chain 1 (about 27 kD). These may be the same as the 30 kD protein reported earlier by others.[13, 15, 16, 27] Such a change in levels of a band of about 36 kd (this study) was observed in HDP-treated samples (Figures 1 - 3) as compared to the control.

Protein components of a high molecular mass protein between 90 and 100 kD form during postmortem aging. For example, the myosin head component, S_1 (light meromyosin), has been recovered in extracts of conditioned beef.[25] Light meromyosin with a lower M_r than α-actinin, does migrate near α-actinin on homogeneous gels where it is difficult to clearly distinguish these two bands particularly on overloaded gels. A protein in this molecular mass range was found in the HDP-treated meat (p2 in Figure 3) and increased as the meat's tenderness level increased or shear force decreased (earlier chapters in this book).

The effect of HDP-treatment on the physical structure of treated strip loins has been well documented.[4] HDP-treatment has been shown to disrupt meat structure at the Z-line and A-I band junctures. The data in Figures 2 & 3 and which have adjacent control samples indicate clearly that there is a relationship between tenderness induced by HDP-treatment and 4 resolvable proteins. These observations are not totally unexpected, since others have also shown a relationship between tenderness and protein composition or cellular distribution.[10,18,19,20,21,22]

The data presented herein (and in the earlier contribution in this book) indicate that the physical disruption and relative level of some proteins in HDP-treated beef is related directly to the HDP-treatment-induced change in the level of tenderness of the strip loin sample. Data suggest that some of these proteins might be related to some changes in proteins in early stages of aging. At this time, the origin/source of these proteins is unknown, but they are most likely result from fragmentation of larger meat proteins[10, 18, 19, 20, 21, 26] such as, but not limited to, titin, nebulin, myosin, connectin, or α-actinin. Alternatively, they may result from bond-disruptions in other high molecular mass protein complexes such as myosin and actin with known functional bonding interaction and banding patterns.[9]

Since HDP-treatment of meat has been shown to lead to visual structural alterations[4] and to chemical alteration of proteins in homogenized meat[5] (Figures 1 – 4 and other data in this book), these proteins or their altered parent proteins may be more susceptible to the physical shearing caused during homogenization and pressure-wave treatment. This change in a meat's susceptibility to perturbations by physical means such as HDP-treatment may also include an enhanced susceptibility to chemical perturbations. For example, proteolytic digestion by endogenous tissue proteases such as cathepsins B, D, H, and L, calpains, and exoproteases or altered reactivity to oxidation could enhance or detract from the initial tenderizing effect of HDP-treatment. This potential response to chemical perturbations must be explored in future experiments because of its potential impact on meat quality and to further examine the relationship, if any, between HDP-treatment and normal postmortem aging.

5 CONCLUSIONS

It is clear from these studies (this manuscript; [28, 29]) of beef that HDP-treatment causes meat to become more tender in responding cuts of meat. The tenderizing effect of HDP-treatment is most likely mediated through alteration of several higher molecular weight meat proteins such as c-protein. Continued research on the effect of hydrodynamic pressure (HDP) on meat proteins will provide information essential for developing HDP or other related technologies into viable, commercially applicable, processes that will attract a high level of consumer confidence. It is also hoped that such research will lead to developing under-exploited markets for new products that produce value-added, high quality meat with desirable palatability, extended shelf-life and improved safety. Last, these data have identified specific protein peaks that can be utilized as indicators of the level of tenderness in a meat and a meat's aging period. Future studies will necessarily have to be performed on various muscle types and on those types obtained from different breeds (future manuscript) to determine which specific proteins may be used to develop a standard curve for rapidly determining the tenderness in specific muscle and specific breed. The documentation and data presented in the four contributions in this issue have already laid the groundwork for a method for the rapid prediction of meat tenderness.

Acknowledgments

The authors graciously acknowledge the technical assistance of several individuals during various portions of this project: Dr. Brad Berry, Mrs. Janice Callahan, Ms. Janet Eastridge, and Mr. Ernie Paroczay. We gratefully acknowledge the assistance of Dr. Bryan Vinyard, Driector, BARC Biometrics Consulting Center. Sincere thanks are also extended to Dr. Morse Solomon who is licensed to handled and detonate the explosives used in this investigation. Sincere gratitude is extended to Dr. Robert (Bob) Schmukler of Pore[2] Bioengineering Inc. for opening his laboratory for this work. This manuscript is dedicated in part to the memory of Dr. Robert (Bob) Romanowski who passed away in February 2000.

References

1. D.G. Morgan, J.W. Savell, D.S.Hale, R.K. Miller, D.B. Griffin, H.R. Cross and S.D. Shackelford, *J. Anim. Sci.*, 1991, **69**, 3274-3283.
2. M.B. Solomon, J.B. Long and J.S. Eastridge, *J. Anim. Sci.*, 997, **75**, 1534-1537.
3. M.B. Solomon, *Proc. Recip. Meat Conf.*, 1998, **51**,171-176.
4. H. Zuckerman and M.B. Solomon, *J. Muscle Foods*, 1998, **9**, 419-426.
5. A.M. Spanier and R.D. Romanowski, *Meat Sci.*, 2000, **56**, 193-202.
6. U.K. Laemmli and M. Favre, *J. Molec. Biol.*, 1973, **80**, 575-599.
7. H. Zuckerman, B.W. Berry, J.S. Eastridge and M.B. Solomon, *J. Muscle Foods*, 2002, **13**, 1-12.
8. A.M. Spanier, *Studies of the Lysosomal Apparatus in Progressive Muscle Degeneration Induced by Vitamin E - Deficiency*. 1977, Ph.D. Thesis, Rutgers, The State University of New Jersey.
9. D.L. Hopkins and J.M. Thompson, *Meat Sci.*, 2000, **57**, 1-12.
10. R.M. Robson, E. Huff-lonergan, F.C. Parrish Jr., C-Y. Ho, M.H. Stromer, T.W. Huiatt, R.M. Bellin and S.W. Sernett, *Recip. Meat Conf.*, 1997, **50**, 43-52.
11. D.G. Olson, F.C. Parrish, Jr. and M. Stromer, *J. Food Sci.*, 1976, **41**, 1036-1041.
12. D.G. Olson, FC. Parrish, Jr., W.R. Dayton and D.E. Goll, *J. Food Sci.*, 1977, **42**, 117-124.
13. M.A. MacBride and F.C. Parish, Jr., *J. Food Sci.*, 1977, **42**, 1627-1629.
14. D.G. Olson, F.C. Parrish Jr., *J. Food Sci.*, 1977, **42**, 506-508.
15. R.D. Culler, F.C. Parrish Jr., G.C. Smith and H.R. Cross, *J. Food Sci.*, 1978, **43**, 1177-1180.
16. R.D. Culler, *J. Food Sci.*, 1979, **44**, 1668-1672.
17. C.L. Davey and K.V. Gilbert, *J. Food Sci.*, 1969, **34**, 69-74.
18. D.L. Hopkins, P.J. Littlefield and J.M. Thompson, *Meat Sci.*, 2000, **56**, 9-22.
19. R.W. Purchas, X. Yan and D.G. Hartley, *Meat Sci.*, 1999, **51**, 135-141.
20. J.D. Fritz and M.L. Greaser, *J. Food Sci.*, 1991, **56**, 607-610.
21. D.E. Goll, M.L. Boehm, G.H. Geesink and V.F. Thompson, *Proc. Recip. Meat Conf.*, 1997, **50**, 60-66.
22. M.L. Greaser, *Proc. Recip. Meat Conf.*, 1997, **50**, 53-59.
23. A. Ouali, *J. Muscle Foods*, 1990, **1**, 126-165.
24. A. Ouali, A. Obled, P. Cottin, N. Merdaci, A. Ducastaing and C. Valin, *J. Sci. Food Agric.*, 1983, **34**, 466-473.
25. L.D. Yates, T.R.Dutson, J. Caldwell and Z.L. Carpenter, *Meat Sci.*, 1983, **9**, 157-179.
26. P.F.M. van der Ven, G. Schaart, H.J.E. Croes, P.H.K. Jap, L.A. Ginsel and F.C.S. Ramaekers, *J. Cell Sci.*, 1993, **106**, 749-759.
27. C.Y. Ho, M.H. Stromer and R.M. Robson, *Biochimie*, 1994, **76**, 369-375.
28. A.M. Spanier, B.W. Berry and M.B. Solomon, *J. Musc. Foods*, 2000, **11**, 183-196.

MULTIQUALITY ENHANCEMENT OF MUSCLE FOOD: A HYPOTHESIS EXPLAINING HOW HYDRODYNAMIC PRESSURE (HDP)-TREATMENT LEADS TO MEAT TENDERNESS. [§]

A.M. Spanier[1] and R.D. Romanowski[2]

[1] Spanier Science Consulting LLC, 14805 Rocking Spring Drive, Rockville, MD 20853
[2] USDA, ARS, BA, Animal & Natural Resources Institute (ANRI), Food Technology and Safety Laboratory (FTSL), Bldg 201, BARC-East, Beltsville, MD 20705

1 ABSTRACT

The tenderness of beef has a profound effect on a consumer's acceptance and purchase decision. While traditional aging of the *muscle* tissue of a slaughtered typically leads to a flavorful and tender *meat* product, the complex biochemical changes and interactions that occur during the conversion of muscle to meat do not account for all of the tenderization that is possible. Variables such as age, breed, sex, muscle cut, feeding regimen, etc. must all be considered. Hydrodynamic pressure (HDP)-treatment of meat has shown promise in tenderizing meat in a more consistent and uniform manner. The HDP-effect on meat tenderness is based on the theory that a pressure shockwave passes through water and disrupts any other object in the water; in the case of muscle the shockwave passes through the meat since meat is >70% water and disrupts all of the solid material in the meat. Data examining HDP-treatment and its effect on tenderness show that HDP-treatment induces changes in meat tenderness in a manner that is different from that of traditional aging. The HDP-treatment effect seems to be for specific meat proteins, principally c-protein. A theory of HDP shock treatment and its relationship to meat tenderness is presented.

2 INTRODUCTION

A major challenge for the meat industry is to market products with acceptable levels of tenderness. Tenderness has proven to be the most difficult quality factor for meat producers and meat packers to manage.[1,2,3] The variation is due not only to factors such as the age, breed, sex, muscle cut, feeding regimen, method of harvesting, etc.,[4,5,6] but also to the complex biochemical changes and interactions that occur during the conversion of muscle to edible meat.[6,7,8] This inconsistency and variation is seen not only among breeds of the same species, but also within the same cut in a given breed, and has been identified

---

[§] Mention of brand or firm names are necessary to report factually on available data and does not constitute an endorsement by the authors over other products or manufacturers of products of a similar nature not mentioned.

as a major problem facing the meat industry.[8] Many procedures have appeared over the years that attempt to address the tenderness issue including, but not limited to, electrical stimulation, proteolytic enzyme treatment, infusion with calcium, pH control, skeletal alteration, blade tenderization, hydrostatic pressure (HPP) and, most recently, hydrodynamic pressure (HDP).

The HDP-treatment process involves use of an explosive to create a pressure shock wave that passes through the water and through any object in the water with a mechanical impedance matching that of water.[9] Since the composition of meat is predominantly water (>75%) the pressure wave passes through the meat and ruptures selected cellular components. For example, sarcomeres in red meat are torn at the Z-line and A-band/I-band.[10] Thus far, evidence indicates that the shock wave affects mainly myofibrillar components; whether HDP-treatment effects the collagen components of meat remains to be answered (Solomon, personal communication). Recent studies of hydrodynamic pressure (HDP)-treated meat have also shown that HDP-treatment not only increases the tenderness of meat but also makes the tenderness more uniform along the entire length of a strip loin.[11,12]

The overall objective of this research was to obtain physicochemical data that might prove useful to understanding the effect of the HDP shock-wave treatment on meat structural and functional properties. Over the course of this work (see other contributions in this volume), HDP-treatment was performed on beef strip loins from multiple sources partially to see if there was a relationship between animals but more importantly to generate samples with different levels of tenderness.

3   MATERIALS AND METHODS

3.1   Meat (Source and Handling)

As indicated above, meat was obtained from many sources and thus complete histories will not be presented here. Some steaks were obtained fresh, some were frozen (to enhance toughness in some cases and to facilitate the number of samples that could be handled at a time). The samples used in this study had a shear resistance value (an indicator of tenderness) ranging from a tender level of 4.0 kg to an extremely tough level of 12.9 Kg.

Strip loins samples to be treated by hydrodynamic pressure (HDP) were typically, but not always, wrapped in Saran® wrap which had been shrunk using a hand-held blower/heat gun at a distance of about 2 cm from the surface. All samples were vacuum packaged in plastic bags (Cryovac®/Sealed Air Corporation; Saddle Brook, NJ) and most shrink-wrapped in a water bath set above 180°C.

3.2   Hydrodynamic Pressure (HDP) – Treatment

Processing of strip loin sections by HDP shock wave technology was performed essentially following the procedure of Solomon[13] using 98.4 liter plastic (26 U.S. gallons) explosive containers (PEC). Packaged strip loin sections were placed fat side up on top of a 1.25 cm thick, 40.6 cm diameter, steel plate. A supersonic shock wave was induced using 100 g of binary explosive (composed of liquid nitromethane and solid ammonium nitrate) submerged from the top of the PEC with wire. Meat samples were placed into the PEC with the explosive charge set to a height of 30.5 to 38.1 cm above the top of the strip loin. The PEC was filled with water to a height of 15.25 cm above the explosive charge. After HPD - treatment the treated strip loin sections as well as their non-HDP treated controls were cut depending upon the experiment into 2.54 cm or 3.18 cm thick steaks for tenderness assessment.

3.3 Cooking of Samples for Shear Resistance Analysis

As described previously,[13] 2.5 cm-thick the strip loin steaks were broiled to an internal end-point cooking temperature of 71EC using Farberware® Open-Hearth broilers (Model T-4850, Hanson Corp., Bronx, NY). Steaks were turned midway between the initial temperature and the final temperature of 71°C. Internal meat temperature was monitored using iron-constantan thermocouples attached to a Speedomax™ multipoint recording potentiometer (Model 1650, Leeds and Northrup, North Wales, PA). Some steaks (3.18 cm thick) were cooked in a similar manner to the same end-point temperature using a George Forman grill (820 watt, Model GR-20, Salton, Inc., Mt. Prospect. IL).

3.4 Measurement of Shear Resistance Values

At least six cores (1.27 cm diameter) parallel to the muscle fiber orientation (Figure 1) were removed from each section of strip loin. Each core from the control (C) and hydrodynamic pressure (HDP) treated strip loin sections was sheared once at right angles to the fiber orientation using a Warner-Bratzler shear test cell mounted on a Food Texture Corp. (FTC) tenderness measurement system (Model TMS-90, FTC, Chantilly, VA) using a 3.18 mm thick flat bottom blade with an inverted V-shape to surround the meat during coring. The cross-head speed was set to 25 cm/min.

3.5 Homogenization and Subcellular Fractionation

Individual 5.0 g samples of HDP meat (and where available control, C meat) were frozen at -20°C until needed. Five (5.0) gram samples of meat were thawed (25°C), finely minced and placed into 45 ml of homogenization medium [HM; consisting of 0.05 M Tris (Trizma acid) with 1.5 mM dithiothreitol (DTT) and 1.5 mM tetrasodium ethylenediaminetetraacetic acid (EDTA) with pH adjusted to 7.0]. In some experiments where the presence of sodium azide would not interfere with low UV analysis the HM also contained 15 mM sodium azide as a bacteriostat.

The procedure used to isolate the subcellular organelles[12] was a modification of that reported previously[14,15,16] all of which were modifications of the classical procedure described for soft tissue by deDuve.[17] A Waring Blender™ with stainless steel holding cup (100 ml capacity) was used for homogenization. Homogenizer speed was controlled using a Powerstat™ rheostat (Superior Electric Co., Bristol, CT). Homogenization was conducted at a rheostat setting of 65 for a duration of 10 seconds.[11] The resulting homogenate was filtered through 2 layers of cheese cloth to remove any fat, connective tissue, and nondisrupted tissue. The filtered homogenate was labeled as total homogenate

(TH) and divided into 2 portions: 30 ml for use in subcellular fractionation studies, and the remaining material for other purposes such as assessment of protein concentration and sodium dodecylsulfate - polyacrylamide gel electrophoresis (SDS-PAGE).

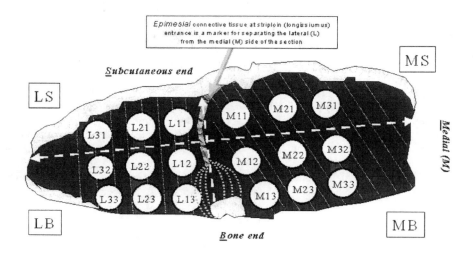

Figure 1 *Diagrammatic representation of beef striploin steak showing the locations that cores were obtained. As many cores as possible were obtained but no less than 6 per steak. Each quarter section (LS, LB, MS, and MB where "L" is lateral end, "M" is medial end, "S" is subcutaneous end and "B" is the bone end of the steak) had at least 1 core taken from it for determination of Warner-Bratzler shear force value (kgf).*

Meat purchased by consumers has typically been aged to permit tenderization and thus the amount and composition of material found in the subcellular fractions can change continuously during postmortem aging as a result of the biochemical and morphological changes.[14,18,19] The letters MF, P and S are used to identify the subcellular fractions and are based on the original scheme of deDuve[17] for non-muscle tissue. This nomenclature will be used for in this manuscript to represent the predominant type of material typically associated with these fractions. Meat samples were obtained from three different breeds and eight different strip loins; each data point represents the mean of at least 3 individual homogenizations of separate samples from a strip loin. An example of tenderness improvement measured by Warner-Bratzler shear force value (kg$_f$) in 56 control screening steaks (C, untreated steak) is compared to the steaks obtained from the matching HDP-treated section (Table 1).

Table 1 Control and HDP-treated shear force values (kg_f) from 56 different striploins obtained from 5 different sources/breeds of cattle.

WBS (kg_f) HDP-treated	Control	% improvement	Source/ID#
2.48	6.25	60.0	Rumsey '96 steers / 5051D
3.39	5.70	35.0	Holstein Select / 58B
3.42	5.60	39.0	Holstein Select / 58D
3.55	6.25	41.0	Rumsey '96 steers / 5051C
3.93	5.60	29.0	Holstein Select / 58C
3.94	5.70	31.0	Holstein Select / 43C
3.97	5.70	41.0	Holstein Select / 43A
4.00	5.70	34.0	Holstein Select / 58A
4.04	6.10	36.0	Holstein Select / 29C
4.24	6.10	33.0	Holstein Select /29D
4.35	6.30	24.0	Holstein Select /29A
4.37	6.34	31.0	Rumsey '96 steers / 5051B
4.44	7.08	37.0	Wye '96 / 7567D
4.51	6.34	39.0	Rumsey '96 steers / 5039D
4.59	6.30	32.0	Holstein Select / 43B
4.62	6.30	26.0	Wye '96 / 7567D
4.89	6.70	14.0	Holstein Select / 43D
4.92	7.33	21.0	Rumsey '96 steers / 5039D
5.01	7.33	32.0	Rumsey '96 steers / 5039B
5.06	7.33	40.0	Rumsey '96 steers / 5029D
5.28	7.92	28.0	Rumsey '96 steers / 5029C
5.34	7.08	25.0	Wye '96 / 7567C
5.44	6.70	5.0	Holstein Select / 29B
5.95	11.25	47.0	Canadian Bullocks / 727DR
6.12	6.42	5.0	Wye '96 / 7558L-C
6.13	7.15	14.0	Wye '96 / 7558L-B
6.17	8.33	5.0	Wye '96 / 7558R-C
6.21	11.01	43.0	Canadian Bullocks / 779BR
6.21	6.42	3.0	Wye '96 / 7558L-D
6.25	8.33	25.0	Wye '96 / 7558R-D
6.51	8.36	36.0	Rumsey '96 steers / 5029B
6.55	8.36	23.0	Rumsey '96 steers / 5027D
6.57	8.36	17.0	Rumsey '96 steers / 5027C
6.69	9.87	32.0	Wye '96 / 7498C
6.73	8.47	20.0	Rumsey '96 steers / 5023D
6.88	9.72	29.0	Wye '96 / 7493D
6.88	9.06	24.0	Wye '96 /7589B
6.90	12.86	46.0	Canadian Bullocks / 770BR
6.98	10.56	32.0	Canadian Bullocks / 779BL
7.08	8.47	16.0	Rumsey '96 steers / 5023C
7.17	12.27	42.0	Canadian Bullocks / 727DL
7.46	10.23	27.0	Rumsey '96 steers / 5013D

Table 1 *(continued)*

WBS (kg$_f$)		% improvement	Breed/ID#
HDP-treated	Control		
7.47	9.27	19.0	Wye '96 / 7558R-B
7.56	10.23	10.0	Rumsey '96 steers / 5013C
7.83	12.82	39.0	Canadian Bullocks / 801BL
8.30	12.82	32.0	Canadian Bullocks / 801DL
8.74	11.01	21.0	Canadian Bullocks / 779DR
8.82	12.82	36.0	Canadian Bullocks / 801DR
8.83	8.06	15.0	Wye '96 / 7589D
8.83	12.86	31.0	Canadian Bullocks / 770BL
9.15	10.80	16.0	Wye '96 / 7498B
9.34	9.72	4.0	Wye '96 / 7493C
9.38	12.86	24.0	Canadian Bullocks / 770DR
9.48	10.56	10.0	Canadian Bullocks / 779DL
9.63	12.86	24.0	Canadian Bullocks / 770DL

Data kindly contributed by Dr. M.B. Solomon (personal communication)

Thirty (30) ml of TH was subjected to differential centrifugation in a refrigerated Sorvall® superspeed RC-2B centrifuge (Sorval Instruments, Inc., Newtown, CT) under the following conditions: thirty (30) ml of TH was centrifuged at 9,500 rpm (10,800 g) at 4EC for 10 min. The supernatant solution (S_1) was saved, the pellet was resuspended to one-half the original volume with cold HM, and again centrifuged for 10 min at 9,500 rpm (10,800 g) at 4°C. The resulting pellet (MF for myofibrillar fraction) was resuspended with HM to its original volume and saved. The supernatant solution from the second centrifugation (S_2) was pooled with the first supernatant solution (S_1) to yield S_3 and was centrifuged for 30 min at 4°C. The resulting pellet (P fraction) was resuspended with HM to one-half (15 ml) its original volume. The final 10,800 g supernatant solution was centrifuged at 20,000 rpm (48,200 g) for 30 min yielding a pellet (P fraction) that was resuspended to one-half its original volume (15 ml) and a supernatant solution (S). The supernatant solution (S fraction) was adjusted to twice its initial volume (2 x 30 ml).

3.6 Analysis of Proteins.

Protein concentration was determined on appropriately diluted samples of homogenized meat or on subcellular fractions of the homogenized meat. Protein level was assessed by reading the absorbency at 230 nm (A_{230}) or when sodium azide was present A_{280} using a Tecan™ SECTRAFLOURPlus™ multiwell plate reader (Tecan, U.S., Inc. Research Triangle Park, NC). Recording absorbency in the low UV range necessitated use of Costar® #3635 UV plates (96-well; Costar is a division of Corning Inc., Corning, New York) for all sample analysis. The protein concentration in the meat was determined by comparison to the absorbency of a standard solution of bovine serum albumin (BSA; 1 mg/ml).

3.7 SDS-PAGE

Proteins were separated electrophoretically on polyacrylamide gels in the presence of sodium dodecyl sulfate (SDS-PAGE) using a slight modification of the discontinuous gel procedure of Laemmli and Favre.[20] Separating gels were either homogeneous or gradient gels with a 37.5:1 cross-link with bis-acrylamide; the upper stacking gel when used was 4.0% acrylamide. Sample combs were placed in the stacking gel prior to polymerization to make wells for sample introduction.

Samples were mixed 1:1 (v/v) with SDS-PAGE solubilization (rendering) media (0.5 M Tris/HCl, pH 6.8, 10% SDS, 5% 2-mercaptoethanol, 0.1% bromphenol blue) in a small, capped, plastic conical centrifuge tube and boiled for 4 min to inactivate and render proteins soluble. Samples were placed in each well by use of a Hamilton syringe (concentrations are noted in each figure). Acrylamide gels (1.5 mm thick) were subjected to a constant voltage of 200 V at room temperature (25°C) for approximately 45 min - 1 hr in a BioRad miniunit powered by a BioRad Power Pac 3000 (BioRad Labs, Hercules, California) power supply or in the BioRad Criterion gel electrophoresis system using the same power supply. The acrylamide gels were removed from the sandwiched position between the glass holding plates and stained with 0.05% Coomassie blue R-250 in 7% acetic: 40% methanol overnight at room temperature. Gels were stained and destained with a mixture of 7% acetic acid and 5% methanol for 48 h or until the gel background was clear. A mixture of proteins standards with a progressive increase of 10 kDa per protein (called a 10 kD protein ladder) were used (Gibco BRL, Grand Island, NY).

4 RESULTS AND DISCUSSION

The data in Table 1 shows the shear force value (Warner-Bratzler Shear, WBS as kg_f) for control, C, and HDP-treated samples used in several HDP experiments. WBS values in C range from ~5 kg_f to almost 13 kg_f while the HDP-treated group (...treated using several modifications of HDP-treatment...) show a range in shear resistance from ~2 to ~9.6 kg_f. All steaks show an improvement in tenderness; however, the improvement is not consistent in each group. Previous reports have indicated that HDP-treatment will improve tenderness and will even make tenderness along a strip loin more consistent,[12] even though tenderness improvement is not necessarily consistent between loins. The authors remind the reader that use of hydrodynamic pressure treatment is a developing technology and optimal HDP-treatment conditions are still under development. Some of the variability appears to be related to breed (Figure 2), with the response of the four animal sources to HDP-treatment seems reasonably similar (as seen by similar slopes); only one the Canadian Bullocks appears slightly different. Thus, the observations that there were differences in the final level of improvement in tenderness led us to ask if some, perhaps consistent, change in utrastructure, physical nature, or chemical constituents could be observed in meat after HDP-treatment. This question has been addressed it various degrees in earlier communications in this issue.

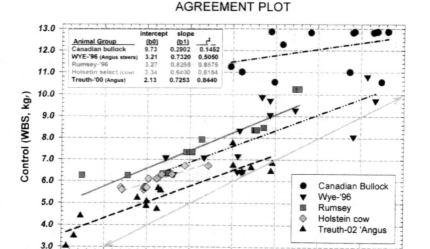

Figure 2 *Agreement plot of data in Table 1 arranged by breed type. The solid line with the arrow at both ends represents the "Line of Equity" at which HDP-treatment has no effect on the tenderness level (kg_f) in the striploin. When HDP-treatment has the effect of tenderizing the meat, the line of agreement will shift to the left.*

4.1 Subcellular Distribututuion of Proteins

4.1.1 Total protein in subcellular fractions

Earlier experiments, which included some of the HDP-treated samples used in this study, showed that HDP-treatment changed the SDS-PAGE profile of HDP-treated samples when compared to their control.[11 and other chapters in this volume] Homogenization of these samples showed that there was a relationship between the texture of the meat and the subcellular distribution of protein such that the more tender (lower shear resistance value) samples have less protein in their myofibrillar fraction and more in their soluble fraction[11] (Figure 3). These data suggested that as samples became more tender, HDP-treatment affected the physical structure of the meat in such a manner as to make the meat more susceptible to the shear stresses of homogenization.

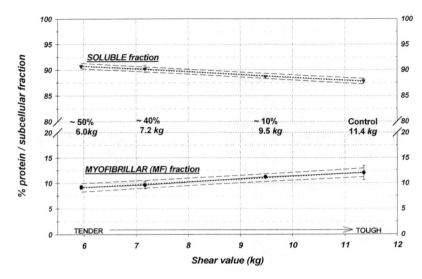

Figure 3. *Effect of HDP-treatment on the protein distribution in beef strip loins. Data is presented as the mean with error bars representing the standard deviation for 3 repeat homogenizations of each HDP-treated sample. Filled, inverted triangles (▼) represent the soluble fraction (post 48,200 g supernatant material) while filled circles (●) represent the nuclear or myofibrillar fraction (MF; 11,000 g pellet). Dotted lines (·····) Are point-to-point lines between data values, solid lines (—) represent the line formed from the linear regression analysis of the data points, and dashed lines (---) represent the 95% confidence limits of the regression analysis. As shear resistance value (kg) increases so does the toughness of the HDP group over its control steak. The slope of the regression analysis of the MF fraction is 0.275% / kg with an r^2 of 0.99 while the slope of the regression analysis of the soluble fraction is -0.298% / kg with an r^2 of 0.99.*

4.1.2 Comparison of protein profiles of control and HDP-treated striploins

Control beef strip loins with a prescreened shear force value of 4, 6, 8, and 10 kg were treated by HDP. Gradient sodium docecyl sulfate polyacrylamide gel electrophoresis (4-20% SDS-PAGE) were run on the soluble fractions obtained from control and HDP-treated samples (Figure 4). The control (C) gels at all four shear force values show similar gel protein profiles. HDP-treated samples from adjacent portions of the control striploin showed some change in protein pattern as the strip loins became more tender (Figure 4). Two bands in particular ... one at about 155 kD and another at ~100 kD ... appear to respond to the HDP-treatment and both show an increase in size as tenderness improves due to the HDP-treatment.

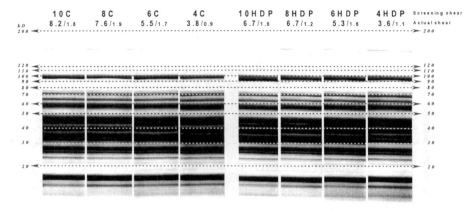

Figure 4 SDS-PAGE of 4 – 20% acrylamide gradient. Each lane consists of 40 µg of BSA equivalent protein per slot of a 10,800 x g soluble fraction from control, C, and HDP-treated beef striploins.

4.1.3 Profile of proteins in beef strip loins having different shear force values after to hydrodynamic pressure treatment

HDP-treated strip loin samples with progressively decreasing shear force values (increasingly tender) were homogenized and the soluble fraction examined by SDS-PAGE analysis. Seventeen major bands were identified (Figure 5) and the molecular mass compared to mass of known standards. The average molecular mass (in kilodaltons) of the 17 visible proteins ± standard error of the mean for (n = 8) ranged from 95.1 to 11.9 kD and is shown in Table 2 along with the reported molecular mass of some known muscle proteins.

SDS-PAGE electrophoresis of soluble fractions from HDP-treated strip loins is shown in Figure 5 and plots of the distribution of individual Coomassie-blue stained, protein bands is shown in Figure 6.

It is clear that there are a few bands that show an increase as a striploin becomes more tender ... presumably a result of HDP treatment as seen in Figure 4. However, we caution the reader to note that since we do not have gels of the controls in this group of samples (Figures 5 & 6), we can not say with absolute surety if the observed change is a result of the meat's level of tenderness or a result of the HDP- treatment. The senior author believes that the change is a result of HDP-treatment (as seen in Figure 5) and experiments planned for the future and those described earlier in this volume will show/have shown this indeed to be the case. No matter what the cause of the decrease in tenderness, whether by HDP-treatment or by normal proteolytic degradation, there are several identifiable protein bands that change with increasing tenderness (decreasing shear resistance values).

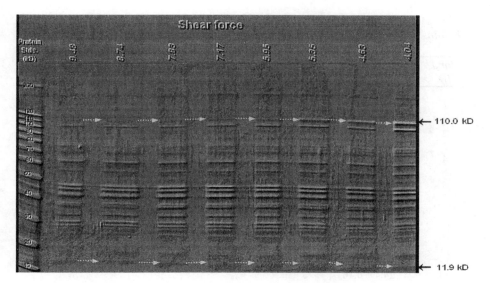

Figure 5 *Homogeneous 20% acrylamide SDS-PAGE gel of soluble fractions from HDP-treated beef striploins. Each lane contained 25μg of protein (based on A230 absorbtion) and were stained with coomassie brilliant blue R-250. The gel was scanned and printed in an embossed format to facilitate viewing.*

Data indicate clearly that as HDP-treated meat becomes more tender, tissue disruption by homogenization yields differing proportions in some protein bands ie., an increase in several protein bands that are associated with increased tenderness (decreased shear resistance values) in HDP-treated meat (Figures 5 and 6 and other chapters in this volume). At this point the origin/source of these proteins is unknown. However, they are likely a result of fragmentation of larger meat structural and functional proteins such as but not limited to titin, nebulin, myosin, connectin, or %-actinin. In an earlier chapter (this issue) the authors have clearly shown that one of the proteins is most likely c-protein. However, we do not rule out that some of the observed fragments may be a result of bond-disruptions in other proteins with known bonding interactions such as myosin and actin.

Since HDP-treatment has been shown to lead to the altered distribution of proteins in homogenized meat samples, these proteins or their altered parent proteins are potentially more susceptible to other physical and chemical perturbations after HDP-treatment. Such changes may include an enhanced susceptibility to proteolytic digestion (by endogenous tissue protease such as cathepsins B, D, H, and L, calpains, exoproteases) or to oxidation which could enhance or detract from the initial tenderizing effect of HDP-treatment. This must be explored in the future.

Table 2 *Mean molecular mass of proteins bands isolated from the soluble fraction of beef strip loins on a 7.5% - 20.0% SDS-PAGE ± standard error of the mean.*

Band ID	MW (kiloDaltons) ± sem	Some meat proteins and their approximate molecular weight (kiloDaltons, kD)
1	110.1 ± 0.87	breakdown products of larger molecular weight beef proteins such as titin, nebulin, or myosin, also breakdown product of c-protein or m-protein ; possibly %-actinin[§]
2	86.7 ± 0.80	
3	65.0 ± 0.45	
4	63.2 ± 0.46	
5	60.9 ± 0.46	
6	53.4 ± 0.33	desmin (tubulin)[1] ....55 kD
7	47.9 ± 0.29	actin[1] ....43-44 kD
8	38.8 ± 0.24	troponin-T (cardiac)1 ....40 kD
9	38.0 ± 0.23	
10	35.4 ± 0.21	tropomyosin[1] .... 35-36 kD
11	31.4 ± 0.17	tropomyosin-I[1] .... 28-30 kD
12	27.2 ± 0.18	myosin-LC_1[1] .... 27 kD
13	24.4 ± 0.16	
14	23.0 ± 0.15	
15	22.0 ± 0.18	
16	19.9 ± 0.22	myosin LC2[1] ....19.0-19.5 kD troponin-c 17-18 kD
17	11.9 ± 0.10	

§ from Schiverick et al., 1975 BBA 393:124

<u>*Other molecular masses / subunit*</u> **(adapted from [19]):**

102,000/subunit (2) α-actinin @ Z-line integral
165,000 M-protein @ M-line
186,000/subunit (2) .Synemin
195,000 Skelemin @ M-line (peripheral)
33,000/submit (2) Cap Z @z-line (integral)

130,000 C-protein (MyBP-C @ thick filament
185000 myomesin; @ M-line
Intermediate filaments at Z-line (peripheral)
280,000/subunit (2) Filamin @ z-line (peripheral)
3,700,000 titin @ longitudinal sarcomeric filaments (Z to M-line)

40,000/subunit (2) creatine kinase @ M-line
69,000 tropomyosin, thin filament
773,000 Nebulin @ parallels (part of) thin filaments to the z-line
178,000/subunit (2) Paranemin / Intermediate filaments.

46,000 Actin @ thin filament
74,000 H-protein (MyBP-H) @ thick filament
53,000/subunits (4) Desmin @ intermediate filaments at Z-line (peripheral)

Quality

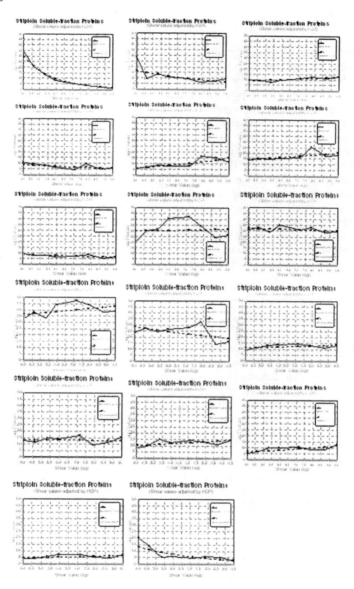

Figure 6 *Seventeen peaks were selected for analysis using the Quantity One software of BioRad. Solid lines go from point to point while dashed lines are a regression analysis of the line. From left to right and from top to bottom each graph represents one of the 17 band numbers assigned as seen in Table 2.*

4.1.3 A working hypothesis

The process of meat tenderization is highly complex and is affected by many factors such as, but not limited to: age, breed, cut of meat, feeding regimen, method of harvesting, animal's stress level during and before slaughter, post-mortem storage, cooking, etc. We have made some excellent initial progress in answering some questions that deal with tenderness development that is not regulated by normal enzymatic activity of the CANP (calcium activated neutral protease, also called calpain, CASF, etc.; see Goll[5] and Koomahraie[21] and summary in Figure 7)

Figure 7 shows the typical enzymatic events that occur during normal tenderization, but does not describe the effect of HDP-on tenderization. Basically, the effect of HDP-treatment is synergistic to the natural effect of aging. All of the data available to us up to this point in time (see other chapters in this issue) seem to indicate that HDP-treatment affects tenderness by direct action on specific higher molecular weight proteins, such as c-protein, that have not been degraded by the enzymatic activity of normal aging. In addition to c-protein, we also believe that HDP-treatment may also target some of the protein fragments of very high molecular weight proteins such as nebulin, titin and others and also has a direct affect on intermolecular bonds such as that between myosin and actin.

5. CONCLUSION

Data show that treatment of tough meat samples (high shear force values) with hydrostatic pressure (HDP) leads to meat that is more tender than before the pressure treatment (Table 1; Figure 2). Data also indicate that the protein profile of more tender cuts of the HDP-treated meat is different than that of tougher cuts. A hypothesis has been developed suggesting that pressure treatment by HDP shock waves specifically targets several of the proteins that are not fragmented by normal enzymatic means during aging and perhaps also affecting some of the proteolytic fragments themselves.

It is our hope that data obtained from this physicochemical research will provide information essential for developing HDP (or other pressure-related) technology into viable, commercially applicable processes that will attract a high level consumer confidence and should lead to developing under-exploited markets for new products that produce value-added, high quality meat with desirable palatability, extended shelf-life and improved safety.

6. ACKNOWLEDGEMENTS

The manuscript is dedicated to the memory of my co-author, Dr. Robert (Bob) Romanowski, who passed away on February 4, 2000; much of this work was a result of Bob's efforts in support of our research. The authors also acknowledge the technical assistance and support by other individuals at various stages of this work (alphabetically): Dr. Brad W. Berry, Dr. Anisha Campbell, Dr. Zvi Holzer, Ms. Janet Eastridge, Mr. Timothy Fahrenholz, Dr. Patti Nedoluha, Mr. Ernie Paroczay, Dr. Robert Schmukler, and Dr. Morse B. Solomon.

Quality

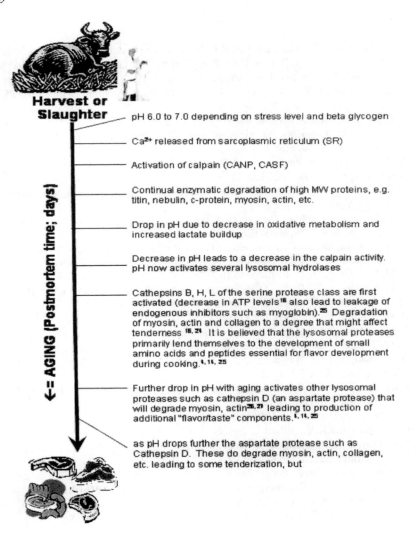

Figure 7 *Flow chart of the normal enzymatic events that occur during aging, i.e. as physiological muscle changes to edible/flavorful meat.*

References

1. R.L. Alsmeyer, J.W. Thornton and R.L. Hiner, *J. Anim. Sci.*, 1965, **24**, 526-530.
2. J.B. Morgan, J.W. Savell, D.S. Hale, R.K. Miller, D.B. Griffin, H.R. Cross and S.D. Shackelford, *J. Anim. Sci.* 1991, **69**, 3274-3283.
3. S.D. Shackelford, T.L. Wheeler and M. Koohmaraie, *J. Anim. Sci.*, 1997, **75**, 175-176.
4. E. Dransfield, *Meat Sci.* 1998, **48**, 319-321.
5. D.E. Goll, M.D. Boehm, G. Geesink and V.F. Thompson, *Proc. 50th Ann. Recip. Meat Conf.*, 1997, **50**, 60-67.
6. A. Asghar and A.M. Pearson, In, *Advances in Food Research*, Vol. 26, Academic Press, New York, 1980, pp. 153-213.
7. C. Faustman, In, *Muscle Foods, Meat poultry and Seafood Technology*, Chapman & Hall, New York, 1994, pp. 63-78.
8. C. Valin and A. Ouali, In, *New Technology for Meat and Meat Products*, Audet Tijdschriften B.V., The Netherlands, 1992, pp. 163-179.
9. K. Kolsky, In, *Stress Waves in Solids*. Dover Publications Inc., New York, 1980
10. H. Zuckerman and M.B. Solomon, *J. Muscle Foods*, 1998, **9**, 419-426.
11. A.M. Spanier and R.D. Romanowski, *Meat Sci.*, 2000, **556**, 193-202.
12. A.M. Spanier, B.W. Berry, M.B. Solomon, *J. Muscle Food*, 2000, **11**, 183-196.
13. M.B. Solomon, 51st *Annu. Recip. Meat Conf.*, 1998, **51**, 171-176.
14. A.M. Spanier, In *Studies of the Lysosomal Apparatus in Progressive Muscle Degeneration Induced by Vitamin E-deficiency*, Ph.D. Dissertation, Physiology Dept., Rutgers, The State University of New Jersey, New Brunswick, NJ, 1977.
15. A.M. Spanier, W.B. Weglicki, D.L. Stiers and H.P. Misra, *Am. J. Physiol.* 1985, **249** (*Cell Physiol.* **18**), C379-C384.
16. A.M. Spanier, B.F. Dickens and W.B. Weglicki, *Am. J. Physiol.* 1985, **249** (*Heart Circ. Physiol.* **18**): H20-H28.
17. C. deDuve, *Science* 1975, **189**, 186-194.
18. C.L. Davey and K.V. Gilbert, *J. Food Sci.* 1969, **34**, 69-74.
19. D.G. Olson, F.C. Parrish, Jr. and M. Stromer, *J. Food Sci.* 1976, **41**, 1036-1041.
20. U.K. Laemmli and M. Favre, *J. Molec. Biol.*, 1973, **80**, 575-599.
21. M. Koohmaraie, *Proc. Recip. Meat Conf.*, 1994, **36**, 93-104.
22. R.M. Robson, E. Huff-Lonergan, F.C. Parrish, Jr, C-Y. Ho, M.H. Stromer, T.W. Huiatt, R.M. Bellin and S.W. Sernett, *Reciprocal Meat Conf. Proc.*, 1997, **50**, 43-52.
23. D.G. Olson, F.C. Parrish, Jr., W.R. Dayton, D.E. Goll, *J. Food Sci.* 1997, **42**, 117-124.
24. A.M. Spanier, M. Flores, K.W. McMillin and T.D. Bidner, *Food Chem.* 1997, **59**, 531-538.
25. A.M. Spanier and J.W.C. Bird, *Muscle & Nerve.* 1982, **5**, 313-320.
26. J.W.C. Bird, W.N. Schwartz and A.M. Spanier, *Acta Biol. Med. Germ*, 1977, **36**, 1587-1604
27. J.W.C. Bird, A.M. Spanier and W.N. Schwartz, In, *Protein Turnover and Lysosome Function.*, H.L. Segal and D.J. Doyle, (eds), Acad. Press, NY, 1978, pp 589-604.

FLAVOR AND QUALITY CHARACTERISTICS OF BAKERY PRODUCTS FROM FROZEN DOUGH WITH VARIOUS ADDED INGREDIENTS

V. Giannou and C. Tzia

National Technical University of Athens, Laboratory of Food Chemistry and Technology, 5, Iroon Polytechniou str, Athens, Greece

1 ABSTRACT

Bakery products are considered to be basic part of human diet in almost every culture. Even though they exhibit various formulations and characteristics they have a common appealing feature: flavor, which comprises the total sensation evoked during their consumption. Besides their formulation ingredients, the flavor generating steps in breadmaking are fermentation and baking during which several compounds are formed due to the constantly modified conditions of moisture, pH, and temperature. The most important flavor generating mechanism is the nonenzymatic Maillard reaction. However, immediately after baking, especially in frozen dough products, volatile materials begin to diffuse and oxidation of the unstable aromatic compounds occurs resulting in the degradation of the overall sensory perception of these products. The aim of the present work is to investigate flavor degradation in bread made from frozen dough with various ingredients as well as to correlate them with the characteristics of appearance and texture.

2 INTRODUCTION

Bread has been regarded for centuries as one of the most popular and appealing food products both due to its relatively high nutritional value and to its unique sensory characteristics. The main sensorial characteristics of bread can be discriminated into appearance, texture and especially taste and flavor. Appearance basically comprises of products color, shape, size and gloss. Texture is primarily the response of the tactile senses to physical stimuli that result from contact between some part of the body and the food. However, it can be further evaluated by kinesthetics (sense of movement and position) and sometimes sight and sound, which is associated with crisp or crackly characteristics. Bread flavor is a complex phenomenon involving psychological and physiological senses and can be defined as the reactions to the stimuli produced by a multitude of chemical compounds present in both the crumb and crust of bread. It includes the total sensations experienced by the consumer, such as aroma, taste, warmth, and freshness.[1-4]

2.1 Formation of Flavor and Texture

Although breadmaking has an extremely long history it still remains one of the most complicated and impressive flavor generating reactions in food preparation. The production of bakery products begins by mixing of raw material (mainly flour, yeast and salt) with water and occasionally with various other ingredients (sugars, shortening, oxidizing agents, etc.). Mixing initiates a long series of complex changes and interactions between those diverse components and finally results to the formation of the gluten network and the development of a cohesive and viscoelastic dough.[2, 5, 6]

However, breadmaking cannot be completed unless dough is baked into bread. Baking entails the development of the crust and crumb texture and color and the formation of bread flavor. Dough is transformed into a porous and resilient protein-starch-lipid matrix with honeycomb structure that encloses minute gas cells, which are created during mixing and yeast fermentation and subsequently dispersed during sheeting and molding. Cell structure is primarily influenced by the processing conditions before baking.[2]

Bread flavor formation is basically governed by a number of different factors such as the following:
- ingredients incorporated in dough formulation,
- the mixing, hydration, enzymatic modifications and hydrolysis of proteins and starch,
- method of leavening (yeast, sourdough, or chemical system),
- chemical and enzymatic reactions occurring during fermentation and baking and
- changing conditions of moisture, pH, and temperature of the baking process.

All of these factors can independently or synergistically contribute to the generation of a great number of volatile and nonvolatile organic compounds.[1, 2, 7, 8]

2.2 Sensory Characteristics of Bakery Products

Bakery products, especially when they are freshly baked, usually possess extremely attractive sensory characteristics, which are initially affected by the fermentation method applied in dough processing. Therefore, bread prepared from a naturally fermented sponge or sourdough is expected to exhibit a light cream-colored crumb with grayish tones, a very definite and distinctive odor and taste, a sharp acetic acid flavor and a wholesome, rustic flavor and aroma. It should be pleasant to chew and have especially delightful eating qualities. Bread leavened with baker's yeast should present a golden crust, a creamy white crumb, and an attractive aroma/flavor.

Bakery products should also possess adequate volume and symmetrical expansion, appealing and uniform crust and crumb appearance. Such products must be proportioned according to product specification and be appropriately shaped with a well-rounded, smooth top and without excessive cracks bulges or streaks. Crust should have an even and pleasant brownish color and proper thickness while crumb cells should preferably be small, fairly thin-walled, slightly elongated, uniform in size and free from large holes.

Additionally, bakery products should present a pleasing, wheaty and sweet taste without off-flavors and fine roasty aroma with a mild yeast overtone. They should exhibit soft, tender, smooth and slightly moist mouthfeel with fine grain. Finally, crumb should be satisfactorily elastic and cohesive and present decreased adhesiveness and a smooth, soft, velvety feel when touched lightly with the tip of the fingers.[2]

All of the above characteristics are very important to the quality and palatability of bakery products. This explains the significant research activity towards the investigation of the mechanisms that contribute to their formation, to the identification of the compounds, which compose bread flavor, and to the development of properly designed sensory evaluation methods and objective methods for their measurement.

Apart from the sensory evaluation of bakery products, which can be conducted by a properly trained test panel, crust and crumb color can be measured with colorimeters and flavor can be assayed through the use of gas chromatography or electronic nose. Bread texture characteristics, mainly firmness, stickiness, elasticity and chewiness, can be measured with texture analyzers. However, firmness is the characteristic most often measured due to the strong correlation between crumb firmness and consumer perception of the freshness of bread. Many instruments can be used for its evaluation, but most use some type of plunger or probe to deform a slice of bread and then measure either the compression distance resulting from an application of a fixed force (softness), or the force needed to compress the sample to a specific distance (firmness).[9]

2.3 Effect of Ingredients

Many of the ingredients, which are used in breadmaking, make a special contribution to the flavor of the final products that is more extensively described below.[8]

2.3.1 Flour. Flour tends to have a fairly bland flavor with most of its contribution coming from the oils of the germ and bran particles present. Thus, wholemeal, wholewheat and bran and germ-enriched flours are expected to deliver enhanced flavor characteristics. Other factors which may influence the sensory characteristics of bakery products are flour type, and proteins, amino-acids and ash content.[8, 10]

2.3.2 Yeast. Metabolic activities of yeast during fermentation may result in the production of aromatic substances some of which are relatively unstable and these may be transformed into other flavor compounds or diffuse during the last stages of fermentation and baking. The most important compounds formed during normal dough fermentation are ethanol, carbon dioxide, and various nonvolatile and volatile organic acid and carbonyl compounds. However, yeast varieties differ markedly in their ability to produce acids, alcohols, and esters. Therefore, yeast genera, species, and variety and the level of yeast used in the recipe are important in determining final bread flavor.[1, 8]

2.3.3 Lipids. Fats or oils added to bread formula can serve as flavor carriers, make crust more tender and crumb softer, moisten mouthfeel, improve palatability and slicing properties, enhance and enrich taste. Furthermore, their oxidation during baking can generate the production of desirable carbonyl compounds.[1]

2.3.4 Salt. Salt is added to the bread formula, to impart and enhance bread taste both due to its characteristic salty taste and its ability to increase the perception of other flavors, which may be present in the bakery products. It also exerts a controlling effect on enzymes and on microbial action during fermentation, and strengthens and tightens the dough gluten proteins.[1, 8]

2.3.5 Sugars. Sugars in bread dough come from three sources: those originally present in the flour; produced from oligosaccharides or polysaccharides by enzyme action during fermentation; and dough ingredients intentionally added. Sugars are added chiefly to obtain the desired sweetness in baked products, to maintain the fermentation process at a desirable level, and to bring about the most distinct flavor of baked products through the reaction with free amino groups in the nonenzymatic browning reaction.[1]

2.4 Effect of Processing

The most important processing steps, especially for the formation of flavor, are fermentation and baking.

2.4.1 Fermentation. Dough fermentation aims at the saturation of dough with carbon dioxide, the formation of the gas bubbles which will form the basic cellular structure of the product, the generation of fermentation products such as lactic acid, which soften the gluten and make it more extensible and the production of flavor compounds which highly contribute to the enhancement of bread aroma and taste.[11]

The flavor constituents of dough resulting from fermentation may be affected by several factors, including fermentation duration and temperature, sugar and salt concentration, strain of yeast, and bacteria. Usually several hours of fermentation are required in order to produce more flavorful bread unless flavor is developed in a pre-ferment, brew or sponge which is later mixed with the remaining ingredients to form the dough for the final processing. The preservation of the fermentation products after baking depends on the temperature rise of bread loaf during baking.[1, 8]

2.4.2 Baking. By far the most important contribution to bread flavor and texture comes from the process of baking. During baking, dough undergoes numerous changes that may be grouped into three important stages:

The first stage involves moisture evaporation from the exterior layers of the dough. This results in decreased temperature rise and permits dough surface to retain a degree of elasticity that aids further loaf development and expansion. The rise in bread internal temperature stimulates fermentation and carbon dioxide gas production which rapidly expands gas cells in the dough. Fermentation continues until the internal temperature of the dough reaches 50 to 60°C and yeast cell are killed.

At the beginning of the second stage, dough still presents plastic behavior and continues to expand. As the internal temperature rises, starch gelatinization and gluten coagulation occur. When the temperature reaches about 70 °C, dough plasticity and development are rapidly ceased.

During the third baking stage, the evaporation from the external loaf surface diminishes, the surface temperature increases and the crust is formed and thickens. The residual sugars that remain in the dough are caramelized, and nitrogen-bearing substances undergo a Maillard reaction. Both of these surface reactions contribute to the formation of crust color which turns from pale yellow, amber, orange, brown to finally dark brown.[12]

During crust formation several organic compounds are formed, some are lost and carried away with oven gases, and others trapped within crust structure and cells. The categories of flavor compounds that have been detected in white bread are:[1, 2, 10]

- Alcohols:
  ethanol, isobutanol, n-propanol, isoamyl alcohol and amyl alcohol.
- Acids:
  acetic, propionic, butyric, isobutyric, valeric, lactic, isovaleric, caproic, heptanoic, octanoic, nonanoic, capric, pyruvic, hydrocinnamic, benzilic, itaconic and levulinic.
- Esters:
  ethyl acetate, ethyl lactate, ethyl succinate, ethyl itaconate, ethyl pyruvate, ethyl levulinate, and ethyl hydrocinnamate.

- Aldehydes:
formaldehyde, propionaldehyde, n-valeraldehyde, n-hexanal, 2-methyl-butanal, 2-ethylhexanal, benzaldehyde, furfural, 2-butanal, hydroxymethyl furfural, acetaldehyde, isobutanaldehyde, isovaleral, crotonaldehyde, and pyruvaldehyde.
- Ketones:
acetone, methyl n-butyl, ethyl n-butyl, diacetyl, acetoin, and maltol.

Even though the lengthening of baking time favors the browning reactions and may result in increased flavor compounds production, it also reduces the nutritive value of the proteins and can cause vitamins destruction. This is primarily attributed to a loss in lysine, which is especially susceptible to side reactions and crosslinkings, or to its reduced nutritional availability that results from the browning reaction in baking.[2, 10]

2.5 Baking – Mechanisms

The thermal browning reactions that occur during baking and are known to be responsible for the development of the crust color and flavor can be characterized as: caramelization and Maillard browning. These reactions are both nonenzymatic and nonoxidative in nature, and heat generated, with caramelization being by far the more energy intensive.

2.5.1 Caramelization. Caramelization is the transformation of sugars when heated above their point of fusion, from colorless, generally sweet substances into compounds varying in color from pale yellow to dark brown, and in flavor from mild and pleasant caramel to burnt, bitter and acrid. During caramelization a significant number of degradation products, which have an acid, or even a slightly bitter or astringent taste (such as hydroxymethyl furfural or furfuraldehyde) are formed as well as a series of carbonyls, aldehydes, and ketones with agreeable volatile odors. The most basic difference between caramelization and Maillard reaction is that the activation temperature for the caramelization reaction is respectably higher.[2, 12]

2.5.2 Maillard browning. The Maillard reaction was first described in 1912 and has been extensively studied ever since due to its significant contribution to the flavor of baked products. It can be described as the interaction of free amino groups of amino acids, peptides, or proteins, with free reducing sugars resulting in the production of melanoidins as end products. The nonenzymatic browning reaction takes place through three stages:[1, 12, 13]

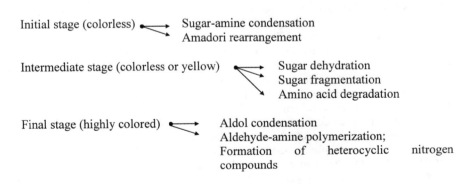

2.6 Frozen Bakery Products

After baking, the sensory characteristics of bakery products degrade rapidly resulting in the disappearance of flavor compounds, the deterioration of textural characteristics and finally staling; thus the main problem with these products is their limited shelf life. The application of freezing in breadmaking has been proposed as a promising method of extending bakery products shelf life. Freezing enhances flexibility and effectiveness of the production since the products can be stored, thawed, proofed and baked in quantities proportional to daily demand, at in-store bakeries, restaurants, institutions and supermarkets, with limited requirements in labor and equipment.

2.7 Role of Additives Used in Frozen Dough

In order to exploit the advantages of freezing (shelf life extension, quality preservation) the selection of raw materials as well as the freezing conditions (temperature level, fluctuation, duration) should be properly controlled. In the case of bread made from frozen dough, it is important to preserve its quality characteristics and overcome some of the problems caused by prolonged storage under freezing conditions. Due to parameters such as increased proof times, decreased inflation ability and variable textural properties, it is important to incorporate into the dough formula appropriate additives and substances apart from the basic ingredients used in conventional breadmaking. Some of these ingredients are trehalose, monoglycerides, vital wheat gluten, whey proteins and soy protein isolate, which are examined in the present study.

Trehalose protects molecules and structure and functionality of cells from destruction caused by freezer dehydration. Because of its decreased sweetness (40-45% of sucrose sweetness) it does not influence foods flavor, it is safe, non-toxic and stable at a great range of pH and temperature values. It has the ability, when added in several foods, to increase glass transition temperature and influences the point of fusion. Finally, it has been proved that it impedes degradation of proteins and contributes in the protection of their three-dimensional configuration during freezing and drying.[14, 15]

Monoglycerides are used in breadmaking because they are considered to strengthen dough by increasing gas holding ability. They increase dough stability during inflation and improve volume and textural characteristics of the final products. When incorporated in dough, they immediately form bonds with hydrated proteins. Thus, the gluten network, which is created, is more cohesive, stable and elastic.[2, 16]

Vital wheat gluten is the dried insoluble gluten protein of wheat flour. Gluten is the principal key to dough utility and its viscoelastic properties make wheat unique among cereal grains. It improves the mixing tolerance and stability of dough during fermentation, makeup and final proofing resulting in increased loaf volume, improved grain, texture and softness of crumb, and a prolonged shelf life of the baked bread products. Being a protein, gluten improves the nutritional value of bakery products.[10, 17]

Whey proteins provide the fortification of bakery products with calcium, protein and essential amino-acids, such as lysine, methionine and tryptophane. Due to their functional characteristics they reinforce dough handling ability, mainly by increasing water absorption, and improve crust color, crumb texture and structure, as well as the bread flavor and aroma. The incorporation of these proteins can also slow down humidity loss resulting in increased final products storage time and maintenance duration.[18, 19]

Soy protein products deliver high amino-acids content (methionine, cysteine and threonine). They also contain important quantities of lysine, which is found in low

Quality

percentage in other cereal. Soy protein isolates are used in breadmaking due to their high protein content and solubility. They provide increased water binding and improve final products maintaining their humidity and remaining soft.[20]

The aim of this work is to examine the effect of trehalose, mono and di-glycerides, vital wheat gluten, whey proteins and soy protein isolate addition on the sensory characteristics of bread prepared from frozen dough.

3 MATERIALS & METHODS

3.1 Raw Materials

The raw materials used for dough preparation were hard wheat flour T.70% (Sarantopoulos Flourmill, Keratsini, Greece), ascorbic acid and trehalose suitable for foods (Merck) and sugar, salt, active dry yeast, plant shortening, vital wheat gluten, mixture of mono and di-glycerides, whey proteins and soy protein isolate of commercial origin.

3.2 Production Process

Breadmaking process initially included weighing of raw materials according to Table 1. Dough ingredients were placed in a "Kenwood" domestic blender and mixed for 2 min at a low speed (speed 2) and for 8 min at a medium speed (speed 4). As soon as dough is formed, it was divided in samples of 80 g, which were round shaped by hands. Samples were placed in aluminum pans, wrapped with plastic membrane, weighed and placed in the freezer. The process flow diagram is presented below.

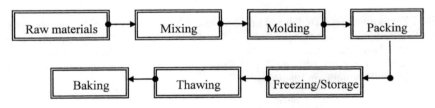

Figure 1 *Process flow diagram*

The formulation of dough samples without additives (blank samples), expressed on flour basis, was the following:

Table 1 *Dough formulation*

Component	Flour	Water	Yeast	Sugar	Salt	Shortening	Ascorbic Acid
Percentage (%)	100	60	2	4	2	3	0.01

The following series of experiments were conducted:

- Series 1: Addition of trehalose in concentration of 0.02% and 0.2% or mixture of mono and di-glycerides (MDG) in concentration of 1 and 2% on flour basis
- Series 2: Addition of vital wheat gluten in concentration of 2, 4, 5 and 6% on flour basis
- Series 3: Addition of whey protein in concentration of 2 and 4% on flour basis
- Series 4: Addition of soy protein isolate in concentration of 1, 2.5 and 4% on flour basis.

3.3 Experimental Process

Samples were withdrawn from the freezer every 6-15 days. They were weighed and placed in a room of constant temperature (25 °C) for 195 min in order to thaw and proof. One dough sample was used for texture characteristics examination, while the rest were placed into the oven (Thermawatt with air circulation) at the temperature of 210 °C, for 35 min. After baking, bread samples were allowed to cool down for about 30min in ambient temperature. Then they were weighed, their specific volume was estimated (rapeseeds displacement) and their crust/crumb color was measured. The samples were also subjected to sensory evaluation and texture examination with texture analyzer.

3.4 Analytical Measurements

3.4.1 Color measurement. Crust and crumb color of baked samples were measured using a Minolta CR/200 colorimeter, which displays the L, a, b color parameters for every sample. Color variation was estimated according to the following equation:

$$E = \sqrt{L^2 + a^2 + b^2}$$

3.4.2 Texture analysis. Dough and bread texture characteristics were determined using a TA-XT2i (Stable Micro Systems, Osaka, Japan) Texture Analyser. Dough samples were subjected to a two cycles compression test using the SMS P/45C probe. Bread samples were subjected to a cut test using the TA-45 Craft Knife and a two cycles compression test using the Sris P/75 Aluminum Platen probe.

4 RESULTS AND DISCUSSION

4.1 Series 1: Addition of Trehalose and MDG

Samples with 2% of MDG presented the best rising ability during the first days of storage followed by blank samples and samples with 0.2% of trehalose. However, after 82 days of storage the rising ability of blank samples deteriorated seriously and their volume was the lowest of all samples. As far as crust color is concerned, samples with 0.02% of trehalose and 1% of MGD presented the lightest while those with 2% of MDG the darkest crust color. Samples with 2% of MDG were the only samples whose crust color became darker during prolonged storage. The measurement of crumb hardness indicated that blank samples were minimally affected by frozen storage while all other samples showed a decrease in crumb hardness. This decrease was higher for samples with 2% of MDG and 0.2% of trehalose. The above findings were confirmed by the sensory evaluation of the samples. Sensory evaluation also showed that blank samples presented more extensive cracks and bulges and that blank samples along with MDG samples had the more intense aroma.

4.2 Series 2: Addition of Vital Wheat Gluten

In this series the best rising ability was observed at samples with 4% of gluten while those with 2% of gluten had similar behavior to the blank samples. During the first days of storage, all samples presented similar crust color. However after prolonged storage, the crust color of blank samples became lighter and the opposite trend was observed for samples with gluten. The greatest color variation was noted for samples with 6% of gluten. Except for the samples with 5% of gluten, all the others demonstrated a decrease in crumb firmness during frozen storage. Firmness decrease was higher for samples with 6% of gluten. Sensory evaluation showed that samples with 2 and 4% of gluten had more extensive cracks and bulges. Most of the samples developed spots and blisters on their crust especially during the first weeks of storage. Samples with 4% of gluten possessed the most appealing aroma and the blank ones had the worst.

4.3 Series 3: Addition of Whey Proteins

The incorporation of whey proteins increased bread volume and samples with 4% of whey presented the best rising ability. Whey proteins also considerably altered the crust color, which became darker as the percentage of whey protein in dough formula increased. Consequently, samples with 4% of whey protein had significantly dark crust color. However, the storage under freezing conditions did not seriously affect crust color of the samples crust color. Crumb and crust firmness increased during storage and blank samples presented the greatest change. According to sensory evaluation, samples with whey proteins had a rather desirable appearance and crumb structure. Furthermore, as the percentage of whey proteins increased, samples presented a more roasted appearance and a characteristic dairy flavor, which was excessive for samples with 4% of whey proteins.

4.4 Series 4: Addition of Soy Protein Isolate

In this series, all samples demonstrated similar rising ability except for those with 4% of soy protein isolate which were slightly better especially during the first days of storage. Blank samples and those with 2.5% of soy isolate had similar behavior as far as crust color was concerned and all samples retained their relatively light crust color during prolonged frozen storage, except for those with 4% of soy isolate. Furthermore, all samples possessed an increase in crumb firmness, which was significantly higher for those with 2.5 and 4% of soy isolate. The sensory evaluation of the samples demonstrated that the incorporation of soy protein isolate in dough formula had a desirable effect on the appearance and crumb structure of the samples. Blank samples presented white spots and blisters on their crust much earlier compared to the other ones. However samples with the highest percentage of soy isolate developed a slightly unpleasant aroma and imparted the characteristic bitter taste of soy proteins to the flavor of bread.

Some characteristic diagrams with the above findings are presented below:

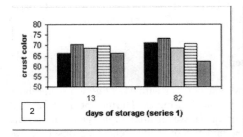
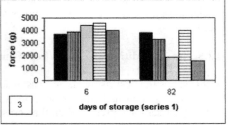

Figures 2-3 *Crust color variation – crumb firmness during compression for series 1*
blank ■ *trehalose 0.02%* ⊞ *0.2%* ▨ *MDG 1%* ⊟ *MDG 2%* ▦

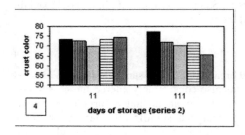
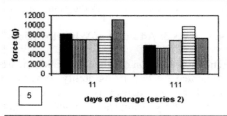

Figures 4-5 *Crust color variation – crumb firmness during compression for series 2*
blank ■ *vital wheat gluten 2%* ⊞ *4%* ▨ *5%* ⊟ *6%* ▦

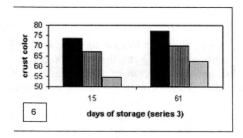
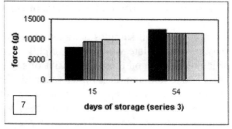

Figures 6-7 *Crust color variation – crumb firmness during compression for series 3*
blank ■ *whey proteins 2%* ⊞ *4%* ▨

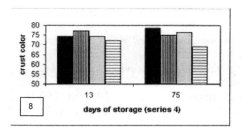 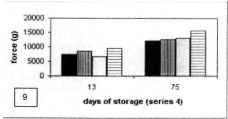

Figures 8-9 *Crust color variation – crumb firmness during compression for series 4*
blank ■ *soy protein isolate* *1%* ▥ *2.5%* ▫ *4%* ▤

5 CONCLUSIONS

After careful examination of the result from all the aforementioned experimental series it could be concluded that:
- Samples with 2% of MDG presented the best rising ability while those with 4% of gluten had the best storage stability during prolonged storage under freezing conditions.
- Samples containing whey proteins exhibited the darkest crust color and the more roasted appearance. The opposite was observed for samples with soy protein isolate.
- Samples with trehalose and MDG presented significantly lower crumb firmness. The incorporation of whey proteins and soy protein isolate in dough formula at excessive percentages per flour weight can adversely affect samples aroma and taste and destroy the characteristic flavor of bread.

Acknowledgements

The present project is been supported by grants from Greek GSRT (General Secretariat of Research and Technology).

References

1. J.A. Johnson and A.A. El-Dash, *J. Agric. Food Chem.*, 1969, **17**, 740.
2. E.J. Pyler, *Baking Science and Technology*, Vol 2, 3rd edn., Sosland Publishing Company, Kansas City, 1988, pp. 589-590, 752-763.
3. A.G. Gaonkar, *Ingredients Interactions – Effects on Food Quality*, Marcel Dekker, Inc., New York, 1995, pp. 110-111
4. Y. Linko and J.A. Johnson, *J. Agric. Food Chem.*, 1963, **11**, 150.
5. C.R. Hoseney and E.D. Rogers, *Crit. Rev. Food Sci.*, 1990, **29**, 73.
6. M.A Rao and J.F. Steffe, *Viscoelastic Properties of Foods*, Elsevier Applied Science London and New York, 1992, pp. 77-83.
7. K. Lorenz and J. Maga, *J. Agric. Food Chem.*, 1972, **20**, 211.

8   S.P. Cauvain and L.S. Young, *Technology of Breadmaking*, Aspen Publishers, Maryland, 1999, pp. 6-8.
9   H. Faridi and J.M. Faubion, *Dough Rheology and Baked Product Texture*, Avi Books, New York, 1990, pp. 356-359.
10 C.A. Stear, *Handbook of Breadmaking Technology*, Elsevier Applied Science, London and New York, 1990, pp. 590-595.
11 S.A. Matz, *Equipment for Bakers*, Pan-Tech International, Texas, 1988, pp. 159-160.
12 R. Calval, R. Wirtz and J.J. MacGuire, *The Taste of Bread*, Aspen Publishers, Maryland, 2001, pp. 69-73.
13 N. De Kimpe & C. Stevens, *Tetrahedron*, 1995, **51**, 2387.
14 T.H. Grenby, *Advances in Sweeteners*, Chapman and Hall, London, 1996, pp. 188-189.
15 M.L. Stecchini, E. Maltini, E. Venir, M. Del Torre and L. Prospero, *J. Food Sci.*, 2002, **67**, 2196.
16 S.E. Friberg and K. Larson, *Food Emulsion*, Marcel Dekker, Inc., New York, 1997, pp. 141-148.
17 P.D. Ribotta, A.E. León, and M.C. Añón, *J. Agric. Food Chem.*, 2001, **49**, 913.
18 E. Mannie and E.H. Asp, *Cereal Food World*, 1999, **44**, 143.
19 S. Kenny, K. Wehrle and E.K. Arendt, *Eur. Food Res. Technol.*, 2000, **210**, 391.
20 M.N. Riaz, *Cereal Food World*, 1999, **44**, 136.

SHELF-LIFE PREDICTION AND MANAGEMENT OF FROZEN STRAWBERRIES WITH TIME TEMPERATURE INTEGRATORS (TTI)

E.D. Dermesonlouoglou, M.C. Giannakourou and P.S. Taoukis

Laboratory of Food Chemistry and Technology, School of Chemical Engineering, National Technical University of Athens, Iroon Polytechniou 5, 15780 Zografou, Athens, Greece.

1 ABSTRACT

Temperature conditions in distribution chain determine the quality of frozen products. These conditions can be monitored with Time Temperature Integrators (TTI), simple devices that show a visible time- temperature dependent response, that can be translated to the food's quality status. For the development of a TTI based management system, quality indices of frozen strawberry, including vitamin C (HPLC), color (CIELab scale) and flavor (sensory evaluation) were kinetically studied in the range -5 to –18°C and their temperature dependence was modelled by the Arrhenius equation. The activation energy, Ea, ranged from 28 to 40 kcal/mol. The response of enzymatic TTIs was also kinetically studied, with Ea calculated at 24 kcal/mol, and correlated to the quality level of the strawberries. The results showed that the TTI provided effective indication of the product's remaining shelf-life, in the real chill chain.

2. INTRODUCTION

Freezing of fruit and vegetables is one of the most widely applied technologies for effective preservation of the quality of these commercially and nutritionally valuable, sensitive products. Strawberries (*Fragaria* x *ananassa*) are rich in nutrients, such as amino acids and vitamins,[1] but are highly perishable due to their high physiological post-harvest activity, making their effective management a challenge.[2,3] Among the berry species, strawberries have been in the center of commercial interest, due to the variety of their potential applications (juices, liqueurs, sorbets, ice cream, concentrates, dairy products and lip colorings).[4,5] Additionally, fresh strawberries are a good source of ascorbic acid.[6]

Frozen fruit and vegetables are microbiologically stable due to the combined effect of low temperature[7] and low effective moisture content.[8] The most important changes in frozen foods are related to their sensory and nutritional value as a result of physicochemical phenomena and chemical and enzymatic reactions during freezing and subsequent frozen storage.[9] According to Mohammad *et al* (2004),[10] who mostly studied the effect of freezing rates on quality reduction of frozen strawberries, most changes during storage are related to vitamin C loss and color change. Polyphenoloxidase (PPO) present in strawberry tissues, causes loss of red color because of deterioration of anthocyanin pigments and browning.[11] In order to preserve red color during subsequent storage, blanching is recommended before minimally processing the strawberry, but is also responsible for the reduction of thermosensitive nutrients.

A considerable body of work on the modes of quality degradation of several frozen fruit and vegetables has been published in the earlier literature[12]. More recent studies take a more systematic kinetic and modelling approach of quality determining indices such as color,[13,14] vitamin C content,[13,14,15,16] chlorophyll content[17] and selected sensory parameters.[18,19] Among these parameters, vitamin C loss was found to be a well-established measure of frozen products degradation.

During processing, distribution, and storage of frozen vegetables, ascorbic acid (vitamin C) oxidizes to dehydroascorbic acid, which is irreversibly hydrolyzed to 2,3-diketogulonic acid, which possesses no vitamin C activity. This oxidation is enhanced by temperature abuse during frozen storage. The retention of ascorbic acid in frozen products is thus strongly dependent on their temperature history.[13] Recent studies report vitamin C contents of several frozen fruit and vegetables at designated time points of their isothermal storage at low temperatures, e.g at -20 and $-30°C$[15,20,21,22,23] without assuming a full, systematic kinetic approach. The temperature range in most studies does not cover the -3 to -10°C range which is very detrimental and does frequently occur.[14,24] Additionally, the applicability of shelf-life models, under possible temperature fluctuations, has not been fully addressed. In order to be able to predict, in a reliable way, at any point of its life cycle, the nutritional level of a product, based on its temperature history, it is important that the established kinetic equations cover the whole relevant range of temperatures.

Recent temperature surveys have shown that significant deviations from specified conditions during frozen distribution, handling and storage often occur. Temperature conditions in the frozen distribution chain are of central importance for the shelf life and final quality of optimally processed and packed frozen fruit and vegetables. Thus, it is necessary to monitor and record temperature. This can be achieved by a structured quality and safety management system through the entire frozen product's life cycle based on prevention that aims at product quality optimisation at the consumer's end.[25] A cost-effective way to individually monitor the temperature conditions of frozen food products throughout distribution is Time Temperature Integrators or Indicators (TTI). TTI are simple, inexpensive devices that can show an easily measurable, time and temperature dependent change that cumulatively indicates the time-temperature history of the product from the point of manufacture to the consumer.[26] Based on reliable models of the shelf-life and the kinetics of TTI response, the effect of temperature can be monitored and quantitatively translated to food quality from production to the point of consumption.[27,28,29,30] At the early stages of TTI development, several studies correlating frozen food quality with TTI were published.[31,32,33,34,35] Recently, however, due to difficulties relating to systematic kinetic modelling of frozen foods and response of TTI in the subfreezing range, few studies are available in this field. Prerequisite to a successful application of TTI is a thorough knowledge of kinetics of frozen food quality loss as well as systematic study of TTI response in the temperatures of interest.

The objective of this work is (a) the systematic kinetic study of the quality of frozen strawberries, focusing on the nutritional deterioration in the whole temperature range of practical interest, (b) the development of response kinetics for suitable TTI in the same temperature range. The aim is to integrate all this kinetic data in an application scheme, where TTI will serve as temperature monitoring devices and prediction tools for the quality status of frozen strawberries, at any point of their life cycle. Finally, the requirements for obtaining a sufficient accuracy in this TTI implementation algorithm are also discussed.

3 MATERIALS AND METHODS

3.1 Sample Preparation

Strawberries (*Fragaria* × *Ananassa*), obtained directly from a producer from Southern Greece, were washed and quick frozen at –40°C with forced air convection [h=11 W/(m^2*K) - Sanyo MIR 553, Sanyo Electric Co, Ora-Gun, Gunma, Japan]. Strawberries of the same size were selected for the experiments. Half of them were blanched before their freezing. Blanching was carried out in hot water of 80°C for 80 s.

Approximately 100 g of frozen strawberries, untreated and blanched, were packed in a laminate film (20 μm BOPP – 48 μm PE) used for commercial frozen vegetables. Packed strawberries were distributed into 4 controlled temperature cabinets (Sanyo MIR 153 and 253, Sanyo Electric Co, Ora-Gun, Gunma, Japan) at the temperatures of -5, -8, -12 and -16°C (±1°C), constantly monitored by type T thermocouples and a multichanel datalogger (CR10X,Campbell Scientific, Leicestershire, UK). Samples were obtained at appropriate time intervals for each storage temperature based on ASLT (Accelerated Shelf Life Testing) methodology (detailed by Taoukis et al[28]) and a rough estimate of expected temperature-dependence of vitamin C loss and color change rate from previously published work.

3.2 Vitamin C Determination

Vitamin C was determined by a high performance liquid chromatography method (HPLC), which was compared and standardized with the 2,6-dichlorophenolindophenol titrimetric method (AOAC, 1984, 43.064).[36] All analyses were carried out in duplicate on fruit tissue, homogenized, using a pestle and mortar as described in the method used by Oruna-Concha and others.[15] 5 g of homogenate were mechanically stirred in 15 ml of a 4.5% (wt/vol) solution of metaphosphoric acid (Merck, Darmstadt, Germany) for 15 min. The mixture was vacuum-filtered and diluted with HPLC grade water (Merck); the total final volume was measured and an aliquot was filtered through a 0.45-μm Millipore (Millex®; Bedford, U.S.A.) filter prior to injection into the chromatographic column. The instrumentation details were: HP Series 1100 (quarternary pump, vacuum degasser, a Rheodyne 20-μl injection loop and a Diode-Array Detector, controlled by HPChemStation software); Hypersil ODS column (250×4.6 mm) of particle size 5 μm; mobile phase: HPLC grade water with metaphosphoric acid to pH 2.2; detection at 245 nm; calibrated by external standard method. The results from chromatographs were translated in mg vitamin C/100 g raw material and plotted against time. All measurements were conducted in duplicate.

3.3 Color Measurement

A continuous objective instrumental quantitation of the color change was done based on measurement of CIELab values (CIE, 1978) with a CR-200 Minolta Chromameter® (Minolta Co., Chuo-Ku, Osaka, Japan) with an 8 mm measuring area. A standard white plate (Calibration plate CR-200) was used to standardize the instrument under "C" illuminant condition according to the CIE (Commission International de l' Eclairage). At predetermined times of storage, according to the experimental design, measurements were conducted on the same numbered strawberries, representative of the group, in order to obtain consistent information.

3.4 Modelling of TTI Rresponse

The color response of enzymatic TTI (VITSAB AB, Malmö, Sweden), Type $M_{2-102140}$ was kinetically studied. These TTI are based on a color change caused by a pH decrease, due to a controlled enzymatic hydrolysis of a lipid substrate. Before activation, the lipase and the lipid substrate are in two separate compartments (mini-pouches). At activation, the barrier that separates them is broken, enzyme and substrate are mixed, and the color gradually changes from deep green to bright yellow. The tested TTI give a triple response via three separate transparent "windows" allowing view of change in three different double-pouches with different enzyme concentrations in each. Color change can be visually graded on a 6 point reference scale constructed from TTI inactivated at a certain level or quantitatively measured on the CIELab scale with the Minolta CR-200 Chroma Meter. Kinetic modelling was based on measurements, at appropriate time intervals for each "window" of the two types, of the response of multiple TTI samples, isothermally stored in the controlled cabinets at temperatures from –20 to 0°C. Response models were assessed against measured values at variable temperature profiles.

The basic principles of TTI modelling and application for quality monitoring are detailed by Taoukis and Labuza.[28-30] Similarly to loss of shelf life of a food which is usually expressed as:

$$f(A) = k(T)t = k_{A_{ref}} \exp\left(-\frac{E_A}{R}\left(\frac{1}{T} - \frac{1}{T_{ref}}\right)\right)t \qquad (1)$$

where f(A) is the quality function, A a characteristic quality index of the food, k is the reaction rate constant, $k_{A_{ref}}$ is the rate constant at a reference temperature T_{ref}, E_A is the activation energy of the reaction that controls quality loss and R the universal gas constant, a response function F(X) can be defined for TTI such that $F(X) = k_I t$, with k_I an Arrhenius function of T.

The value of the functions, $f(A)_t$ at time t, after exposure at a known variable temperature exposure, T(t), can be found by integrating equation 1. Introducing the term of the effective temperature T_{eff}, which is defined as the constant temperature that results in the same quality value $f(A)_t$, as the variable temperature distribution over the same time period, equation 1 gives:

Quality

$$f(A) = \int_0^t k(T)dt = k_{A_{ref}} \exp\left(-\frac{E_A}{R}\left(\frac{1}{T_{eff}} - \frac{1}{T_{ref}}\right)\right) t \qquad (2)$$

For a TTI exposed to the same temperature fluctuations, T(t), as the food product, and corresponding to an effective temperature T_{eff}, the response function can be expressed as:

$$F(X) = k_{I_{ref}} \int_0^t \exp\left(-\frac{E_{A_I}}{R}\left(\frac{1}{T} - \frac{1}{T_{ref}}\right)\right) dt = k_{I_{ref}} \exp\left(-\frac{E_{A_I}}{R}\left(\frac{1}{T_{eff}} - \frac{1}{T_{ref}}\right)\right) t \qquad (3)$$

where $k_{I_{ref}}$ and E_{A_I} are the Arrhenius parameters of the TTI.

Thus, the basic elements for a TTI based food quality monitoring scheme are (a) a well established kinetic model to describe quality loss of the food, (b) the response function of the TTI and (c) the temperature dependence of both food quality loss and TTI response rate, expressed by the respective values of the activation energies. The essence of the TTI implementation algorithm lies in the calculation of the T_{eff} of the exposure (Equation 3), based on the TTI response reading. This T_{eff} value is also assumed to describe the integrated effect of temperature history on food quality loss, allowing for the estimation of the quality function value and the value of index A from equations (1) and (2). This assumption requires that food quality degradation and TTI response rate are similarly affected by temperature, i.e. the activation energies of the two phenomena do not differ by more than 25 kJ/mol. Under these conditions the application scheme would reliably provide the extent of the quality deterioration of the food and a prediction of the remaining shelf life at any assumed average storage temperature.

4. RESULTS AND DISCUSSION

4.1 Kinetic Study of Vitamin C Loss

The results of Vitamin C content, obtained for strawberry samples, untreated and blanched, were plotted vs time for all temperatures studied, and the apparent order of vitamin C loss (L-ascorbic acid oxidation) was determined.

The average retention of vitamin C is expressed relative to an initial, average value of day 0 of the experiment (Figure 2a), where VitC represents the concentration of vitamin C in 100 g of raw material. In all cases, Vitamin C loss was found to be adequately described by an apparent first order reaction (Equation 4):

$$VitC = VitC_0 e^{-k_{vit} t} \text{ or } \ln\frac{VitC}{VitC_0} = -k_{vit} t \qquad (4)$$

where VitC and $VitC_0$ are the concentrations of L-ascorbic acid at time t and zero respectively and k_{vit} is the apparent reaction rate of Vitamin C loss, estimated by the slope of the linearized plot of $\ln(VitC/VitC_0)$ vs t.

The temperature dependence of the deterioration rate k_{vit} was then modelled by the Arrhenius equation (Equation 5):

$$k_{vit} = k_{vit,ref} \exp\left[\frac{-E_A}{R}\left(\frac{1}{T} - \frac{1}{T_{ref}}\right)\right] \tag{5}$$

where $k_{vit,ref}$ is the reaction rate of the Vitamin C oxidation at a reference temperature T_{ref}, E_A is the activation energy of the chemical reaction and R is the universal gas constant. By linearly correlating $\ln k_{vit}$ vs $(1/T_{ref}-1/T)$ (Arrhenius plot), the E_A of L-ascorbic oxidation was estimated from the slope of the fitted line.

After the freezing/blanching process, during the subsequent isothermal frozen storage, strawberries exhibited a significant loss of Vitamin C at all temperatures studied (Figure 1a). Rates of loss in unblanched and blanched frozen samples did not differ significantly, with the blanched samples showing slightly reduced rates (Table 1). Compared to other published work, the vitamin C content of frozen strawberries in this study is better retained. For example, Mallet[37] measured 100% loss of vitamin C after 30 days of storage time at $-5°C$. In our case, according to Figure 1a, a 30% retention was observed at the same time.

Temperature dependence of Vitamin C deterioration was expressed with the Arrhenius equation (Figure 1b). The estimated activation energies, E_A, are 118.9 and 126.2 kJ/mol, for the case of untreated (R^2= 0.962) and blanched (R^2= 0.976) frozen strawberries, respectively.

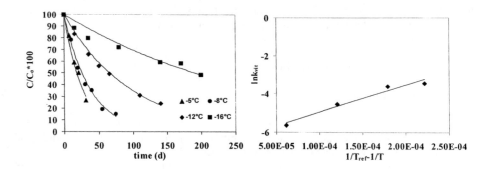

Figure 1 (a) Results for Vitamin C loss vs time at 4 storage temperatures and (b) Arrhenius plot of the Vitamin C loss rate for frozen strawberries, unblanched (with $T_{ref} = -20°C$).

Table 1 *Rates of Vitamin C loss and initial Vitamin C concentration, for blanched and unblanched frozen strawberries.*

Temperature	k_{VitC} (d^{-1})	
	Blanched strawberries	Unblanched strawberries
-5°C	0.0353	0.0326
-8°C	0.0213	0.0273
-12°C	0.0107	0.0106
-16°C	0.0030	0.0030
C_o (mg VitC/100g raw material) (mean value of 8 samples±standard deviation)	72.0±14.2	98.0±17.7

4.2 Kinetic Study of Color Change

For frozen strawberries, the color change, expressed as shown in Equation 6 was modelled by a first-order reaction.

$$\frac{DC_{max} - DC}{DC_{max}} = \exp(-k_{col} * t) \qquad (6)$$

where DC ($= \sqrt{(a-a_o)^2 + (b-b_o)^2}$) is the chroma change, DC_{max} is the asymptote value of DC observed after a long period of storage, a_o and b_o the values of a and b color parameters at zero time, k_{col} is the kinetic rate at temperature T. Color change for untreated strawberries at all temperature studied is shown in Figure 2a.

The temperature dependence of the color change rate is also modelled by the Arrhenius equation (similar to Equation 5).

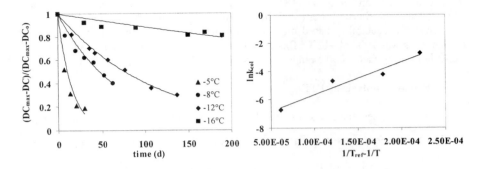

Figure 2 *(a) Results for color change vs time at 4 storage temperatures and (b) Arrhenius plot of the color change rate for frozen strawberries, untreated and blanched (with $T_{ref} = -20°C$).*

Temperature dependence of color change was expressed with the Arrhenius equation (Figure 2b). The estimated activation energies, E_A, are 191.5 and 169.8 kJ/mol, for the case of untreated (R^2= 0.944) and blanched (R^2= 0.985) frozen strawberries, respectively.

Decrease in red (CIE a) and yellow (CIE b) intensity in frozen strawberry color was the characteristic color change during frozen storage. Color parameters indicated that color change of untreated frozen strawberry samples was more intense than the change of color of blanched samples. Indeed, the estimated rates of color change (k_{col}) had lower values for the blanched samples. In Figure 3, this improved retention, accomplished with the application of blanching, is illustrated.

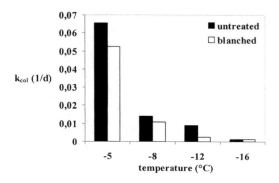

Figure 3   *Rates of color change for untreated and blanched strawberries.*

A notable observation is that the two alternative modes of quality deterioration, vitamin C vs. color loss, of frozen strawberries have a different temperature sensitivity with color change exhibiting an elevated E_A. As a result frozen strawberries stored at low temperatures (below –12° C) exhibit good color stability for long storage times (shelf life of 122d at -12°C and 973d at -18°C). Acceptable Vitamin C loss (40% of initial) is surpassed at much shorter times (Shelf life of 54d at -12°C and 194d at -18°C).

Sensory acceptability based on limited results from a 5 member panel at the same temperatures was 100d at –12°C. Based on the above measurements and observations, the shelf life limiting factor in the range of –5 to –20°C is Vitamin C degradation.

4.3 Kinetic Analysis of TTI

Based on Lab values, the index that was found to quantify effectively the color change of TTI was the chroma value $C = \sqrt{a^2 + b^2}$. The normalized chroma $X_C = \dfrac{C - C_{min}}{C_{max} - C_{min}}$ was used as the response X of the TTI, which, when plotted as a function of time, had a sigmoidal shape, rather similar to a Gaussian function (expressed as $X = 1-\exp[-(kt)^2]$). Accordingly, rearrangement of the previous general equation, leads to the following form of a linearized response function (Equation 7):

Quality

$$F(X_C) = \sqrt{\ln\left(\frac{1}{1-X_C}\right)} = k_I t \qquad (7)$$

From the F(X$_C$) vs. time plots the value of k_I, the response rate of the tested TTI, was determined at each temperature by linear regression analysis. In Figure 4a such a plot for the second of the 3 windows of enzymatic TTI type M, model $M_{2\text{-}102140}$ is shown. For this type of TTI, nominally the response times from window 1 to 3 are set at a ratio of 10:21:40 respectively. Temperature dependence is the same for the 3 windows. The temperature dependence of the response rate was modelled by the Arrhenius equation. The value of activation energy ± 95% confidence range, of TTI type M were calculated as 92.2±18.7 kJ/mol (R^2=0.955).

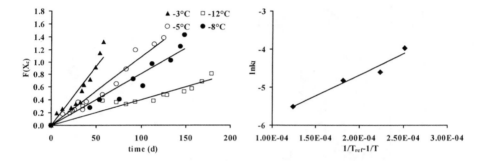

Figure 4   *TTI enzymatic Type M (model $M_{2\text{-}102140}$, window 2) (a) response vs time plot (b) Arrhenius plot of the response rate (T_{ref} = -20°C)*

4.4   Prediction of the Quality Level of Frozen Strawberries with TTI

In order to test the accuracy and the reliability of the application scheme of TTI, a representative scenario of distribution of frozen strawberries was studied, using realistic time-temperature data.[14] This scenario includes 10 days of storage in the producer warehouse, 20 days of stocking in a central distribution center, 20 days of retail display and finally 30 days of storage in the domestic freezer, taking also into consideration all intermediate transports. The temperature profile at each respective stage is illustrated in Figure 5. When the time-temperature history of products is constantly monitored, it is possible to calculate, at designated points of their marketing route, the extent of their nutritional deterioration, and, therefore their remaining Shelf Life (SLR). SLR is based on nutritional criteria. 40% vitamin C loss is set as the acceptability limit. The SLR is estimated from the respective time-temperature integral, or equivalently, by the calculation of the corresponding T_{eff} (Equation 2). These estimations can then be compared to the predictions of a TTI of Type M, that would be attached on the food throughout its circuit from the producer to the final consumer. In Figure 5, the comparison between SLR calculated from the TTI response and SLR based on the actual time-temperature data is illustrated for studied distribution scenario. In this case, at any point of the 80 days cycle, the TTI prediction was in good agreement to the actual remaining shelf life. To put that in terms of current practice, the "presumed" SLR, at any point of the distribution, would be solely based on its expiration date label, not taking into account its real time-temperature handling. In the case studied in our example, at the end of the 80 days distribution, the "presumed" SLR would be 222 days (shelf life at –20°C: 302d), compared to the 68d signaled by the attached TTI and the 88d of actual SLR. Similar differences were calculated, when applying different distribution scenarios, ranging from ideal to abusive ones. The calculations confirm that the studied TTI type M can acceptably serve as a practical monitoring tool for nutritional degradation of frozen strawberries.

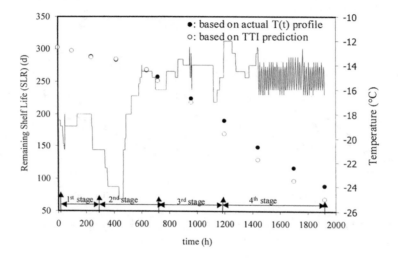

Figure 5   *Remaining shelf life of blanched frozen strawberries, based on 40% vitamin C loss, estimated either by the actual time-temperature profile, or by the TTI prediction, at designated points of the illustrated distribution scenario. Remaining Shelf Life (SLR) is the time the product would remain acceptable after the consumption time if stored at –20°C.*

A similar approach was used for assessing the TTI potential for color loss prediction of frozen strawberries. Color loss rate has a much higher activation energy and for all temperatures below -5°C color change will be preceded by unacceptable vitamin C degradation. At -12°C unacceptable color is reached only after 122 d. If nutritional criteria are not considered then color can be used as a shelf life index. Different scenarios of storage temperature at different marketing stages were studied (Table 2). At the end of each assumed scenario, the remaining shelf life was calculated based on the actual T(t) profile and color loss kinetics, and subsequently compared to the prediction obtained by the TTI response reading. From the calculated differences in SLR, it is seen that the particular type of TTI is not appropriate for SLR monitoring of frozen strawberries, when color change is considered as the limiting factor. This is mainly due to the significant difference between the activation energies of the color loss rate and the TTI response. Alternatively, a different type of enzymatic TTI, with a higher E_A, would significantly improve the prediction accuracy of the TTI systems. Enzymatic TTI of Type L (VITSAB AB, Malmö, Sweden), has an activation energy of 163 kJ/mol at temperature ranges of 0-12°C. If the same temperature sensitivity is confirmed, as expected, in the frozen range, from testing that is in progress, the differences between SLR_{ACTUAL} and SLR obtained by TTI of Type L, in cases 1-3 of Table 2 would be significantly decreased. Further improvement of the TTI prediction could be achieved by using a double TTI system of different temperature sensitivity (i.e. combination of Type M and L on the same package), instead of a single TTI tag, by applying a correction algorithm.[38]

Table 2 *Remaining shelf life (SLR) signaled by color change of $DC_{fin} = 6$, for frozen strawberries 165 d after production, for 3 alternative storage scenarios.*

Scenarios	Temperature (1st stage : 60 days at the producer warehouse)	Temperature (2nd stage: 45 days at the distribution center)	Temperature (3rd stage: 30 days at the retail outlet)	Temperature (3rd stage: 30 days at the domestic freezer)	SLR (actual) (d)	SLR (TTI Type M) (d)	SLR (TTI Type L) (d)
1	-20°C	-12°C	-11°C	-13°C	165	568	283
2	-15°C	-16°C	-13°C	-12°C	616	712	648
3	-14°C	-15°C	-12°C	-11°C	60	193	105

Based on the above, it was demonstrated that TTI with the appropriate response E_A value can be reliably used as time- temperature recorders and quality loss (primarily nutritional deterioration) monitors for frozen strawberries. Consequently, when addressing the problematic distribution chain, it would be possible to control and improve the management of such frozen products, by introducing a TTI based monitoring system, coded LSFO (Least Shelf Life First Out). This novel approach, detailed in Giannakourou and Taoukis,[14] is based on the classification of products at designated points of their marketing route, according to their quality status, as it is predicted by the attached TTI. The principles that lie behind this management system are illustrated in Figure 6, showing a case study where 60 cases of frozen strawberry products of the same production date arrive at the retail warehouse where they are supposed to be split into 3 groups for successive stocking of the retail freezer cabinets, every 10 days. TTI application allows for a classification, based on real quality (Vitamin C) criteria, instead of a random split, according to the First In First Out approach. With LSFO, products that have suffered the higher vitamin C loss due to worse temperature handling would be advanced first to the Super Market shelves (cabinets). They would be followed by the two other groups, in order of decreasing vitamin C loss, in 10 and 20 days respectively. Overall, this system would optimize the current inventory management system leading to products of more consistent quality and nutritional value at the time of consumption.

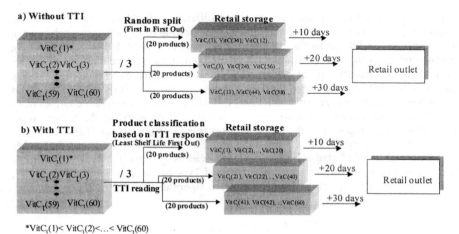

Figure 6 *Schematic illustration of the decision-making principles of the TTI based management system, at the point of retail display of frozen strawberry products, compared to the traditional FIFO practice.*

References

1. S.W. Souci, W. Fachmann and H. Kraut, *Food Composition and Nutrition Table*, Sixth Ed. *Medpharm Scientific Publishers*, Stuttgart, 2000, 908-916.
2. C. Han, Y. Zhao, S.W. Leonard and M.G. Traber, *Postharvest Biol. Technol.* 2004, **33**, 67-78.
3. M.A. Garcia, M.N. Martino and N.E. Zaritzky, *J. Agric. Food Chem.* 1998, **46**, 3758-3767.
4, J. Moreno, A. Chiralt, I. Escriche and J.A. Serra, *Food Res. Internat.* 2000, **33**, 609-616.
5. A. Garcia-Palazon, W. Suthanthangjai, P. Kajda and I. Zabetakis, *Food Chem.* 2004, In press.
6. K.P. Wright and A.A. Kader, *Postharvest Biol. Technol.* 1997, **10**, 39-48.
7. H.D. Goff, *In: M.C. Erickson, Y.C. Hung, editors. New York: Chapman & Hal,*. 1997, p.29.
8. O.R. Fennema, *In: Fennema, OR, editor. 3rd ed. New York: Marcel Dekker Inc.*, 1996, p.17-94.
9. M.C. Giannakourou and P.S. Taoukis, *J. Food Sci.* 2002, **67**, 2221-2228.
10. A.S. Mohammad, B.F. Mohsen and H.E. Zohreh, *Food Chem.* 2004, **86**, 357-363.
11. P. Markakis, *Food Technol.* 1974, **4**, 437.
12. T.P. Labuza, *In: Labuza TP. Westport, Connecticut, USA: Food & Nutrition Press, Inc.*, 1982, p.289-340.
13. M.C. Giannakourou and P.S Taoukis, *Food Chem.* 2003, **83**, 33-41.
14. M.C. Giannakourou and P.S. Taoukis, *J. Food Sci.* 2003, **68**, 201-209.

15. M.J. Oruña-Concha, M.J. Gonzalez-Castro, J Lopez-Hernandez and J Simal-Lozano, *J. Sci. Food Agric.* 1998, **76**, 477-480.
16. D.J. Favell, *Food Chem.* 1998, **62**, 59-64.
17. R.C. Martins and C.L.M. Silva, *J. Food Engineering,* 2004, **64**, 481-488.
18. M. Martens, *J. Food Sci.* 1986. 51, 599-617.
19. Y.C. Hung and D.R. Thompson, *J. Food Sci.* 1989, **54**, 96-101.
20. L.A. Howard, A.D. Wong, A.K. Perry and B.P Klein, *J. Food Sci.* 1999, **64**, 929-936.
21. M. Haag, S. Ylikoski and J. Kumpulainen, *J. Food Comp. Anal.* 1995, **8**, 12-20.
22. W. Kmiecik and Z. Lisiewska, *Food Chem.* 1999, **67**, 61-66.
23. Z. Lisiewska and W. Kmiecik, *Food Chem.* 1997, **60**, 633-637.
24. E.U. 1080/94/000069, *Report from the research on the temperatures of frozen products.* 1995.
25. J.H. Wells and R.P. Singh, *J. Food Proc. Preser.* 1989, **12**,:271-292.
26. P.S. Taoukis, B. Fu and T.P. Labuza, *Food Technol.* 1991, **45 (10)**, 70-82.
27. B. Fu and T.P. Labuza, *Food Control*, 1993, 4, 125-133.
28. P.S. Taoukis and T.P. Labuza,. *In: Valentas KJ, Rotstein E, Singh RP. Editors. Handbook of food Engineering Practice, New York: CRC,* 1997, 361.
29. P.S. Taoukis and T.P. Labuza,. *J. Food Sci.* 1989, **54**, 783-788.
30. P.S. Taoukis and T.P. Labuza,. *J. Food Sci.* 1989, **54**, 789-792.
31. J.H. Wells, R.P. Singh and A.C. Noble, *J Food Sci.* 1987, **52**, 436-439, 444.
32. J.W. Farquhar, *In: proceedings of XVIth International Congress of Refrigeration, Tome III,* 1983, p811-819.
33. K.D. Dolan, R.P. Singh and J.H. Wells, *J. Food Processing,* 1985, **9**, 253-271.
34. R.P. Singh and J.H. Wells, *Food Technol.* 1985, **39**(12), 42-50.
35. S.H. Yoon, C.H. Lee, D.Y. Kim, J.W. Kim and K.H. Park, *J. Food Sci.* 1994, **59**, 490-493.
36. AOAC Official Methods of Analysis, 43.064-43.068, 1984, 844-845.
37. C.P. Mallet, *Frozen Food Technology, Glasgow, U.K.: Chapman and Hall,* 1993.
38. M.C. Giannakourou and P.S. Taoukis, In proceedings of ICEF9: International Congress on Engineering and Food, Montpellier, France, 2004.

STABILITY OF METHANOLIC EXTRACT ACTIVITY FOR LEAVES, PEELS AND CITRUS SEEDS UNDER UV IRRADIATION

Emad S. Shaker

Agricultural Chemistry Department, Faculty of Agriculture, Minia University, Minia, Egypt

1. ABSTRACT

Grapefruit (*Citrus paradise* Macf.) **(gr)**; Valencia orange (*C. sinensis* Osbeck) **(vi)**; Sour orange (*C. aurantium* L.) **(lr)**; Lemon (*C. limon* L.) **(ad)**; Balady mandarin (*C. reticulate* Blanco) **(y)**; Balady orange (*C. sinensis* Osbeck) **(or)** methanolic HCl extracts have been analyzed for antioxidative properties in their leaves **L**, peels **P** and seeds **S**. For the last two citrus, only leaves were extracted and evaluated.

The results showed that grapefruit leaves, lemon seeds and valencia peel extracts have the highest antioxidative property (AOA). Comparing phenol content, lemon leaves, grapefruit seeds and sour orange peel extracts have the highest phenolic amounts of all analyzed samples. In spite of that citrus seeds contain the lowest mean phenols amounts, they showed mean AOA more than 93% comparing to that for control.

The chemical composition was investigated using GC-MS for the citrus seed methanolic extracts **(SME)**. Three major peaks were detected and identified as hexadecanoic acid methyl ester (I), flavanone derivatives specifically a pentahydroxy dimethoxy flavone (II), and Kaempferide (III) a methoxy derivative. The grapefruit seed methanolic extract has the highest amount of compounds (I, III). Valencia seed extract has the highest amount of compound (II). Valencia seed extract has the lowest content of compounds (I, III) and the lemon seed extract has the lowest content for compound (II).

2. INTRODUCTION

Citrus fruits are important for their nutritional and antioxidant properties. Tons of peel and rag citrus are generated every year by industries which produce citrus fruit juices. These by-products are used either for the production of flavonoids or as substrate for flavor production. Orange peel oil is used extensively as an approved flavor enhancer in fruit drinks, carbonated beverages and as a scenting agent in soaps and cosmetics. Much research has been published that describes how to extract valuable compounds, which could be very useful in food science and medicine.

The medical importance of Orange peel[1] has been demonstrated in that it reduces diabetes. Tangerine peel extract and citrus flavonoids (naringin and hesperidin) showed cholesterol lowering effects,[2] comparable to that of pectin extracted from citrus peels.[3] Naringin (a bioflavonoid in citrus fruit peel) has an antioxidative and an antiatherogenic effect.[4] D-limonene from orange peel oil have chemopreventive activity against rodent mammary cancer and pancreatic tumors.[5,6] Hesperidin (the most important flavanone of Citrus sp.) has been found in green tangerine peel (*citrus reticulate*),[7] and orange peel and possesses antihypertensive and diuretic effects.[8] Hesperidin, neohesperidin, nobiletin and tangeritin were isolated from the green tangerine peel (*Citrus reticulata*).[9]

An antibacterial effect of citrus peel has also proved. Different dilutions of lemon peel have a bactericidal ability.[10,11] It was also found that natural and concentrated lemon juice and fresh and dehydrated lemon peel have similar activity. *Barringtonia asiatica* (leaves, fruits, seeds, stem and root barks) and their fractions exhibited a very good level of broad spectrum antibacterial and antifungal activity.[12]

Citrus fruit and juice were especially protective for decreasing the risk of an ischemic stroke in the Health Professionals Follow-up Study.[13] Sweet orange (*Citrus sinensis*) seed flour has been evaluated and found to contain 54.2% lipid, 28.5% carbohydrate, 5.5% crude fiber, 3.1% crude protein and 2.5% ash.[14] Flour samples were found to contain high levels of calcium and potassium. Accumulations of trace elements in the fruit peels have been also found.[15] Two major limonoid glucosides have been identified in the methanol extract of the peel of *Citrus grandis* L. Osbeck.[16]

Citrus fruit seeds are a rich source of flavonoids and their structures show antioxidant and antiproliferative activity.[17] Their effect for reducing inflammation, heart disease and cancer are described in traditional Eastern medicine. The concentration of polyphenols, phenolic acids and flavonoids has been shown to account for the antioxidative activities for apple and pear peel and pulp.[18] It was proved the effect of flavanone glycosides, such as hesperidin, narirutin and naringin, the most important phenols in the water-soluble fraction.[19]

Our research has concentrated on the value importance of citrus byproducts (leaves, peels and seeds), and their stability under UV irradiation conditions. The destructive power of the short-wavelength ultraviolet photons with many cellular molecules is known. The antioxidative activity under these conditions, as well as the total phenolic content have been measured in the methanolic extract of leaves, peels and seeds extracts. Identification of some of the valuable components using GC-MS has been carried out for the potent seed extracts.

3. MATERIALS AND METHODS

Grapefruit (*Citrus paradise* Macf.) **(gr)**; Valencia orange (*C. sinensis* Osbeck) **(vi)**; Sour orange (*C. aurantium* L.) **(lr)**; Lemon (*C. limon* L.) **(ad)**; Balady mandarin (*C. reticulate* Blanco) **(y)**; Balady orange (*C. sinensis* Osbeck) **(or)** (family Rutaceae) have been grown by the Food Technology Institute Giza (Giza, Egypt). The fruits were harvested at a commercial stage.

3.1 Sample Preparation

Fruits and samples were taken in May 2003, washed by distilled water and left at room temperature until dry. The dry leaves, seeds for the first four fruits; and dry peels for all the six fruits were all well crushed. Dry seeds, dry leaves or dry peels (1.0 g) were mixed and stirred with 10 ml methanolic HCl (1%). The solutions were kept in refrigerator for further experiments.

3.2 Antioxidative Activity Assay (AOA)

The determination of antioxidative activity was carried out by using the linoleic acid system. Each sample (0.5 ml) was added to a solution mixture of linoleic acid (0.33 ml), 99% distilled ethanol and 50 mM phosphate buffer, pH 7; the total volume was adjusted to 10 ml, with distilled water. These mixtures containing either the citrus antioxidants, the chemical antioxidants or the control were exposed to a UV lamp.

3.2.1 *UV irradiation.*
A Rayonet RPR-100 chamber reactor with UV lamps (254 nm) was used for irradiating the samples with linoleic acid. The results of the antioxidative activity for the mixtures were determined at hour intervals. The exposure distance was 11 cm between the sample beakers and the light source. The temperature of the equipment was maintained at 25 °C.

Aliquots of 0.1 ml were taken at different time intervals (until 17 hrs) during UV-exposure for measuring the degree of oxidation using ferric thiocyanate (FTC) method[20] by sequentially adding ethanol (47 ml 75%), ammonium thiocyanate (0.1 ml, 30%) and ferrous chloride (0.1 ml, 20 mM in 3.5% HCl). Absorbance at 500 nm was measured using a spectrophotometer. A control was run using linoleic acid without citrus extracts. 0.5 ml (500 ppm) of BHT, vit E or vit C were used as standards in the antioxidative determination.

Antioxidative activity was expressed[21] as % inhibition relative to the control using:

$$AA=[(\text{degradation rate of control} - \text{degradation rate of sample}) / \text{degradation rate of control}]\ 100$$

3.3 Determination of Total Phenolic Compounds (TPC)

The concentration of phenolic compounds was measured[22] and calculated using tannic acid as a standard. Different volumes of the dry samples in methanolic HCl (0.1; 0.05; 0.01 ml) were added to 2 ml 2% Na_2CO_3. After 2 min, 50% Folin-Ciocalteu reagent (0.1 ml) was added to the mixture, then it was left for 30 min. Absorbance at 750 nm was measured using a spectrophotometer. Results were expressed as milligrams percentage of tannic acid equivalent.

3.4 Identification of Seed Methanolic Extract (SME) Components using GC-MS

Characterization of the seed extracts components was carried out using a Shimadzu-Japan gas chromatograph 14A equipped with a flame ionization detector (FID) and a 50m x 0.25mm diameter polar (Carbo-wax^{TM}) bonded phase capillary column was used. The injector and detector temperatures were 250 °C and 250 °C, respectively. The oven temperature was programmed from 100 °C to 210 °C at 10 °C/min and held for 20 min. The linear velocity of the helium carrier gas flow was 35.5 cm/s with a split ratio of 1:42. The volume of the sample injected was 1 µl.

Gas chromatographic peaks were integrated by a Spectra Physics Integrator (SP 4290). Components were identified by their mass spectra utilizing MS in the commercial library (Wiley 138K, Mass Spectral Database, Wiley, 1990).

4. RESULTS AND DISCUSSIONS

4.1 The Antioxidative Activity for the Citrus Waste Components

Leaves and seeds of grapefruit (*Citrus paradise* Macf.) **(gr)** extracts showed higher antioxidative activity than that for peels, especially after 8 and 17 hrs. The activity of grapefruit leaves (grL), and grapefruit seeds (grS) exceeded at certain exposure times the activity of BHT as shown in Figure (1a). After 17 hrs exposure to UV, grapefruit leaves showed antioxidative activity greater than that for BHT. It was proved that UV absorption of natural phenolics induced the free radical scavenging capability of antioxidants.[23]

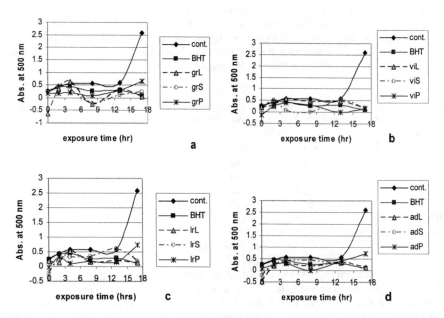

Figure 1 *Antioxidative activity for (a) grapefruit gr; (b) Valencia vi; (c) sour orange lr; (d) lemon ad leaves (L), seeds (S) and peels (P) methanolic extracts at different times under exposure to UV irradiation.*

It has been reported that grapefruit among other food extracts has antioxidant effects with cytochrome P450 containing microsomes.[24] On the other hand, orange peels contain beta-carotene with nutritional evaluation higher than that for the synthetic beta-carotene.[25]

Approximate similar antioxidative activity for leaves, seeds and peels of valencia orange (*C. sinensis* Osbeck) **(vi)** extracts is shown in Figure (1b). Peel extract exhibited satisfactory activity up to 13 hrs of exposure time. After 17 hrs exposure, the AOA for valencia leaves, seeds and peels were similar to BHT.

Strong activity has shown for seeds and peels of some orange extracts comparing to that for the synthetic antioxidant at 4 hrs exposure. On the other hand, the activity of sour orange (*C. aurantium* L.) **(lr)** peel extract was lower than that for leaves and seeds extracts after 17 hrs exposure as shown in Figure (1c). Sour orange peel has a high efficiency in relation with a high content of glycosylated flavanones and phenolic acids.[26] The effect of Taiwanese orange peels has proved on the oxidation stability of soybean oil under light storage.[27] Yen and Chen found that various pigments in orange peels have the antioxidant effectiveness dependent upon their concentration.[27]

The antioxidative effect of extracts of Lemon (*C. limon* L.) **(ad)** leaves and seeds were positive especially after 13 and 17 hrs exposure, and much higher than that for peel extract as shown in Figure (1d). After 17 hrs, lemon seeds activity was higher than that for BHT. The chemical composition from peel and leaf oils of lemons and limes was investigated by Lota, et al.[28] Three major chemotypes were formed for lemon peel oils: limonene; limonene/β-pinene/γ-terpinene; and limonene/linalyl acetate/linalool. Two chemotypes were identified for lemon leaf oils: limonene/β-pinene/geranial/neral and linalool/linalyl acetate/α-terpineol.

The AOA percentages have been calculated comparing to the control as shown in Table 1 for the 0.5 ml citrus fruit extracts and the total phenolic compounds (TPC) amount (mg in 100 mg) as tannic acid. AOA for the tested standard antioxidants for BHT, vit E and vit C were 95.8, 23.8 and 34.4 respectively. The highest AOA value in the study was for grapefruit leaves (97.9%) and then for lemon seeds (97.6%).

Table 1 *The antioxidative activity (AOA) percentages compared to the control; and the total phenolic compounds (TPC) as tannic acid (mg / 100 mg) for citrus extracts.*

Leaves		Seeds		Peels		
AOA	TPC	AOA	TPC	AOA	TPC	
97.9	0.92	91.4	0.48	74.8	0.57	gr
93.8	1.44	92.9	0.44	92.7	0.73	vi
95.0	1.10	92.6	0.47	71.1	1.00	la
94.7	1.57	97.6	0.39	71.6	0.53	ad
97.2	1.23	Nt*	Nt*	Nt*	Nt*	Y
94.0	0.90	Nt*	Nt*	Nt*	Nt*	or

Nt* : not tested.

The antioxidant activity (AOA) for different phenolic extracts in citrus extracts wants further investigation. In general, the mean activity (AOA) for leaves was higher (more than 95%) than that for seeds (more than 93%). AOA for phenolic extract in peels was much lower (less than 78%) than that for leaves and seeds, compared to the control. The antioxidant activity order for citrus peels is not the same as of that for their seeds as reported.[26] The AOA results did not correlate with the result of total phenolic compounds in the extracts. The order of mean AOA was; leaves > seeds > peels, while mean phenolic content order was; leaves > peels > seeds. The mean phenolic content for leaves was the highest (6%), while the phenolics in seeds were the lowest (2.2%). The phenolics in peels showed a moderate content (3.5%). That may indicate the phenolics in seeds are of high quality. In agreement, it was mentioned[26] that seeds have more antioxidtive activity than peel but their flavanone content is lower.

The highest antioxidant activity for leaves was in grapefruit, followed by Balady mandarin, sour orange, lemon, Balady orange and Valencia. The highest AOA for seeds was in lemon, followed by Valencia, sour orange and grapefruit. The highest AOA for peels was in Valencia, followed by grapefruit, lemon and sour orange. Under UV exposure, rapid loss of vitamins E and C was deserved (Figure 2a).

The methanolic extracts of Balady mandarin and Balady orange leaves were 97.2, 94.0 % respectively (Figure 2b). It is reported the methanolic extracts of mandarin and sweet orange seeds have superior antioxidant properties on the oxidation of citronella oil under strong oxidizing conditions.[26] From *Citrus reticulata* peel, some potent polymethoxylated flavone bio-preservatives can be obtained for food applications.[29] The potent activity for the orange peel oil butanol fraction was showed to be greater than that for vitamin C.[30]

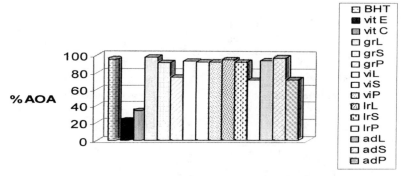

Figure 2a *Antioxidative activity for methanolic extracts of citrus leaves, seeds and peels compared to standard antioxidants (BHT, vit E and vit C) after 17 hrs exposure to UV irradiation.*

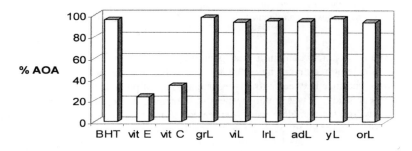

Figure 2b *Antioxidative activity for methanolic extracts citrus leaves after 17 hrs exposure to UV irradiation.*

4.2 The Phenolic Contents in the Citrus Extract Components

The total phenolic content has been studied as methanolic extracts in three concentrations from citrus materials. Increased sample amount of the methanolic citrus extracts (from 0.01 to 0.1 ml) lead to increase of total phenolic content. The phenolic content in leaves were the highest, followed by that for peels and finally the seeds as shown in Figure 3 a,b,c,d. The phenolic content is considered one of the major factors for the antioxidative activity. Vitamin C has also been shown to account for 65-100% of the antioxidant potential of beverages derived from citrus fruit.[31]

Grapefruit extracts contain total phenolic compounds between 0.48-0.92 mg/ 100 mg extract, while Valencia extracts have 0.44-1.44 mg/100 mg extract. Cranberries and red grapes have the largest amount of both free and total phenols among the fruits, while citrus fruits have a very low concentration.[32] Vinson, et al. stated that fruits contain better quantity and quality of antioxidants (that can enrich lower density lipoproteins LDL) compared to vegetables.[32]

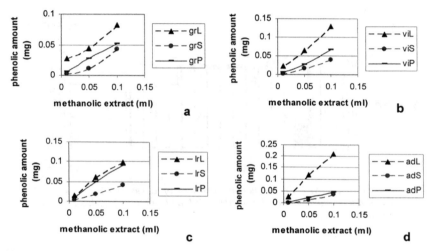

Figure 3   *Total phenolic compounds percentages in 0.01, 0.05, and 0.1 mls methanolic (a) grapefruit gr; Valencia vi(b); sour orange lr(c); lemon ad(d) extracts.*

Sour orange extracts contain a total phenolic content between 0.47-1.10 mg/100 mg extract, while lemon extracts have 0.39-1.57 mg/100 mg extract. The lemon extracts showed wide variance between extracts in total amount of phenols (Figure 3d).

In a recent study, it was found hesperidin, naringin, polymethoxylated flavones and numerous hydroxycinnamates were generated from fruit processing.[33] The highest concentrations occurred in peel molasses of orange (*C. sinensis*) and tangerine (*C. reticulate* Blanco.) compared to grapefruit (*C. paradise* Macf.) and lemon (*C. limon*). Limonene (92.6%), and polymethoxylated flavones[34] are the main components have been identified in the volatile and non-volatile fractions of maltese sweet orange oil.

The highest total phenolics for leaves were in lemon (adL) as shown in Figure 4, followed by Valencia, Balady mandarin, sour orange, grapefruit and Balady orange. The highest phenolic values for seeds were in grapefruit, followed by sour orange, Valencia and lemon. The phenolic values for peels were highest in sour orange, followed by Valencia, grapefruit and lemon.

Quality

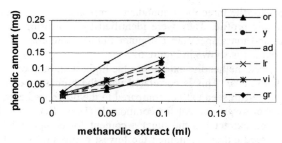

Figure 4 *Total mgs phenolic compounds in 0.01, 0.05, and 0.1 ml methanolic citrus leaf extracts.*

Generally, average phenol quantities in citrus seeds were the lowest, while their mean AOA were 93% compared to that for control. The results were in agreement with that seeds that possess greater antioxidant activity than peels.[26] The lack of correlation observed between the antioxidant activity and the quantity of phenolic compounds in the seeds extracts, shows that other substances must be responsible for the activity of the extracts. It is known that different phenolic compounds have different responses in the Folin-Ciocalteu method. The molecular antioxidant response of phenolic compounds varies widely, depending on their chemical structure.[35]

4.3 Identification of Seed Methanolic Extract (SME) Components by GC-MS

Information concerning citrus seed active compounds is limited in the literature. The relation between the AOA and phenolic content is the aim of this research. A high antioxidative activity for seed extracts from one side, and the poor phenolic content from the other side has been found from Table 1. Retention time (Rt) and percentages (%) for the most abundant compounds are presented in Table 2.

Table 2 *The retention times (R_t), the MS fragments for the most abundant compounds in the seed methanolic extracts (SME) and their percentages (%).*

R_t	Lemon %	Grapefruit %	Valencia %	Sour orange %	MS (fragments)
9.71-9.81	7.8	0.2	3.1	3.6	<u>63</u>,78,133
12.41-12.46	5.4	0.2	5.0	4.0	55,<u>74</u>,143,242
14.71-14.76 (I)	17.7	18.9	14.1	16.9	55,<u>74</u>,87,143,185,227, 270
15.06-15.11	4.6	1.6	8.8	5.3	<u>55</u>,74,96,123
16.71-16.76	3.8	2.3	5.9	4.2	<u>55</u>,83,111,196
17.86-17.91 (II)	24.2	26.4	27.1	27.0	<u>55</u>,83,149,177,222,264, 355
18.66-18.76 (III)	10.4	32.8	4.6	13.6	55,<u>67</u>,81,109,135,194, 263,294
19.96-20.11	0.7	0.6	0.8	2.5	<u>79</u>,95,121
20.96-21.01	2.1	2.6	3.5	3.2	55,<u>74</u>,87,143
21.36-21.41	1.8	1.9	6.7	2.5	69,<u>97</u>,126

The numbers above lines in MS have Intensities 100% in the spectrum.

The peak areas for the three most abundant compounds were selected on the basis of their chromatographic profile. They were identified by measuring their mass fragments, the literature data and mass spectrum computer library matching. The chromatogram of sour orange seed extract is shown in Figure 5.

The grapefruit seed methanolic extract has the highest amount of compound I (18.9%) and compound III (32.8%). This is correlated with the highest phenolic content and the lowest AOA for grapefruit seed extract compared to the other seed extracts. Valencia seed extract possessed the lowest content for compounds (I and III), 14.1, 4.6% respectively. While, Valencia seed extract has the highest content for compound II (27.1%). On the other hand, lemon seed extract contained the lowest compound II (24.2%). Lemon seed contains principally eriocitrin and hesperidin, whereas the peel is rich in neoeriocitrin, naringin and neohesperidin. Sour orange is a source of naringin and neohesperidin.[26]

From mass spectral fragments and computer matching, compound (I) proved to be hexadecanoic acid methyl ester. The mass spectrum for compound I in grapefruit seed extract is shown in Figure 6. High content of carbonyl compounds and alkyl acetates[34] and polymethoxylated flavones have been identified in *C. sinensis* (L.) osbeck cv. Maltese.

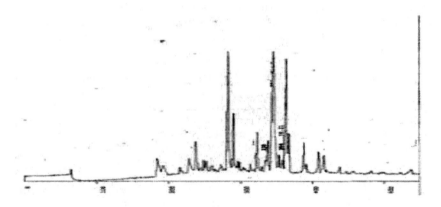

Figure 5 *GC Chromatogram for Sour Orange Seed Extract*

The expected fragments for compound I found to be as follows;
.CH.CH$_2$.CO. (**55**), CH$_3$.CO.OCH$_3$ (**74**), .CH$_2$.CH$_2$.COOCH$_3$ (**87**), .(CH$_2$)$_5$.CH$_2$.COOCH$_3$ (**143**), .(CH$_2$)$_8$.CH$_2$.COOCH$_3$ (**185**),
.(CH$_2$)$_{11}$.CH$_2$.COOCH$_3$ (**227**), .CH$_3$.(CH$_2$)$_{13}$.CH$_2$.COOCH$_3$ (**270**).

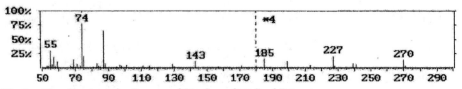

Fig 6 *Mass Spectrum for Compound I in Grapefruit Seed Extract*

Compounds (II and III) proved to be flavonoid; penta hydroxyl- dimethoxy flavone and methoxy-kaempferide, respectively. The mass spectral fragmentation pattern for compound II in sour orange seed extract and for compound III in lemon seed extract are shown in Figures 7 and 8 respectively. The fragment (355) of compound II contains protonated penta hydroxyl groups.

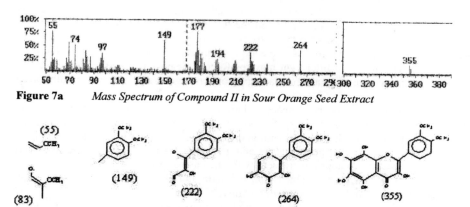

Figure 7a   *Mass Spectrum of Compound II in Sour Orange Seed Extract*

Figure 7b   *Mass Fragments for Compound II*

Flavonoids having fewer than four hydroxyl groups show very little antioxidant activity.[36] Myricetin with three adjacent hydroxyl groups, is one of the most active antioxidants among flavonoids. On the other hand, flavonols, with a free hydroxyl group at the C-3 position of the flavonoid skeleton, have high inhibitory activity to B-carotene oxidation.[37] Antiradical activity depends on the free hydroxyl group at the C-4′ position, free hydroxyl at C-3 and/or C-3′ and a C2-C3 double bond.

Figure 8a   *Mass Spectrum of Compound III in Lemon Seed Extract*

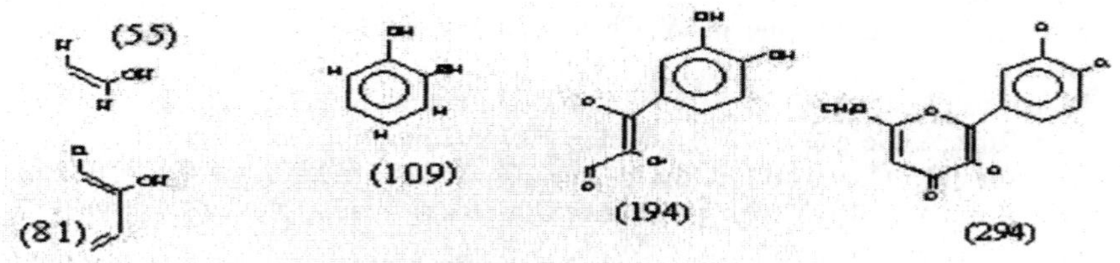

Figure 8b   *Mass Fragments for Compound III*

Kaempferide has high antioxidative activity[37] (60%), and the methoxyl at certain positions can increase the antiradical activity. Morin and kaempferol have the highest activity[38] in inhibiting the autooxidation of emulsified linoleic acid and methyl linoleate in the dark at room temperature. The activity was attributed to their ability to donate a hydrogen atom to the peroxy radical derived from the autooxidizing fatty acid derivative.

In explanation to our result, kaempferol proved to undergo photobleaching in illuminated chloroplasts.[39] Takahama suggested inhibition O^{-2} promoted redox reactions within the chloroplast. Quercetin inhibited the $^1O_2$-induced[40] bleaching of the carotenoid pigment, crocin. Flavonoids, with their strong absorption in the 300-400 nm UV region, act as internal light filters[41] for the protection of chloroplasts and other organelles from UV damage. The ability to filter light may reinforce their powerful antioxidant effects to provide high protection level against damaging oxidants generated either thermally or by light.

A GC-MS study of the head space volatiles of *Citrus taitensis* oil showed (60%) linalool, N-containing compounds; benzyl cyanide, indole and methyl anthranilate (16.9%) and flavonoids, dihydrorobinetin and genistein.[42] Limonin glucoside and phlorin were extracted from orange peel residues and grapefruit seeds.[43] Linalool concentrations in juice of Valencia oranges increased with increasing peel oil levels.[44] Limonoids (triterpene derivatives), such as sinensetin, nobiletin and phenylpropanoids showed high antioxidant potential and health promoting ability.[45]

Fruit juice factories produce a huge amount of product each year from seeds and peels. The aim of this study to evaluate the importance of citrus extracts as food preservatives and protective effects against harmful irradiation. The high content and quality of flavanones, their synergistic effects, tocopherols, ascorbic acid and other non identified substances are responsible for the high antioxidative effect.

References

1   D. Mahabir and M. Gulliford, *Rev. Panam. Salud. Publica.*, 1997, **1**, 174.
2   S. Bok, S. Lee, Y. Park, K. Bae, K. Son, T. Jeong and M. Choi, *J. Nutr.*, 1999, **129**, 1182.
3   A. Terpstra, J. Lapre, H. deVries and A. Beynen, *Eur. J. Nutr.*, 2002, **41**, 19.
4   S. Choe, H. Kim, T. Jeong, S. Bok and Y. Park, *J. Cardivasc. Pharmacol.*, 2001, **38**, 947.
5   P. Crowell, *Breast Cancer Res. Treat.*, 1997, **46**, 191.
6   P. Crowell, *J. Nutr.*, 1999, **129**, 775.
7   T. Wang, X. Guo and J. Zhang, *Zhongguo*, Z., 1997, **22**, 156.
8   E. Galati, A. Trovato, S. Kirjavainen, A. Forestieri, A. Rossitto and M. Monforte, *Farmaco.*, 1996, **51**, 219.
9   X. Guo, T. Wang, M. Guo and Y. Chen, *Zhongguo. Zhong. Yao. Za.Zhi.*, 2000, **25**, 146.
10   M. deCastillo, C. deAllori, R. deGutierrez, O. deSaab, N. deFernandez, C. deRuiz, C. deRuiz, A. Holgado and O. deNader, *Biol. Pharm. Bull.*, 1997, **20**, 1033.
11   M. deCastillo, C. deAllori, R. deGutierrez, O. deSaab, N. deFernandez, C. deRuiz, C. deRuiz, A. Holgado and O. deNader, *Biol. Pharm. Bull.*, 2000, **23**, 1235.
12   M. Khan and A. Omoloso, *Fitoterapia.*, 2002, **73**, 255.
13   K. Joshipura, A. Ascherio, J. Manson, M. Stampfer, E. Rimm, F. Speizer, C. Hennekens, D. Spiegelman and W. Wllett, *J. Am. Md. Assoc.*, 1999, **282**, 1233.

14   M. Akpata and P. Akubor, *Plant. Food Hum. Nutr.*, 1999, **54**, 353.
15   M. Selema and M. Farago, *Phytochem.*, 1996, **42**, 1523.
16   Q. Tian, J. Dai and X. Ding, *Se-Pu.*, 2000, **18**, 291.
17   E. Middleton, C. Kandaswami and T. Theoharides, *Pharmacol. Rev.*, 2000, **52**, 673.
18   M. Leontowicz, S. Gorinstein, H. Leontowicz, R. Krzeminski, A. Lojek, E. Katrich, M. Ciz, B. Martin, F. Soliva, R. Haruenkit and S. Trakhtenberg, *J. Agric. Food Chem.*, 2003, **51**, 5780.
19   A. Gil-Izquierdo, M. Gil, F. Ferreres and F. Tomas-Barberan, *J. Agric. Food Chem.*, 2001, **49**, 1035.
20   H. Mitsuda, K. Yasumoto and K. Iwami, *Eiyoto Shakuryo*, 1966, **19**, 210.
21   M. Ogata, M. Hoshi, K. Shimotohno, S. Urano and T. Endo, *JAOCS*, 1997, **74**, 557.
22   M. Taga, E. Miller and Pratt, *JAOCS*, 1984, **61**, 928.
23   X. Chen and D. Ahn, *JAOCS*, 1998, **75**, 1717.
24   G. Plumb, S. Chambers, N. Lambert, S. Wanigatunga and G. Williamson, *Food Chem.*, 1997, **60**, 161.
25   A. Ghazi, *Nahrung.*, 1999, **43**, 274.
26   A. Bocco, M. Cuvelier, H. Richard and C. Berset, *J. Agric. Food Chem.*, 1998, **46**, 2123.
27   W. Yen and B. Chen, *Food Chem.*, 1995, **53**, 417.
28   M. Lota, S. deRocca, F. Tomi, C. Jacquemond and J. Casanova, *J. Agric. Food Chem.*, 2002, **50**, 796.
29   G. Jayaprakasha, P. Negi, S. Sikder, L. Rao and K. Sakariah, *Z.Naturforsch.[C]*, 2000, **55**, 1030.
30   E. Shaker and M. Ghazy, First Conference of Role of Biochemistry in Environment and Agriculture, Feb. 6-8, 2001, Biochem. Dept. Faculty of Agric. Cairo Univ. Giza, Egypt.
31   P. Gardner, T. White, D. McPhail and G. Duthie, *Food Chem.*, 2000, **68**, 471.
32   J. Vinson, X. Su, L. Zubik and P. Bose, *J. Agric. Food Chem.*, 2001, **49**, 5315.
33   J. Manthey and K. Grohmann, *J. Agric. Food Chem.*, 2001, **49**, 3268.
34   A. Trozzi, A. Verzera and G. Lamonica, *J. Essent. Oil Res.*, 1999, **11**, 482.
35   M. Satue-Gracia, M. Heinonen and E. Frankel, *J. Agric. Food Chem.*, 1997, **45**, 3362.
36   D. Pratt, "Phenolic, Sulfur and Nitrogen Compounds in Food Flavors", G. Charalambous and I. Katz eds, ACS Sympos. Ser., Washington, DC., 1976, 26., p.1.
37   S. Burda and W. Oleszek, *J. Agric. Food Chem.*, 2001, **49**, 2774.
38   J. Torel, J. Cillard and P. Cillard, *Phtochem.*, 1986, **25**, 383.
39   U. Takahama, *Plant Physiol.*, 1983, **71**, 59.
40   U. Takahama, R. Youngman and E. Elstner, *Photobiochem. Photobiophys.*, 1984, **7**, 175.
41   M. Caldwell, R. Robberecht and S. Flint, *Physiol. Plant*, 1983, **58**, 445.
42   M. Saleh, F. Hashem and K. Glombitza, *Food Chem.*, 1998, **63**, 397.
43   R. Braddock and C. Bryan, *J. Agric. Food Chem.*, 2001, **49**, 5982.
44   R. Bazemore, R. Rouseff and M. Naim, *J. Agric. Food Chem.*, 2003, **51**, 196.
45   C. Kaur and H. Kapoor, *International J. Food Sci. Technol.*, 2001, **36**, 703.

EFFECT OF DIFFERENT INITIAL AND SUPPLEMENTARY BRINING TREATMENTS ON THE FERMENTATION OF GREEN OLIVES CV. CONSERVOLEA

E.Z. Panagou, C.C. Tassou and K.Z. Katsaboxakis

National Agricultural Research Foundation, Institute of Technology, Quality and Safety of Foods, 1. Sof. Venizelou str., Lycovrissi, Greece, GR-141 23.

1 ABSTRACT

The effect of different brining conditions on the fermentation of green olives cv. Conservolea was studied. Different treatments included: (a) brine acidification with 2% (v/v) lactic and 1% (w/v) citric acids (control), (b) addition of 25% (v/v) 1N HCl, (c) substitution of the initial brine by 20% (v/v) with a brine from a previous fermentation. Brine re-use decreased the survival period of enterobacteria to 24 days followed by the HCl treatment and the control. However, after 35 days, pH reached a plateau above 4.8 in all treatments. Addition of 1.5% (w/v) glucose in the HCl treated and brine re-use processes as well as 5% (v/v) lactic acid in the control reduced the pH to 4.3-4.5. Glucose increased the concentration of lactic acid in brine re-use and HCl treated processes (73.4 mM and 67.8 mM) compared with the control that was lacking in acidity (44.7 mM).

2 INTRODUCTION

Table olive fermentation is one of the oldest uses of biotechnology in food processing that historically appeared in the Mediterranean region and subsequently spread over to other countries. Every year approximately 95,000-100,000 tonnes of table olives are processed in Greece either as Spanish-style green or naturally black olives in brine. The traditional Spanish-style processing of Conservolea green olives involves a lye treatment step (1.8-2.0, w/v, NaOH) to hydrolyze the bitter agent oleuropein, followed by a washing step to remove the excess of alkali. The fruits are then submerged in brine (6-8%, w/v, NaCl) where they undergo a spontaneous fermentation mainly by lactic acid bacteria and yeasts.[1] The microbial succession in green olive fermentation is characterized by three distinct phases.[2] In the first phase, the microbial groups that dominate the process are Gram-negative bacteria, mostly belonging to the *Enterobacteriaceae* family. The second phase is characterized by the progressive growth of lactic acid bacteria and yeasts and the gradual decrease of Gram-negative bacteria. During the third phase, an abundant growth of lactobacilli is observed, mainly from species of *Lactobacillus plantarum*, that coexist with a yeast flora and become the dominant microflora of the fermentation. The main driving forces of the process are the availability of fermentable substrates, salt content, free and combined acidity, degree of aerobiosis or anaerobiosis, application of starter cultures, pH and temperature control.[3-9] Process control of these parameters can improve fermentation and produce consistent and high quality final products. A successful green olive fermentation is characterized by a rapid production of high acidity, low pH, short survival period of spoilage bacteria and high population density of the desirable lactic acid bacteria. The initial brining conditions can affect the aforementioned parameters and determine the quality attributes of the final product.

In bulk fermentation processes, the duration of the first and second phases must be reduced to a minimum, thus favouring rapid dominance of lactic acid bacteria over potential spoilage microorganisms. Acidification of the initial brine with lactic, acetic or hydrochloric acid, inoculation with a pure culture of lactobacilli or use of a previous well-fermented brine are some ways to minimize the first two phases of the process, thus favouring rapid initiation of a vigorous lactic acid fermentation. It must be emphasized however, that due to the high buffering capacity of the brines, a complete lactic acid fermentation, and hence the development of high acidity/low pH, may be difficult to be attained. To overcome this problem, supplementary treatments may be necessary at the end of fermentation such as the addition of fermentable material and/or lactic acid in the brines. These treatments will improve quality attributes (pH/acidity) and enhance storage stability of the final product.

The aim of the present research was to study (a) the physico-chemical and microbiological changes during spontaneous fermentation of cv. Conservolea green olives, in brines supplemented at the onset of fermentation with (i) lactic acid, (ii) HCl and (iii) brine from a previous fermentation, and (b) to evaluate the effectiveness of two of the most common supplements (lactic acid vs. glucose) at the end of fermentation to enhance the final characteristics (acidity/pH) of the process.

3 METHODS AND RESULTS

Green olives cv. Conservolea were harvested in mid-September and subjected to traditional Spanish-style treatment. A total of 30 kg of olives were immersed in a 2% (w/v) NaOH solution for 7.5 h at room temperature (20-25°C) until the alkali reached 2/3 of the flesh as measured from the epidermis to the pit. A washing step was followed by replacing the NaOH solution with tap water and changing it at intervals of 4 and 10 h, respectively.

Fermentation took place in 8 l total capacity screw-capped PVC fermenters containing 5 kg of olives and 3 l of freshly prepared 6% (w/v) NaCl. Three initial brining treatments were investigated: (a) acidification with 2% (v/v) lactic acid of the brine volume (control treatment), (b) acidification with 25% (v/v) 1N HCl. (HCl treatment), (c) partial substitution of 20% of the initial brine's volume (ca. 600 ml) with brine from a previous fermentation (pH 4.2; acidity 0.6%) (brine re-use treatment). Each set of conditions was repeated twice. The addition of lactic acid and HCl was made 24 h after brining. At the end of fermentation (approximately one month after brining), the following supplementary treatments were investigated: control process was supplied with 5% (v/v) lactic acid while in the other two treatments glucose was added at a quantity equal to 1.5% (w/v) of the brine volume contained in each fermenter.

Brine samples were analyzed at regular time intervals throughout fermentation. The samples (1 ml) were aseptically transferred to 9 ml sterile ¼ Ringer's solution. Decimal dilutions in the same Ringer's solution were prepared and duplicate 1 or 0.1 ml samples of the appropriate dilutions were mixed or spread on the following agar media: de Man-Rogosa-Sharp medium (MRS; Merck 1.10660, Darmstadt, Germany) for lactic acid bacteria, overlaid with the same medium and incubated at 25°C for 72 h; Rose Bengal Chloramphenicol agar (RBC; Oxoid CM 549, Basingstoke, Hampshire, England, supplemented with selective supplement SR 78) for yeasts and moulds, incubated at 25°C for 48 h; Violet Red Bile Glucose agar (VRBGA; Oxoid CM 458, Basingstoke, Hampshire, England) for enterobacteria counts, incubated at 37°C for 24 h.

Growth data from plate counts were enumerated as $\log_{10}$ values. The Baranyi model was fitted to the logarithm of the cell concentration in order to estimate the maximum specific growth/death rates of the various groups of microorganisms.[10,11] For curve fitting, the in-house program DMFit (Institute of Food Research, Norwich, UK) was used. The death rate of enterobacteria was estimated by linear regression of the straight segment of the curve.

Brine samples were routinely analyzed for pH, using an Orion model 940 pH meter (Orion Research Inc., Boston, MA, USA). Titratable acidity was determined according to IOOC.[12] Organic acids (lactic, acetic, succinic, citric) were analyzed by HPLC. Details of the method are given elsewhere.[13]

The population dynamics of the different microbial groups in the brines during fermentation are presented in Figure 1. Enterobacteria counts showed a rapid increase within the first 10 days in all treatments, followed by a sharp decline thereafter. No viable counts were enumerated after 24 and 28 days of fermentation in the brine re-use and HCl treated processes, respectively (Figures 1b and 1c), compared with the control where the same group survived for 35 days (Figure 1a). Enterobacteria showed higher death rates in HCl treated brines (-0.532 days^{-1}) than in the control and brine re-use process where the corresponding values were –0.206 and –0.259 days^{-1}, respectively (Table 1).

Lactic acid bacteria and yeasts increased steadily and became the dominant members of the microflora during fermentation in all treatments. In the HCl treated and brine re-use processes, lactic acid bacteria reached their maxima within the first 10 days (7.8-8.1 $\log_{10}$cfu/ml) compared with control brines in which a maximum was attained 4 days later (7.8 $\log_{10}$cfu/ml). It must be stressed that although the final population density was the same, however, different treatments affected the growth rate of lactic acid bacteria (Table 1). Thus, the use of brine from a previous fermentation served as a starter culture inoculum, resulting in an accelerated fermentation process as concluded by the higher specific growth rate (0.605 days^{-1}), followed by HCl process (0.583 days^{-1}) and the control (0.443 days^{-1}).

Yeasts co-existed with lactic acid bacteria in all processes but their counts were lower by approximately 3-4.5 log cycles. They grew in similar populations with lactic acid bacteria during the first week and then remained almost stable until the end of fermentation ranging from 2.8-4.2 $\log_{10}$ cfu/ml.

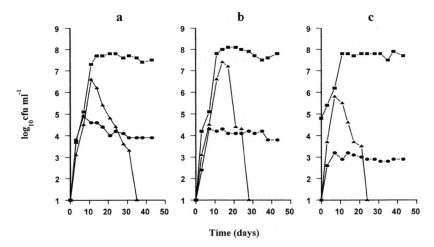

Figure 1 *Changes in lactic acid bacteria (■), yeasts (●) and enterobacteria (▲) during fermentation of cv. Conservolea green olives: (a) brine acidified with 2‰ (v/v) lactic acid; (b) brine acidified with 25‰ (v/v) 1N HCl; (c) original brine was substituted by 20% (v/v) with brine from a previous fermentation*

Curves for titratable acidity and pH are presented in Figure 2. The acidity in the brine re-use and HCl process was, on the average, some 0.1% higher than in the control throughout the first 30 days. Acidity increased slowly in the control process, particularly during the first 20 days. A better response was obtained in both HCl treated and brine re-use treatments although no differences were observed between them throughout fermentation from a practical point of view. In the case of pH, results were similar to that of acidity. Specifically, pH decrease was noticeable in all treatments within the first 28 days reaching a plateau thereafter. Overall, the process seemed to progress faster in the case where the brine was supplemented with HCl and when brine from a previous fermentation was used. It is noteworthy that pH did not drop below 4.8 in any treatment after 35 days of fermentation and in the case of the control it was even above 5. The addition of 1.5 % (w/v) glucose in the HCl treated and brine re-use processes as well as 5‰ (v/v) lactic acid in the control on day 35 resulted in a further decrease of pH near 4.3-4.5 in all processes. The pH decrease was followed by a corresponding increase in titratable acidity, keeping the difference below 0.1% among different processes.

Lactic acid was the major metabolic product with a significant presence in the brines. An increase in lactic acid concentration was observed in the HCl treated and brine re-use processes ranging from 30.9 to 56.3 mM after 28 days of fermentation (Table 2), compared with the control which was continuously lacking in lactic acid, not exceeding 21.6 mM in the same time. The supplementary addition of 1.5% (w/v) glucose in the two former processes as well as 5‰ (v/v) lactic acid in the control on day 35 resulted in an increase in lactic acid concentration in all processes. By the end of fermentation the highest concentration was observed in the brine re-use process (73.4 mM) followed by the HCl treated (67.8 mM) and control fermentation (44.6 mM). An accelerated production of acetic acid was also noticeable in all treatments ranging from 18.8 to 22.6 mM. Succinic acid was also measured but in no case its concentration exceeded 13 mM.

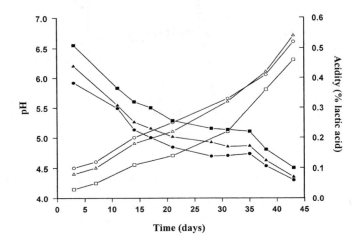

Figure 2 *Changes in acidity (open symbols) and pH (solid symbols) during fermentation of cv. Conservolea green olives: (■) brine acidified with 2‰ (v/v) lactic acid; (▲) brine acidified with 25‰ (v/v) 1N HCl; (●) original brine was substituted by 20% (v/v) with brine from a previous fermentation*

Citric acid was also present at the beginning of the process, but was metabolized in all cases. No citric acid was detected after 28 days of fermentation regardless of the treatment (Table 2).

The use of HCl in the brines has been suggested by other researchers.[14,15] The concentration of this inorganic acid should not exceed certain levels, otherwise a strong inhibition of the normal lactic acid fermentation may occur due to very low initial pH. Moderate levels of HCl have enhanced the process and can actually help to inhibit Gram negative and other spoilage microorganisms originally present in the brine environment.[16] In the present work the amount of added HCl had been predetermined to neutralise the residual NaOH present in olive flesh after lye treatment and subsequent washings. The addition of this inorganic acid reduced drastically the survival period of enterobacteria by 7 days compared to the control (Figure 1b) and also favoured spontaneous lactic acid fermentation as judged by the higher estimated growth rate of lactic acid bacteria (Table 1). In addition, HCl treated brines presented higher acidity by approximately 0.1% during the first 30 days in comparison with the control process (Figure 2), which is precisely the period where the risk of spoilage from non-lactic acid bacteria is highest.

Brine re-use process supplies the brine with a population of lactobacilli that will enhance the dominance of lactic acid bacteria over the indigenous flora and achieve an accelerated lactic acid fermentation.[17] In this way, the probability of spoilage is minimized and a more predictable fermentation process is feasible. Experimental data obtained in this study suggest that this treatment had a positive effect in minimizing the survival period of enterobacteria that were not countable after 24 days (Figure 1c). The positive contribution of brine re-use is also depicted in the estimated growth rate of lactic acid bacteria that presented the highest value than in any other fermentation vessel.

Table 1 *Effect of different brining conditions on the growth/death rates of selected microbial groups during fermentation of cv. Conservolea green olives*

Fermentation	Organism	Maximum specific growth/death rate (day^{-1})	SE of fit	R^2
Control[a]	Lactic acid bacteria	0.443	0.027	0.98
	Yeasts	0.534	0.123	0.86
	Enterobacteria	-0.206	0.047	0.93
HCL process[b]	Lactic acid bacteria	0.583	0.043	0.95
	Yeasts	0.519	0.018	0.96
	Enterobacteria	-0.532	0.074	0.91
Brine re-use[c]	Lactic acid bacteria	0.605	0.018	0.96
	Yeasts	0.543	0.013	0.94
	Enterobacteria	-0.259	0.167	0.85

[a] Brine acidified with 2‰ (v/v) lactic acid
[b] Brine acidified with 25‰ (v/v) 1N HCl
[c] Original brine was substituted by 20% (v/v) with brine from a previous fermentation

It must be emphasized that a pH value of 4.5 or less was not reached in any of the three processes after 35 days of fermentation. Consequently, the evolution of pH and acidity cannot reach values that ensure microbiological stability of the final fermented product as well as the expected organoleptic characteristics by the consumer. To overcome this situation, supplementary treatments are needed such as addition of fermentable material (e.g. glucose) or organic acids (e.g. lactic acid). In our work, the addition of 1.5% (w/v) glucose in HCl treated and brine re-use processes on day 35, which was assumed to be the end of fermentation, resulted in further drop in pH near 4.3 and also caused a slight increase in the population of lactic acid bacteria by 0.2-0.3 log cycles (Figures 1b, 1c). A positive effect was also observed in the control process where the supplementary addition of 5% (v/v) lactic acid reduced the pH to about 4.5. Although the final pH values were not different among the processes from a practical point of view, there was a significant difference in lactic acid concentration at the end of fermentation (Table 2). The addition of glucose increased the concentration of lactic acid to 73.4 and 67.8 mM, respectively, in brine re-use and HCl treated processes compared with 44.6 mM in the control, indicating that the use of glucose as a supplementary treatment is better than lactic acid.

Table 2 *Changes in organic acids concentration in brine during fermentation of Conservolea green olives*

Fermentation	Time (days)	Organic acids (mM)			
		Lactic	Acetic	Succinic	Citric
Control[a]	14	14.2	nd[d]	6.4	9.5
	28	21.6	13.4	8.1	Nd
	43	44.6	18.8	8.9	Nd
HCL process[b]	14	16.3	nd	nd	3.1
	28	30.9	12.3	5.4	Nd
	43	67.8	16.1	5.3	Nd
Brine re-use[c]	14	18.1	nd	nd	3.5
	28	56.3	13.1	12.1	Nd
	43	73.4	18.6	12.2	Nd

[a] Brine acidified with 2‰ (v/v) lactic acid
[b] Brine acidified with 25‰ (v/v) 1N HCl
[c] Original brine was substituted by 20% (v/v) with brine from a previous fermentation
[d] Not detected

4 CONCLUSIONS

The initial brining conditions can affect the titratable acidity/pH evolution as well as the survival period of spoilage enterobacteria. The best brining treatments were substitution of the original brine with a proportion of brine from a previous fermentation (brine re-use) followed by the addition of HCl in the brine to neutralize the excess alkali from the washing step. The process that is currently used by local industry, i.e. addition of lactic acid, gave the lowest titratable acidity and the longest survival period of enterobacteria. However, none of the treatments could produce the desired amounts of lactic acid that could ensure preservation of the final product. Addition of glucose at the end of fermentation in HCl treated and brine re-use processes resulted in a further decrease in pH as was the case in the control when lactic acid was added. However, the concentration of lactic acid produced in glucose supplemented brines was higher than simple lactic acid addition, thus denoting a clear advantage of glucose as a supplement.

References

1. G.D. Balatsouras, *Table Olives*, 3rd edn., Agricultural University of Athens, Athens, 2004, pp 211-279.
2. A. Garrido Fernández, M.J. Fernández Díez, and M.R. Adams, *Table Olives: Production and Processing*, Chapman & Hall, London, 1997, pp 155-157.
3. M. Bobillo and V.M. Marshall, *J. Appl. Bacteriol.*, 1992, **73**, 67.
4. M. Bobillo and V.M. Marshall, *Food Microbiol.*, 1991, **8**, 153.
5. P. García García, C. Durán Quintana, M. Brenes Balbuera and A. Garrido Fernández, *J. Appl. Bacteriol.*, 1992, **73**, 324.
6. M.J. Fernández González, P. García García, A. Garrido Fernández and M.C. Durán Quintana, *J. Appl. Bacteriol.*, 1993, **75**, 226.
7. A. Montaño, M. Bobillo and V.M. Marshall, *Lett. Appl. Microbiol.*, 1993, **16**, 315.
8. M.C. Durán Quintana, P. García García and A. Garrido Fernández, *Int. J. Food Microbiol.*, 1999, **51**, 133.
9. K.E. Spyropoulou, N.G. Chorianopoulos, P.N. Skandamis and G.-J.N. Nychas, *Int. J. Food Microbiol.*, 2001, **66**, 3.
10. J. Baranyi, T.A. Roberts and P. McClure, *Food Microbiol.*, 1993, **10**, 43.
11. J. Baranyi and T.A. Roberts, *Int. J. Food Microbiol.*, 1994, **23**, 277.
12. IOOC, International Olive Oil Council, *Table Olive Processing*, Madrid, 1990.
13. N. Kakiomenou, C.C. Tassou and G.-J.N. Nychas, *Int. J. Food Sci. Technol.*, 1996, **31**, 359.
14. F. González Cancho, M. Rejano Navarro, C. Durán Quintana, F. Sánchez Roldán, P. García García, A. de Castro Gómez Millán and A. Garrido Fernández, *Grasas y Aceites*, 1983, **34**, 375.
15. A. Sánchez, P. García García, L. Rejano, M. Brenes and A. Garrido Fernández, *J. Sci. Food Agric.*, 1995, **68**, 197.
16. F. González Cancho, M. Rejano Navarro, C. Durán Quintana, F. Sánchez Roldán, P. García García, A. de Castro Gómez Millán and A. Garrido Fernández, *Grasas y Aceites*, 1984, **35**, 155.
17. W.H. Holzapfel, *Food Control*, 1997, **8**, 241.

SHELF-LIFE DETERMINATION OF UNTREATED GREEN OLIVES CV. CONSERVOLEA PACKED IN POLYETHYLENE POUCHES UNDER DIFFERENT MODIFIED ATMOSPHERES

E.Z. Panagou and C.C. Tassou

National Agricultural Research Foundation, Institute of Technology, Quality and Safety of Foods, 1. Sof. Venizelou str., Lycovrissi, Greece, GR-141 23.

1 ABSTRACT

Microbiological, physicochemical and organoleptic changes were studied in non-pasteurized samples of untreated green olives cv. Conservolea, stored in polyethylene pouches (HDPE, 60 μm thickness) at 20°C for 180 days under different modified atmospheres (air, vacuum and 40%CO_2/30%O_2/30%N_2). The dominant microbial flora consisted of lactic acid bacteria and yeasts. Olives presented a progressive loss in fruit firmness and color with time as judged by texture readings and L*, a* and b* values respectively. In the end of the storage period, olives packed in aerobic conditions and 40%CO_2/30%O_2/30%N_2 presented the lower firmness, acidic taste and overall eating quality. The best quality characteristics were maintained in vacuum packed olives.

2 INTRODUCTION

Table olives are probably the most popular fermented food worldwide with an annual production of approximately one million tonnes, the majority of which is produced in the EU, mainly Spain, Italy and Greece.[1] Green olives are the most important commercial preparation with nearly 400,000 tonnes/year, with Spain being the major producer with 120,000 tonnes/year.[2] During the preparation of this product, olives are subjected to lye treatment (1.8 – 2.6 %, w/v NaOH) followed by 2-3 washings with tap water to remove the excess alkali. The fruits are then immersed in brine (6-8 %, w/v NaCl) where they undergo lactic acid fermentation. The final product is then packed, together with freshly prepared acidified brine, in glass or plastic containers or tins with a capacity ranging from 250 g to 13 Kg.[3] However, there are other traditional preparations less well known in the international market. One of those involves the direct brining of green olives without prior debittering with NaOH solution. In this case, the fruits undergo a spontaneous fermentation by a mixed microflora of lactic acid bacteria and yeasts.[4] This preparation is known as "untreated green olives in brine" or "naturally green olives."[5]

Traditionally, olives are packed in glass and plastic containers as well as in tins. However, there is a new tendency, particularly in retail outlets, to use other packaging materials, such as polyethylene and aluminium pouches filled with brine and in some cases with gases. The shelf life of these products is not yet clearly defined, although the majority of Greek industries determine a shelf life of 1-2 years on the label, based on the fact that olives are a fermented product of high acidity. However, this length of time is not supported by relevant studies. It is evident that shelf life determination should take into account quality parameters and try to establish lower quality limits, beyond which the product is unacceptable for consumption.[6] For instance, in the case of the Manzanilla variety processed with the Spanish method, it is mentioned that a color index (i) < 23.7 is related with a poor color of the fruit[7] and that a value of A_{440}-A_{700} in the packing brine > 0.23 AU is considered unacceptable.[8] It must be emphasized that the majority of literature is focused on green olive fermentation,[9-14] while there is little information about the quality changes of olives during storage.

The aim of the present work was to study the effect of modified atmosphere packing on the physicochemical and microbiological characteristics of untreated green olives of the Conservolea cultivar.

3　METHOD AND RESULTS

Conservolea green olives were supplied by a local processor in Volos, Central Greece. The fruits were harvested in September, processed at the factory to produce "naturally green olives in brine" and transported to the laboratory within 24 hours. On arrival, olives were hand-selected to remove defective fruits and the remainder was washed thoroughly under pressurized tap water to remove any impurities and left to dry before given any treatment.

Packaging was performed by placing a sample of 250 g of green olives in polyethylene pouches (HDPE, 80 μm thickness). The atmosphere of the packages was modified prior to sealing in a two-line sealing machine (Tecnovak, Milan, Italy) creating the following atmospheres: (i) 40%CO_2/30%N_2/30%O_2, (ii) vacuum and (iii) air (control samples). All packages were stored at controlled temperature (20°C±1°C) in a thermostatic chamber for 180 days. The experiment was repeated twice. Duplicate packages from each sampling occasion were analysed at regular time intervals during storage.

Olive samples (25 g) were aseptically transferred to 225 ml sterile ¼ Ringer's solution respectively. Decimal dilutions in the same Ringer's solution were prepared and duplicate 1 or 0.1 ml samples of the appropriate dilutions were mixed or spread on the following agar media: Plate Count Agar (PCA; Merck 1.05463, Darmstadt, Germany) for total viable counts, incubated at 25°C for 48h; de Man-Rogosa-Sharp medium (MRS; Merck 1.10660, Darmstadt, Germany) for lactic acid bacteria, overlaid with the same medium and incubated at 30°C for 72h; Pseudomonas Agar (Oxoid CM 559, Basingstoke, Hampshire, England, supplemented with selective supplement SR 103) medium for pseudomonads counts, incubated at 25°C for 48h; Rose Bengal Chloramphenicol agar (RBC; Oxoid CM 549, Basingstoke, Hampshire, England, supplemented with selective supplement SR 78) for yeasts and moulds, incubated at 25°C for 72 h; Violet Red Bile Glucose agar (VRBGA; Oxoid CM 458, Basingstoke, Hampshire, England) for enterobacteria counts, incubated at 37°C for 24h. Results were expressed as $\log_{10}$ values of colony forming units per gram (cfu g^{-1}) of olives.

The pH of the fruits was measured using an Orion model 940 digital pH meter (Orion Research Inc., Boston, USA). The pH was measured in a sample (50 g) of olive flesh homogenised previously at room temperature in an Ultra Turrax T25 blender (IKA Labortechnik, Staufen, Germany) in 100 ml of distilled water until obtaining a fluid slurry. Fruit firmness was determined using a Kramer shear compression cell mounted to a texturemeter T-2100G (Food Technology Corp., Maryland, USA) with cross-head speed of 200 mm min^{-1}. The firmness of the fruits was expressed as the mean of 5 replicated measurements, each of which was performed on 5 olives. The fruits were previously depitted using a manual machine. Shear compression force was expressed as newtons per gram of product.

The measurement of fruit color was carried out using a Minolta CR-300 Chroma meter (Minolta Co., Osaka, Japan) with a 8 mm diameter measurement area and 0° observation angle. The CIE (Commission Internationale de l' Éclairage) L* a* b* colorimetry system was used for color determination. The apparatus was calibrated with a standard white tile (L_s^* = 96.10, a_s^* = +0.98, b_s^* = +7.27). Color values are expressed as the mean of three replicate readings on ten olives.

In the end of the storage period, olives were evaluated organoleptically by a 10 member taste panel. The quality attributes selected for evaluation were the following: fruit color, firmness, aroma, saltiness, bitterness, crispness, acidic taste, pit detachment and overall eating quality. The intensity of each attribute was rated by marking a 10 cm scoring scale. At both ends of the scale, an anchor term was placed indicating low and high scores respectively. More specifically, in the color scale, the anchor terms were "dark green" on the left (0) and "green-yellow" on the right (10); in the aroma scale, the anchor terms were "poor" on the left (0) and "excellent" on the right (10); in the saltiness, bitterness, firmness, crispness and acidic taste scales, the anchor terms were "high" on the left (0) and "low" on the right (10); in the pit detachment scale, the anchor terms were "very difficult" on the left (0) and "very easy" on the right (10); finally, in the overall eating quality scale, the anchor terms were "dislike" on the left (0) and "like very much" on the right (10). Sensory profiles were graphically presented by the QDA (quantitative descriptive analysis) method.

The population dynamics of the prevailing microflora, lactic acid bacteria and yeasts, during storage in modified atmospheres is presented in Figure 1. The initial population of lactic acid bacteria presented slight changes and oscillated around 7 log cfu g^{-1} regardless of the modified atmosphere applied. This is possibly due to the fact that lactic acid bacteria have entered the stationary phase of growth and thus the growth rate is balanced by the death rate,[15] resulting thus in a stable number of lactic acid bacteria on the fruits. In addition, since the fruits are not immersed in brine, there is hardly any leaching of fermentable material from the flesh, making the growth of lactic acid bacteria difficult. In this case, lactic acid bacteria are located on the epidermis and sub-stomata spaces with limited access to the interior of the fruit.[16]

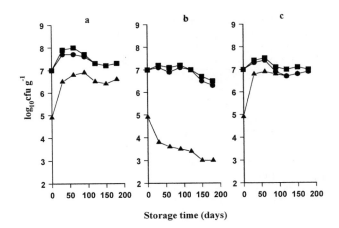

Figure 1 *Changes in total viable counts (■), lactic acid bacteria (●) and yeasts (▲) in green olive samples of cv. Conservolea packed in modified atmospheres at 20°C for 180 days (a: aerobic storage; b: vacuum; c: 40 %CO_2/30 %O_2/30 %N_2).*

The effect of modified atmosphere composition was very pronounced on yeast survival (Figure 1). In vacuum packed olives, the population of yeasts started to decline from the initiation of experiments and after 150 days of storage reached the lowest count (3 log cfu g^{-1}) presenting an overall decrease of 2 log cycles. On the contrary, the samples packed under aerobic conditions (control samples) and 40 % CO_2/ 30 % N_2/ 30 % O_2 presented a slight increase in yeast counts of about 1-1.5 log cfu g^{-1} during the first 30 days of storage, remaining almost stable thereafter.

The effect of different modified atmospheres on the physico-chemical characteristics of packed olives is shown in Table 1. The pH of the flesh presented a slight increase of 0.2-0.3 units without any significant changes among the various treatments. The most notable changes were observed on fruit firmness. In all packages, there was a progressive decrease of firmness regardless of the composition of the atmosphere. Firmness loss was significantly lower in vacuum packed olives, followed by 40%CO_2/30% N_2/30%O_2. This effect is due to the activity of yeasts, in particular oxidative yeasts, which have the ability to produce pectinolytic enzymes, mainly polygalacturonases and pectin-methyl esterases.[17] This observation is in agreement with previous researchers who reported that yeasts from the indigenous microflora of olives can seriously deteriorate flesh firmness by excreting pectinolytic enzymes.[18-20] The same researchers noted that the activity of these yeasts can be restricted by anaerobic conditions in the packages.

According to Sánchez et al.[21] the decrease in fruit firmness with time follows first order kinetics. The slope of the straight line is called softening rate constant, k (months^{-1}). For a degradation with first order kinetics it is possible to estimate the shelf life (t_s) of the product from the expression: $t_s = \ln(F_0/F_1)/k$,[41] where F_0 is the firmness at the beginning of storage (t = 0) and F_1 the limit of firmness below which the fruits are not acceptable. Although this lower limit has not yet been determined for the Greek table olive varieties,

Table 1 Changes[a] in the physico-chemical characteristics of untreated green olives of the Conservolea cultivar packed in modified atmospheres at 20°C for 180 days

Package	Time (days)	Firmness ($N\ g^{-1}$)	k[b]	t_s[c]	pH fruit	Fruit color		
						L^*	a^*	B^*
Air (control)	0	42.3 a			4.1 a	91.64 a	5.96 a	30.15 a
	30	37.1 b			4.3 b	76.62 b	3.35 b	19.46 b
	60	32.6 c			4.2 c	41.25 c	3.19 c	18.66 c
	90	28.1 d	0.1488	9	4.3 b	42.34 d	3.69 d	19.49 b
	120	23.5 e			4.3 b	39.80 e	3.70 d	16.12 d
	150	21.6 f	$r^2 = 0.989$		4.3 b	39.18 f	3.38 b	14.10 e
	180	14.0 g			4.3 b	39.15 f	3.25 e	13.50 f
Vacuum	0	42.3 a			4.1 a	91.64 a	5.96 a	30.15 a
	30	40.2 b			4.2 b	90.52 b	0.41 b	29.15 b
	60	38.7 c			4.2 b	49.21 c	1.07 c	28.80 c
	90	36.3 d	0.0615	23	4.3 c	49.00 d	0.96 d	29.36 d
	120	34.1 e			4.4 d	47.23 e	0.31 e	26.33 e
	150	32.7 f	$r^2 = 0.990$		4.4 d	46.27 f	0.18 f	25.44 f
	180	31.6 g			4.4 d	45.16 g	0.75 g	25.30 g
40%CO_2/30%O_2/30%N_2	0	42.3 a			4.1 a	91.64 a	5.96 a	30.15 a
	30	38.5 b			4.4 b	77.18 b	5.26 b	21.29 b
	60	35.7 c			4.2 c	40.33 c	4.74 c	16.29 c
	90	32.1 d	0.0919	15	4.3 d	39.92 d	3.59 d	16.96 d
	120	30.8 e			4.2 c	41.43 e	3.37 e	18.14 e
	150	27.1 f	$r^2 = 0.985$		4.4 b	38.51 f	4.48 f	14.83 f
	180	25.8 g			4.4 b	37.55 g	4.25 g	13.54 g

[a]: Means with different letters within a column for each packing treatment are significantly different ($p < 0.05$)
[b]: softening rate constant, k (month^{-1})
[c]: shelf life in months

the value $F_1 = 10$ N g^{-1} was used as an approximation. This value has been previously used[21] for the Manzanilla cultivar which has many similarities with Conservolea. The softening rate constants (k) and the estimated shelf life for each packing treatment are presented in Table 1. Vacuum packed olives presented the lowest softening rate constant and hence the longest shelf life (23 months), followed by 40%CO_2/30%N_2/30%O_2 and control samples (aerobic storage) with a shelf life of 15 and 9 months respectively.

The composition of the modified atmospheres had a pronounced effect on the lightness (L*) and the yellowness (b*) color parameters (Table 1). Fruits packed under vacuum presented the highest L* and b* values throughout storage compared with the other two packing treatments. It is worth noting that the b* values (yellowness) were almost double in vacuum packed olives after 6 months of storage, indicating a much slower color degradation. According to Spanish researchers,[5,22,23] color degradation is due to the oxidation of phenolic compounds present on the epicarp of the fruits. Storage in an atmosphere with elevated CO_2 concentration can retard this phenomenon. However, the presence of O_2 in the package can act as an antagonist to CO_2 minimizing the effect of the latter.[24] For this reason, the 40%CO_2/30%N_2/30%O_2 atmosphere was not very effective in retarding fruit color degradation during storage, due to the high concentration of O_2 present in the packages.

The results of the organoleptic evaluation for selected quality attributes is presented in Figure 2. In general, vacuum packed olives were considered acceptable by the panelists after 6 months of storage, as indicated by the higher quality attribute scores. The most important parameters which influenced the judgment of the panelists were the color and firmness of the fruits. The initial green-yellow color of the fruits packed in air and 40 % CO_2/ 30 % N_2/ 30 % O_2 was turned into dark green after 6 months of storage. This color was unacceptable by the panelists as indicated by the low color scores. The same results were observed for fruit firmness where the lowest ratings were given again for the samples packed in air and 40%CO_2/30%N_2/30%O_2.

4 CONCLUSIONS

Olives packed in modified atmospheres (vacuum, 40%CO_2/30%O_2/30%N_2 and air) presented a progressive loss of fruit firmness and color with time. At the end of the storage period, olives packed in aerobic conditions and 40%CO_2/30%O_2/30%N_2 presented the lowest firmness and skin color and were unacceptable by the panelists. The best quality characteristics were maintained in vacuum packed olives as indicated by the higher organoleptic scores. Firmness degradation followed first-order kinetics and the expected shelf life was 23, 15 and 9 months for vacuum packed, 40%CO_2/30%O_2/30%N_2 and air respectively.

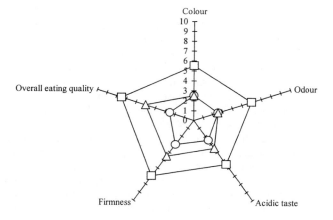

Figure 2. *Sensory profile of untreated green olives of the Conservolea cultivar after 180 days of storage in modified atmospheres at 20 °C (o: air; □: vacuum; △: 40%CO_2/30%O_2/30%N_2)*

References

1 Anonymous, *Olivae*, 2001, **89**, 32.
2 ICAP, *Sector report for table olives*, ICAP, Athens, 2003, p. 84.
3 G.D. Balatsouras, *Table Olives*, 3rd edn., Agricultural University, Athens, 2004, p. 326.
4 J.M.R. Borbolla y Alcalá, F. González Pellissó and F. González Cancho, *Grasas y Aceites*, 1971, **22**, 455.
5 A. Garrido-Fernández, M.J. Fernández Díez and M.R. Adams, *Table olives: Production and Processing*, Chapman & Hall, London, 1997, p. 134.
6 A.A. Jones and C.M.D. Man, 'Ambient-stable sauces and pickles' in *Shelf Life Evaluation of Foods*, eds., C.M.D. Man and A.A. Jones, Chapman & Hall, London, 1994, pp 275-295.
7 A.H. Sánchez Gómez, L. Rejano and A. Montaño, *Grasas y Aceites*, 1985, **36**, 258.
8 A. Montaño, A.H. Sánchez Gómez and L. Rejano Navarro, *Alimentaria*, 1988, **193**, 78.
9 P. García García, M.C. Durán Quintana, M. Brenes Balbuera and A. Garrido Fernández, *J. Appl. Bacteriol.*, 1992, **73**, 324.
10 J.L. Ruiz-Barba, D.P. Cathcart, P.J. Warner and R. Jiménez-Díaz, *Appl. Environ. Microbiol.*, 1994, **60**, 2059.
11 M.C. Durán Quintana, P. García García and A. Garrido Fernández, *Int. J. Food Microbiol.*, 1999, **51**, 133.
12 K.E. Spyropoulou and G.-J.N. Nychas, 'Addition of fermentable substrates and thiamine during the fermentation of green olives with or without starter cultures' in *17th International Symposium of the International Committee on Food Microbiology*

and Hygiene (ICFMH), eds., A.C.J. Tuijtelaars, R.A. Samson, F.M. Rombouts and S. Notermans, Veldhoven, The Netherlands, 1999, pp 685-689.
13 K.E. Spyropoulou, N.G. Chorianopoulos, P.N. Skandamis and G.-J.N. Nychas, Int. J. Food Microbiol., 2001, **66**, 3.
14 E.Z. Panagou, C.C. Tassou and C.Z. Katsaboxakis, J. Sci. Food Agric., 2003, **83**, 1.
15 M.J. Pelczar, R.D. Reid and E.C.S. Chan, 'The growth of bacteria' in Microbiology, eds., M.J. Pelczar, R.D. Reid and E.C.S. Chan, McGraw-Hill, New-York, 1997, pp 118-135.
16 G.-J.N. Nychas, E.Z. Panagou, M.L. Parker, K.W. Waldron and C.C. Tassou, Lett. Appl. Microbiol., 2002, **34**, 173.
17 D. Marquina, C. Peres, F.V. Caldas, J.F. Marques, J.M. Peinado and I. Spencer-Martins, Lett. Appl. Microbiol., 1992, **14**, 279.
18 E.M. Mrak, R.H. Vaughn, M.W. Miller and H.J. Phaff, Food Techn., 1956, **10**, 416.
19 R.H. Vaughn, T. Jakubczyk, J.D. MacMillan, T.E. Higgins, B.A. Davé and V.M. Crampton, Appl. Microbiol., 1969, **18**, 771.
20 R.H. Vaughn, K.E. Stevenson, B.A. Davé and H.C. Park, Appl. Microbiol., 1972, **23**, 316.
21 A.H. Sánchez, L. Rejano and A. Montaño, J. Sci. Food Agric., 1991, **54**, 379.
22 M.J. Ellis, 'The methodology of shelf life determination' in Shelf Life Evaluation of Foods, eds., C.M.D. Man and A.A. Jones, Chapman & Hall, London, 1994, pp 27-39.
23 L. Rejano, M. Brenes, A.H. Sánchez, P. García and A. Garrido, Sci. Aliments, 1995, **15**, 541.
24 A.H. Sánchez, A. Montaño and L. Rejano, J. Agric. Food. Chem., 1997, **45**, 3881.
25 R. Maestro Durán, J.M. García and J.M. Castellano, Hortsci., 1993, **28**, 979.

MICROBIOLOGICAL, PHYSICOCHEMICAL AND SENSORY CHANGES OF MARINATED FISH PRODUCTS

J. S. Arkoudelos, C. C. Tassou, P. Galiatsatou, and F. J. Samaras

National Agricultural Research Foundation, Institute of Technology, Quality and Safety of Foods, S. Venizelou 1, Lycovrissi, Attiki GR-141 23

1. ABSTRACT

Commercial packages of ready to eat marinades of anchovies (*Engraulis enchrasicolus*) and octopus (*Octopus vulgaris*) were obtained immediately after production. The changes in microbial flora, physicochemical parameters and sensory characteristics were assessed during storage at 15 and 25°C. Lactobacilli were the main spoilage microorganisms. TBA values, pH, total volatile basic nitrogen (TVB-N) and TMA were determined. Sensory characteristics such as color, odor, rancidity, texture and overall appearance of samples stored at 25°C reached at un-acceptable levels after 50 days of storage. Samples stored at 15°C were acceptable for 100 days of storage. There was a good correlation among microbial population, physicochemical parameters and sensory characteristics.

2 INTRODUCTION

Marinated fish products are produced by the combined action of organic acids and salt and this is a typical example of the hurdle concept. The objective of the marination process is to prevent the growth of both pathogenic and the majority of spoilage microorganisms.[1] Marination is also used to change taste or to tenderize, textural and structural properties of fish meat.[2] The quality of the final product is affected by the initial quality of the raw material[3] while the shelf-life of marinated products depends largely upon the storage temperature.

Anchovies (*Engraulis enchrasicolus*) and octopus (*Octopus vulgaris*) are common in Mediterranean diets. In Greece, these products are consumed fresh or salted (only for anchovies). Recently pickling has been applied to anchovy as an alternative process to traditional salting. Pickling of octopus is a usual way of consuming it at restaurants and home. Lately, some marinated octopus and anchovy, 'ready-to-eat' products have been developed and introduced into the market. These products, supplemented with herbs and spices, are highly appreciated by the consumers and used as appetizers.

The microbiology of the marination process of anchovies and the quality and stability of the final products have been studied to some extent.[1,2] Recently, the chemical, microbiological and sensory changes occurring during marination of sardine fillets[4] as

Quality

well as the shelf-life of marinated sardine at 4° C have been reported[3]. However, there is no information on the microbiological, physicochemical and sensory changes during shelf-life of other marinated products like anchovies and octopus.

This work was part of a shelf-life study of marinated anchovy and octopus products and refers to the determination of the changes during storage at the abusive temperatures of 15 and 25° C. The aim of storing the products at these above freezing temperatures was to speed up the spoilage process in order to determine the level of microbiological (specific spoilage organisms) or chemical indicators of spoilage.

3 METHODS AND RESULTS

3.1 Preparation and storage of samples

Packages (plastic pouches), approximately 200 g each, of marinated 'ready-to-eat' anchovies and octopus were obtained from a fish company in Greece. Marinated anchovies contained anchovy fillets, sunflower oil, sliced green olives, salt, vinegar, red peppers, parsley, garlic, spices, citric acid and sugar. Marinated octopus consisted of octopus arms, sunflower oil, vinegar and oregano. Immediately after production the products were delivered to the laboratory within a few hours under refrigeration. Upon arrival, packed fish were stored at 15 and 25° C, in order to determine the shelf-life. On each sampling occasion, two samples of each product were taken after 0, 14, 42, 55, 70, 84, and 99 days and 0, 8, 28, 55, 70 and 84 days of storage at 15 and 25° C, respectively, and analyzed as describes in 3.2, below.

3.2 Microbiological analysis

Twenty-five grams of each sample were aseptically placed into a sterile stomacher bag with 225 ml of sterile ¼ Ringer's solution and homogenized for 1 min in a Seward 400 Stomacher (Seward Medical UAC House, London, UK). Decimal serial dilutions in 1/4 Ringer's solution were prepared and duplicate 0.1 or 1 ml samples of the appropriate dilutions were spread or poured in the following agar media: Plate Count agar (Merck 1.05463, Darmstadt, Germany) for total viable count, incubated at 25° C for 48h; Baird - Parker agar (Merck 1.05406) for staphylococci counts, incubated at 37° C for 48 h; Pseudomonas agar base with CFC selective supplement (Oxoid CM 559, Basingstoke Hampshire, UK) for *Pseudomonas* spp. counts, incubated at 25°C for 48 h; Rose-Bengal Chloramphenicol agar with SR 78 supplement (Oxoid CM 549) for yeasts and moulds incubated at 25°C for 72 h; STAA agar (Biolife 402079, Milano, Italy) for *Brochothrix thermosphacta* counts incubated at 25°C for 72 h; de Man Rogosa Sharpe agar (Merck 1.10660) for lactic acid bacteria, incubated at 25°C for 96 h; Violet Red Bile Glucose agar (Oxoid CM 458) for enterobacteria, incubated at 37°C for 24 h; Iron agar (bacteriological peptone 20g/l, "lab lemco" powder 3.0, yeast extract 3.0, ferric citrate 0.3, sodium thiosulfate 0.3, sodium chloride 5.0, L-cysteine 0.6, agar 12.0) for hydrogen sulphide producing bacteria especially *Shewanella putrefaciens*, incubated at 25°C for 96 h.

Results of the microbiological analyses of marinated anchovies and octopus are presented in Tables 1 and 2, respectively. Only the changes of the lactic acid bacteria counts are given, as they were the dominant microorganisms on both kinds of marinated products during storage at 15 and 25°C. No growth of other genera was observed on the referred above selective media, except of some low counts of yeasts observed occasionally. Intrinsic factors such as the low pH due to acetic and citric acids present, the occurrence of antimicrobial agents of spices[5] (garlic, oregano, etc.) and the salt and sugar content, create hurdles for microbial growth. Lactic acid bacteria are able to grow in such an environment. The initial load of lactic acid bacteria was higher in marinated anchovies than octopus (ca 5 and 2 logcfu/g, respectively). However, it has been reported[1,4] that at the end of the marination process of fish, the microbial association of the raw material had been completely inhibited while others have stated that the selection of the final flora of fish at the end of marination depends on the type and concentration of the organic acid used.[2] It has been referred also that lactic acid bacteria found in marinades may be originated from the added spices.[6] In marinated anchovies, after 30 days storage at 25° C and 40 days storage at 15°C, lactic acid bacteria reached their maximum level (ca 7 logcfu/g). This population survived for 100 days of storage at 15°C while at 25°C gradually decreased. In marinated octopus, lactic acid bacteria, although in low initial level, increased rapidly at 25°C and reached high levels (ca 8 logcfu/g) in 30 days of storage, while at 15°C the growth was gradual throughout the storage period of 100 days.

Table 1 *Changes in lactic acid bacteria (LAB) counts, pH, total volatile basic nitrogen (TVB-N) and thiobarbutiric acid (TBA) values during storage of marinated anchovies at 15 and 25° C*

15°C				
Storage days	LAB (logcfu/g)	pH	TVB-N (mg/100g)	TBA (mg/Kg)
0	4.8	4	20	22.9
14	5.2	4	22	15.4
42	7.4	4.2	26	11.8
55	6.8	3.9	26	7.5
70	7.2	4.1	28	9.2
84	7.1	4.3	37	7.1
99	7.0	4.1	38	7.9

25°C				
0	4.8	4.0	20	22.9
8	5.5	4.0	27	20.4
14	6.4	3.8	26	16.4
28	7.0	3.9	25	11.0
42	6.3	3.9	26	3.8
69	5.8	4.2	36	5.2
84	5.4	3.9	41	7.9

Table 2 Changes in lactic acid bacteria (LAB) counts, pH, total volatile basic nitrogen (TVB-N), thiobarbutiric acid (TBA) and trimethylamine (TMA) values during storage of marinated octopus at 15 and 25° C

15°C

Storage days	LAB (logcfu/g)	pH	TVB-N (mg/100g)	TBA (mg/Kg)	TMA (mg/100g)
0	2.5	4.0	15	22	1.9
14	2.3	4.0	13	19	1.1
42	4.8	4.0	15	22	1.9
55	5.4	4.1	28	20	3.0
70	6.4	4.1	24	16	1.1
84	6.0	4.0	34	18	3.0
99	6.7	4.0	39	20	1.1

25°C

Storage days	LAB (logcfu/g)	pH	TVB-N (mg/100g)	TBA (mg/Kg)	TMA (mg/100g)
0	2.5	4.0	15	22	1.9
8	3.1	3.8	14	18	0.7
14	4.6	4.0			
28	7.5	4.0	28	13	1.0
42	8.2	4.1	32	12	3.5
69	7.7	4.0	33	8.0	6.0
84	8.1	4.2	34	7.0	6.3

3.3 Physico–chemical analysis

3.3.1 Determination of pH. The pH of the homogenised fish slurry (1:10) used for microbiological analysis was measured with a pH meter (Metrohm 691, Herisa, Switzerland), immediately after microbiological sampling.

3.3.2 Determination of total volatile basic nitrogen (TVB-N). Total volatile basic nitrogen was measured using the method of Fagan.[7] The quantity of TVB-N was determined by the formula TVBN = n x 16.8 vmg of N/100 g, where n was the amount (cm^3) of sulphuric acid required.

3.3.3 Determination of thiobarbituric acid reactive substances (TBARs). The extent of lipid oxidation was determined by the 2-thiobarbituric acid (TBA) method of Pearson[8]. The TBA number is expressed as milligrams of malonaldehyde equivalents per kilogram of sample using a conversion factor of 7.8 for absorbance to TBA values.

3.3.4 Determination of trimethylamine (TMA). The TMA content was determined using the method of AOAC[9].

Results of the physicochemical analyses of marinated anchovies and octopus are shown in Tables 1 and 2, respectively. The pH of marinated products was 4.0 due to organic acids present (acetic and citric acids). It did not change significantly throughout the storage period. Similar observations have been made in the pH values of marinated sardines during storage at 4°C.[3] However, there is a possibility that the pH of marinades may rise during storage as a result of growth and metabolism of hetero-fermentative lactic acid bacteria: the degradation of amino acids, the formation of carbon dioxide and other decarboxylation products may bind acetic acid.[10]

TVB-N values showed an increase in marinated products at both temperatures, being more intense in marinated octopus at 25° C. According to Sikorski, Kolakowska & Burt,[11] a level of 30 mg/100 g has been considered the upper limit above which fishery products are considered unfit for human consumption while according to others the level of 35mg/100g is the upper limit.[12,13] In marinated octopus TVB-N values reached that level in 40 and 60 days of storage at 25 and 15°C, respectively. In marinated anchovies, although the initial TVB-N values were high (ca 20 mg/100 g), they reached and exceeded the level of 30 mg/100 g in 70 days of storage at both temperatures. It has been stated that the increase in TVB-N values depends on fish species and the amount of salt and acetic acid used.[3]

TBA is usually used as an indicator of fresh fish quality as it reflects the degree of lipid oxidation. In this study (Tables 1 and 2), the initial TBA values were high and decreased during storage. It has been stated[14,15] that TBA values may not reveal the actual lipid oxidation, since malonaldehyde and other aldehydes can interact with other components of the fish flesh.

TMA values (Table 2) reflect the trimethylamine oxide (TMAO) in fish, as TMA is produced by the decomposition of TMAO and is used also as a quality and freshness indicator. TMA was not detected in marinated anchovies but in marinated octopus it increased slightly during storage at 15°C and reached high levels (6 mg/100 g) after 70 days at 25°C, but still within acceptable limits. TMA values of 5-10 mg/100 g have been reported as the limit for acceptability of fish.[11] On the other hand, TMA could not be used as an objective quality indicator even when it is related to fresh fish.

3.4 Sensory evaluation

A trained 10 member sensory panel evaluated the fish at each sample storage period according to the Quality Index Method (QIM) developed by the Netherlands Institute for Fisheries Research (RIVO-DLO). Panelists were Institute staff with long-term training and experience in fish evaluation. Samples stored at 15 and 25° C were drawn on the 0, 14, 42, 55, 70, 84, 99 and 0, 8, 28, 55, 70 and 84th day of storage, respectively, and evaluated. The attributes examined were color, odor, texture and overall appearance of the products (Figure 1). The scale used for color, odor and texture evaluation was: excellent (0), good (1), fair (2) and poor (3). Though, for overall appearance a different scale was used: very poor (1-2), poor (3-4), fair (5-6), good (7-8) and excellent (9-10).

The quality deterioration was accelerated during storage at 25°C and detected by the panelists. In the case of anchovies, panelists pointed out the beginning of deterioration as early as from the 42nd of storage for the attributes; color, texture and overall appearance of the product. The scoring of these attributes had a very good correlation with TVB-N content (Table 1) increase, r=0.937, 0.802 and 0.853, respectively. In octopus samples the panellists started to detect color deterioration and overall appearance changes at the borders of acceptance on the 55th day of storage at 25°C, which were correlated with TVB-N content increases, r=0.929 and 0.911, respectively. Although, panelists had not scored any quality attributes below the acceptance limits, the TBV-N values had been above the acceptance limits (35 mg/100 g) until the 84th day of storage at 15°C.

Quality

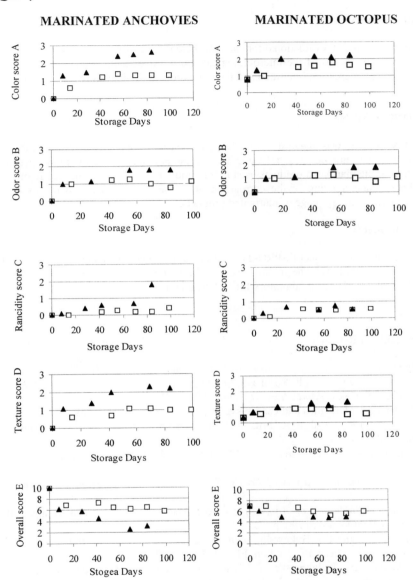

Figure 1 *Changes in color (A), odor (B), rancidity (C), texture (D) and overall appearance (E) scores of marinated anchovies and marinated octopus stored at 15° (□) and 25°C (▲)*

3.5 Data Analysis

The means ratings (of 10 evaluations) for each attribute for each fish were determined and correlated with the means of the microbiological and physicochemical analyses (of two experiments). The regression line fitted and the regression coefficient were determined.

4 CONCLUSIONS

The TVB-N content was the only objective parameter in shelf-life determination of marinated anchovies and octopus. TBA and TMA could not be correlated with the quality deterioration of the marinated products used in this study. Further work is needed in order to gain better understanding of the correlation of the sensory attributes with possible chemical indices of the spoilage of marinated products.

Acknowledgements

This work is part of the project ESPRO-TR-24 funded by the Greek General Secreteriat of Research and Technology. The authors would like to thank the technical staff of the microbiological and chemical laboratory for technical assistance.

References

1. S.R. Fuselli, M.R Casales, R. Fritz and M.I. Yeannes, *Lebensm.-Wiss. u-Technol.*, 1994, **27**, 214.
2. I. Poligne and A. Collignan, *Lebensm.-Wiss. u-Technology*, 2000, **33**, 202.
3. N. Gökoğlu, E. Cengiz and P. Yerlikaya, *Food Control*, 2004, **15**, 1.
4. B. Kilinc and S. Cakli, *Food Chemistry*, 2004, **88** (2), 275.
5. C.C. Tassou, E.H. Drosinos and G.J.E. Nychas, *J. Food Protec.*, 1996, **59**, 31.
6. S.R. Fuselli, M.R. Casales, R. Fritz and M.I. Yeannes, *J. Aquatic Food Product Technol.*, 2003, **12**, 55.
7. J.D. Fagan, T.R. Gormley and M.Ui Mhuircheartaigh, *Lebensm.-Wiss. u-Technol.*, 2003, **36**, 647.
8. D. Pearson, *Chemical Analysis of Foods*, 7th ed. Churchill, Livingstone, Edinburgh, 1976.
9. AOAC *Official Methods of Analysis*, 15th ed. Association of Official Analytical Chemists, Arlington, VA, 1990, p. 123.
10. V.I. Shenderyuk and P.J. Bykowski, *Seafood: resources, nutritional composition and preservation*, CRC Press, Boca Raton, FL, 1989.
11. Z.E. Sikorski, A. Kolakowska and J.R. Burt, *Seafood: resources, nutritional composition and preservation*, CRC Press, Boca Raton, FL, 1989.
12. W. Ludorff and V. Meyer, Fische und fischerzeugnisse, Paul Parey Verlag, Hamburg, Berlin, 1973.
13. J. Schormuller, Handbuch der lebensmittelchemie, Springer Verlag, Berlin, Heidelberg, New York, 1968.
14. D. Taliadourou, V. Papadopoulos, E. Domvridou, I. N. Savvaidis and M.G. Kontominas, *J. Sci. Food Agric.*, 2003, **83**, 1373.
15. S.P. Auburg, *Int. J.Food Sci. Technol.*, 1993, **28**, 323.

Subject Index

ABTS radical ion, 294, 298
Agroresource refining concept, 105
3-Alkyl-2-phenyl-2-
 hydroxymorpholinium cations,
 335-339
Allergens, 351
Ananas comosus, 231, 235
Angtiotention converting enzyme, 352,
 353
Anoca analysis, 387, 388, 412
Anthocyanins, 72, 75
Antifungal activity
 of white truffle, 320, 321
Antioxidant capacity, 293, 294
Antioxidative activity
 of citrus seeds, 472, 476
 of cruciferous vegetables, 304,
 305, 307
 of white truffle, 312, 319
Arazá pulp fruit, 164, 165
Aroma active compounds
 of fresh orange juice, 256
 of fresh truffle, 269
Aroma compounds
 effect of storage in wine, 66
 glycosidically-bound, 145, 147,
 169, 170
 of Arazá pulp fruit, 167-169
 of saffron flowers, 108
 of truffle, 120-123
Aroma profile
 of saffron flowers, 113
Arthritis, 332
Atomic emission detector, 95
Autoxidation, 309

Black truffle, 115, 260-268
Blueberry,
 Fruits, 293
 leaves, 293
Blood pressure, 355

Broccoli, 304
Buckwheat, 350
 sprout sour juice, 351

Cancer, 333
Cadiovascular disease, 332
Calcium caseinate, 42-49
Capillary electrophoresis, 409-410
Carotenoids, 252
Cheese,
 flavor of Piacentinu ennese, 39
 making, 15
 Piacentinu Ennese, 23, 26-28
 Ricotta, 23, 29
Chemesthetic, 94
Chenopodium quinoa Willd, 278
Chiral isomers, 280
Cinnamic acid
 glycoconjugates, 170
Citral, 248
Citric seeds, 472, 473
Citric acid, 85, 88, 488
Cocona, 156, 157
Combinatorial chemistry program, 175-207
Comprehensive two-dimentional GC,
 243-250
Consevolea green olive, 472
Crocus sativus L., 104
Cruciferous vegetables , 304, 305
p-Cymen-8-ol, 248
Cystamine, 215

2,4-Decadienal, 219
Diabetes, 333
Dienamides, 93, 101, 102
Discriminant analysis, 79
DPPH radical scavenging, 294, 297
Dynamic headspace analysis, 106
 of black truffle flavor, 265

Electrophoresis, 421
Emulsion, 44, 45
Enantiomer,
    of linalool, 273
Eugenia stipitata Mac Vaugh, 164, 165
Extraction, solvent, 62-63

Fatty acid composition
    of duck meat, 138
Fish products
    marinated, 492-494
Flavorants
    from mango, 234
    from pineapple, 235
Flavor analysis, 243
Flavor perception, 13
Flavor release, 13-15
    of flavor compounds of cheese, 17
Frozen dough, 452
Furanones,
    biosynthesis, 219
Furfural, 248

GC-Olfactometry, 255
    of black truffle, 263-265
    of duck meat, 143
    of fresh saffron flower, 106
    of Piacentinu Ennese, 24
    of Piacentinu Ennese, 36
GCxGC, 243-250
Germinated brown rice, 356-358
Glucosidase activity, 146
    of vanilla beans, 147
Glucoside
    of vanilla beans, 151
Glucovanillin, 147
Glycosidically bound volatiles, 236
    of cocona, 161, 162
Grape, 271, 273

Halal, 3-9
    certification, 9
    demographics, 8
    markets, 8
    status, 5, 6

Halal foods, 4
    flavor issues, 6
    use of alcohol, 7
Heat challenge, 248, 249
Heat coagulation time, 42, 46, 47
Hepatic acute-phase response, 335, 337, 341, 343-345
HRGC, 166
HS-SPME, 166, 233
Human nutrition
    of structured lipids, 329
Hydrodistillation, 106
Hydrodynamic pressure, 375, 391, 393, 405-407, 418, 420, 431
Hydrodynamic pressure - treatment,
    of striploin steaks, 376, 393, 407-408, 420, 431-433
5-Hydroxymethylfurfural, 248

Immobilization supports, 51
Immobilized cells, 53
Indian long pepper, 93-102
Ingredients,
    effect on bakery quality, 449, 545-457

Kosher, 6

Lactaldehyde, 220
Lactic acid bacteria, 487, 494
Lemon juice, 246
Lemon-flavored soft drink, 246-248
Limonene, 248
Linalool, 248, 271-276
    glycosidically bound, 275
Lipid autoxidation,
    in olive oil, 224-225
Lipolysis-resistant lipids, 345-347
Lycopene, 252

Maceration, 107
Maillard reaction, 213
    of bakery products, 451
Malic acid, 85, 87, 88
Mangifera indica, 231, 234
Mango, 231, 234

Subject Index

Medium-chain fatty acid, 327
p-Methylacetophenone, 248
Methylobacterium extorquens, 220, 221
Microbiological analysis, 501-502
Migration, 286, 288
Modified atmosphere, 493
Morpholine, 335
Multidimensional gas chromatograph, 158
Muscat de Frontignan, 271

Norisoprenoids, 252, 256, 257
    aroma impact of, 257, 258
Nutrition bioassay
    of white truffle, 314, 318

Off-flavor, of duck meat, 136, 142
    of olive oil, 224
Olfactory intensity devise, 107
Olive,
    brining treatment of, 489-490
    Coservolea, 492, 496
    green, 492
Olive oil, 224-229
Omega-3 fatty acids, 324, 325, 328
Organic acids, 83, 85, 88, 490
Osmodehydrated fruit, 237
Osmotic dehydration, 232
Oxidation index values
    of olive oil, 228

Packaging materials, 283, 288
Pellitorine, 95
Peroxide value, 306
Phenolic compounds,
    in germinated brown rice, 358, 359
Phenolic content
    of white truffle, 314, 319
Phenolic glycosides
    from quinoa, 278, 280
Pineapple, 231, 232
Piper longum Linn., 93-102
Pollutants, 283
Principal component analysis, 78
Propanediol, 222

Quality prediction, 467
Quinoa, 278, 279

Reducing power
    of white truffle, 320
Reverse phase HPLC, 378, 380-385
RTD low-alcoholic beverage, 250

Saccharomyces cerevisiae, 51, 53
Saffron flowers, 104-108
SDS-PAGE, 395-397, 400, 422-427, 437
Sensory analysis
    of fish products, 504, 505
    of olive oil, 226
Sensory characteristics
    of bakery products, 448-449, 456-457
Sensory profile
    of green olive, 498
Sensory rancidity
    of olive oil, 228
Serum cholesterol, 341, 347
Shear force determination, 376
Shear resistance values, 433, 435-436
Shelf-life determination
    of green olive, 492, 497
Solanum sessiliflorum Dunal, 156, 157
Soy protein isolate, 455
SPME, 24, 36, 118, 119, 254, 262, 265
Stereochemical analysis, 274
Storage stability
    of olive oil, 224-226
Strawberry
    flavor, 219
    frozen, 459
Strecker aldehydes, 213-216
Structured lipids, 323-333
    enzymatic synthesis of, 323, 324
Superoxide scavenging activity, 364
    of germinated brown rice, 363-365
    of phenolic compounds, 367
Synthase,
    linalool, 275, 276

Taste, of phenolic compounds, 361-362
Temporal perception, 19, 20
Trehalose, 454
TBARS, 331, 503
TEAC, 294, 300
Tenderness, meat, 431
Thiamine, 175-176
    thermal degradation of, 175-176
Thiazolidine derivatives
    of aldehydes, 214
    of methional, 214
    of phenylacetaldehyde, 214
Thrombosis, 332
Time temperature integrators, 461, 462
Tina, 29, 32
    Cocci community of, 32
Tingle character, 94, 101
    compounds possessing, 101, 102
Tingle sensation, 94
Total polyphenols
    of blueberry, 294, 295
    of citrus seeds, 474
Trimethylamine, 503-504
Truffles, principal species of, 116
Tuber borchii, 312-321
Tuber magnatum, 116
Tuber melanosporun, 115, 116
Tyrosinase inhibitory activity, 368
    of germinated brown rice, 369

Ume fruit, 83
    organic acids in, 85
Ume liqueur, 82-89
    manufacturing of, 83
    organic acids in, 86

Vanilla beans, 145-153
Vanillin, 147
Vitamin C, 461
    kinetic of loss, 463
Vitis vinifera L., 271
Volatile compounds
    of Arazá pulp fruit, 167-169
    of black truffle, 266
    of cocona, 159-160
    of duck meat, 139-140
    of Greek wines, 66-68
    of Indian long pepper, 97-100
    of olive oil, 225, 226
    of Piacentinu Ennese cheese, 27-29, 35-41
    of *Piper longum* Linn., 97-100
    of Ricotta cheese, 30, 31
    of wines, 57

Warner-Blatzler shear, 422
Wheat gluten, 455
Whey, 44
Whey proteins, 455
White truffle fungus, 312-321
Wines,
    classification, 76
    Tempranillo, 76
    Zitsa, 65

Zanthxylum, 101
Zocor, 341, 347
Zocor-induced acute-phase response, 347